Library of
Davidson College

DIFFUSION IN CRYSTALLINE SOLIDS

This is a volume in the
Materials Science and Technology series.
Editors: *A. S. Nowick and G. G. Libowitz*

A complete list of the books in the series appears at the end of the volume.

DIFFUSION IN CRYSTALLINE SOLIDS

Edited by GRAEME E. MURCH

Materials Science and Technology Division
Argonne National Laboratory
Argonne, Illinois

ARTHUR S. NOWICK

Henry Krumb School of Mines
Columbia University
New York, New York

 1984

ACADEMIC PRESS, INC.
(Harcourt Brace Jovanovich, Publishers)
Orlando   San Diego   San Francisco   New York
London   Toronto   Montreal   Sydney   Tokyo

COPYRIGHT © 1984, BY ACADEMIC PRESS, INC.
ALL RIGHTS RESERVED.
NO PART OF THIS PUBLICATION MAY BE REPRODUCED OR
TRANSMITTED IN ANY FORM OR BY ANY MEANS, ELECTRONIC
OR MECHANICAL, INCLUDING PHOTOCOPY, RECORDING, OR ANY
INFORMATION STORAGE AND RETRIEVAL SYSTEM, WITHOUT
PERMISSION IN WRITING FROM THE PUBLISHER.

ACADEMIC PRESS, INC.
Orlando, Florida 32887

United Kingdom Edition published by
ACADEMIC PRESS, INC. (LONDON) LTD.
24/28 Oval Road, London NW1 7DX

Library of Congress Cataloging in Publication Data
Main entry under title:

Diffusion in crystalline solids.

(Materials science and technology series)
Includes bibliographical references and index.
1. Diffusion. 2. Solids. I. Murch, G. E. II. Nowick,
Arthur S. III. Series: Materials science and technology.
QC176.8.D5D54 1984   530.4'1   84-355
ISBN 0-12-522662-4 (alk. paper)

PRINTED IN THE UNITED STATES OF AMERICA

84 85 86 87   9 8 7 6 5 4 3 2 1

Contents

LIST OF CONTRIBUTORS ix
FOREWORD xi
PREFACE xv

1. The Measurement of Tracer Diffusion Coefficients in Solids

   S. J. Rothman

   I. Introduction 2
   II. Preparation of Diffusion Samples 10
   III. Annealing of Diffusion Samples 15
   IV. Sectioning and Microsectioning 22
   V. Counting of Radioactive Sections 42
   VI. Determination of D from a Penetration Plot 48
   VII. Conclusions 55
   References 56

2. Diffusion in Silicon and Germanium

   Werner Frank, Ulrich Gösele, Helmut Mehrer, and Alfred Seeger

   I. Introduction 64
   II. Basic Features of Bulk Diffusion in Crystalline Solids 66
   III. Self-Diffusion and Related Phenomena 71
   IV. Survey of Foreign-Atom Diffusion 88
   V. Oxidation-Influenced Diffusion
      of Group III and Group V Elements in Silicon 93
   VI. A Barrier against Vacancy–Interstitial Recombination 105
   VII. Diffusion of Group III and Group V Elements
      and Its Dependence on Doping 108
   VIII. Anomalous Diffusion Phenomena 110
   IX. Substitutional–Interstitial Interchange Diffusion
      and Application to Gold and Nickel in Silicon
      and to Copper in Germanium 116
   X. Concluding Remarks 136
   References 137

3. **Atom Transport in Oxides of the Fluorite Structure**

   A. S. Nowick

I.	Introduction	143
II.	Diffusion Studies	145
III.	Conductivity and Relaxation	152
IV.	Comparison of Ionic Conductivity and Oxygen Diffusion	183
	References	185

4. **Tracer Diffusion in Concentrated Alloys**

   H. Bakker

I.	Introduction	189
II.	Theoretical Background	191
III.	Empirical Rules	193
IV.	Theoretical Considerations of the Kinetics of Diffusion in Random Alloys	197
V.	Tracer Diffusion Experiments in Primary (Terminal) Phases	200
VI.	Theoretical Considerations of Diffusion in Ordered Structures	213
VII.	Tracer Diffusion Experiments in Intermediate Phases	235
VIII.	Conclusions	253
	References	253

5. **The Mathematical Analysis of Diffusion in Dislocations**

   A. D. Le Claire and A. Rabinovitch

I.	Introduction	259
II.	The Dislocation Model	261
III.	Solutions of the Diffusion Equations	266
IV.	Properties of the Solutions	274
Appendix A.	Derivation of Eq. (39)	313
Appendix B.	The Poles of Eq. (39) for the Case of the Dislocation Array	314
Appendix C.	Numerical Considerations in the Calculation of $Q(\eta, \varepsilon/\alpha)$	314
Appendix D.	Numerical Considerations in the Calculation of $Q(\eta)$	315
Appendix E.	An Order of Magnitude Estimate of Δ	316
	References	316

6. **Grain Boundary Diffusion Mechanisms in Metals**

   R. W. Balluffi

I.	Introduction	320
II.	The Diffusion Spectrum	321
III.	Present Knowledge of the Structure of Grain Boundaries and Their Line and Point Defects	322
IV.	Review of Experiments Relevant to the Atom Jumping Mechanism in Boundaries	348

V.	Model for Atom Jumping in Boundaries	355
VI.	Influence of Boundary Structure on Boundary Diffusion	360
VII.	Diffusion along Migrating Boundaries	365
VIII.	Model for Grain Boundaries as Point Defect Sources and Sinks and Comparison with Experimental Observations	367
IX.	Conclusions	372
	References	374

7. Simulation of Diffusion Kinetics with the Monte Carlo Method

Graeme E. Murch

I.	Introduction	379
II.	Tracer Diffusion	381
III.	Ionic Conductivity	407
IV.	Chemical Diffusion	412
V.	Conclusions	423
	References	424

8. Defect Calculations beyond the Harmonic Model

Gianni Jacucci

I.	Introduction	431
II.	Lattice Dynamics	436
III.	Vacancy Formation in fcc Lennard–Jones Crystals	442
IV.	Vacancy Migration	453
V.	Analytical Treatment of Anharmonic Jump Frequency	460
VI.	Conclusions	473
	References	474

INDEX   477

List of Contributors

Numbers in parentheses indicate the pages on which the authors' contributions begin.

H. BAKKER (189), Laboratory of Natural Sciences, University of Amsterdam, 1018 XE Amsterdam, The Netherlands

R. W. BALLUFFI (319), Department of Materials Science and Engineering, Massachusetts Institute of Technology, Cambridge, Massachusetts 02139

WERNER FRANK (63), Max-Planck-Institut für Metallforschung, Institut für Physik, and Universität Stuttgart, Institut für Theoretische und Angewandte Physik, Stuttgart, Federal Republic of Germany

ULRICH GÖSELE* (63), Max-Planck-Institut für Metallforschung, Institut für Physik, Stuttgart, Federal Republic of Germany

GIANNI JACUCCI (429), Center of Studies of the National Research Council and Department of Physics, University of Trento, Povo, Italy, and Department of Physics and Materials Research Laboratory, University of Illinois at Urbana–Champaign, Urbana, Illinois 61801

A. D. LE CLAIRE (257), Materials Development Division, Atomic Energy Research Establishment, Harwell, Oxon OX11 0RA, United Kingdom

HELMUT MEHRER† (63), Universität Stuttgart, Institut für Theoretische und Angewandte Physik, Stuttgart, Federal Republic of Germany

GRAEME E. MURCH (379), Materials Science and Technology Division, Argonne National Laboratory, Argonne, Illinois 60439

A. S. NOWICK (143), Henry Krumb School of Mines, Columbia University, New York, New York 10027

A. RABINOVITCH‡ (257), Materials Development Division, Atomic Energy Research Establishment, Harwell, Oxon OX11 0RA, United Kingdom

* Present address: Siemens AG, Zentrale Technik, Forschungslaboratorien, Munich, Federal Republic of Germany.

† Present address: Universität Münster, Institut für Metallforschung, Münster, Federal Republic of Germany.

‡ Present address: Department of Physics, Ben Gurion University of the Negev, Beer-Sheva, 84105 Israel.

S. J. ROTHMAN (1), Materials Science and Technology Division, Argonne National Laboratory, Argonne, Illinois 60439

ALFRED SEEGER (63), Max-Planck-Institut für Metallforschung, Institut für Physik, and Universität Stuttgart, Institut für Theoretische und Angewandte Physik, Stuttgart, Federal Republic of Germany

Foreword

Arthur Nowick and I both started research in the field of diffusion in solids some 35 years ago. The intervening period has made us, with certainty, much older, only arguably much wiser. This area of research has proved remarkably durable for its fundamental interest to both condensed-matter physicists and materials scientists, as evidenced by the contents of this volume, which contains contributions from sexagenarians (like Nowick and me) along with those from young research workers a third our age. It is perhaps in the nature of the beast that the elementary diffusional process is so very fundamental and ubiquitous in the art and science of dealing with matter in its condensed phase that it never ceases to be *useful* but, at the same time, is a problem which is never really *solved*. It remains *important* by any measure.

Interest in diffusion is as old as metallurgy or ceramics, but the scientific study of the phenomenon may probably be dated from the classic papers of H. B. Huntington, which appeared in the *Physical Review* some four decades ago. These papers were the first to attempt to identify the basic underlying atomistic mechanisms responsible for mass transport through solids by a quantitative theoretical analysis of the activation energies required for diffusion by exchange, interstitial, and vacancy mechanisms in copper. Prior to this time, there had been little concern with treating diffusional phenomena on a microscopic basis, and most research was concerned with fairly crude observations of overall bulk transfer processes at junctions between regions with strong compositional differences.

Although large masses of experimental data were prevalent in the literature—to the extent that whole books dealing with diffusion had already appeared by the late 1940s—experimental techniques were largely limited to optical microscopic observations of changes in color and texture and wet chemical analysis of layers near diffusion boundaries. Most numerical values of diffusion coefficients were deduced from the familiar (but painfully imprecise) Matano–Boltzmann method of analysis. It was not until the late 1940s, with the availability of reactor-produced radioisotopes of high purity and high specific activity, that measurements of real precision became possible. The scientific developments since that time have been extraordinary. With precise data available, it became possible to find answers to precise questions.

Steve Rothman, who wrote the first chapter of this volume, is one of the pioneers who developed the optimal experiment methods for measuring the diffusion coefficient by means of radiotracer techniques. No one can expound with greater authority on this vital aspect of the subject. Since so much of the past progress, as well as the future, of this field of research is reliant on acquisition of data of high precision, it is entirely fitting that the book begin with a review of measurement techniques.

For the first few decades of *scientific* study of the subject, attention was focused on identifying the atomistic mechanism or mechanisms which could be invoked to explain diffusion in the simplest of systems: pure, monatomic, monocrystalline metals, simple cubic salt crystals, etc. The lattice vacancy emerged as the dominant defect in most substitutional lattices, with the singular exception of the silver halides. However, even the simple vacancy turned out to be more than the ideal *point* defect calculated by Huntington; there were relaxation effects which involved a large number of its lattice neighbors. Neither was it clear that the vacancy acted only singly; there was good evidence for pairs and higher-order clusters of vacancies in some systems.

More recently, interest has centered on extending the earlier studies to the diffusional behavior of more complex systems and structures. Some of the most exciting recent work is reported in this volume by some of the world's leading experts: elemental semiconductors, by Frank, Gösele, Mehrer, and Seeger; oxides, by Nowick; concentrated alloys, by Bakker; diffusion along grain boundaries, by Balluffi.

Progress in theory has been somewhat less dramatic than that in experiment over the four decades. It has been difficult to expand following Huntington's original models, which involved *ab initio* calculations of the total energies of ground and saddle-point states with and without defects. Only the differences between these two large energies could be compared with the activation energies actually measured for diffusion. The *theoretical errors* in the calculated differences were far larger than the experimental uncertainties in the activation energies, even if no relaxation effects were considered. If these are included—to be consistent with experimental findings—lattice symmetry is lost in the vicinity of the defects, and even ground-state energies are nearly impossible to calculate with any degree of precision. Certain critical areas, however, have yielded to the ingenuity of theorists. I am pleased to see the review by Le Claire and Rabinovitch of the mathematical analysis of diffusion along dislocations.

FOREWORD

The advent of modern, high-speed computers has made possible a new kind of theory, inconceivable four decades ago. It is now possible to perform large-scale simulations of diffusional motions inside the mathematical "lattice" of a computer's memory. There, one can follow individual atomic jumps and sequences of jumps and try to correlate this microscopic behavior with observed macroscopic tracer motion. Murch's chapter brings us up to date on these methods.

One of the bases for all diffusion theories is the so-called theory of absolute reaction rates, initially introduced by Wigner and Eyring as a means to describe chemical reactions between systems *in equilibrium*. Activation energies and entropies are equated to differences in Gibbs free energies between ground and excited saddle-point states, and the manipulations of normal thermodynamics are used to relate calculated parameters to experiment. A basic question, most often avoided by theorists, is why should such parameters, deduced for an equilibrium system, be valid descriptions for diffusion, which is patently a nonequilibrium process? The connection between the kinetic theory of the process, which involves consideration of the enormous spectrum of lattice modes which must enter into both formation and motion of defects, and the reaction-rate parameters has only very recently been brought into focus by the exciting work of C. P. Flynn and some of his co-workers, including G. Jacucci. It is fitting that the final chapter in this volume gives us Jacucci's up-to-date account of these new developments.

Department of Physics DAVID LAZARUS
and Materials Research Laboratory
University of Illinois
Urbana, Illinois

Preface

This book is a sequel to "Diffusion in Solids: Recent Developments," edited by A. S. Nowick and J. J. Burton and published in 1975 by Academic Press. By following the aims of the original work, we have chosen to focus upon some of the most active areas of diffusion research. Although the editors' choice must inevitably be somewhat subjective, we have endeavored to select those subjects which, in our opinion, have matured to the extent that there is general agreement on their scope and interpretation.

The backbone of diffusion is undoubtedly the precise measurement of diffusion coefficients. In the first chapter, S. J. Rothman has compiled extensive information on the measurement of diffusion coefficients with radioisotopes. The following three chapters consider diffusion in materials of substantial technological importance for which, in addition, considerable basic understanding has developed. W. Frank, U. Gösele, H. Mehrer, and A. Seeger deal with diffusion in silicon and germanium and A. S. Nowick with atomic transport in oxides of the fluorite structure, while H. Bakker analyzes diffusion in concentrated alloys, including intermetallic compounds.

The next two chapters delve into diffusion along short-circuiting paths. A. D. Le Claire and A. Rabinovitch analyze the effect of diffusion down dislocations on the form of the tracer concentration profile, while R. W. Balluffi deals with the mechanisms of diffusion in grain boundaries in metals by invoking considerable work done on grain-boundary structure.

In recent years computer simulation has made a substantial contribution to diffusion theory. The last two chapters are concerned with the two main streams of such activities. In the first, G. E. Murch describes the application of the Monte Carlo method to the calculation of random-walk-related quantities. In the final chapter, G. Jacucci focuses on machine calculations of the fundamental atomic migration process by reviewing some state-of-the-art calculations for defect energies and the topology of the saddle surface.

DIFFUSION IN CRYSTALLINE SOLIDS

1

The Measurement of Tracer Diffusion Coefficients in Solids*

S. J. ROTHMAN

MATERIALS SCIENCE AND TECHNOLOGY DIVISION
ARGONNE NATIONAL LABORATORY
ARGONNE, ILLINOIS

I.	Introduction	2
	A. General Remarks	2
	B. The Relation of Diffusion Experiments to the Mathematics of Diffusion	3
	C. The Reproducibility of Diffusion Measurements	8
	D. Some Useful Formulas for Thin Layer Geometry	9
II.	Preparation of Diffusion Samples	10
	A. Polishing, Etching, and Preannealing	10
	B. Making the Diffusion Couple	10
III.	Annealing of Diffusion Samples	15
	A. Low-Temperature Anneals ($T < 200°C$)	16
	B. Intermediate-Temperature Anneals ($200 < T < 1850°C$)	16
	C. High-Temperature Anneals ($T > 1850°C$)	17
	D. Temperature Measurements	18
	E. Calibration of Temperature Measuring Instruments	19
	F. Accuracies Obtainable in Temperature Measurements	20
	G. Correction for Heat Up and Cool Down	20
IV.	Sectioning and Microsectioning	22
	A. Mechanical Sectioning	22
	B. Other Techniques	29
	C. Microsectioning	30
V.	Counting of Radioactive Sections	42
	A. Counting of Gamma Rays	44
	B. Counting of Alpha or Beta Particles or X Rays	45
	C. The Measurement of the Simultaneous Diffusion of Several Radioisotopes	46

* Work supported by the U.S. Department of Energy.

Copyright © 1984 by Academic Press, Inc.
All rights of reproduction in any form reserved.
ISBN 0-12-522662-4

VI.	Determination of D from a Penetration Plot	48
	A. Gaussian Plots	48
	B. Nongaussian Plots	50
VII.	Conclusions	55
	References	56

I. Introduction

A. GENERAL REMARKS

The first measurement of diffusion in the solid state was made by Roberts-Austen (1896). Many measurements, especially of chemical diffusion in metals, were made in the 1930s; the field was reviewed by Mehl (1936), Jost (1952), and Seith (1955). Diffusion research increased after World War II; the increase was motivated by the connection among diffusion, defects, and radiation damage and helped by the availability of many artificial radiotracers. It was at this time that suggestions on how to carry out high-precision, highly reproducible diffusion experiments were first put forward (Slifkin et al., 1952; Tomizuka, 1959).

The three major factors that determine the quality of a diffusion measurement are

1. the method used,
2. the care taken in the measurement, and
3. the extent to which the material is specified (Nowick, 1951).

The most accurate method has, in general, been considered to be radiotracer sectioning (Tomizuka, 1959), and most of this article is devoted to this method, especially to points for which special care must be taken; these are the measurement of temperature, the accuracy of sectioning, and the reproducibility of counting the radioactivity.

The importance of specifying the material cannot be overstated. The measured diffusion coefficient depends on the chemistry and structure of the sample on which it is measured. Impurities, nonstoichiometry of compounds, grain boundaries, and dislocations can give apparent values of the diffusion coefficient that are different from, and usually larger than, the true value.

The objective of this chapter is to describe some experimental techniques that are useful in carrying out diffusion measurements. We have organized the chapter around general principles that are applicable to all materials, and then listed the particulars. The materials we consider are mainly inorganic solids, especially metallic materials; however, organic solids are also mentioned. The effect of pressure on diffusion is omitted. Previous reviews covering mainly metals and inorganic crystals have been given by Hoffman (1951), Tomizuka (1959), Čadek and Janda (1957), Adda and Philibert (1966, Chapter 4), Lundy (1970), and Benière (1983). Chadwick and Sherwood (1975) have reviewed techniques for organic crystals.

Radioactive tracers are essential to many of the experiments described in this article. Radioactive tracers are hazardous materials, and the experimenter who uses them is under the strongest moral obligation to avoid exposure of his colleagues and contamination of his environment.

B. The Relation of Diffusion Experiments to the Mathematics of Diffusion

For measurable diffusion to take place a gradient of some kind is necessary. Diffusion is a consequence of the hopping motion of atoms through a solid. The diffusion coefficient D is defined in Fick's first law (Fick, 1855; Manning, 1968),

$$\mathbf{J} = -D\,\nabla C + C\mathbf{V} \tag{1}$$

where $\mathbf{J}$ is the flux of atoms, C their concentration, and $\mathbf{V}$ the velocity of the center of mass, which moves due to the application of a force such as an electric field or a thermal gradient. A number of different diffusion coefficients exist, e.g., for the diffusion of a radioactive tracer in a chemically homogeneous solid in the absence of external forces,

$$\mathbf{J}^* = -D^*\,\nabla C^* \tag{2a}$$

where the asterisk denotes the radioactive species. For diffusion in a chemical gradient,

$$\mathbf{J} = -\tilde{D}\,\nabla C \tag{2b}$$

where $\tilde{D}$ is the interdiffusion or chemical diffusion coefficient. Any of these equations can be combined with the equation of continuity

$$\partial C/\partial t = -\nabla \cdot \mathbf{J} \tag{3}$$

to yield Fick's second law

$$\partial C/\partial t = \nabla \cdot (D\,\nabla C) \tag{4a}$$

where the mass flow term has been omitted. For a tracer in a homogeneous system,

$$\partial C^*/\partial t = D^*\,\nabla^2 C^* \tag{4b}$$

Equations (4a) and (4b) describe the types of diffusion experiments discussed in this article.

The tracer diffusion coefficient is given also in the atomistic form

$$D^* = \gamma a^2 \Gamma f \tag{5}$$

where γ is a geometric factor, a the jump distance, Γ the atomic jump frequency, and f the correlation factor (Bardeen and Herring, 1951; Manning, 1968). It is thus possible, in principle, to measure D^* by measuring Γ in a resonance experiment of some kind (Nowick and Berry, 1972; Wolf, 1979). This kind of experiment will not be treated here.

We are concerned here with diffusion measurements where the diffusion coefficient is obtained via Fick's second law, i.e., from a solution of the diffusion equation. Fick's second law is used rather than his first because concentrations are easier to measure than fluxes and because the magnitudes of D in the solid state are so small that the required steady state is seldom reached.

In order to obtain a solution of the diffusion equation, the initial and boundary conditions (IC and BC) must be known. The IC correspond to the distribution of the diffusing substance in the sample before the diffusion anneal, and the BC describe what happens to the diffusing substance at the boundaries of the sample during the diffusion anneal. If the experimental IC and BC correspond to the mathematical conditions, the mathematical solution to the diffusion equation $C(x, y, z, t)$ will describe the distribution of the diffusing substance as a function of position in the sample and of annealing time. The diffusion coefficient is finally obtained by fitting the experimentally determined $C(x, y, z, t)$ to the appropriate solution of the diffusion equation with D as a parameter. This chapter describes some methods for setting up the IC, maintaining the BC as well as the implicit conditions mentioned later in this section, and determining $C(x, y, z, t)$.

Most laboratory experiments are arranged so that diffusion takes place in one dimension. The solution of the diffusion equation is then $C(x, t)$. One most often determines $C(x)$ at constant t, i.e., the concentration distribution along the diffusion direction after a diffusion annealing time t. It is also possible to determine $C(t)$ at a constant x (e.g., the concentration at a surface) or $\iint C(x, t)\, dx\, dt$ (e.g., the weight gain of a sample as a function of time).

The IC, BC, and solutions to the diffusion equation (for $D = \text{const}$) for some common geometries are described below. These, and solutions for other cases, are given by Crank (1975) and Carslaw and Jaeger (1959).

(i) Thin Layer or Instantaneous Source Geometry (Fig. 1a). An infinitesimally thin layer ($\ll (Dt)^{1/2}$) of diffusing substance is deposited on

1. TRACER DIFFUSION COEFFICIENTS IN SOLIDS

one surface of a semi-infinite ($\gg (Dt)^{1/2}$) solid. The initial condition is

$$C(x, 0) = M\delta(x) \tag{6}$$

where δ is the Dirac delta function and M the strength of the source in atoms per unit area. The boundary condition is

$$\frac{\partial C}{\partial x}(0, t) = 0 \tag{7}$$

i.e., there is no flux through the surface (impermeable boundary). The solution is

$$C(x, t) = (M/\sqrt{\pi Dt}) \exp(-x^2/4Dt) \tag{8}$$

One determines $C(x)$ for constant t.

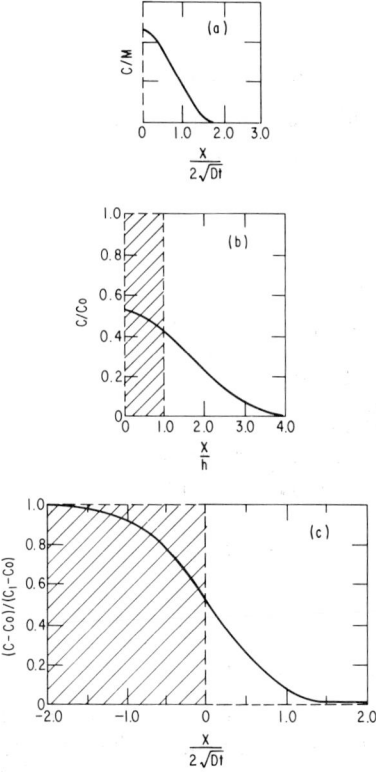

Fig. 1. Concentration distributions for different initial conditions. Dotted line is for $t = 0$, solid line is for a finite t. (a) Thin layer geometry [case (i)]; (b) thick layer geometry [case (ii)], solid curve for $Dt = h^2$; (c) infinite couple [case (iii)]. [After Crank (1975).]

(ii) Thick Layer Geometry (Fig. 1b). Similar to the above, except that the layer thickness h is of order of the diffusion distance:

IC: $C(x, 0) = C_0$, $h \geq x \geq 0$
     $C(x, 0) = 0$, $x > h$ (9)

BC: $\dfrac{\partial C}{\partial x}(0, t) = 0$ (10)

solution: $C(x, t) = \dfrac{C_0}{2}\left[\text{erf}\left(\dfrac{x+h}{2\sqrt{Dt}}\right) - \text{erf}\left(\dfrac{x-h}{2\sqrt{Dt}}\right)\right]$ (11)

where

$$\text{erf}(\lambda) = \dfrac{2}{\sqrt{\pi}}\int_0^\lambda \exp(-\eta^2)\, d\eta \quad (12a)$$

Measure $C(x)$ for constant t. [Note:

$$\text{erfc}(\lambda) \equiv 1 - \text{erf}(\lambda)] \quad (12b)$$

(iii) Infinite Couple (Fig. 1c). A sample of uniform concentration C_0 is welded to a sample of uniform concentration C_1. The weld plane is situated at $x = 0$.

IC: $C(x, 0) = C_1$, $x < 0$
     $C(x, 0) = C_0$, $x > 0$ (13)

Solution: $C'(x, t) \equiv \dfrac{C(x, t) - C_0}{C_1 - C_0} = \dfrac{1}{2}\left[1 - \text{erf}\left(\dfrac{x}{2\sqrt{Dt}}\right)\right]$ (14)

Measure $C(x)$ for constant t.

(iv) Vapor–Solid Couple. A semi-infinite couple containing a volatile component is placed into a dynamic vacuum at $t = 0$:

IC: $C(x, 0) = C_0$, $x > 0$ (15a)
BC: $C(0, t) = 0$, $t > 0$ (16a)

Solution: $C(x, t) = C_0\, \text{erf}(x/2\sqrt{Dt})$ (17a)

Exposing a sample initially devoid of volatile component to a vapor of the volatile component at a pressure in equilibrium with C_0 gives the analogous mathematics:

IC: $C(x, 0) = 0$, $x > 0$ (15b)
BC: $C(0, t) = C_0$, $t > 0$ (16b)

1. TRACER DIFFUSION COEFFICIENTS IN SOLIDS

Solution: $C(x, t) = C_0[1 - \mathrm{erf}(x/2\sqrt{Dt})] = C_0\,\mathrm{erfc}(x/2\sqrt{Dt})$ \hfill (17b)

The same equations apply to isotopic exchange between solid and vapor. Measure either $C(x)$ at constant t or integral weight gain (loss)

$$\int_0^\infty \int_0^t C(x, t)\, dt\, dx$$

(v) *Grain Boundary Diffusion.* The mathematics are more complicated (Le Claire, 1963), owing to the coupled lattice diffusion, but one still measures $C(x)$ at constant t.

(vi) *Exchange Experiment* (Carman and Haul, 1954). This technique is used for materials for which a massive sample cannot be prepared. It involves diffusion exchange between an assembly of powder and a gas of limited volume, from which very small aliquots are drawn at different times.

In the first three sample configurations two bodies of widely different composition are brought into contact. The assumption implicit in the BC is that diffusing material passes across the resulting interface without hindrance, i.e., it is not held up by surface oxides, low solubility, chemical reactions, etc. Nonfulfillment of this condition leads to deviation of the experimental $C(x)$ from the solution of the diffusion equation, as discussed in Section VI.B.

In the vapor–solid couple and the exchange experiment, the assumption implicit in the BC is that the surface of the solid equilibrates with the gas phase instantaneously. However, optical measurements of the change of the surface concentration at low temperature have indicated that the attainment of solid–gas equilibrium can be a slow process (Swanson et al., 1962).

The solutions of the diffusion equation given above rest on the assumption that D is the same everywhere in the sample, except for grain boundary diffusion, where $D' \gg D$ (D' is the grain boundary D), or in a chemical concentration or temperature gradient, where D may vary along the gradient. The condition of uniform D implies that D is independent of position in the sample, i.e., that there are no short-circuiting paths, stress gradients, voids or other defects, second phases, anisotropy, etc., that would make D a function of position. The condition of a well specified material implies that its chemistry and structure remain unchanged during the diffusion anneal, i.e., pickup of impurities, change of stoichiometry, devitrification, etc., do not take place. Violation of these implicit assumptions will cause the form of the measured $C(x)$ to deviate from the calculated.

Such deviations are most easily detected when using the thin layer geometry [case (i)]. Since the gaussian solution [Eq. (8)] lends itself easily to linearization, the deviations from linearity due to violations of the BC or of the above assumptions are especially easy to detect.

The thin layer geometry has several other advantages. The thin layer can be deposited without straining the sample, which is essential for single crystal samples. A thin layer also allows the use of high specific activity radioisotopes, and thus measurements of diffusion without a chemical gradient. Diffusion under large chemical gradients can lead to deformation of the sample and generation of defects (Seith and Kottman 1952; Queisser, 1961; Ayres and Winchell, 1968, 1972).

For the above reasons, the thin layer geometry is most often used in experiments in which diffusion is measured in order to study the fundamentals of diffusion and defect behavior in solids. Such experiments usually concern diffusion as a function of temperature, pressure, or concentration, and small differences in D are important, in contrast to engineering experiments in which the magnitude of the penetration of one material into another is of interest.

C. The Reproducibility of Diffusion Measurements

In the best of scientific diffusion studies, reproducibility of a few percent in D can be obtained. Simultaneously annealed samples should yield Ds reproducible within 1%, while D values obtained from separately annealed samples are reproducible within $\pm 3\%$ (Rothman and Peterson, 1969) or better (Rothman et al., 1970). Values for D from easily duplicable materials (e.g., 99.999% pure Ag single crystals) are reproducible to $\pm 10\%$ between laboratories for $D \geq 10^{-12}$ cm^2 s^{-1} (Hoffman and Turnbull, 1951; Slifkin et al., 1952; Tomizuka and Sonder, 1956; Rothman et al., 1970) or to $\pm 30\%$ for much lower values of D for Ag (Lam et al., 1973; Backus et al., 1974; Bihr et al., 1978) and for W (Mundy et al., 1978; Arkhipova et al., 1977).

The most important factor in the accurate measurement of D is the accurate measurement of the temperature T. Scatter in D is most often due to uncertainty in T (Mallard et al., 1963; Rothman et al., 1980). As D more or less follows an Arrhenius relationship,

$$D = D_0 \exp(-Q/RT) \tag{18}$$

(Q is the activation enthalpy and D_0 is prosaically called the preexponential factor), an error ΔT in T causes a fractional error in D,

$$\Delta D/D = (Q/RT^2)\,\Delta T \tag{19}$$

or, for $Q = 160$ kJ mole^{-1} and $T = 1000$ K, $\Delta T = 2$ K causes an error of $\sim 4\%$ in D.

The other important causes of lack of reproducibility in D among laboratories are either improperly specified materials or annealing conditions, in the case of materials in which the atmosphere affects the stoichiom-

1. TRACER DIFFUSION COEFFICIENTS IN SOLIDS

etry. The presence of grain boundaries in metals, divalent impurities in alkali halides, and nonstoichiometry in oxides can cause immense differences in D, and these factors must be controlled.

D. Some Useful Formulas for Thin Layer Geometry

It should be noted that all the solutions to the diffusion equation in Section I.B are expressed in terms of the dimensionless variable $x/(2\sqrt{Dt})$. The length $2\sqrt{Dt}$ is a kind of mean penetration distance, and this has to be the same order of magnitude as the *characteristic distance* associated with an experiment. For a sectioning experiment, the characteristic distance is the section thickness. For ion-beam depth profiling, it is the ion range, etc.

In the ordinary thin-layer sectioning experiment, one wishes to measure diffusion over a drop in specific activity C of $\sim 10^3$; any effects due to diffusion along short-circuiting paths are likely to show up as curvature in the penetration plot over such a range, while they may not be visible if the measurement is only over a factor of 6 in C (Wajda, 1954). Twenty sections suffice to define a penetration plot; from Eq. (8), the section thickness required to get a drop of 10^3 in C over 20 sections is

$$\theta \approx \sqrt{Dt}/3.8 \tag{20}$$

A preliminary estimate of D is useful in planning an experiment.

The amount of activity for deposition is calculated as follows. The minimum activity that can be counted conveniently is about one-half the background. The background is ~ 400 counts min^{-1} (cpm) for a 3 × 3-in. NaI–Tl well crystal, counting in the integral mode. Thus the activity of the last section is ≥ 200 cpm, and therefore that of the first section is about 2×10^5 cpm. Now from Eqs. (8) and (20), the amount of activity in the first section is given approximately by

$$A_1 = \frac{M}{\sqrt{\pi Dt}} \frac{\sqrt{Dt}}{3.8} \tag{21}$$

so $M \approx 1.5 \times 10^6$ cpm, or, since the counter efficiency under these conditions is $\sim 50\%$, $M \approx 5 \times 10^4$ Bq or 1.3 μCi. If the isotope decays significantly during the time of the experiment, more radioisotope has to be deposited. Under the conditions of $\theta \approx \sqrt{Dt}/3.8$, the specific activity drops by a factor of ~ 2 per section at the twentieth section. These last points on the penetration plot have the greatest weight in determining D, so the counting statistics must be maintained and the penetration plot extended as far as possible. This implies use of an intense source of radioisotope; on the other hand, too much activity poses an unnecessary health hazard as well as increasing the dead-time correction.

The radiotracer may rapidly reach the side surfaces of the sample by surface diffusion or evaporation, and then diffuse inward. To keep the diffusion one-dimensional, one removes $\approx 6\sqrt{Dt}$ from the sides of the sample before sectioning.

II. Preparation of Diffusion Samples

A. Polishing, Etching, and Preannealing

Preparation of a diffusion sample usually involves preparation of flat (irregularities $<2\sqrt{Dt}$), strain-free surfaces, and joining the parts with differing concentrations. Polishing of metals and oxides is usually by standard metallographic techniques, followed by etching or electropolishing to remove cold worked material. Alkali halides are cut with a microtome, on a plane a few degrees from (100) to prevent cleavage chipping on subsequent microtome sectioning. Soft materials such as organic crystals and some metals are also cut with a microtome. Surface preparation involves a compromise between flatness and removal of strain. Mechanical methods (Samuels, 1982; Kehl, 1953) produce the best flatness but introduce strain. Chemical or electrochemical methods (Tegart, 1959) do not introduce strain, but they tend to destroy flatness by causing rounding, selective removal of material from grain boundaries, etc.

The criterion roughness $<2\sqrt{Dt}$ cannot be applied to samples to be sectioned by microsectioning techniques, and this problem is discussed further in Section IV.C. Samples to be sectioned by sputtering should be sputtered for a long time before depositing the tracer so that any microstructure developed by sputtering develops before, not during, the sectioning (Mundy *et al.*, 1981). Materials that deform plastically should be annealed to remove cold work, and then lightly polished or etched. Materials whose stoichiometry is determined by the annealing atmosphere should be preannealed at the diffusion temperature in the appropriate atmosphere for a time long enough to establish the desired stoichiometry via chemical diffusion.

B. Making the Diffusion Couple

After a smooth surface is prepared on the base material, the diffusion couple must be fabricated, i.e., the IC must be established. In a gas–solid couple this is done by introducing the sample into the appropriate atmosphere. In a solid–solid couple, either the two parts must be welded together [case (iii), infinite couple] or the layer of tracer must be deposited on the prepared surface.

1. TRACER DIFFUSION COEFFICIENTS IN SOLIDS

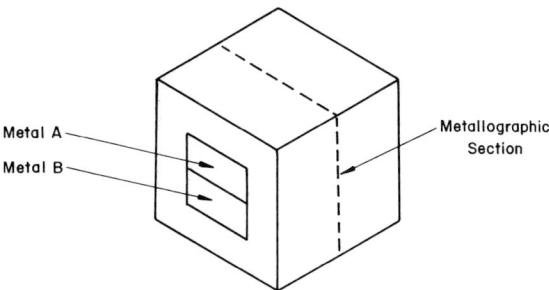

Fig. 2. Sketch of "window frame" for welding diffusion couples. [After Kittel (1948), quoted by Rothman (1962).]

1. *Infinite Couples*

Infinite couples are usually used to study chemical diffusion, although this geometry is used for measuring self-diffusion when the specific activity of the diffusing isotope is very low, or when both halves of the couple contain the diffusing isotope. In the former case, e.g., self-diffusion in aluminum (Lundy and Murdock, 1962), a thick layer of the material containing the radioisotope is required. In the latter case, e.g., self-diffusion in uranium (Adda and Kirianenko, 1958), the concentration of diffusing isotope in the richer half of the couple must be maintained at the original level, so an infinite couple is used.

The chief requirement in such a couple is good contact between the two halves of the diffusion couple. In the case of metals and alloys, such contact is obtained by pressure welding, i.e., pressing the two halves of the couple together until they deform plastically, and then heating to a temperature below the diffusion annealing temperature. One reason for plastic deformation is to break up oxide films. Jigs for welding have been described by Lindemer and Guy (1968) and Adda and Philibert (1966, p. 237). A "window frame" of a material with lower thermal expansion (e.g., Ta) than the materials of the couple can also be used (Fig. 2; Rothman, 1962). On the other hand, good contact has apparently been obtained on oxide samples (Wuensch and Vasilos, 1961) by simply placing them in contact.

2. *Requirements for Deposition of Thin Layers*

In the deposition of thin layers of tracers, the primary objective is to deposit the tracer in a chemical form that allows it to dissolve in the base material in a time $t_s \ll t$, the annealing time. If t_s is not much less than t, the BC is not obeyed and an upward concave plot of log C versus x^2 (penetration plot) is found (Fig. 3). Attempts to fit this with a gaussian plot may yield an incorrect value of D (Mundy et al., 1974). The solution of the diffusion equation for $t_s \leq t$ (Malkovich, 1963) involves both the solubility limit and t_s, which must be treated as an unknown parameter in fitting the data, introducing an element of uncertainty into the result. If $t_s > t$, the BC becomes constant flux at $x = 0$; the solution for this case is treated in Section VI.B.1.

The second objective is to obtain a good yield of radioisotope on the sample with minimum contamination of equipment. A less important requirement is that the deposit be uniformly distributed over the surface. If the irregularities in the x direction are $< \sqrt{Dt}$, homogeneity of the deposit is not important as long as the same section in the y–z plane is counted in each section (Tannhauser, 1956). However, a homogeneous deposit is desirable if the progress of sectioning is to be monitored by autoradiography, as it is difficult to distinguish between uneven blackening of the film due to an inhomogeneous tracer deposit or more serious causes.

3. *Problems of Tracer Solubility*

One reason for a problem with solubility is that the tracer is deposited in a form that is insoluble in the base material. An example of this is the chloride of a radiotracer, a common chemical form for purchased radiotracers, deposited on a metallic substrate. Chloride ions are usually insoluble in the metal. If exchange of the radiotracer with atoms of the base material is rapid enough to fulfill the condition $t_s \ll t$ [$CrCl_3$ on Cr (Mundy et al.,

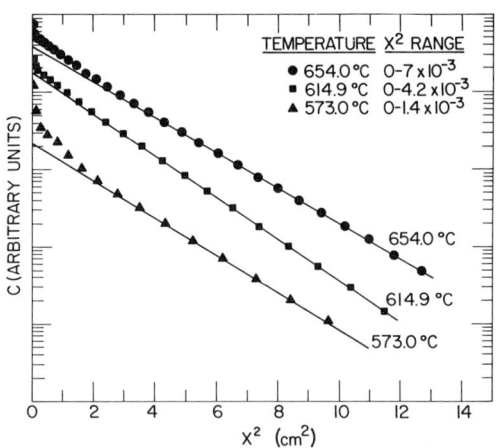

Fig. 3. Surface hold up due to slow dissolution of the tracer Fe in Al. Note that extent of hold up increases with decreasing temperature. [From Peterson and Rothman (1970).]

1976)], or if the base metal quickly reduces the tracer [Nb_2O_5 on U (Peterson and Rothman, 1964)], a good penetration plot with the correct value of D is obtained. Otherwise, the chemical form of the tracer has to be changed. For low temperature diffusion work, elemental radiotracers are essential for metallic diffusion samples.

4. *Electrodeposition of Thin Layers*

In the case of an inorganic compound of the tracer and a metallic base material, the tracer can be simultaneously reduced and deposited by electroplating, if it is not too high in the electromotive series. As the tracer is usually of high specific activity, the concentration of the tracer element in the solution is low, and industrial electroplating recipes (e.g., Canning Handbook on Electroplating, 1978) do not work. One way to increase the concentration and keep the amount of tracer high is to decrease the volume. A drop of the appropriate solution is placed on the sample face (up to 200 μl on a 1-cm diameter sample can be held by surface tension), and a Pt loop anode is just touched to the top of the drop (Fig. 4). As much as 50% of the tracer in the drop can be plated out. After plating, the remaining liquid can be pipetted off the surface or rinsed into a beaker. The sample is rinsed with water, followed by alcohol. Both are collected in a beaker which is then left to dry, since it is easier to dispose of dry, rather than liquid, active waste.

Most cations plate out better when they are present in the solution as a complex ion. The ammonium ion is a useful complexing agent for electroplating Ag, Zn, Ni, Co, and Cu. These cations plate out well from a neutral or slightly basic (pH 7–8) solution. Gold plates well from an alkaline cyanide solution. Iron plates best from $FeCl_3$ dissolved in an ammonium oxalate–oxalic acid solution at pH 5–6 (Mullen, 1961). Lead has been plated from a nitrate–$HClO_4$ solution at pH 2 (Rothman, 1954), and Pd from an acidic chloride solution containing NH_4Cl (Peterson, 1964). A recipe for plating Sn on Pb has been given by Gupta and Campbell (1980).

High vapor pressure metals such as Zn should be plated on the entire sample, not just on the prepared face. Otherwise, the Zn redistributes over the entire surface by vapor-phase transport during the anneal, leading to a large decrease of activity at the prepared face.

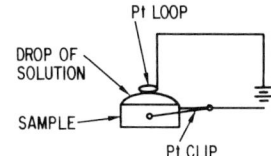

Fig. 4. Sketch of setup for electroplating radiotracers.

5. Other Techniques for the Deposition of Thin Layers

Some metals are so high in the electrochemical series that they cannot be electroplated. If the radioisotope can be obtained as the metal, it can be evaporated in vacuum (see Holland, 1956). However, the yield is usually $<2\%$; the yield is equal to the sample area, typically ~ 1 cm^2, divided by $2\pi r^2$, where r, the distance of the sample from the boat, is usually ≥ 3 cm so the sample does not get overheated by thermal radiation. Thin layers of radioactive salts, such as halides, can also be deposited by evaporation. Tantalum boats are a preferred evaporation dish, being less brittle than tungsten. The biggest problem is attack of the boat by the material to be evaporated; We have found that rapid heating to high temperature sometimes avoids this problem. A chimney or large beaker should be placed over the sample and boat to prevent contamination of the belljar. Sputtering is seldom used for depositing radioisotopes, because sputtering targets are usually much too big to be made of high-specific activity radioisotopes; however, sputtering has been used to deposit thin layers of uranium (Weil et al., 1959), and ^{110m}Ag was sputtered onto Pd–Si metallic glass from an electrodeposit on Pd (Gupta, 1975; Gupta et al., 1975).

Some metals such as Al cannot be plated on because of the oxide layers on their surfaces (Peterson and Rothman, 1970); the tracer can then be deposited by vacuum evaporation of the metal. Tracers can be successfully electroplated onto Ti from a dimethyl sulfoxide electrolyte (Santos and Dyment, 1973). Metallic tracer can also be electroplated onto a Pt (but not Ta or W) filament and evaporated from there. The tracer deposit to be evaporated from the filament must be perfectly dry or vaporization of the water content will blow the deposit all over the belljar. Some chemical reactions to produce metallic tracer can be carried out on the filament just prior to evaporation, e.g., exchange of metallic ^{23}Na with ^{22}Na in NaCl (Mundy, 1971) or the reduction of ^{49}VCl$_3$ by Ca (Stanley and Wert, 1961). In some cases, when the tracer cannot penetrate through an oxide layer on the sample, the oxide layer is removed by sputtering and the tracer is evaporated onto the sample without breaking vacuum (Rothman, 1961).

The most elegant and undoubtedly the most expensive method of depositing radiotracers is ion implantation (Hood, 1970). The yield is $\leq 1\%$, the accelerator is contaminated, and some radiation damage is caused in the sample. On the other hand, this method can be made very clean, surface oxides can be penetrated, and because the implantation is done in a mass separator a carrier-free deposit is obtained, unless the carrier combines with a residual gas as in the codeposition of ^{59}CoH and ^{60}Co (Hood et al., 1983). The deposit occurs at a depth that depends on the implantation energy as well as on the atomic masses of the ions and base material (Littmark and Ziegler, 1980); for 50 keV ions, the implantation depth is a few tens of nanometers, which cannot be considered as $x = 0$ in microsectioning experiments. The profile of the deposit is roughly gaussian, with FWHM equal to about one-half the depth of the maximum concentration.

A tracer can be deposited on a less noble substrate, e.g., Cu on Pb (Mundy et al., 1974), by cementation. Chemical vapor deposition can be an advantageous technique for refractory materials such as tungsten (Mundy et al., 1978).

6. *Deposition of Thin Layers on Nonmetallic Samples*

In the case of oxides and oxide glasses, the radiotracer is dried on from aqueous solution. It is preferable first to convert to hydroxide by passing the tracer solution through a hydroxide-form anion exchange resin[1] (Rothman et al., 1982), or to convert to the sulfate or oxalate by taking to dryness with sulfuric or oxalic acid, respectively (Peterson and Chen, 1982), and picking up with H_2O. A wetting agent such as methyl alcohol or tetraethylene glycol can be added to spread the tracer solution uniformly over the sample surface. Of course, this technique cannot be used on materials, such as sodium silicate glasses or alkali halides, that are attacked by H_2O; drying an organic solution (e.g., ^{22}NaCl in ethanol) or vacuum evaporation is probably the best method for depositing the layers of radiotracer on such materials.

III. Annealing of Diffusion Samples

The objectives of a diffusion anneal are to hold the sample at a constant temperature T for a time t in an atmosphere that does not change the chemistry of the sample, and to measure T accurately.

Most laboratory furnaces are electrically heated. A sensor such as a thermometer or thermocouple feeds a signal to the controller, which increases or decreases the current. A wide variety of controllers are available and they will not be described here. The sample temperature is measured by a sensor attached to the sample. The exact type of heating apparatus depends on the desired temperature range.

---

[1] Analytical grade anion exchange resin, Ag 1 × 4, hydroxide form, Bio-Rad Laboratories, Richmond, California.

A. Low-Temperature Anneals ($T < 200°C$)

For $T < 200°C$, i.e., below the flash point of a number of silicone oils, a circulating bath can be used, with a thermometer as sensor. The fluid used in the bath depends on the temperature desired; oil, water, or alcohol are possible. If the sample could react with the fluid, it is encapsulated. Gupta and Kim (1980) have annealed Pb samples in an oil bath, using a Cu container tube filled with H_2 for heat conduction.

B. Intermediate-Temperature Anneals ($200 < T < 1850°C$)

In this temperature range, an electric resistance furnace is used, commonly a tube furnace with an alumina tube on which the heating element is wound (Fig. 5). For temperatures up to 1000°C, Nichrome elements are used. Kanthal is good to 1200°C, above which Pt–30% or 40% Rh windings are used; these materials need no protective atmospheres. Other types of heating elements are also possible. High-purity recrystallized Al_2O_3 tubes fail at 1800–1850°C, or at ~1650°C under vacuum. The control thermocouple should be as close to the windings as possible to increase sensitivity. A separate couple is used to measure the sample temperature, which is not necessarily the same as the temperature set on the controller. Up to perhaps 1200°C, a Pt-lined stainless steel or Monel metal heat sink can be used to level out temperature gradients and increase the thermal inertia of the system.

Samples to be annealed in vacuum at $T \leq 1100°C$, or at higher temperatures for short times, can be encapsulated in fused quartz. To prevent reaction with the SiO_2, the sample can be placed in an Al_2O_3 cup or wrapped

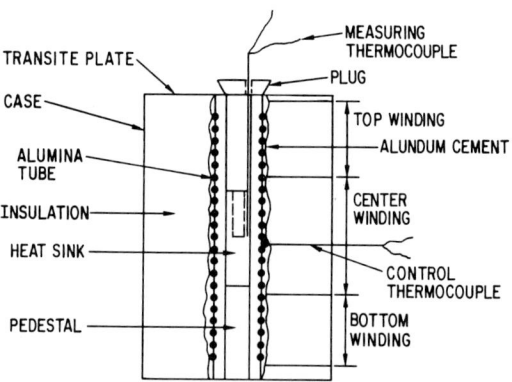

Fig. 5. Schematic diagram of vertical tube furnace. The windings are held on the alumina tube with Alundum cement. Power to the windings is adjusted separately to minimize the gradient. Only the center winding is controlled. The plug and pedestal are made of fire brick.

in Ta or Pt foil. Samples to be annealed in vacuum at $T > 1200°C$ can be electron-beam welded into Ta or Pt capsules (Peterson, 1964), but Ta capsules must be protected by an inert gas atmosphere. An advantage of a capsule is the relatively small volume (~ 10 cm^3)—even with a poor vacuum, the amount of reactive gas present is small, and can be gettered with Zr chips or Ta sheet. Metallic samples can be annealed in flowing purified Ar or He. If oxidation of the samples is to be avoided, a tight all metal and glass system must be used, as well as leak tested regulators on the gas cylinder. Commercial room-temperature oxygen traps, followed by Ti chips maintained at 800°C, lower the O_2 concentration in a tight system to below the parts per billion range from the nominally 99.999% pure noble gas (Mundy, 1982). Possible radioactive contamination due to sweeping of the radioisotope by the flowing gas must be carefully controlled.

Oxides are annealed under the oxygen pressure appropriate to the defect concentration of interest, as obtained from the usual mass action laws (Kofstad, 1972). The desired oxygen pressure over the range 1 to $\sim 10^{-4}$ atm ($10-10^5$ Pa) O_2 is obtained by mixing pure O_2 or an Ar–O_2 mixture with high-purity Ar; Peterson (1982) and co-workers have gotten as low as 10^{-7} atm (10^{-2} Pa) O_2 with Ar–O_2 mixtures. The oxygen pressure in this range is measured with a ZrO_2 cell; these are available commercially.[2] Lower oxygen pressures are obtained from CO–CO_2 or H_2–H_2O mixtures using special gas mixers; the former are more convenient. The $p(CO)/p(CO_2)$ ratio can be varied from 10^{-3}–10^3, which corresponds to a 10^{12} change in $p(O_2)$. The exact pressure range depends on the temperature in the manner depicted in Fig. 6 (Muan and Osborne, 1965). The $p(CO)/p(CO_2)$ ratio in the exit gas can be measured with a gas chromatograph, or a ZrO_2 cell can again be used. Standard mixtures of CO/CO_2 (100:1, 1:1, 1:100) are available for calibration purposes.

C. High-Temperature Anneals ($T > 1850°C$)

The only containers for a controlled atmosphere that can sustain temperatures above 1850°C are ZrO_2 tubes, and these are expensive and sensitive to thermal shock. Otherwise, one uses tungsten heaters or electron cloud furnaces (Flinn, 1962) in vacuum, or a graphite heater in vacuum or inert gas; the latter is thoroughly deoxidized by the hot graphite. The sample can be encapsulated in a Ta can by electron beam welding: encapsulation of the sample can prevent contamination of the furnace by evaporating radiotracer. Temperature gradients in the sample are possible under these conditions and must be guarded against.

---

[2] Thermox Co., Pittsburgh, Pennsylvania.

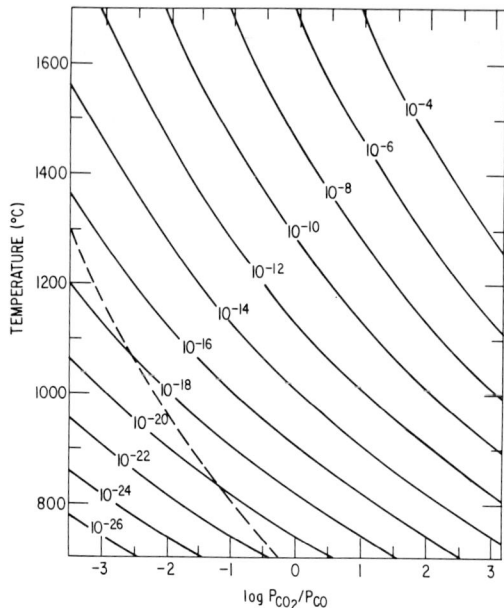

Fig. 6. Oxygen pressure (atm) as a function of temperature and CO_2/CO ratio at a total pressure of 1 atm. Region below dashed line is metastable. [After Muan and Osborne, "Phase Equilibria Among Oxides in Steelmaking," © 1965. Addison-Wesley, Reading Massachusetts. Figure 11. Reprinted with permission.]

D. Temperature Measurements

For a diffusion measurement to have any meaning, the temperature of the diffusion anneal must be known accurately in terms of the 1968 International Practical Temperature Scale (IPTS, 1976). A temperature measurement thus consists of two parts, the measurement of an apparent temperature, and calibration of the instrument according to points of the IPTS. These two parts will be discussed separately. The temperature of the sample may differ from the temperature of the sensor (thermometer or thermocouple), and a calibration of this difference may also be necessary. Also, there may be severe temperature gradients across the sample.

Temperatures of samples in fluid baths are measured with thermometers touching, or near to, the sample, or inside the sample container (Gupta and Kim, 1980). Temperatures from 200 to 1500°C are measured with thermocouples. The thermocouple can touch the sample when neither is encapsulated. An encapsulated sample can be placed, together with the thermocouple, into a heat sink, which tends to level out thermal gradients, or the thermocouple can be attached to the outside of the capsule. The type of thermocouple used depends on the temperature and atmosphere of the anneal. Chromel–Alumel couples (Type K) are useful in oxidizing atmospheres to perhaps 1000°C. Noble metal thermocouples are the freest from drift and least likely to become contaminated; the Pt–Pt 10% Rh (Type S) thermocouples can be used in oxidizing atmospheres to 1500°C, but must be recalibrated frequently if used in reducing or neutral atmospheres above 1000°C. The Pt 6% Rh–Pt 30% Rh thermocouples can be used up to 1700°C. Tungsten–rhenium couples (W 5% Re–W 26% Re) can be used in vacuum or a reducing atmosphere to 2750°C, but the calibration changes due to loss of Re. As the thermocouple EMF depends on the difference in temperature between the hot and cold junction of the thermocouple, the cold junction is held at a fixed temperature, usually 0°C. This is done more conveniently with an electronic cold junction compensator than with an old-fashioned ice bath. The EMF can be measured with either a digital voltmeter or a potentiometer. If the thermocouple does not touch the sample, a separate calibration run should be made with a second thermocouple touching a dummy sample in order to determine the true sample temperature.

High sample temperatures ($T > 1200$°C) can also be measured with an optical pyrometer (Kostkowski and Lee, 1962). This is essential in an electron cloud furnace, in which the sample is at a potential of several kV with respect to ground. In such a case, blackbody holes (Quinn, 1980) should be drilled in the sample so that emission corrections are small and easily calculated; a deep tube furnace is in fact a blackbody hole. Our preference is to use thermocouples whenever possible, as they require less subjective operator judgment than a disappearing filament pyrometer. Scattered light from the electron-emitting filament in an electron cloud furnace can give false optical pyrometer readings (Einziger and Mundy, 1976).

E. Calibration of Temperature Measuring Instruments

By calibration, we mean relation to IPTS-68. This scale consists of primary and secondary fixed points as well as equations defining the temperature between and outside these points.

The calibration of thermocouples has been authoritatively described by Roeser and Lonberger (1958). Our practice has been to calibrate thermocouples at fixed points up through the Ni or Pd points, and to interpolate and extrapolate these differences on a linear basis. Calibrations up to the Cu point (Sn, Zn, Al, Ag, and Cu) are done in a National Bureau of Standards (NBS) type apparatus (Roeser and Lonberger, 1958), in which the thermocouple, protected by an alumina tube, is immersed in a bath of NBS-certified pure metal. Calibrations at the Au, Pd, or Ni points are carried out by the wire method (Roeser and Lonberger, 1958), in which a few millimeters of wire of the metal is spot welded between the legs of the thermocouple to be calibrated. The EMF is recorded while the assembly is heated, until the metal melts.

Another method of calibrating thermocouples is to heat them together with either a thermocouple that has been calibrated at the IPTS fixed points or a calibrated thermocouple purchased from NBS (Roeser and Lonberger, 1958). The hot junctions of the thermocouples should be close together in a massive heat sink. Many laboratories keep a "primary standard" thermocouple calibrated against IPTS or by NBS; a "working standard" thermocouple is calibrated against the primary standard, and is used to calibrate the furnace thermocouples.

Optical pyrometers are more difficult to calibrate than thermocouples because more individual judgment is involved. The calibration should be carried out by the same person who will use the pyrometer, and under very similar conditions. Strip lamps or blackbodies and sectored disks from NBS can be used; details are given in the paper by Kostkowski and Lee (1962).

F. Accuracies Obtainable in Temperature Measurements

The uncertainty in a temperature measurement is the square root of the sum of the squares of the uncertainties in the measurement of the apparent temperature, in the calibration to IPTS-68, in the IPTS standard, and in the sample–sensor temperature difference. Temperatures in fluid baths can be measured to $\pm 0.1°C$. Estimates of obtainable accuracies have been given by Roeser and Lonberger (1958) for thermocouples, by Kostkowski and Lee (1962) for optical pyrometers, and by Gray and Finch (1971). Our estimates of the uncertainty of a sample temperature are considerably more conservative, e.g., $\pm 1.5°C$ with a thermocouple near $1000°C$. [This estimate is obtained from the $\pm 3\%$ uncertainty in D (see Section I.C).] With an optical pyrometer, a skilled operator under ideal conditions can read the temperature of a blackbody hole with an uncertainty of a few $°C$, but in a vacuum at $T > 2000°C$, the temperature gradients can be large, and the uncertainty in the temperature of the diffusion zone can be an order of magnitude greater. However, Eq. (19) indicates that a given value of ΔT leads to a smaller value of ΔD at high temperature; e.g., $\Delta T = 30°C$ at $3000°K$ leads to $\Delta D/D = 0.2$ for $Q = 480$ kJ mole^{-1}.

G. Correction for Heat Up and Cool Down

A sample placed into a furnace at temperature takes some time to heat up, and to cool down when removed from the furnace. In addition, there are temperature fluctuations during the anneal, giving a temperature (T) versus time (t) plot for the anneal as shown schematically in Fig. 7. This plot is converted to the appropriate values of T and t, i.e., T_0 and t_0, by integrating D over the anneal as

$$D(T_0)t_0 = \int_0^{t_f} D[T(t)]\, dt$$

$$= D_0 \exp\left(-\frac{Q}{RT_0}\right) \int_0^{t_f} \left[\exp -\frac{Q}{R}\left(\frac{1}{T(t)} - \frac{1}{T_0}\right)\right] dt \quad (22)$$

Then t_0 is just the integral on the right-hand side. The value of T_0 is first determined by averaging the temperature readings in the flat region (say between 10 and 45 min in Fig. 7). If the fluctuations are not too great, the thermocouple EMF can be averaged and converted to temperature. A graph of $\exp\{(-Q/R)[(1/T(t)) - (1/T_0)]\}$ versus t (Fig. 7) is then constructed by using a preliminary value of Q, and the desired area determined. The calculation can be done on a computer, using a polynomial to convert millivolts to temperature, or by hand and graphical integration. The calibration corrections are added to T_0 later. When final values of the Ds are obtained, the value of the activation enthalpy Q can be refined. The heat up correction can be very important if Q or the time at temperature is small, or if the heat up and cool down are prolonged because the sample is sensitive to thermal shock.

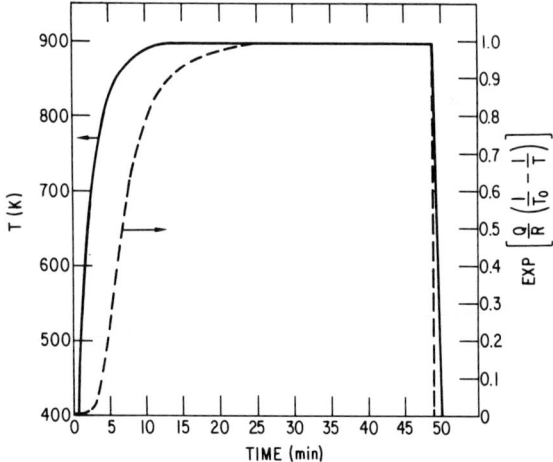

Fig. 7. Plots for calculating heat up correction for Al sample heated to 625°C; t_f is 50 min; (—) temperature in °K, (---) $\exp\{(Q/R)[(1/T_0) - (1/T)]\}$.

IV. Sectioning and Microsectioning

The sectioning technique consists of dividing the diffusion zone into sections and determining both the x coordinate of the section and the concentration in it of diffusing material, or the specific activity of diffusing radiotracer. From fitting the resulting $C(x)$ to the solution of the diffusion equation, D is obtained. It is in principle possible to determine $C(x)$ without sectioning, e.g., with an electron microprobe (Adda and Philibert, 1966, p. 266), from a hardness traverse (Lévy et al., 1959), etc., but in the case of a radiotracer experiment, sectioning allows a more precise determination of $C(x)$ because radiation from neighboring sections does not interfere.

To understand sectioning, the reader should think in terms of the isoconcentration contours. These are parallel to the original surface on which the thin layer is deposited, and perpendicular to the diffusion direction. The most important criterion of correct sectioning is the parallelness of the sections to the isoconcentration contours.

In this discussion, we make an arbitrary distinction between sectioning and microsectioning. In the former, the original surface is flat to better than $2\sqrt{Dt}$, and is therefore used as a reference surface. The sections are removed parallel to it, and the sample is remounted, if need be, with the normal to that surface as the reference. For microsectioning, the roughness of the original surface is $>2\sqrt{Dt}$, so the isoconcentration contours are not flat on the scale of $2\sqrt{Dt}$ and the surface is not a reference surface for alignment. Another difference is that in sectioning, the thicknesses of the sections are determined individually, usually from their weights. In microsectioning, the entire thickness removed is determined. Microsectioning is discussed separately in Section IV.C; however, the mathematics described in Section I.D remain the same: one removes $\sim 6\sqrt{Dt}$ from the sides and takes ~ 20 sections, each $\sim \sqrt{Dt}/3.8$ thick.

A. Mechanical Sectioning

Most sectioning is done mechanically, the exact technique depending on the material and the thickness desired (Table I).

1. Lathe Sectioning

Lathe sectioning is the standard technique for ductile metals and alloys with $D > 10^{-12}$ cm^2 s^{-1}. The sample, cylindrical in shape, is placed in an adjustable chuck which is then aligned so that the front face of the sample is accurately perpendicular to the lathe axis. The requisite $6\sqrt{Dt}$ is removed from the radius, and the sample is then sectioned.

TABLE I
Instruments for Mechanical Sectioning

Materials and characteristics	Instrument	Section thickness (μm)
Ductile, but not too soft (metals and alloys, e.g., Ni, Cu, Brass)	Lathe	5–100[a]
Ductile (metals and alloys, e.g., W, stainless steel, Cu)	Grinder	3–100
Ductile, but not too soft (metals and alloys, Al, Pb, Fe)	Microtome with carbide blade	1–3[a]
Plastic and very soft (organic solids, Pb, Al, Sn)	Microtome with steel blade	1–10[a]
Soft but brittle (alkali and silver halides)	Microtome with steel blade	3–10[a]
Hard and brittle (ceramics, silicate glasses, semiconductors)	Grinder	3–100

[a] For the lathe and microtome, these are thicknesses of one cut. Several cuts can be put into the same section.

The essential equipment for lathe sectioning is a precision lathe, an adjustable chuck, preferably with a chip catcher, and a properly shaped tool. The lathe must have excellent bearings with no play in the headstock. Adjustable chucks for aligning samples have been described by Rothman and Sobocki (1959) (Fig. 8), Adda and Philibert (1966, p. 254), and Layer and Meyer (1962). Chip catchers are described in the first two references. A good chip catcher prevents loss of machined material, which is important because the section weight enters into the value of the x and C coordinate of the section, and the x coordinate of all subsequent sections (see Table II). The type of tool used depends on the material. Soft, pure materials (e.g., noble metals and their alloys) are sectioned with a tool steel bit, while tough materials with hard inclusions (e.g., uranium) require a carbide bit. The tool bit must leave a surface smooth to a fraction of the section thickness and cut chips without making dust, which causes chip loss. A different tool shape has to be developed for each material. Common characteristics are a large top rake angle to let the chip curl, a small radius, and a finely polished cutting surface. The tool bit must be mounted rigidly; we use a $\frac{3}{4}$-in. boring bar, which also goes easily through a hole in the chip catcher. The lathe speed and feed must also be optimized for each material.

If the sample is held directly in the adjustable chuck, care must be taken that it is neither deformed in tightening the chuck nor held so loosely that it falls out. The sample can also be mounted with epoxy or Eastman 910 on a platen which is held in the adjustable chuck. The sample is aligned with a 0.0001-in. (2.5-μm) dial gage against the face. If this process moves the sample off the lathe axis, it is recentered using a dial gage on the cylindrical surface and the four-jaw lathe chuck which holds the adjustable chuck. The alignment is then rechecked. Alignment may take more than an hour, as much time as the rest of the sectioning, although Layer and Meyer (1962) claim alignments in a few minutes with their chuck. The sides are next removed, with the tool moving out, i.e., opposite to the diffusion direction,

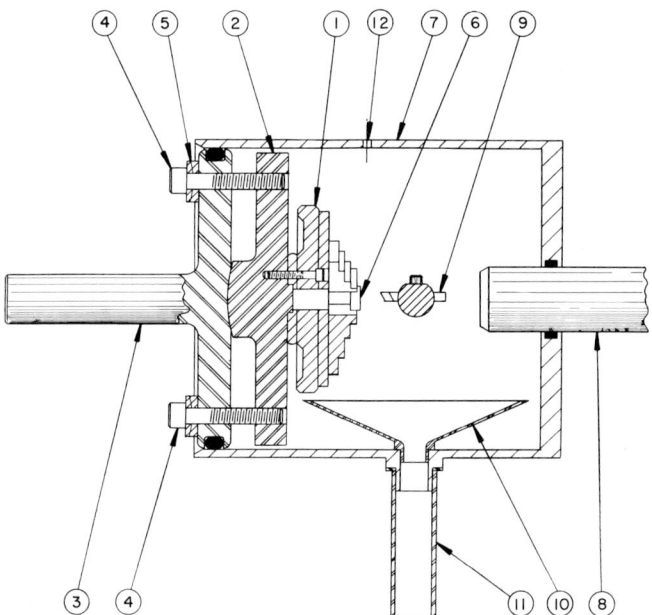

Fig. 8. Chuck and chip catcher for lathe sectioning: (1) three-jaw chuck for a jeweler's lathe; (2) chuck adapter plate (note spherical radius machined on back so that center of sphere is in sample face); (3) headstock bar; (4) screws for lining up; (5) spherical washers; (6) specimen; (7) acrylic plastic chip catcher; (8) tailstock bar; (9) tool; (10) funnel; (11) tared bottle; (12) hole for admitting gas. The last item, as well as the o-ring between the headstock bar and the chip catcher, were not used.

on the last cuts, to avoid smearing hot material over cold. Light cuts are always taken to avoid deforming the sample. The sample is then removed, weighed, and realigned. An error in alignment of angle α causes a fractional error in D of $d^2\alpha^2/32Dt$, where d is the diameter of the sample (Shirn et al., 1953). For very soft materials such as Au single crystals, the sides are removed during sectioning by running the tool in for the desired distance and discarding the chips, and then taking the proper section (Makin et al., 1957).

Finally, the sections are cut from the front face. The advance of the tool rest is measured with a dial gage mounted on the ways of the lathe. The chips are swept with an artist's brush from the catcher into a tared bottle; the bottles are handled with cotton gloves to avoid pick up of perspiration. A fresh brush is used for each section and one looks carefully for chips stuck in the brush. Brushes can be washed and reused many times. The section thickness can be determined from the section weight, or by means of a capacitor device (Meyer and Sekizawa, 1968). After sectioning, the sample is weighed again and the weight removed in sectioning is compared to the sum of the section weights; the difference should be $<1\%$. Occasional sections can be counted to see when two times background is reached; the proper counting procedures are described in Section V.

Radioactive contamination of the lathe is no problem if there is no chip loss. Nonetheless, the lathe bed should be covered with plastic sheeting.

2. Sectioning by Grinding

Sectioning by grinding is the standard method of sectioning brittle materials, and for metals if one wants sections 3–100 μm thick. The main instrument needed is a custom-made precision grinder: designs have been

TABLE II

PROBLEMS IN SECTIONING

Problem	Effect on plot of log C versus x^2	Remedy
Weight loss or incorrect weight for a section	A bend	Use average section thickness as the thickness of affected section to determine x for that and subsequent sections
Incorrect remounting of sample on grinder	Alternate low and high points	Combine counts and weights of low and high sections into one section

published by Letaw *et al.* (1954), Fisher and McSkimmin (1958), Goldstein (1957), Schamp *et al.* (1959), Deiss (1963), De Bruin and Clark (1964), Le Blans and Verheijke (1964), Adda and Philibert (1966, p. 256), and Hong *et al.* (1978). The principle of these instruments is the same (Fig. 9): The sample is mounted in a reproducible manner on a piston that moves accurately parallel to the normal to the front face, and is rubbed against an abrasive surface held accurately parallel to this face. The exact nature of the abrasive and how it and the sample are held depend on the material and the radiation characteristics of the isotope under study. The x coordinate can be determined by weighing the sample after each section; this involves removal and remounting of the sample after each section. The other possibility is to mount the sample holder in an interferometric device; however, because grinding removes the entire front surface, the reference surface must be the platen on which the sample is mounted. We weigh the sample to get x, and mount it on the platen with double-faced tape. This makes for quick removal and remounting, but there is a certain amount of slop involved, and as a result one may encounter alternately low and high points in the penetration plot for section thickness <3 μm (Table II, Fig. 10). The thickness of each section can be checked approximately by a micrometer measurement of the sample thickness.

Before sectioning, $6\sqrt{Dt}$ must be removed from the edges. For brittle samples, grinding must be used, hence a rectangular sample shape is preferred. Ductile samples can be machined on a lathe, between pads if they are too small for a chuck or collet, the tool again moving opposite to the diffusion direction to avoid smearing activity. Electrically conducting samples can also be trepanned by spark machining.

Alignment of the sample on the grinder is accomplished by grinding the

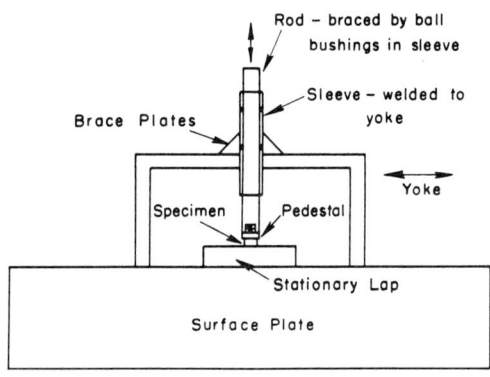

Fig. 9. Schematic diagram of grinder. [After Fisher and McSkimmin (1958).]

1. TRACER DIFFUSION COEFFICIENTS IN SOLIDS

front and back faces of the sample parallel on the same grinder before the isotope is deposited. After the edges are taken off, the sample is mounted on the grinder with the back face toward the abrasive and the back face is reground parallel to the front face. Special care must be taken with samples the sides of which were taken off on a lathe, because there can be a burr left. When taking the first section, the sample face is inspected after a very few turns of the grinder to see if the entire face is being cut; this is sometimes hard to observe on transparent materials such as silicate glass.

For studying the diffusion of gamma-emitting radioisotopes in metals, medium hard ceramics, and semiconductors, emery or SiC abrasive papers are excellent for section thicknesses 1 μm and up. The paper is held flat on a vacuum holder in our grinder. Grinding of nonradioactive samples to determine section thickness removed from new materials under a set of grinding conditions is strongly recommended. To remove 2 μm, a few figure eights on 600-grit paper will suffice for most materials.

After cutting a section, the sample is cleaned and weighed. The abrasive paper is folded and put in a counting bottle, along with the cleaning wipes and the double-faced tape on which the sample was mounted. This works for isotopes that emit hard gamma radiation, as long as the grinding papers are folded reproducibly, but the radiation from beta emitters would clearly be absorbed. Such samples can be ground with diamond powder on Ta foil, and the activity counted on the Ta foil with an end window counter (Mundy et al., 1978). If analytical chemistry needs to be done on the material removed, one can use diamond powder in a viscous but easily destroyed organic carrier such as butyl alcohol on flat glass plates (Rothman et al., 1966b). Preliminary counting of the samples during sectioning allows one to monitor the progress of the penetration plot, and to change the section thickness when this is needed.

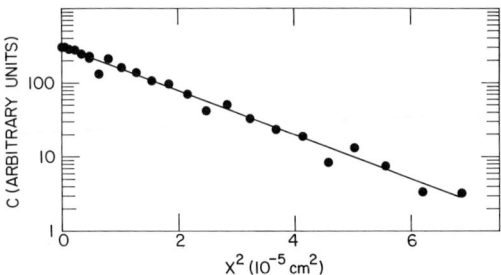

Fig. 10. Penetration plot for the tracer diffusion of Cr in a 15 at. % Cr–20 at. % Ni–1.4 at. % Si–Fe alloy, annealed 2560 s at 1363°C, showing scatter due to inaccurate remounting of sample. [From Rothman et al. (1980).]

3. Microtomes

Microtomes are more often used to prepare thin sections of tissue in biology laboratories. The sample moves against the edge of a very sharp blade to cut the section. The sample can be advanced very small distances (~ 1 μm), usually by means of a fine-threaded screw. Microtome blades must be kept very sharp to cut well, and present a safety hazard to the fingers.

A microtome with a steel blade is the sectioning instrument of choice for soft metals, e.g., Sn (Coston and Nachtrieb, 1964), Pb (Miller et al., 1973), Al (Peterson and Rothman, 1970), for organic materials (Brissaud-Lancin, 1978; Chadwick and Sherwood, 1975) alkali halides (Rothman et al., 1966a; Benière, 1970), and silver halides (Peterson et al., 1973). Use a D-profile blade for metals. A rugged instrument in which the sample is held firmly and the blade does not flex is needed. A sledge-base microtome, in which the sample holder slides on ways, is usually used. A carbide bladed, hydraulically powered sledge-base model[3] has been used to cut Fe and Zr, leaving optically flat surfaces (Graham, 1969), and for some superb work on alkali halides (Benière, 1970).

The sides are removed as described in Section 2, or by cleaving in the case of alkali halides. Alignment is by means of an optical system (mirror on sample, prism, graph paper on a convenient wall 2 m or so away, Rothman et al., 1966a) a feeler gage (Graham, 1969), or use of a demountable sample holder, on which the surface is originally prepared (Gupta and Kim, 1980). A special sample tray is fabricated with a hole under which the tared bottle is held, and the chips are swept in with a brush. Most microtomes have an automatic sample advance on which the thickness of the cut can be set. Certain materials (alkali halides, Al) cannot be cut too thin as the cuttings come apart and are difficult to collect. Cuts that are too thick cause the knife to dig into the sample. Alkali halide samples need to be moistened slightly to keep the chip from breaking into powder; the sample bottles are baked to drive off the H_2O before weighing.

4. Some Special Problems in Sectioning

The sectioning of reactive materials is carried out under a protective atmosphere, e.g., alkali metals can be sectioned on a hand microtome in a glove box filled with dried nitrogen (Mundy, 1971). An apparatus for sectioning samples by grinding at liquid N_2 temperature has been described by Weil et al. (1979).

---

[3] Model K, R. Jung, A.G., Heidelberg, West Germany.

B. Other Techniques

A few sectioning experiments have been carried out by chemical or electrochemical sectioning (Section IV.C), e.g., ^{22}Na in Na silicate glass by Barr et al. (1972), ^{60}Co in Co metal by Lange et al. (1962), Fe in Ni by Neiman and Shinyayev (1954).

The depth profile of radiotracers has also been determined nondestructively by moving the diffusion sample parallel to the diffusion direction past a collimating slit which is in front of a detector (Kim et al., 1977; Dejus et al., 1980). This technique is suitable only for $D \geq 10^{-7}$ cm^2 s^{-1}. Especially good resolution is obtained by coincidence counting the 0.511 MeV annihilation radiation from positron emitting radioisotopes with two counter–slit combinations located at exactly 180° from each other on opposite sides of the sample (Kim et al., 1977).

Instead of measuring the activity of the sections, one can measure the residual activity of the sample (Gruzin, 1952; Seibel, 1964). The residual activity after removing a length L is given by

$$A(L) = \int_L^\infty C(x) \exp[-\mu(x - L)] \, dx \qquad (23)$$

where collimated radiation and exponential absorption with linear coefficient μ have been assumed. Seibel (1964) has given the general solution, which is independent of the functional form of C:

$$C(L) = A(L)(\mu - d \ln A(L)/dL) \qquad (24a)$$

We consider the Gruzin method to be less desirable than counting the sections, except in the limiting cases $\mu \gg d \ln A(l)/dl$ and $\mu \ll d \ln A(l)/dl$. The first corresponds to radiation so weak that most of it is absorbed in the thickness of one section; examples are the weak β radiation from ^{3}H, ^{14}C, or ^{63}Ni. In this case the two methods are equivalent, except that the Gruzin technique obviates the tedious preparation of sections for counting. The second case corresponds to gamma radiation so hard that absorption is negligible, and then reduces to a subtraction technique

$$A(\text{section } n) = A(L_n) - A(L_{n+1}) \qquad (24b)$$

which can be less precise than a straightforward measurement. In the case in which the two terms in Eq. (24a) are comparable, μ must be measured accurately in the same geometry as the sample is counted. Also, the Gruzin technique is useful only when the specimen can be moved to the counter repeatedly without losing alignment.

There is an "integral activity" method that does not require measurement of the absorption coefficient; activity is measured as a function of time on both the front and back surfaces of the sample (Zhukovitskii et al., 1955). Obviously, the thickness of the sample must be of the order of μ^{-1}, the reciprocal of the linear absorption coefficient. For example, gold foils 10^{-3} cm thick have been used with the radioisotope ^{195}Au, which emits x rays (Gainotti and Zecchina, 1965). The L x rays (about 10 keV) have an absorption coefficient of 2320 cm^{-1}, while the K x rays ($\sim$67 keV) are not absorbed at all and are counted to make sure no radioisotope is lost by evaporation. The solution to the diffusion equation is, for a thin layer of thickness h and for t large and specimen thickness l small,

$$C = (C_0 h/l)\{1 + 2\exp[-(\pi/l)^2 Dt]\cos \pi x/l\} \tag{25a}$$

where C_0 is the activity of the layer. If $I_1(t)$ and $I_2(t)$ are the activities from the front and back sides, respectively, then

$$\ln\left(\frac{I_1 - I_2}{I_1 + I_2}\right) = \ln k - \frac{\pi^2 Dt}{l^2} \tag{25b}$$

where the constant k contains all the absorption terms (Zhukovitskii et al., 1955). An application to grain boundary diffusion as well as a detailed description of the experimental procedure have been given by Gupta and Campbell (1980).

C. Microsectioning

We define *microsectioning* (Gupta, 1975) as the cutting of sections a few hundred nm or less in thickness, so that the surface on which the tracer is deposited, the "front" surface, is not necessarily flat on the scale of $\sqrt{Dt}$, and so the thicknesses of the individual sections are not determined separately. The isoconcentration contours then follow the contour of the front surface, and one must remove sections parallel to this nonflat surface, rather than parallel to a flat surface. If this condition is met and if the undulations in the front surface are gentle (radius of curvature $\rho \gg \sqrt{Dt}$), one can treat the sections as if they were flat; see, however, Wuttig (1969).

If the irregularities in the surface are sharp, e.g., small pits, so that the sectioning proceeds across rather than along them, they can behave as a Fisher (1951) grain boundary, i.e., a thin region of high-diffusivity material, and thus give rise to a tail on the penetration plot (Fig. 11) apparently due to a "short-circuiting path" (Rothman et al., 1976; Mundy et al., 1981; Atkinson and Taylor, 1981). Even careful surface preparation can fail to eliminate all surface irregularities; the tails due to the ones that remain are eliminated in the data analysis, but at the cost of added uncertainty in the values of D.

This problem is especially serious for measurements of grain boundary diffusion, where both the artifact and the grain boundary diffusion are in the tail (Atkinson and Taylor, 1979).

Another resolution problem found in microsectioning is the apparent spreading of the thin layer (Rupp *et al.*, 1969; Rothman *et al.*, 1976). That is, even though the thickness of the deposited isotope layer, as calculated from the activity of the sample and the specific activity of the isotope, is less than one section, the penetration plot for a blank, i.e., undiffused sample is spread over several sections (Fig. 12). This spreading, which is more or less gaussian, is treated as an instrument resolution function and the *apparent* $4Dt$ is subtracted from the $4Dt$ obtained from the penetration plot.

The thickness of sections removed in microsectioning is too small to be determined accurately either by weighing or optical means; one therefore determines the total thickness removed and assumes that the section thickness is uniform if the sectioning conditions are maintained. This assumption has been checked by sectioning irradiated single crystals (McCargo *et al.*, 1963; Davies *et al.*, 1964; Lam *et al.*, 1972a; Mundy *et al.*, 1981) and found to be good to a few percent. Note that this only means that the same amount of material has been removed in each section; it does not guarantee that the sections follow the isoconcentration contours. However, the uniformity of sufficiently thick anodized oxide films can be seen from the interference colors, and the uniformity of removed layers can be checked interferometrically.

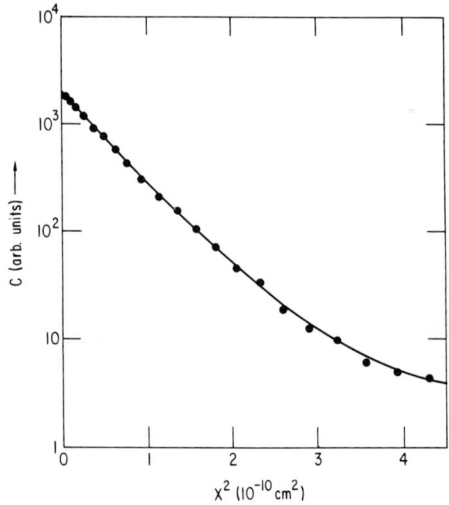

Fig. 11. Penetration plot for self-diffusion in Ag at 424.9°C. Nongaussian behavior is believed to be due to surface artifacts. [From Lam *et al.* (1973).]

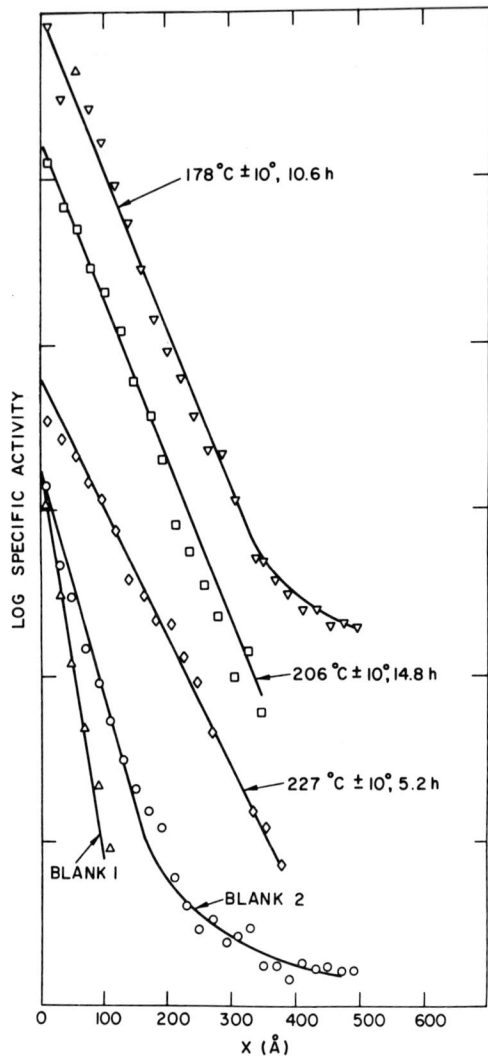

Fig. 12. Plot of log C versus x for self-diffusion in Ag single crystals, two blank samples, and three samples annealed under electron irradiation. Note that slopes of penetration plots from blank samples are (1) irreproducible, (2) significant with respect to slopes of annealed samples [From Rothman *et al.* (1976).]

The main techniques for microsectioning are chemical or electrochemical dissolution and sputtering. Vibratory polishing has also been proposed (Whitton, 1965), but this method can be used only with the residual activity (Gruzin) technique, and works only on fairly hard materials. The main advantage of chemical and electrochemical techniques is their low cost. When the conditions have been worked out, e.g., for anodizing tungsten (McCargo *et al.*, 1963), the method is simple and requires only beakers, chemicals, and a small dc power source. On the other hand, working out the proper conditions for a new material can be time consuming, expensive, and not always successful. Sputtering is applicable to most materials that do not decompose under ion bombardment, but requires an expensive apparatus to carry out.

It is appropriate to mention that many microsectioning techniques were developed not for diffusion experiments but for the determination of the ranges of energetic ions in crystals (McCargo *et al.*, 1963; Anderson and Sørensen, 1968; Davies *et al.*, 1960; Whitton and Davies, 1964). Table III lists the details of the chemical or electrochemical methods for several materials.

1. *Chemical Techniques*[4]

Simple chemical dissolution followed by counting the solution has been used for metals (Barr *et al.*, 1962), alloys (Styris and Tomizuka, 1963), and silicate glasses (e.g., Barr *et al.*, 1972; Kelly and Tomizawa, 1980). One uses a constant volume of solvent in the counting vial, rinses the sample so that the rinse flows into the same vial, and then adjusts the total volume of solution so that it is the same for all sections. Alternatively, the sample can be rinsed by dipping it into water in a second counting vial, then counting both vials and adding the activities. The amount of metal in each section can be determined spectrophotometrically (Barr *et al.*, 1962). Exact control of the dissolution time is important. The absence of preferential dissolution at grain boundaries, etc., should be checked under an optical microscope. The sides and back of the sample are masked with lacquer or paint. It is also possible to grow a corrosion layer on a metallic sample and dissolve this layer chemically; this is the chemical analog of the anodization described in Section IV.C.2 (Andersen and Sørenson, 1968).

2. *Anodizing*

Anodizing applies to metals. An anodic oxide film is grown on the front face of the sample and is subsequently dissolved in a solution that does not

---

[4] See also Section IV.B.

TABLE III

Chemical and Electrochemical Techniques for Microsectioning Metals

Metal	Technique	Solution for forming corrosion layer	Conditions for layer formation	Solution for stripping corrosion layer	Thickness removed	Remarks	Reference
Ag	Chemical	I_2 in chloroform or Br_2 vapor	—	1M $Na_2S_2O_3$	Tens of nm, Depends on I_2 conc. and time	AgI is photosensitive—strip immediately after forming layer; excellent polish required on original surface	Andersen and Sørenson (1968, 1969)
Ag	Anodize and strip	5 g liter^{-1} KOH	0.035–1.7 mA cm^{-2}, <2 min, Pt cathode, room temp.	1% (vol) NH_4OH	<100 nm, rate in nm s^{-1} = 0.91 (mA cm^{-2})$^{-1}$	Anodization follows Faraday's law; layer thickness is independent of crystal orientation or KOH conc.	Lam et al., (1972a)
Al	Anodize and strip	30 g liter^{-1} NH_4 citrate, 25°C	Al beaker serves as cathode, 10 min	30 g liter^{-1} Cr_2O_3, 50 g liter^{-1} H_3PO_4, 90°C	Thickness in nm = 0.1 (V + 1.8)	Voltage-limited anodization	Davies et al. (1960)
Au	Chemical	Br_2 vapor	Parabolic with time	Chloroform	Tens or hundreds of nm, increases with temp.	—	Andersen and Sørenson (1968, 1969)
Au	Anodize and strip	1M H_2SO_4	Gold cathode	5N HCl	2–400 nm	Anodization follows Faraday's law; efficiency increases with H_2SO_4 conc.; 1.5% of removed Au in anodizing solution; rate independent of crystal orientation	Whitton and Davies (1964)
Cu	Chemical	I_2 in chloroform	Immersion	1% HCl in ethyl alcohol	≤10 nm	Thickness increases with I conc. and parabolically with time	Andersen and Sørenson (1968, 1969)
Cu	Anodize and strip	0.1 g liter^{-1} Na_2SO_4	0.1–2 mA cm^{-2}, Room temp., Pt cathode	0.1% HCl in ethyl alcohol, 10 s immersion	Rate in nm s^{-1} = 0.42 mA cm^{-2}	Anodization follows Faraday's law, oxide dissolves during anodization	Lam et al., (1972b)

Metal	Method	Solution	Conditions	Thickness/Voltage info	Notes	Reference
Mo	Anodize and strip	Glacial acetic acid base, 0.02 M $Na_2B_4O_2 \cdot 10 H_2O$, 1.0 M H_2O	25°C, Pt cathode, 2 mA cm^{-2} for 6 min, dry with compressed air	1.0 g $liter^{-1}$ KOH, 30 s immersion	—	Voltage dependent anodization	Arora and Kelly (1972)
Nb and Ta	Anodize and strip	0.2% KF	10–200 V	Strip with Scotch tape, 200 Å	0.95 nm metal V^{-1} (Nb), 0.65 V^{-1} (Ta) thickness, 200 Å	Voltage dependent anodization, excellent surface needed for successful stripping	Pawel and Lundy (1964), Pawel (1964), Lundy et al. (1965)
Nb	Anodize and strip	H_2SO_4-3 $(C_2H_5)_2SO_4$	Pt cathode, current < 5 mA cm^{-2}, allow current to drop 2 min after maximum V is reached	Sat. KOH	Thickness removed in nm: $1.4 + 0.48\,V$, $15 < V < 50$	Voltage dependent anodization, stripping time is important as metal is attacked	Arora and Kelly (1974)
Si	Anodize and strip	0.04M KNO_3 in n-methylacetamide	Pt cathode; Immerse sample in bath; Stir bath at 34–35°C, 6 mA cm^{-2}, until desired voltage is reached	1.0 HF, immerse for 2–3 min	3–70 nm (see references for equations)	Voltage dependent anodization, oxide does dissolve in anodizing solution, there are problems with very pure Si	Davies et al. (1964); Przyborski et al. (1969)
V	Anodize and strip	Same as Mo	Same as Mo	Same as Mo	Thickness removed in nm = $9.0 + 1.0\,V$ for $V \geq 4$	Same as Mo	Arora and Kelly (1973)
V	Anodize and strip	Acetone containing 22 g $liter^{-1}$ benzoic acid and 10 cc $liter^{-1}$ H_2O supersaturated with $Na_2B_4O_7$	Cathode is the stainless steel beaker, hermetically closed, 3 min time, rinse with acetone	10% NH_4OH, 3 min	Thickness removed in nm is $10.3 + 1.25\,V$, 20–60 nm	Voltage dependent anodization, metal not attacked by stripping solution, oxide is dissolved by water	Pelleg (1974)
W	Anodize and strip	0.4 M KNO_3, 0.04 M HNO_3	Pt foil cathode 1.5–70V, to max V, then hold at that V for 60s	1.0 g $liter^{-1}$ KOH, 1 min	1–100 nm, thickness removed in nm is $0.44 + 0.67\,V$ $(V < 5)$, $1.2 + 0.55\,V$ $(V > 5V)$	—	McCargo et al. (1963)

attack the base metal. The amount of metal in the film thus equals the amount of material in the section.

There are two kinds of anodization, voltage- and charge-dependent. In the former, the thickness of the anodic film is determined by the voltage applied. In the latter, the thickness of the film is proportional to the charge passed, according to Faraday's law. The latter films can blister or dissolve in the anodizing solution, so anodizing times should be kept short.

Our practice has been to attach the current lead to the back of the sample with silver paint, and to paint the back and sides of the sample as well as the masking ring on the front face with a stripable paint such as Tygon. The sample is anodized in a counting vial using a strip of Pt foil as the cathode. The oxide film is stripped in another counting vial, and rinsed with a measured volume of H_2O into the latter vial. The activity in both vials can be counted and added to give the total activity of the section. Other special techniques are described in the papers cited.

Another technique for microsectioning involves electrolytic dissolution of a moving metallic sample on a filter paper soaked in electrolyte (Blackburn et al., 1961). The activity gradient on the filter paper can be determined autoradiographically or by counting, and the amount of material removed is determined from the electrical current.

3. *Sputtering*

a. *Principles of Sputtering.* In sputtering, material is removed by ion bombardment owing to the transfer of momentum from the bombarding ions to the atoms of the target. A depth profile can be constructed by (i) analyzing the sputtered-off material in a mass spectrometer (SIMS), (ii) collecting and analyzing the sputtered-off material, (iii) determining the concentration of the diffusing material in the remaining surface by, e.g., Auger electron spectroscopy (Holloway, 1975; Mayer and Poate, 1978; Stein and Joshi, 1981; Lea, 1983), or (iv) counting the residual activity of the entire sample (Lutz and Sizmann, 1964). As a rule, noble gas ions, especially Ar^+, are accelerated to a few hundred eV or more, with current densities ≤ 1 mA cm^{-2}. This is called physical sputtering, in contrast to bombardment with reactive ions, which is called chemical sputtering. Typical removal rates are of the order of 10 nm min^{-1} for 1 mA cm^{-2} of 500-eV Ar^+ ions. There are two excellent reviews of the subject of sputtering (Behrisch, 1981, Chapman, 1980), and the reader is referred to these for an understanding of the process. The present discussion is limited to the problems of precision sectioning by sputtering with noble gases.

The amount of material removed per unit time is equal to the current density of bombarding ions times the sputtering yield Y per ion (Sigmund, 1969, 1981). To obtain uniform removal of material, both values must be uniform over the bombarded surface. Problems in sectioning by sputtering most often occur because of spatial or temporal variations of Y. The sputtering yield Y depends on the mass and energy of the bombarding ions, as well as their direction with respect to the crystal axes of the target, because of channeling effects. The yield also depends on material parameters, mainly the surface binding energy, but in a complicated and not yet quantitatively understood way. Other things being equal:

(i) Chemically different species in a target are removed at different rates (preferential sputtering). However, a steady state is reached in the sputtering of a homogeneous sample with the ratio of the surface concentrations of elements A and B given by

$$C_A^S/C_B^S = C_A^V Y_B / C_B^V Y_A \tag{26}$$

where the superscripts refer to surface and volume. The material sputtered off has composition C_A^V and C_B^V. Thus catcher experiments are more direct than ones that measure surface concentration.

(ii) Different crystal faces of the same materials may be removed at different rates, as may grain boundaries (anisotropic surface binding energy). Experimental evidence on this point is conflicting. Cathodic vacuum etching, used as a metallographic technique, does give good grain contrast [see, e.g., Stroud (1959)] and shows grain boundaries. However, Gupta and Asai (1974) claim that in their sectioning experiments on polycrystals, the different grains were eroded at the same rate, perhaps because the ion energy was <1 keV (Gupta, 1975).

(iii) Different materials sputter at different rates—e.g., Cu and Cu–30% Zn.

The first two processes suggest that sectioning by sputtering is applicable only to monatomic single crystals, but this limitation is not borne out by experiment. In addition to the self-diffusion experiments on single crystals (Gupta and Tsui, 1970; Gupta, 1973a; Maier et al., 1976; Maier, 1977; Bihr et al., 1978; Mehrer et al., 1978; Atkinson and Taylor, 1979; Mundy et al., 1981), successful volume and grain boundary self-diffusion experiments have been carried out on polycrystals (Gupta, 1973a,b; Gupta and Asai, 1974; Atkinson and Taylor, 1981) polycrystalline alloys (Perkins, 1973; Perkins et al., 1973), silicon (Mayer et al., 1977), and metallic glass [see, e.g., Gupta et al. (1975)]. Grain boundary diffusion of an impurity has also been measured (Atkinson and Taylor, 1982). A number of alloys have been depth profiled by sputtering and Auger electron spectroscopy (Holloway, 1975), but values of D were obtained in only one study (Wildman et al., 1975), perhaps because, as the third caveat above suggests, the section thickness is not uniform for a sample with a finite concentration gradient. (Many such depth profiles are plotted versus sputtering time, not distance.)

The resolution of the sputtering technique is limited by ion bombardment mixing [see, e.g., Anderson (1979)]. A zeroth order estimate of this resolution limit is the size of the collision cascade that gives rise to atomic displacements in the target. This is ≤ 10 nm for ions of energy ≤ 1 keV, and increases with increasing ion energy. Thus better resolution is obtained by bombarding with low-energy ions; the yield is increased by increasing the current density.

b. *Equipment for Sputtering.* All equipment for sputtering includes a vacuum chamber, pumping equipment, and a controlled gas leak such as a micrometer needle valve. A high-speed pumping system is needed as gas is passed continuously and there are bursts of desorbed gases to cope with. Cold-trapped diffusion pumps, cryopumps, or turbopumps have all been used.

All sputtering equipment has a gaseous discharge in it. Common glow discharges are not suitable, as too high a gas pressure is required, with resulting low mean-free paths and back diffusion of the sputtered atoms. Therefore, either an ion gun or an rf power source is used.

In an ion gun system, a plasma is created inside the gun and positive ions are accelerated out of the plasma through an aperture in the gun. The gas is introduced into the gun and leaks into the pumped chamber, so the pressure in the latter is reasonably low ($\leq 10^{-4}$ Torr or 10^{-2} Pa). Most guns contain an electron-emitting filament, a potential to accelerate the electrons through the gas and ionize the gas atoms, a magnet to make the electrons travel in a longer spiral path, and a potential to accelerate positive ions out of the gun. Two types of ion sources have been used in sectioning experiments, the custom-modified duoplasmatron of Maier and Schüle (1974), and a commercially available Kauffman-type gun[5] (Reader and Kauffman, 1975; Mundy and Rothman, 1983a); almost any ion source used in ion milling should be usable. The main requirement is that the source put out ion currents ≥ 1 mA cm^{-2} at ~ 1 kV over ~ 4 cm^2 area in a reasonably uniform beam ($\pm 10\%$ except at the very edge), and that the current stay constant over a period of several hours. The length of a run is limited by the life of the filament. Some of the Kauffman guns have an electron-emitting neutralizer filament in front of the gun. The electrons emitted by this filament

---

[5] Ion Tech Co., Fort Collins, Colorado.

keep the ion beam from blowing up due to Coulomb repulsion and also prevent the buildup of positive charge on an insulating sample.

In an rf system (Gupta and Tsui, 1970; Atkinson and Taylor, 1977) the plasma is maintained by the rf power even at pressures $\leq 10^{-3}$ Torr, and because the frequency is so high (13.56 MHz), no charge buildup occurs. Radio-frequency systems are also available commercially.[6]

In addition to the ion source, chamber, pumps, and valving, one needs a collector and a sample holder. These are usually custom made. Designs have been given by Gupta and Tsui (1970) and Atkinson and Taylor (1977) for rf systems, and by Maier and Schüle (1974) and Mundy and Rothman (1983a) for ion gun systems. The collector is either a carousel, with six Al planchets, which allow six sections to be taken before the chamber is opened (Gupta and Tsui, 1970), or a device like a camera back, on which polyester film is rolled; the latter allows 32 sections to be taken. The film is cut up and folded into counting bottles for counting. The sample holder in an ion gun system should allow

(i) the sample to be rotated to homogenize the bombardment,
(ii) the current to be measured, and
(iii) the heat of the bombardment (~ 0.5 W cm^{-2}) to be conducted away.

The edges of the sample are either removed in advance, as described in Section IV, or the sample is masked; the latter can lead to problems with removal of material from the side of the hole. An ion gun apparatus should be equipped with a movable shutter between the sample and gun, which allows the gun to be conditioned without damaging the sample.

4. *Secondary Ion Mass Spectrometry (SIMS)*

In the SIMS technique, the sample is bombarded by reactive ions, and the sputtered-off molecules are ionized in a plasma and fed into a mass spectrometer. The mass spectrum is scanned and the ion current for tracer and host atoms can be recorded simultaneously. The beam is swept over the sample and, in effect, digs a crater, the bottom of which is more or less flat; an aperture prevents ions originating from the edges of the crater from reaching the mass spectrometer (Fig. 13). The penetration plot is constructed from the plots of instantaneous tracer/host atom ratio versus sputtering time and of distance versus sputtering time; the distance is obtained by using a Talysurf[7] or interferometric measurement of the total crater depth

---

[6] Nordiko Ltd., Havant, United Kingdom.
[7] Rank Organization, Leicester, United Kingdom.

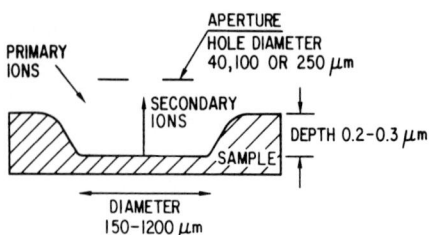

Fig. 13. Schematic diagram of crater caused by sputtering in a SIMS apparatus. [After Dorner et al. (1982).]

under the assumption that material is removed uniformly as a function of time. Large changes of chemical composition along the diffusion direction can invalidate that assumption.

The limitations of the SIMS technique have been discussed by Liebl (1975) and Reuter and Baglin (1981), and a detailed description of its application to diffusion has been given by Seran (1976) and Macht and Naundorf (1982). A recent paper (Dorner et al., 1982) shows the quality of results that can be obtained. In general, the resolution is no worse than that obtained by sputtering and counting the sections, and the sensitivity is no worse than that of counting techniques.

The great advantage of SIMS is that it can be used to measure the tracer diffusion of elements which have no suitable radioisotopes. One of the most important such elements is oxygen, and a number of oxygen self-diffusion measurements in oxides have been made by this technique (e.g., Contamin and Slodzian, 1968a, 1968b; Marin and Contamin, 1969; Arita et al., 1979; Reed and Wuensch, 1980). As these measurements do not involve a chemical gradient, there are fewer difficulties involved in estimating removal rates. However, a background of stable isotope is always present in such a self-diffusion experiment, so a radiotracer study is preferred when a convenient radioisotope is available.

The major disadvantage of SIMS is its cost. The SIMS apparatus is commercially made[8] but represents a large capital investment. Not withstanding the cost of the apparatus, careful controls must be applied to the measurements and artifacts (Slusser and Slattery, 1981) must be avoided.

5. *Depth Measurements and Related Problems*

If the entire area of the sample is sectioned, the depth of material removed is best determined by weighing on a microbalance. With care, a sample can be weighed to ± 3 μg, which corresponds to ± 150 nm for a cross-

---

[8] Cameca, Courbeudic, France; Hitachi, Tokyo, Japan.

sectional area of 0.1 cm^2, which is about the minimum useful area, and a density of 2 g cm^{-3}. For larger areas or densities, even better sensitivities are obtained, down to perhaps ± 10 nm.

If sections are so thin that the total amount of material removed in 20 sections is too small to be determined with the required accuracy, the weight removed can be increased by taking more sections but not counting them. Or, in the case of sputtering, the last section can be sputtered for a much longer time and not be counted. The area can be calculated from the sample dimensions as measured by a micrometer on a regularly shaped sample, or an enlarged photograph of the area and a scale can be copied onto graph paper, and the area measured by counting squares. The same techniques apply to a crater dug into a sample by SIMS or by sputtering or anodizing a masked face, as long as the crater is a right cylinder. If the sides of the crater are inclined, the depth must be determined by a Talysurf apparatus or interferometrically. The resolution of the former depends heavily on the skill of the operator and is perhaps 4% (Wildman et al., 1975). The reslution of the latter is perhaps one-quarter of a fringe, or ~ 75 nm for sodium light.

Sloping sides on the crater pose a serious problem because the sections are no longer parallel to the isoconcentration contours. This is avoided in SIMS by the use of physical or electronic apertures that restrict the analysis to the center of the crater (see Liebl, 1975). However, in sputtering and anodizing, the removal of higher activity material from the sides of the crater leads to a tail on the penetration plot. Without knowing the exact details of the evolution of the shape of the crater as a function of time, one cannot model this tail mathematically and must use approximate expressions that increase the χ^2 of a least-squares fit.

In addition to interferometry, another useful technique for monitoring the progress of sectioning is autoradiography. The density on an autoradiograph should be uniform if the sectioning is following the isoconcentration contours and if the original isotope layer was uniform. Deviation from uniform photographic density in the sectioned area indicates that a tail should be expected on the penetration plot. There are several possible sources of uneven darkening that can be recognized on the film, such as sloping crater walls, sectioning at an angle to the original interface, insufficient material removed from the edges, or dark spots due to tracer accumulation along short circuiting paths. In the case of self-diffusion in chromium, the dark spots were shown to correlate with inclusions in the base metal (Mundy et al., 1981). All of these phenomena give rise to a concave penetration plot.

Autoradiographs of samples containing gamma- or hard beta-emitting isotopes can be taken by simply laying the sample on x-ray film, and developing and fixing the film. Exposures take from a few minutes for an unsectioned sample to overnight for one that has been sectioned. Soft beta-emitting

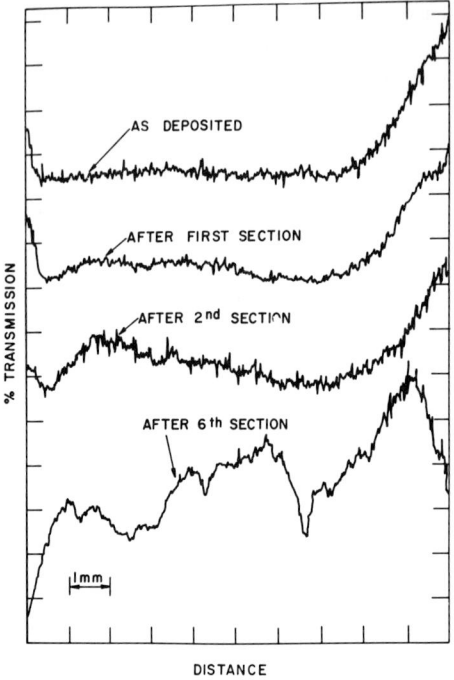

Fig. 14. Density scans of autoradiographs of an Ag self-diffusion sample as a function of section number. Note development of irregularities (down is more activity). [Beck et al., 1975].

isotopes require nuclear track emulsions such as Kodak NTA.[9] Density scans of autoradiographs of a silver sample at different stages of sectioning by anodization and stripping are shown in Fig. 14. Note how the irregularities increase with depth.

V. Counting of Radioactive Sections

The ordinate on a penetration plot is the logarithm of the specific activity of the radioisotope in the various sections. The specific activity is the net counting rate from that radioisotope, corrected for background and counter dead time, divided by the weight of the section. Since the ordinate is a logarithm and D is obtained from the slope, multiplication of the specific activities for all the sections by the same number, such as a counter efficiency, does not change the value of D. By the same token, these multipliers must

---

[9] Eastman-Kodak, Rochester, New York.

be maintained constant for all sections. That is, the sample-counter geometry, and the absorption of the radiation in the sample, must be kept the same for all sections from a sample. This is the primary requirement for precise counting.

The second requirement for precise counting is adequate counting statistics. Radioactive disintegrations follow first-order Poisson statistics [see, e.g., Price (1958, Chapter 3)] so that the standard deviation σ of a number of counts N is given by

$$\sigma = \sqrt{N} \tag{27}$$

Thus to obtain a statistical accuracy of 1%, one must collect 10^4 counts above background. For low activity samples, good statistics must be obtained for the background as well. The standard deviation σ_R of the difference of two counting rates R_s and R_b is

$$\sigma_R = (R_s/t_s + R_b/t_b)^{1/2}, \tag{28}$$

where the ts are counting times and the subscripts s and b refer to sample (which is really sample + background) and background, respectively.

In most simple single-isotope experiments, enough radioisotope can be deposited so that even after a drop in activity by a factor of 10^3 the net activity in the last section is still $\sim 10^3$ cpm. Then two or three 15-min background counts during the counting of the sections suffice and the standard deviation σ_b of R_b is calculated from the scatter of these values. If the sample activity is lower, longer background counts are needed.

For samples with a high counting rate, a correction is required for the counter dead time [see, e.g., Price (1958)]. To first order,

$$R_{\text{true}} = R_{\text{meas}}/(1 - \tau R_{\text{meas}}) \tag{29}$$

where τ is the counter dead time, which is assumed to be independent of the count rate. The dead time can be determined by the two-source method. A holder that can hold two samples is mounted on the counter, and the two samples, call them 1 and 2, are counted as follows. First 1 is counted alone, then 1 and 2 are counted together with 2 being placed in the second hole without disturbing 1, then 1 is removed without disturbing 2 and 2 is counted alone. Neglecting decay during the counting, τ is obtained from the measured counting rates via the equation

$$\frac{R_{12}}{1 - \tau R_{12}} = \frac{R_1}{1 - \tau R_1} + \frac{R_2}{1 - \tau R_2} - R_b \tag{30}$$

Because this method deals with small differences, very good counting statistics ($N \approx 10^7$ counts) must be obtained. The usual scatter in dead time,

$\sigma_\tau \approx 0.1\tau$, causes a $\sim 10^{-3}$ uncertainty in the counting rate, since the dead time correction τR is typically $\sim 2\%$ ($\tau \approx 1$ μsec, $R \approx 10^6$ cpm). In some isotope effect experiments $R \approx 10^7$ cpm; then τ must be determined to a few percent, and the count-rate dependence of τ must be taken into account.

We emphasize that in straightforward diffusion experiments a 1% level of precision in the net counting rate is more than adequate. Attempts to refine the experiment beyond that are a waste of time and money.

A. Counting of Gamma Rays

Gamma rays of energy ≥ 100 keV are not appreciably absorbed in a section. Thus the only problem in reproducibility is geometrical, i.e., positioning the section reproducibly in the counting vial, and positioning the vial reproducibly in the counter. The first is taken care of by folding the grinding paper or catcher foil in a reproducible manner, by using the same volume of solution in the counting vial, or by dissolving the lathe turnings in the same volume of acid. The second is accomplished by using either a well-type detector, with the counting vial in the well, or making a holder that positions the sample on the detector. The former has the advantage of nearly 4π solid angle, resulting in a factor of two increase in counting rate. The usual counting vials are either glass (e.g., 3.7 ml shell vials are a good fit into a $\frac{3}{4}$-in. diameter well) or plastic, and are cheap enough to be discarded after one use.

Two types of detectors are used for counting gamma rays: NaI–Tl scintillation detectors and solid-state detectors [Ge(Li), intrinsic Ge]. The former are more efficient at lower cost, while the latter have an energy resolution that is better by a factor of ~ 40. For diffusion measurements with a single radioisotope, a NaI–Tl detector is perfectly adequate, but for some determinations of the simultaneous diffusion of several isotopes (Herzig and Eckseler, 1979; Bussman et al., 1981), solid state detectors are necessary. The latter are also useful for identifying traces of radioactive impurities (Mundy and Rothman, 1983b).

The signal from the detector is shaped by a preamplifier and amplifier. The gamma-ray spectrum can then be displayed on a multichannel analyzer (MCA). The actual counting can be done either by using the MCA as a scaler, or by passing the signal from the amplifier into a lower level discriminator or single channel analyzer (SCA) and counting the pulses from the latter in a scaler. An MCA must be used with a solid-state detector to take advantage of the high detector resolution. An MCA is also useful for displaying spectra, and setting SCA windows. Most scalers and MCAs have timers built in for the determination of the counting time; MCAs have live time meters that obviate the calculation of the MCA dead time.

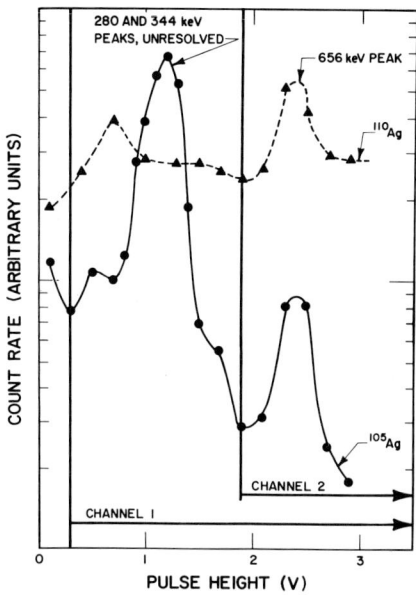

Fig. 15. Gamma-ray spectra of ^{110}Ag and ^{105}Ag showing lower level discriminators set in the Compton valleys. [After Rothman et al. (1970).]

When scintillation counting a single radioisotope, one can take an integral count, i.e., all gamma rays above a certain energy, which is usually chosen in the Compton valley below the lowest energy peak (Fig. 15), or one can set an SCA window around a single peak. For hot sources the former is better, as the higher background over the entire spectrum (400–500 cpm in a 3 × 3-in. well-type counter) is overwhelmed by the sample activity. For lower activities, the reduced background inside a narrow window (~100 cpm for a 3 × 3-in. well-type counter) is more advantageous.

The most important criterion for performance of the electronics is stability. For counting rates $< 10^6$ cpm, count rate effects are usually negligible, and the long-term stability of most off-the-shelf equipment is adequate for ordinary experiments. Samples of a long-lived radioisotope can be counted periodically to check the stability of the counter.

B. Counting Alpha or Beta Particles or X Rays

Alpha or β particles or x rays are absorbed in the sections. The most important aspect of reproducibility is, therefore, preparation of the sections for counting. The technique used depends on the type of radiation, the type of detector, the sectioning technique, and the chemistry of the system.

A useful apparatus for counting easily absorbed radiation is a liquid scintillation counter. The section is placed in the scintillating medium, which is a commercially available solution or "cocktail," and the vial containing the section plus cocktail is counted in a commercial counting unit [see, e.g., Neame and Homewood (1974)]. The section must be transparent to the photons emitted by the cocktail, otherwise quenching takes place and erroneous data are obtained.

For samples that are sectioned by sputtering, the polyester catcher film can be rolled around the inside of the liquid scintillation vial. Sections removed by anodizing and stripping (use the liquid scintillation vial for the stripping solution), or sections cut on a lathe or microtome and dissolved, can also be counted this way. The scintillation cocktail must be compatible with the solution so that quenching is avoided, and about the same amount of sample material should be present in each section.

The counting technique used for samples sectioned by grinding depends on how much the radiation is absorbed. If the isotope emits energetic beta particles (≥ 300 keV) or x rays, the sample is ground with diamond powder on a metal planchet (Mundy et al., 1978) and the planchets are counted with a thin, end-window counter (proportional or plastic scintillator). if the radiation is very weak (e.g., the β particles from ^{14}C or ^{63}Ni), the sample itself can be counted in an internal gas flow proportional counter, and analyzed by the Gruzin technique.

Various radiochemical techniques can also be applied. Weightless deposits can be prepared from the sections for use in α spectrometry or fission counting (Gray and Hagemann, 1962). In a heterodiffusion experiment, the radioisotope can be extracted into an organic solvent that is miscible with the scintillation cocktail, e.g., ^{63}Ni separated from uranium (Peterson and Rothman, 1964) or ^{55}Fe and ^{59}Fe from V (Coleman et al., 1968).

Other techniques have been used to prepare samples for end-window counting: electroplating (Mullen, 1961) of ^{55}Fe and ^{59}Fe in Cu, precipitation (Peterson et al., 1973) of ^{105}Ag and ^{111}Ag in AgBr and AgCl, and just arranging chips of ^{71}Ge in microtome-sectioned Al in a dish (Peterson and Rothman, 1970). The important thing is to keep the yield and the absorption constant from section to section.

C. The Measurement of the Simultaneous Diffusion of Several Radioisotopes

Much work in preparing, annealing, and sectioning samples can be saved by measuring the diffusion of the radioisotopes of several different tracers simultaneously, if the radiation from the different tracers can be separated. More important, the errors in D due to temperature measurement and sectioning are the same for each diffusing element; and the Ds are measured in the identical environment. If two tracers are isotopes of the same element, simultaneous measurement is essential, and it can also be useful in studies of solute–vacancy interaction or of diffusion measurements in compounds (Gupta et al., 1967). Or, sometimes a radioactive impurity is present in a solution [^{95}Nb daughter of ^{95}Zr, ^{95}Zr impurity in ^{95}Nb (Einziger et al., 1978), ^{54}Mn in pile irradiated Fe (Peterson and Rothman, 1964)]; the wise experimentalist goes for the extra set of data rather than the headaches of radiochemical purification.

The diffusion of radioisotopes of different elements presents somewhat different problems from the measurement of the isotope effect. In the former case, the Ds are likely to differ by more than the few percent commonly encountered in the latter. If the difference is more than a factor of 10, a simultaneous measurement is difficult as the appropriate section thicknesses differ by too much, unless one uses Stolwijk's (1980) trick of depositing the slow diffuser for a long anneal, removing the sample from the furnace, then depositing the faster diffuser, and annealing at the same temperature for an additional short time. It is also possible to cut 20 thin sections followed by 20 thick ones. Also, the absolute values of the Ds are the desired quantities, rather than their ratio, which is of primary interest in the isotope effect. Another problem is to work out the separation scheme so that the signal from the slower diffusing isotope at the end of its penetration plot is not overwhelmed by the radiation from the faster diffuser, the concentration of which has barely begun to decrease.

The separation schemes can be based on half-life, on type of radiation, or on energy discrimination, or a combination of these. None of these separations are complete—either some short-lived isotope still remains even after many half-lives of decay, or Compton-scattered radiation always exists below a γ-ray peak. These contributions may not cause a serious error in a 1% experiment, but they have to be calibrated with pure radio-isotopes, or otherwise taken into account in an isotope effect experiment. For instance, the simultaneous measurement of the diffusion of ^{59}Fe (γ rays at 1.095 and 1.292 MeV) and ^{51}Cr (γ ray at 0.31 MeV) (Rothman et al., 1980) is easily carried out by scintillation spectrometry but the Compton contribution of the ^{59}Fe under the ^{51}Cr peak must be subtracted.

The problem in isotope effect measurements is somewhat different in that the desired quantity, the fractional difference in D_1 and D_2, $\Delta D/D \equiv (D_1 - D_2)/D_2$ is small, typically, $0.0 < \Delta D/D < 0.05$. Therefore the relative activity of isotopes 1 and 2 changes only by a few percentage points over the entire penetration plot. The small size of this difference indicates that the counting must be done with a statistical precision of $\sim 0.1\%$ (10^6 counts for each measurement) in order to measure $\Delta D/D$ with sufficient precision.

The separation techniques used in isotope effect experiments are half-life, energy spectroscopy, or the use of different kinds of radiation. In the first method (Rothman and Peterson, 1967), one requires a short-lived isotope with $8h < \tau_{1/2}(\text{short}) < 5d$ and a long-lived isotope with $\tau_{1/2}(\text{long}) > 3\tau_{1/2}(\text{short})$. In the second method, scintillation spectroscopy can be used if the gamma spectra are favorable (see Fig. 15), but a high-resolution, high-efficiency Ge(Li) detector (Herzig and Eckseler, 1979) allows measurements on a large number of isotope pairs. The corrections for Compton-scattered radiation must be made. The third method takes advantage of differences in absorption (Mullen, 1961) or counter efficiency (Peterson et al., 1973) for the two kinds of radiation, or the different paths for positrons and electrons in a magnetic field (Hehenkamp and Schleit, 1977). All of these methods require careful monitoring of radioactive impurities by either half-life measurements or spectroscopy. In addition, the following tricks are useful:

(i) Count sections in random order. Count a long-lived isotope between sections to check stability of the counting system.

(ii) Dilute the most active sections, near $x = 0$, to decrease the dead time correction, maintaining the same volume of solution. This is not possible for sections obtained by grinding.

(iii) Check for count-rate effects by running a "null effect," i.e., about 5 aliquots from the same isotope mixture, covering the same range of count rates as the sections.

For suitable isotope pairs, we prefer the half-life separation technique.

The measurement of isotope effects for diffusion in organic crystals usually involves one molecule labeled with tritium, one with ^{14}C, and separation by liquid scintillation counting (Freer et al., 1982; Brissaud-Lancin, 1978). The masses of the molecules are changed not so much by labeling as by deuterating.

VI. Determination of D from a Penetration Plot

A. Gaussian Plots

The values of C and x for the Nth section are calculated from

$$C_N = [R_s/(1 - \tau R_s) - R_B]/W_N \tag{31}$$

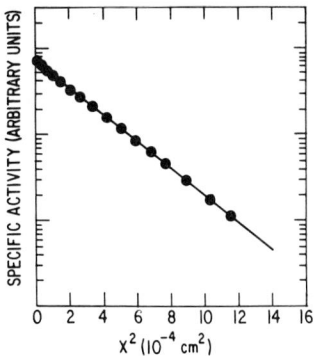

Fig. 16. Penetration plot for self-diffusion in potassium at 35.5°C. [From Mundy *et al.*, (1971).]

and

$$x_N = \left(\frac{W_N}{2} + \sum_{i=1}^{N-1} W_i\right) \bigg/ \rho A \qquad (32)$$

where ρ is the density of the material, A the cross-sectional area, and W_i the weight of the ith section. The other quantities were defined in Section V.

If all the points on the plot of log C versus x^2 fall on a straight line drawn with a sharp pencil (Fig. 16), the value of D can be obtained graphically within 1%. However, it is more elegant to obtain D from a least-squares fit to Eq. (8) or its linearized version

$$\ln C = \text{const} - x^2/4Dt \qquad (33)$$

From a strict statistical point of view, these equations are not equivalent, and the rigorous procedure is to fit Eq. (8), weighting each point with the reciprocal of the estimated variance of C.

$$\text{var } C = \frac{1}{W^2} \frac{R_s}{t_s(1 - R_s\tau)^4} + \frac{1}{W^2} \frac{R_b}{t_b} + \left(\frac{R_s}{1 - \tau R_s} - R_b\right)^2 \frac{\sigma_W^2}{W^4}$$

$$+ \frac{1}{W^2}\left(\frac{R_s}{1 - R_s\tau}\right)^4 \sigma_\tau^2 \qquad (34)$$

where σ_W^2 and σ_τ^2 are the variances of the section weight and dead time, respectively and the other components of the variance were obtained in Section V. A properly weighted least-squares analysis should yield a value of χ^2 close to 1, where χ^2 is the mean-square deviation

$$\chi^2 = \sum_{i=1}^{N} \frac{[C_i(\exp) - C_i(\text{calc})]^2}{\text{var } C_i} \bigg/ (N - 2) \tag{35}$$

where N is the number of sections in the penetration plot. Computer programs for weighted nonlinear least squares are available commercially.[10] Note that the uncertainties in the density and cross-sectional area introduce an uncertainty in x and hence in D, that has to be included in the variance of D in the fit D to T^{-1}.

If the points are scattered about a straight line, e.g., for the reasons mentioned in Table II, computer fitting of the data is necessary to avoid bias. The value of χ^2 in the case of Fig. 10 is $\gg 1$.

B. Nongaussian Plots

If the penetration plot is not straight, an appropriate model must be found before a computer fit can be made, and this requires a thorough understanding of the chemistry, physics, and microstructure of the system. The criteria for the validity of a model are (i) physical reasonability of the postulated processes, (ii) reasonable fit to the data, and (iii) Occam's razor. The fitting of curved penetration plots is a rather subjective gray area where the experimenter must exercise the most scrupulous honesty.

The physical or chemical processes that give rise to curved penetration plots can all be viewed as violations of the IC, BC, or implicit assumptions mentioned in Section I.B. Some of the problems we have encountered are (1) violation of the surface BC $[(\partial C/\partial x)(0, t) = 0]$ (surface hold up of the tracer, evaporation of the tracer, or evaporation of the base material), (2) violation of the assumption of homogeneity of the material (diffusion along short circuiting paths, trapping of the tracer, or presence of impurities), and (3) presence of two independent diffusion mechanisms.

1. Violation of the Zero Flux Condition at the Surface

a. *Surface Hold Up of the Tracer.* Sometimes the tracer remains on the surface because it is held in a stable chemical compound, such as an oxide or intermetallic compound (Preston, 1972), or in the oxide layer on the base metal. If a fraction of the tracer enters into the base metal instantaneously while the rest remains on the surface for the duration of the experiment, as happens with ^{235}U deposited on U by sputtering (Rothman *et al.*, 1959), only the first point on the penetration plot is high, and is omitted from the

---

[10] For example, the SAS Statistical Analysis System. SAS Institute, Raleigh, North Carolina.

least-squares fit. On the other hand, the tracer can dribble slowly into the base material during the entire duration of the diffusion anneal as is the case for Fe evaporated onto Al. In this case, stable Fe_4Al_{13} is formed (Preston, 1972), which decomposes over a period of days even at 600°C. Diffusion samples treated in an analogous fashion have yielded penetration plots which are curved for more than half the investigated diffusion zone and are then more or less gaussian (Peterson and Rothman, 1970) (Fig. 3). If one assumes that the BC is $\partial C/\partial x\,(0, t) = -K$, i.e., is independent of time, the solution to the diffusion equation is (Carslaw and Jaeger, 1959, p. 75)

$$C(x, t) \equiv 2K \left\{ \left(\frac{Dt}{\pi}\right)^{1/2} \exp\left(-\frac{x^2}{4Dt}\right) - \frac{x}{2} \text{erfc}\left[\frac{x}{2(Dt)}\right]^{1/2} \right\} \quad (36)$$

a plot of which is shown as the line in Fig. 3. At large x, the erfc term approaches a gaussian (Jahnke and Emde, 1945), so the correct value of D can be obtained from the gaussian portion of the penetration plot, ignoring the curved part of the penetration plot near $x = 0$.

b. *Evaporation of the Tracer.* A high vapor pressure element (e.g., Zn) will simultaneously evaporate and diffuse from the surface. If the tracer vapor just fills the capsule, not much material evaporates and a gaussian plot is obtained. If, however, there is continuous removal of the tracer, e.g., by condensation at the cold end of the capsule or reaction with the capsule, a penetration plot with a peak (Fig. 17) before a gaussian portion is obtained. The proper BC is (Carslaw and Jaeger, 1959, p. 358)

$$\frac{\partial C}{\partial x}(0, t) = hC \quad (37)$$

where h is the proportionality constant. The solution is

$$C(x, t) = \frac{M}{2\sqrt{Dt}} \left\{ 2 \exp\left(-\frac{x^2}{4Dt}\right) \right\}$$
$$- Mh \exp(Dth^2 + hx) \,\text{erfc}\left(\frac{x}{2\sqrt{Dt}} + h\sqrt{Dt}\right) \quad (38)$$

where M is the source strength. The solutions given by Reimers (1967) and by Trakhtenberg (1974) are based on an incorrect BC.

c. *Evaporation of the Base Material.* This process proved troublesome in our measurements on self-diffusion in chromium (Mundy et al., 1981). The above considerations on continuous removal also apply here, and a solution has been given by Ghoshtagore (1967) as

$$C(x', t) = M[(\pi Dt)^{-1/2} \exp(-\eta^2) - (V/2D) \,\text{erfc}\,\eta] \quad (39)$$

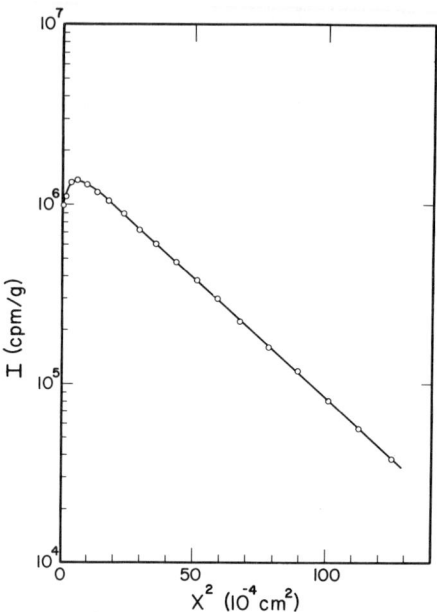

Fig. 17. Penetration plot for the diffusion of ^{65}Zn in NaCl doped with 107 mole ppm Zn at 639°C for 4620 s. [From Rothman *et al.* (1966a).]

where $\eta \equiv (x' + Vt)/2\sqrt{Dt}$, x' is the distance from the surface after diffusion, V the rate at which the surface recedes due to evaporation.

2. *Structurally Inhomogeneous Materials*

a. *Diffusion along Short-Circuiting Paths.* Distortion of profiles by "grain boundary tails" is one of the best known reasons for nongaussian penetration plots. In addition to grain boundaries, dislocations and surface artifacts can also act mathematically as a Fisher (1951) grain boundary, causing an upward curvature following the gaussian portion of the penetration plot (Fig. 11). To use the gaussian part indiscriminately can lead to disaster, i.e., a D an order of magnitude high (Wajda, 1954). One must fit to the sum of a volume diffusion and a short-circuit diffusion term. The latter takes a mathematical form for large x of

$$C = \text{const} \exp(-Ax^{6/5}) \qquad (40)$$

with A defined by

$$D'\delta \approx 0.66 A^{-5/3}(4D/t)^{1/2} \qquad (41)$$

(Le Claire, 1963; Suzuoka, 1961). Here D' is the diffusion coefficient and δ

1. TRACER DIFFUSION COEFFICIENTS IN SOLIDS

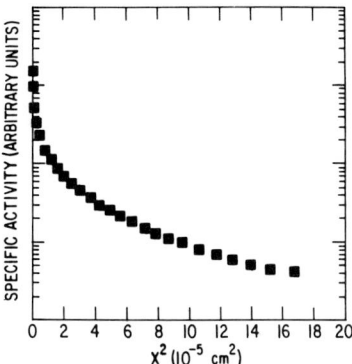

Fig. 18. Penetration plot for the diffusion of ^{64}Cu in Armco iron at 901.6°C, showing the effect of trapping of tracer on the penetration plot. [From Rothman *et al.* (1968).]

is the width of the short-circuiting paths. One can fit the data to the sum of (40) and a gaussian plot, ignoring the fact that the grain boundary term makes a negative contribution to C near $x = 0$, due to depletion of tracer near the surface by rapid flow down the short-circuiting paths. This fit can only be done if the volume term is dominant near $x = 0$, otherwise the iterative procedure of Suzuoka (1961) has to be used to obtain the correct value of D. In any case, it is difficult to obtain accurate values of both D and $D'\delta$ from the same experiment. If the experiment is designed to measure D, the value of $\beta = D'\delta/2D(Dt)^{1/2}$ (Le Claire, 1963) is so small that Eq. (41) is not valid, and the $x^{6/5}$ term is just an empirical correction containing no physics, but needed to get a fit to the data. If the experiment is designed to measure $D'\delta$, a fine-grained sample is used and the annealing time and temperature are chosen so that the volume part extends over only about three to four sections, so only an imprecise value of D is obtained. Criteria for separating volume and short-circuiting diffusion have been given by Harrison (1961); a simultaneous measurement of lattice, dislocation pipe, and grain boundary diffusion has been carried out by Gupta (1975).

b. *Trapping of the Tracer.* If the tracer is trapped, at impurities or precipitates, oddly shaped penetration plots (Fig. 18) or odd values of D are obtained. One situation discussed by Crank (1975, p. 327) is for the concentration of trapped species

$$S = RC \qquad (42)$$

where R is a trapping rate constant. Then the diffusion equation becomes

$$\frac{\partial C}{\partial t} = \frac{D}{R+1}\frac{\partial^2 C}{\partial x^2} \qquad (43)$$

so one obtains gaussian behavior but with an apparent $D = D/(R + 1)$. We believe that the very low values of D obtained for tracer diffusion of rare earths in Ag (Williams and Slifkin, 1963) and in Cu (Badrinarayanan and Mathur, 1970) are due to trapping of the tracer by oxygen dissolved in the base metal.

c. *Presence of Impurities.* If impurities significantly affect the diffusion of the tracer, and the impurities are distributed nonhomogeneously in the sample, warped penetration plots are obtained. This can be especially bad in the case of divalent impurities in alkali halides, the impurities being present either in the radioisotope solution or coming from the capsule. Mathematical solutions have been given (Benière *et al.*, 1972; Feit *et al.*, 1973), but it is easier to keep the system clean.

3. *Two Independent Diffusion Mechanisms*

Assume that a tracer can diffuse by two mechanisms. If a tracer atom can jump at random via both mechanisms, then application of the Einstein

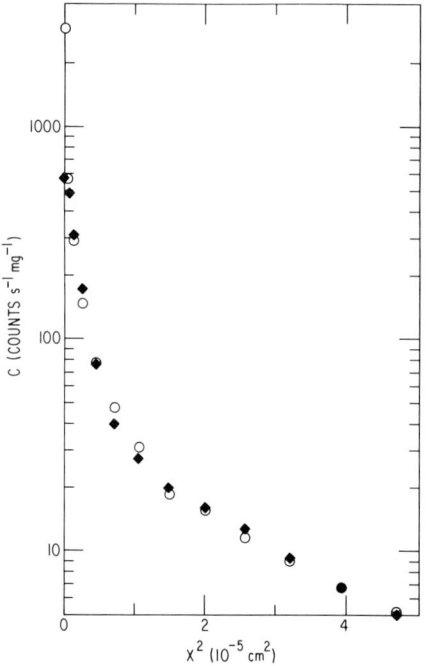

Fig. 19. Plot of the sum of two gaussians (◆). The penetration plot (○) for the diffusion of ^{86}Rb in vitreous SiO_2 at 697°C for 25d (Rothman *et al.*, 1982) is shown for comparison.

(1905) equations (see Bardeen and Herring, 1951) yields a gaussian penetration plot with a D equal to the sum of the Ds for the two mechanisms. If, however, there is no exchange of tracer atoms between mechanisms, the penetration plot is the sum of two gaussian plots, the one with a smaller D being near the surface and the other looking like a tail (Fig. 19). This has been suggested for diffusion of alkali ions in fused quartz (Frischat, 1970), and Fig. 19 shows such a set of data also.

A subclass of this phenomenon is anisotropic diffusion. If diffusion is measured in a polycrystalline sample of noncubic substance with grain size greater than the penetration depth, a curved penetration plot may result (Rothman et al., 1962). The qualitative similarity of the plots on Figs. 3, 11, 18, and 19 emphasizes the necessity of understanding the reason for nongaussian behavior before constructing the model.

4. Mathematical Techniques

All of the mathematical solutions mentioned above can be used in weighted least-squares fitting of the data by using, e.g., the SAS system. The usual criteria to the goodness of fit are used, the value of χ^2, the randomness of deviations. Because the model is seldom exact, i.e., the physical processes are not exactly described, $\chi^2 \gg 1$. A few outlying points on any penetration plot can be left out of the least-squares analysis with a clear conscience, especially ones near $x = 0$, as nongaussian behavior there often does not affect the penetration plot seriously. Discarding points near the tail can lead to incorrect values of D.

VII. Conclusions

Values of D in well-defined systems can be measured with a reproducibility of a few percent to tens of percent by the sectioning technique. Further developments of the technique are likely to consist of adaptations to cope with the properties of new materials as these become of interest. New techniques of analysis developed for the microelectronic industry, most of them involving particle beams and apparatus costing more than \$$10^5$, all have their place in diffusion studies, but none are likely to displace the sectioning technique for precise measurements of tracer Ds.

Acknowledgments

We thank R. Kentra and W. Cann for reminding us of tool bit shapes; N. L. Peterson for telling us how nonstoichiometric oxides are annealed and for other helpful discussion; one of the editors, G. E. Murch, for useful conversations; V. Johnson for finding dimly remembered references; D. Gupta, H. Mehrer, J. N. Mundy, N. L. Peterson, and C. L. Wiley for many constructive comments on the manuscript; and Bonnie Russell for typing the manuscript. In a wider sense, this chapter is made up of contributions from colleagues in the diffusion business. Although they are too numerous to list, a few outstanding debts must be acknowledged: To N. H. Nachtrieb and C. T. Tomizuka, in whose laboratories at the University of Chicago we were first exposed to precise diffusion measurements, and to our collaborators at Argonne: J. J. Hines, N. Q. Lam, J. N. Mundy, L. J. Nowicki, and N. L. Peterson, from whom we have learned so much.

References

Adda, Y., and Kirianenko, A. (1958). *C. R. Acad. Sci.* **247**, 744.
Adda, Y. and Philibert, J. (1966). "La Diffusion dans les Solides." Presses Universitaires de France, Paris.
Andersen, H. H. (1979). *Appl. Phys.* **18**, 131.
Andersen, T., and Sørensen, G. (1968). *Can. J. Phys.* **46**, 483.
Andersen, T., and Sørensen, G. (1969). *Radiat. Eff.* **2**, 111.
Arita, M., Hosoya, M., Kobayashi, M., and Someno, M. (1979). *J. Am. Ceram. Soc.* **62**, 443.
Arkhipova, N. K., Klotsman, S. M., Rabovskii, Y. A., and Timofeyev, A. N. (1979). *Fiz. Met. Metalloved.* **43**, 779.
Arora, M. R., and Kelly, R. (1972). *J. Electrochem. Soc.* **119**, 270.
Arora, M. R., and Kelly, R. (1973). *J. Electrochem. Soc.* **120**, 128.
Arora, M. R., and Kelly, R. (1974). *Electrochim. Acta* **19**, 413.
Atkinson, A., and Taylor, R. I. (1977). *Thin Solid Films* **46**, 291.
Atkinson, A., and Taylor, R. I. (1979). *Philos. Mag. A* **39**, 581.
Atkinson, A., and Taylor, R. I. (1981). *Philos. Mag. A* **43**, 979.
Atkinson, A., and Taylor, R. I. (1982). *Philos. Mag. A* **45**, 583.
Ayres, P. S., and Winchell, P. G. (1968). *J. Appl. Phys.* **39**, 4820.
Ayres, P. S., and Winchell, P. G. (1972). *J. Appl. Phys.* **43**, 816.
Backus, J. G. E. M., Bakker, H., and Mehrer, H. (1974). *Phys. Status Solidi B* **64**, 151.
Badrinarayanan, S., and Mathur, H. B. (1970). *Indian J. Pure Appl. Phys.* **8**, 324.
Bardeen, J., and Herring, C. (1951). "Atom Movements" (J. H. Hollomon, ed.), pp. 87–111. American Society for Metals, Metals Park, Ohio.
Barr, L. W., Blackburn, D. A., and Brown, A. F. (1962). "Radioisotopes in the Physical Sciences and Industry," p. 137. Int. At. Energy Ag., Vienna.
Barr, L. W., Mundy, J. N., and Rowe, A. H. (1972). *In* "Amorphous Materials" (R. W. Douglas and B. Ellis, eds.), p. 243. Wiley, New York.
Beck, C. A., Nowicki, L. J., and Rothman, S. J. (1975). Unpublished work.
Behrisch, R. (1981). *In* "Sputtering by Particle Bombardment I" (R. Behrisch, ed.). Springer-Verlag, Berlin and New York.
Benière, F. (1970). Thesis, Paris.
Benière, F., (1983). "Mass Transport in Solids" (F. Benière and C. R. A. Catlow, eds.). Plenum, London.
Benière, F., Benière, M., and Chemla, M. (1972). *J. Chem. Phys.* **56**, 549.
Bihr, J., Mehrer, H., and Maier, K. (1978). *Phys. Status Solidi A* **50**, 171.
Blackburn, D. A., Morrison, H. M., and Brown, A. F. (1961). *Acta IMEKO, Proc. Int. Meas. Conf.*, 2nd, pp. 525–536.

Brissaud-Lancin, M. (1878). Thèse d'état, Université de Paris.
Bussman, W., Herzig, C., Hoff, H. A., and Mundy, J. N. (1981). *Phys. Rev. B* **23**, 6216.
Čadek, J., and Janda, E. (1957). *Hutn. Listy* **12**, 1008 (*English transl.: AERE Transl.* 840, Harwell).
"Canning Handbook on Electroplating" (1978). Canning, Birmingham, England.
Carman, P. C., and Haul, R. A. W. (1954). *Proc. R. Soc. London, Ser. A* **222**, 109.
Carslaw, H. S., and Jaeger, J. C. (1959). "Conduction of Heat in Solids," Oxford Univ. Press (Clarendon), London and New York.
Chadwick, A. V., and Sherwood, J. N. (1975). *In* "Point Defects in Solids" (J. H. Crawford, Jr., and L. M. Slifkin, eds.), Vol. 2, p. 441–475. Plenum, London.
Chapman, B. (1980). "Glow Discharge Processes." Wiley, New York.
Coleman, M. G., Wert, C. A., and Peart, R. F. (1968). *Phys. Rev.* **175**, 788.
Contamin, P., and Slodzian, G. (1968a). *C. R. Acad. Sci., Ser. C* **267**, 805.
Contamin, P., and Slodzian, G. (1968b). *Appl. Phys. Lett.* **13**, 416.
Coston, C., and Nachtrieb, N. H. (1964). *J. Phys. Chem.* **68**, 2219.
Crank, J. (1975). "The Mathematics of Diffusion." Oxford Univ. Press (Clarendon), London and New York.
Davies, J. A., Friesen, J., and McIntyre, J. A. (1960). *Can. J. Chem.* **38**, 1526.
Davies, J. A., Ball, G. C., Brown, F., and Domej, B. (1964). *Can. J. Phys.* **42**, 1070.
De Bruin, H. J., and Clark, R. L. (1964). *Rev. Sci. Instrum.* **35**, 227.
Deiss, W. J. (1963). Commisariat à l'Energie Atomique Rep. No. 2267. Unpublished.
Dejus, R., Sköld, K., and Granéli, B. (1980). *Solid State Ionics* **1**, 327.
Dorner, P., Gust, W., Lodding, A., Ocklins, H., Predel, B., and Roll, U. (1982). *Z. Metallk.* **73**, 325.
Einstein, A. (1905). *Z. Phys.* **17**, 549.
Einziger, R. E., and Mundy, J. N. (1976). *Rev. Sci. Instrum.* **47**, 1547.
Einziger, R. E., Mundy, J. N., and Hoff, H. A. (1978). *Phys. Rev. B* **17**, 440.
Feit, M. D., Mitchell, J. L., and Lazarus, D. (1973). *Phys. Rev. B* **8**, 1715.
Fick, A. (1855). *Pogg. Ann.* **94**, 59.
Fisher, E. S., and McSkimmin, H. J. (1958). *J. Appl. Phys.* **29**, 1473.
Fisher, J. C. (1951). *J. Appl. Phys.* **22**, 74.
Flinn, P. A. (1962). *Rev. Sci. Instrum.* **33**, 1247.
Freer, R., Salthouse, P. W., and Sherwood, J. N. (1982). *Philos. Mag. A* **45**, 205.
Frischat, G. H. (1970). *Phys. Chem. Glasses* **11**, 25.
Gainotti, A., and Zecchina, L. (1965). *Nuovo Cimento* **40B**, 295.
Ghoshtagore, R. N. (1967). *Phys. Status Solidi* **19**, 123.
Goldstein, B. (1957). *Rev. Sci. Instrum.* **28**, 289.
Graham, D. (1969). *Rev. Sci. Instrum.* **40**, 897.
Gray, J., Jr., and Hagemann, F. T. (1962). *Rev. Sci. Instrum.* **33**, 1258.
Gray, W. T., and Finch, D. I. (1971). *Phys. Today* **24** (September) p. 32.
Gruzin, P. L. (1952). *Dokl. Akad. Nauk. SSSR* **86**, 289.
Gupta, D. (1973a). *Phys. Rev. B* **7**, 586.
Gupta, D. (1973b). *J. Appl. Phys.* **44**, 4455.
Gupta, D. (1975). *Thin Solid Films* **25**, 231.
Gupta, D., and Asai, K. W. (1974). *Thin Solid Films* **22**, 121.
Gupta, D., and Campbell, D. R. (1980). *Philos. Mag. A* **42**, 513.
Gupta, D., and Kim, K. K. (1980). *J. Appl. Phys.* **51**, 2066.
Gupta, D., and Tsui, R. T. C. (1970). *Appl. Phys. Lett.* **17**, 294.
Gupta, D., Lieberman, D., and Lazarus, D. (1967). *Phys. Rev.* **153**, 863.

Gupta, D., Tu, K. N., and Asai, K. W. (1975). *Phys. Rev. Lett.* **35**, 796.
Harrison, L. G. (1961). *Trans. Faraday Soc.* **57**, 1191.
Hehenkamp, T., and Schleit, W. (1977). *Acta Metall.* **25**, 1109.
Herzig, C., and Eckseler, H. (1979). *Z. Metallk.* **70**, 215.
Hoffman, R. E. (1951). *In* "Atom Movements" (J. H. Hollomon, ed.), p. 51. American Society for Metals, Metals Park, Ohio.
Hoffman, R. E., and Turnbull, D. (1951). *J. Appl. Phys.* **22**, 634.
Holland, L. (1956). "Vacuum Deposition of Thin Films." Wiley, New York.
Holloway, D. M. (1975). *J. Vac. Sci. Technol.* **12**, 392.
Hong, J. D., Griffin, W. E., and Davis, R. F. (1978). *Rev. Sci. Instrum.* **49**, 83.
Hood, G. M. (1970). *Philos. Mag.* **21**, 305.
Hood, G. M., Schultz, R. J., and Armstrong, J. (1983). *Philos. Mag. A* **47**, 775.
International Practical Temperature Scale (1976). *Metrologia* **12**, 7.
Jahnke, E., and Emde, F. (1945). "Tables of Functions," p. 24. Dover, New York.
Jost, W. (1952). "Diffusion in Solids, Liquids, Gases." Academic Press, New York.
Kehl, G. (1953). "Principles of Metallographic Laboratory Practice." McGraw-Hill, New York.
Kelly, J. E., III, and Tomozawa, N. (1980). *J. Am. Ceram. Soc.* **63**, 478.
Kim, K. K., Mundy, J. N., and Puri, S. M. (1977). *Rev. Sci. Instrum.* **48**, 1628.
Kittel, J. H. (1948). Unpublished work.
Kofstad, P. (1972). "Nonstoichiometry, Diffusion, and Electrical Conductivity in Binary Metal Oxides," Chaps. 2–3. Wiley, New York.
Kostkowski, H. J., and Lee, R. D. (1962). "Theory and Practice of Optical Pyrometry," Monograph 41. Natl. Bur. Stand., Washington, D.C.
Lam, N. Q., Rothman, S. J., and Nowicki, L. J. (1972a). *J. Electrochem. Soc.* **119**, 715.
Lam, N. Q., Rothman, S. J., and Nowicki, L. J. (1972b). *J. Electrochem. Soc.* **119**, 1344.
Lam, N. Q., Rothman, S. J., Mehrer, H., and Nowicki, L. J. (1973). *Phys. Status Solidi B* **57**, 225.
Lange, W., Hässner, A., and Sieber, K. (1962). *Isotopentechnik* **2**, 42.
Layer, H. P., and Meyer, R. O. (1962). *Rev. Sci. Instrum.* **33**, 1458.
Lea, C. (1983). *Met. Sci.* **17**, 357.
Le Blans, L. M. L. J., and Verheijke, N. L. (1964). *Philips Tech. Rev.* **25**, 191.
Le Claire, A. D. (1963). *Br. J. Appl. Phys.* **14**, 351.
Letaw, H., Slifkin, L., and Portnoy, W. (1954). *Rev. Sci. Instrum.* **25**, 865.
Levy, V., Bouchet, P., Siouffi, J., and Adda, Y. (1959). Commisariat à l'Energie Atomique Rep. No. 1317. Unpublished.
Liebl, H. (1975). *J. Vac. Sci. Technol.* **12**, 385.
Lindemer, T. B., and Guy, A. G. (1968). *Welding (Pittsburgh)* **47**, 2225.
Littmark, U., and Ziegler, J. F. (1980). "Range Distributions for Energetic Ions in All Elements." Pergamon, Oxford.
Lundy, T. S. (1970). *In* "Techniques of Metals Research" (R. A. Rapp, ed.), Vol. 4, Part 2, Chap. 9A. Wiley, New York.
Lundy, T. S., and Murdock, J. F. (1962). *J. Appl. Phys.* **33**, 1671.
Lundy, T. S., Winslow, F. R., Pawel, R. E., and McHargue, C. J. (1965). *Trans. Metall. Soc. AIME* **233**, 1533.
Lutz, H., and Sizmann, R. (1964). *Z. Naturforsch.* **19A**, 1079.
McCargo, M., Davies, J. A., and Brown, F. (1963). *Can. J. Phys.* **41**, 1231.
Macht, M.-P., and Naundorf, V. (1982). *J. Appl. Phys.* **53**, 7551.
Maier, K. (1977). *Phys. Status Solidi B* **78**, 689.
Maier, K., and Schüle, W. (1974). EURATOM Rep. 5234d. Unpublished.

Maier, K., Mehrer, H., Lessman, E., and Schüle, W. (1976). *Phys. Status Solidi B* **78**, 689.
Makin, S. M., Rowe, A. H., and Le Claire, A. D. (1957). *Proc. Phys. Soc. London B* **70**, 545.
Malkovich, R. Sh. (1963). *Fiz. Met. Metalloved.* **15**, 880.
Mallard, W. C., Gardner, A. B., Bass, R. F., and Slifkin, L. M. (1963). *Phys. Rev.* **129**, 617.
Manning, J. R. (1968). "Diffusion Kinetics for Atoms in Crystals." Van Nostrand, Princeton, New Jersey.
Marin, J. F., and Contamin, P. (1969). *J. Nucl. Mater.* **30**, 16.
Mayer, H. J., Mehrer, H., and Maier, K. (1977). "Radiation Effects in Semiconductors" (N. B. Urli and J. W. Corbett, eds.), p. 186. Institute of Physics, London.
Mayer, J. W., and Poate, J. M. (1978). "Thin Film Interdiffusion and Reactions" (J. M. Poate, K. N. Tu, and J. W. Mayer, eds.), pp. 119–160. Wiley, New York.
Mehl, R. F. (1936). *Trans. AIME* **122**, 11.
Mehrer, H., Maier, K., Hettich, G., Mayer, H. J., and Rein, G. (1978). *J. Nucl. Mater.* **69, 70**, 545.
Meyer, R. O., and Sekizawa, H. (1968). *Rev. Sci. Instrum.* **39**, 265.
Miller, J. W., Rothman, S. J., Mundy, J. N., Robinson, L. C., and Loess, R. E. (1973). *Phys. Rev. B* **8**, 2411.
Muan, A., and Osborn, E. F. (1965). "Phase Equilibria among Oxides in Steel Making Processes," p. 26. Addison-Wesley, Reading, Massachusetts.
Mullen, J. G. (1961). *Phys. Rev.* **121**, 1649.
Mundy, J. N. (1971). *Phys. Rev. B* **3**, 2431.
Mundy, J. N. (1982). Private communication.
Mundy, J. N., and Rothman, S. J. (1983a). *J. Vac. Sci. Technol. A* **1**, 74.
Mundy, J. N., and Rothman, S. J. (1983b). *In* "Methods in Experimental Physics, Solid State: Nuclear Methods" (J. N. Mundy, S. J. Rothman, M. J. Fluss, and L. C. Smedskjaer, eds.), Chap. 1. Academic Press, New York.
Mundy, J. N., Miller, T. E., and Porte, R. J. (1971). *Phys. Rev. B* **3**, 2445.
Mundy, J. N., Miller, J. W., and Rothman, S. J. (1974). *Phys. Rev. B* **10**, 2275.
Mundy, J. N., Tse, C. W., and McFall, W. D. (1976). *Phys. Rev. B* **13**, 2349.
Mundy, J. N., Rothman, S. J., Lam, N. Q., Hoff, H. A., and Nowicki, L. J. (1978). *Phys. Rev. B* **18**, 6566.
Mundy, J. N., Hoff, H. A., Pelleg, J., Rothman, S. J., and Nowicki, L. J. (1981). *Phys. Rev. B* **24**, 658.
Neame, K. D., and Homewood, C. A. (1974). "Liquid Scintillation Counting." Wiley, New York.
Neiman, N. B., and Shinyayev, A. Y. (1954). *Dokl. Akad. Nauk SSSR* **96**, 315.
Nowick, A. S. (1951). *J. Appl. Phys.* **22**, 1182.
Nowick, A. S., and Berry, B. S. (1972). "Anelastic Relaxation in Crystalline Solids," Chaps. 7 and 10. Academic Press, New York.
Pawel, R. E. (1964). *Rev. Sci. Instrum.* **35**, 1066.
Pawel, R. E., and Lundy, T. S. (1964). *J. Appl. Phys.* **35**, 435.
Pawel, R. E., and Lundy, T. S. (1968). *J. Electrochem. Soc.* **115**, 233.
Pelleg, J. (1974). *J. Less-Common Met.* **35**, 299.
Perkins, R. A. (1973). *Metall. Trans.* **4**, 1665.
Perkins, R. A., Padgett, R. A., Jr., and Tunali, N. K. (1973). *Metall. Trans.* **4**, 2535.
Peterson, N. L. (1964). *Phys. Rev. A* **136**, 568.
Peterson, N. L. (1982). Private communication.
Peterson, N. L., and Chen, W. K. (1982). *J. Phys. Chem. Solids* **43**, 29.
Peterson, N. L., and Rothman, S. J. (1964). *Phys. Rev. A* **136**, 842.
Peterson, N. L., and Rothman, S. J. (1970). *Phys. Rev. B* **1**, 3264.

Peterson, N. L., Barr, L. W., and Le Claire, A. D. (1973). *J. Phys. C* **6**, 2020.
Preston, R. S. (1972). *Metall. Trans.* **3**, 1831.
Price, W. J. (1958). "Nuclear Radiation Detection." McGraw-Hill, New York.
Przyborski, W., Roed, J., Lippert, J., and Sarholt-Kristensen, L. (1969). *Radiat. Eff.* **1**, 33.
Queisser, H. J. (1961). *J. Appl. Phys.* **32**, 1776.
Quinn, T. J. (1980). *High Temp. High Pressures* **12**, 359.
Reader, P. D., and Kauffman, H. R. (1975). *J. Vac. Sci. Technol.* **12**, 1344.
Reed, D. J., and Wuensch, B. J. (1980). *J. Am. Ceram. Soc.* **63**, 88.
Reimers, P. (1967). *Phys. Status Solidi* **22**, 227.
Reuter, W., and Baglin, J. E. E. (1981). *J. Vac. Sci. Technol.* **18**, 282.
Roberts-Austen, W. C. (1896). *Philos. Trans. R. Soc. London, Ser. A* **187**, 404.
Roeser, W. F., and Lonberger, S. T. (1958). Methods of testing thermocouples and thermocouple materials, Circular 590. Natl. Bur. Stan., Washington, D.C.
Rothman, S. J. (1954). Thesis. Stanford Univ., Stanford, California.
Rothman, S. J. (1961). *J. Nucl. Mater.* **3**, 77.
Rothman, S. J. (1962). *Adv. Nucl. Sci. Technol.* **1**, 116.
Rothman, S. J., and Peterson, N. L. (1967). *Phys. Rev.* **154**, 552.
Rothman, S. J., and Peterson, N. L. (1969). *Phys. Status Solidi* **35**, 305.
Rothman, S. J., and Sobocki, L. J. (1959). *Rev. Sci. Instrum.* **30**, 201.
Rothman, S. J., Lloyd, L. T., and Harkness, A. L. (1959). *Trans. Metall. Soc. AIME* **218**, 605.
Rothman, S. J., Peterson, N. L., and Moore, S. A. (1962). *J. Nucl. Mater.* **7**, 212.
Rothman, S. J., Barr, L. W., Rowe, A. H., and Selwood, P. G. (1966a). *Philos. Mag.* **14**, 501.
Rothman, S. J., Bastar, R., Hines, J. J., and Rokop, D. (1966b). *Trans. Metall. Soc. AIME* **236**, 897.
Rothman, S. J., Peterson, N. L., Walter, C. M., and Nowicki, L. J. (1968). *J. Appl. Phys.* **39**, 5041.
Rothman, S. J., Peterson, N. L., and Robinson, J. T. (1970). *Phys. Status Solidi* **39**, 635.
Rothman, S. J., Lam, N. Q., Nowicki, L. J., and Beck, C. A. (1976). "Fundamental Aspects of Radiation Damage in Metals," Vol. 2, p. 1077. U.S. Atomic Energy Comm., Washington, D.C.
Rothman, S. J., Nowicki, L. J., and Murch, G. E. (1980). *J. Phys. F* **10**, 383.
Rothman, S. J., Marcuso, T. L. M., Nowicki, L. J., Baldo, P. M., and McCormick, A. W. (1982). *J. Am. Ceram. Soc.* **65**, 578.
Rupp, W., Ermert, U., and Sizmann, R. (1969). *Phys. Status Solidi* **33**, 509.
Samuels, L. E. (1982). "Metallographic Polishing by Mechanical Methods." Amer. Soc. Metals. Metals Park, Ohio.
Santos, E., and Dyment, F. (1973). *Plating* **60**, 821.
Schamp, H. W., Jr., Oakes, D. A., and Reed, N. M. (1959). *Rev. Sci. Instrum.* **39**, 1028.
Seibel, G. (1964). *Int. J. Appl. Radiat. Isot.* **15**, 679.
Seith, W. (1955). "Diffusion in Metallen." Springer, Berlin.
Seith, W., and Kottman, A. (1952). *Angew. Chem.* **64**, 379.
Seran, J.-L. (1976). Thesis, Université de Paris (Available as Commisariat à l'Energie Atomique Rep. CEA-R-4717).
Shirn, G. A., Wajda, E. S., and Huntington, H. B. (1953). *Acta Metall.* **1**, 513.
Sigmund, P. (1969). *Phys. Rev.* **184**, 383.
Sigmund, P. (1981). *In* "Sputtering by Particle Bombardment I" (R. Behrisch, ed.), Chap. 2. Springer, Berlin.
Slifkin, L., Lazarus, D., and Tomizuka, C. T. (1952). *J. Appl. Phys.* **23**, 1032.
Slusser, G. J., and Slattery, J. S. (1981). *J. Vac. Sci. Technol.* **18**, 301.
Stanley, J., and Wert, C. (1961). *J. Appl. Phys.* **32**, 267.

Stein, D. F., and Joshi, A. (1981). *Annu. Rev. Mater. Sci.* **11**, 485.
Stolwijk, N. A. (1980). Thesis. University of Amsterdam.
Stroud, P. T. (1959). *Vacuum* **9**, 269.
Styris, D. L., and Tomizuka, C. T. (1963). *J. Appl. Phys.* **34**, 1001.
Suzuoka, T. (1961). *Trans. Jpn. Inst. Met.* **2**, 25.
Swanson, M. L., Mehl, R. F., Pound, G. M., and Hirth, J. P. (1962). *Trans. AIME* **224**, 742.
Tannhauser, D. (1956). *J. Appl. Phys.* **27**, 662
Tegart, W. J. M. (1959). "The Electrolytic and Chemical Polishing of Metals." Pergamon, London.
Tomizuka, C. T. (1959). *In* "Methods of Experimental Physics" (K. Lark-Horovitz and V. A. Johnson, eds.), Vol. 6, Part A, pp. 364–373. Academic Press, New York.
Tomizuka, C. T., and Sonder, E. (1956). *Phys. Rev.* **103**, 1182.
Trakhtenberg, I. Sh. (1974). *Fiz. Met. Metalloved.* **37**, 348.
Wajda, E. S. (1954). *Acta Metall.* **2**, 184.
Weil, Raoul, Lazarus, D., and Witt, F. (1979). *Rev. Sci. Instrum.* **50**, 642.
Weil, Rolf, Rothman, S. J., and Lloyd, L. T. (1959). *Rev. Sci. Instrum.* **30**, 541.
Whitton, J. L. (1965). *J. Appl. Phys.* **36**, 3917.
Whitton, J. L., and Davies, J. A. (1964). *J. Electrochem. Soc.* **111**, 1347.
Wildman, H. S., Howard, J. K., and Ho, P. S. (1975). *J. Vac. Sci. Technol.* **12**, 75.
Williams, G. P., Jr., and Slifkin, L. (1963). *Acta Metall.* **11**, 319.
Wolf, D. (1979). "Spin-Temperature and Nuclear-Spin Relaxation in Matter: Basic Principles and Applications." Oxford Univ. Press, London and New York.
Wuensch, B. J., and Vasilos, T. (1961). *In* "Reactivity of Solids" (J. H. de Boer, ed.), p. 57. Elsevier, Amsterdam.
Wuttig, M. (1969). *Scr. Metall.* **3**, 175.
Zhukovitskii, A. A., Kryukov, S. N., and Geodakyan, V. A. (1955). Symp. 34, Moscow Steel Institute, Moscow (English transl.: AEC Transl. 3100, Part II, p. 3).

2

Diffusion in Silicon and Germanium

WERNER FRANK

MAX-PLANCK-INSTITUT FÜR METALLFORSCHUNG
INSTITUT FÜR PHYSIK
AND UNIVERSITÄT STUTTGART
INSTITUT FÜR THEORETISCHE UND ANGEWANDTE PHYSIK
STUTTGART, FEDERAL REPUBLIC OF GERMANY

ULRICH GÖSELE*

MAX-PLANCK-INSTITUT FÜR METALLFORSCHUNG
INSTITUT FÜR PHYSIK
STUTTGART, FEDERAL REPUBLIC OF GERMANY

HELMUT MEHRER†

UNIVERSITÄT STUTTGART
INSTITUT FÜR THEORETISCHE UND ANGEWANDTE PHYSIK
STUTTGART, FEDERAL REPUBLIC OF GERMANY

ALFRED SEEGER

MAX-PLANCK-INSTITUT FÜR METALLFORSCHUNG
INSTITUT FÜR PHYSIK
AND UNIVERSITÄT STUTTGART
INSTITUT FÜR THEORETISCHE UND ANGEWANDTE PHYSIK
STUTTGART, FEDERAL REPUBLIC OF GERMANY

---

* Present address: Siemens AG, Zentrale Technik, Forschungslaboratorien, Munich, Federal Republic of Germany.
† Present address: Universität Münster, Institut für Metallforschung, Münster, Federal Republic of Germany.

I.	Introduction	64
II.	Basic Features of Bulk Diffusion in Crystalline Solids	66
	A. Diffusion Mechanisms	66
	B. Thermal-Equilibrium Diffusion versus Enhanced Diffusion	69
III.	Self-Diffusion and Related Phenomena	71
	A. General Remarks	71
	B. Experimental Techniques	73
	C. Self-Diffusion in Intrinsic Germanium and Silicon	74
	D. Diffusion of Germanium Tracers in Silicon	80
	E. Effects of Doping	83
	F. Isotope Effects	84
	G. Pressure Effects	86
	H. Self-Diffusion in Si–Ge Alloys	87
IV.	Survey of Foreign-Atom Diffusion	88
V.	Oxidation-Influenced Diffusion of Group III and Group V Elements in Silicon	93
	A. Introductory Remarks and Basic Concepts	93
	B. Influence of Surface Oxidation on the Concentrations of Intrinsic Point Defects	95
	C. Qualitative Features of Oxidation-Influenced Diffusion	96
	D. Analysis of Experiments on Oxidation-Enhanced and Oxidation-Retarded Diffusion	98
	E. Diffusion of Intrinsic Point Defects through Silicon Wafers	103
	F. Conclusions	105
VI.	A Barrier against Vacancy–Interstitial Recombination	105
VII.	Diffusion of Group III and Group V Elements and Its Dependence on Doping	108
	A. Germanium	108
	B. Silicon	108
VIII.	Anomalous Diffusion Phenomena	110
	A. Phenomenological Description	110
	B. The Dislocation Climb Experiments of Claeys *et al.*	112
	C. The Dislocation Climb Experiments of Strunk *et al.*	113
	D. Mechanisms of the Generation of Interstitial Supersaturations	115
IX.	Substitutional–Interstitial Interchange Diffusion and Application to Gold and Nickel in Silicon and to Copper in Germanium	116
	A. Introductory Remarks	116
	B. Theory of Substitutional–Interstitial Interchange Diffusion	117
	C. Comparison with Experiments	125
X.	Concluding Remarks	136
	References	137

I. Introduction

This chapter deals with diffusion processes in silicon and germanium. Considerable progress in this area over the past years makes an up-to-date presentation of both the experimental data and our present-day understanding highly desirable. The relatively large portion of this chapter devoted to silicon reflects the rapid progress achieved for this material as well as its

2. DIFFUSION IN SILICON AND GERMANIUM

great technological importance for electronic devices.

This survey aims at making the reader familiar with the exciting new developments rather than at painstakingly presenting all diffusion data available, a good deal of which may be found in other books and review articles (Seeger and Chik, 1968; Kendall and de Vries, 1969; Sharma, 1970; Hu, 1973; Shaw, 1975; Willoughby, 1978; Fair, 1981a). Nevertheless, a sufficiently broad background of experimental and basic theoretical information will be presented, in order to make the chapter self-contained.

After describing the basic features of bulk diffusion in crystalline solids (Section II), we shall present experimental data on self-diffusion in silicon and germanium (Section III). Following Seeger and Chik (1968), it is suggested that self-diffusion in Ge may be explained in terms of the vacancy mechanism, whereas in Si both self-interstitials and vacancies appear to contribute to self-diffusion. However, it is realized that, particularly in the case of Si, experiments on self-diffusion alone do not permit one to draw definite conclusions about the self-diffusion mechanisms. Therefore, this problem will continue to play a central role in subsequent sections dealing with the diffusion of foreign atoms. Section IV gives a survey of impurity diffusion. In Section V, treating the influence of surface oxidation on the diffusion of group III and group V elements in silicon, it is shown that, in accordance with the Seeger–Chik model of self-diffusion, in Si both vacancies *and* self-interstitials can act as diffusion vehicles. The dynamics of vacancies and self-interstitials in silicon is the subject of Section VI. The discussion of the influence of doping on the diffusion of group III and group V elements in Si and Ge (Section VII) and of the so-called anomalous diffusion phenomena (e.g., the emitter-push effect) of B or P in Si (Section VIII) rounds off our picture of the diffusion of substitutional solutes developed in the preceding sections.

Section IX on the substitutional–interstitial interchange diffusion mechanisms occupies a central position in this chapter. In the authors' opinion, the understanding of the diffusion of gold in silicon, taking place via these mechanisms, is the most remarkable progress in the field of diffusion in semiconductors during the early 1980s. This justifies the detailed mathematical treatment of the two basic interchange diffusion mechanisms, the kick-out mechanism and the dissociative mechanism (Section IX.B), which precedes the discussion of the experimental material (Section IX.C). Among the most important results deducible from investigations of the interchange impurity diffusion is unambiguous information on the self-diffusion mechanisms in host crystals. From studies of the diffusion of Au (and to a lesser extent of Ni) in Si, it was found that indeed both self-interstitials and vacancies contribute to Si self-diffusion, whereas the occurrence of purely dissociative diffusion of Cu in Ge supports the view that Ge self-diffusion occurs exclusively by means of vacancies.

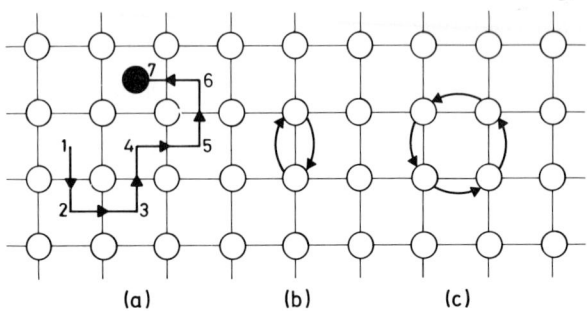

Fig. 1. Direct diffusion mechanisms: (a) Direct interstitial mechanism [foreign atom (●) jumping from interstice 1 to interstice 2, from 2 to 3, etc.], (b) direct exchange of a pair of neighboring lattice atoms [○], (c) ring mechanism.

II. Basic Features of Bulk Diffusion in Crystalline Solids

A. Diffusion Mechanisms

The conceptually simplest mechanisms of bulk diffusion in crystals are the so-called *direct mechanisms* (Fig. 1). Foreign atoms that are—in an otherwise perfect crystal—located exclusively in interstices may jump directly from interstice to interstice (Fig. 1a). This direct interstitial mechanism is presumably responsible for the diffusion of hydrogen in silicon and germanium. The direct diffusion of substitutionally incorporated atoms or of host atoms on regular lattice sites involves the exchange of atoms on two neighboring lattice sites (Fig. 1b) or of a ring of atoms (Fig. 1c). So far, no example of this kind of direct diffusion has been found.[1]

By contrast, *indirect diffusion* of self-atoms or of foreign substitutional atoms requires intrinsic defects as diffusion vehicles. The best known indirect diffusion mechanism is the *vacancy mechanism* (Fig. 2). This mechanism controls self-diffusion in all metals so far investigated in detail (Mehrer, 1978; Peterson, 1978) and in germanium (Section III.C.1) and yields the main contribution to self-diffusion in silicon below about 1270 K (Section IX.C.2). The counterpart of the vacancy mechanism is the *interstitialcy mechanism* (Fig. 3). In silicon, this mechanism dominates self-diffusion above about 1270 K (Section X.C.1.b) and plays a prominent role in the diffusion of the substitutional solutes P, B, Al, and Ga (Sections V and VII).

---

[1] On the basis of calculations using noncentral atomic potentials, but not taking into account lattice relaxations, Glazman and Myaken'kaya (1977) suggested that self-diffusion in silicon above about 1170 K occurs via a direct exchange of two neighboring atoms. This proposal is at variance with the conclusions drawn in this chapter.

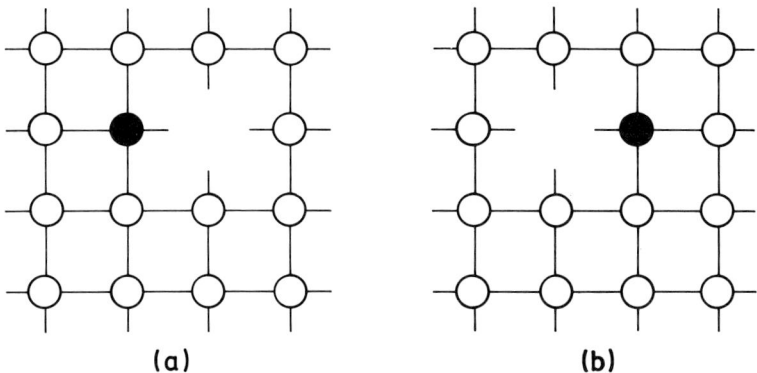

Fig. 2. Vacancy mechanism: The radioactive self-atom in tracer self-diffusion or foreign atom in substitutional–solute diffusion (●) moves, by jumping into the vacancy on its right-hand side (a), to the right (b) by one nearest-neighbor distance of the regular lattice atoms.

Each of the two basic indirect diffusion mechanisms may exhibit a number of complications, some of which will be briefly considered in the following list.

(i) The diffusion vehicles may be multiple defects. A well-studied example of this is the contribution of divacancies to the self-diffusion in metals in the vicinity of the melting temperature (Mehrer, 1978; Peterson, 1978).

(ii) The vacancy and the interstitialcy mechanisms may operate simultaneously [cf. self-diffusion in Si (Section IX.C.1)].

(iii) Complexes of intrinsic and extrinsic point defects may diffuse as entities. Examples are the A center (vacancy–interstitial oxygen pair) and the E center (vacancy–substitutional phosphorus pair) in Si (Watkins *et al.*, 1959; Watkins and Corbett, 1964) as well as the so-called mixed dumbbells [self-interstitial–substitutional foreign-atom pairs (Dederichs *et al.*, 1978)].

(iv) Foreign atoms (A) that under conditions of thermal equilibrium may be incorporated either as substitutional (A_s) or interstitial (A_i) atoms (e.g., Au or Ni in Si; see Sections IV and IX) may diffuse via the *kick-out mechanism* (Gösele *et al.*, 1980) and/or the *dissociative mechanism* (Frank and Turnbull, 1956). What the two mechanisms have in common is that the mobility of foreign atoms is much higher when they are located in interstices than when they are located on regular lattice sites. As a consequence, the long-range transport of A atoms takes place via the direct interstitial mechanism that is undergone by the A_i configuration (Fig. 1a). The two mechanisms differ in the way in which the foreign atoms interchange between the A_s and A_i configurations.

In the kick-out mechanism the interchange involves self-interstitials (I). The interchange may be described by the quasi-chemical reaction

$$A_s + I \rightleftharpoons A_i \qquad (1)$$

From left to right, reaction (1) is identical with the partial step of the interstitialcy mechanism that leads from the configuration of Fig. 3a to that of Fig. 3b. In the opposite direction, reaction (1) agrees with the other step of the interstitialcy mechanism, corresponding to the transition from Fig. 3b to Fig. 3c.

In the dissociative mechanism the transitions between A_s and A_i involve vacant lattice sites (V) according to

$$A_s \rightleftharpoons A_i + V \qquad (2)$$

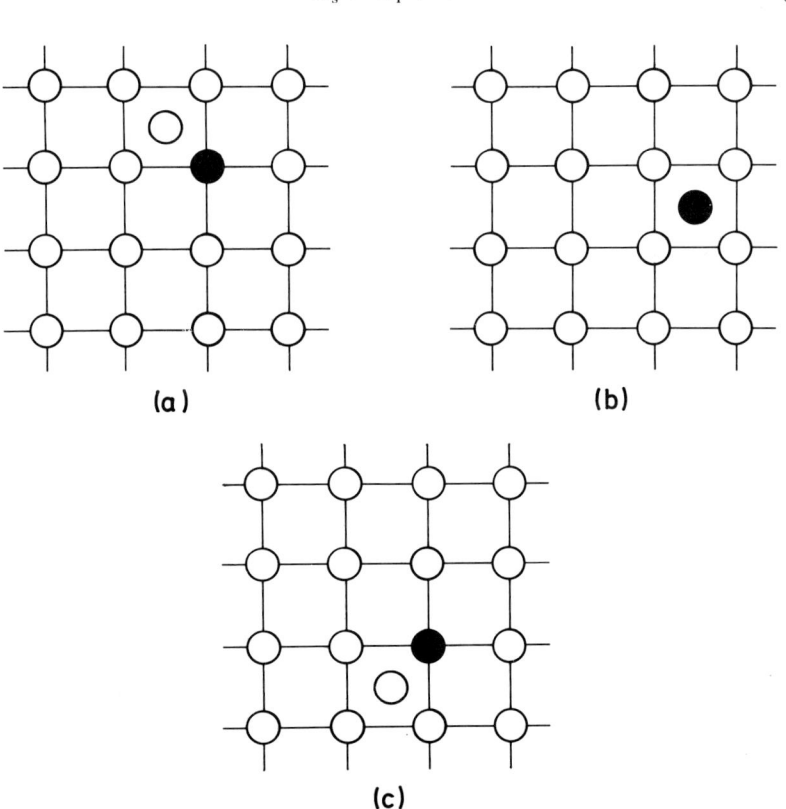

Fig. 3. Interstitialcy mechanism: In (a) a self-interstitial (○ in the center of a lattice cell) has approached a radioactive self-atom or a substitutional foreign atom (●), respectively; in (b) the marked atom has exchanged its original position with the self-interstitial. In this way the marked atom has temporarily become an interstitial, whereas the original self-interstitial has disappeared by occupying a regular lattice site. In (c) the marked atom has re-occupied a regular site by kicking a self-atom into an interstice.

Depending on its direction, Eq. (2) may be considered as the dissociation of a substitutional foreign atom A_s into a Frenkel pair with extrinsic interstitial partner A_i, or as the recombination of such a pair. Note that the reactions (1) and (2) are not symmetric under the exchange of I and V. This has the important consequence that the theoretical predictions for the kick-out and the dissociative mechanism may differ widely (Section IX.B).

Examples of both kick-out and dissociative mechanisms in elemental semiconductors will be thoroughly discussed in Section IX. These mechanisms may play a role in compound semiconductors, too. For instance, the diffusion of Zn in GaAs has been found to occur via kick-out diffusion (Gösele and Morehead, 1981; van Ommen, 1983).

(v) In ordered compounds diffusion by nearest-neighbor jumps of vacancies may lead to interchange of atoms on different sublattices, resulting in the formation or elimination of so-called antisite defects. An example is provided by the diffusion in group III–V semiconductor compounds (Vorob'ev et al., 1981; Weiler and Mehrer, 1983).

B. Thermal-Equilibrium Diffusion versus Enhanced Diffusion

In all diffusion mechanisms the atoms under consideration have to carry out jumps between different sites. If the extreme case of coherent tunneling is left aside, the diffusional jumps are assisted by the thermal movement of the atoms. In the standard situation the jump rate is entirely determined by the temperature T (apart from the effects of hydrostatic pressure, which may be incorporated by formulating the theory in terms of enthalpy and Gibbs free energy). For the purposes of the present chapter, which does not consider the diffusion of hydrogen isotopes or of polarons, we may disregard quantum mechanical contributions to the diffusivity, so that in cubic crystals the diffusion coefficient under standard conditions may be written as an Arrhenius expression

$$D_\alpha = D_{\alpha 0} \exp(-H_\alpha^M/kT) \qquad (3)$$

with the preexponential factor

$$D_{\alpha 0} = g_\alpha a_0^2 v_{\alpha 0} \exp(S_\alpha^M/k) \qquad (4)$$

Here H_α^M denotes the enthalpy and S_α^M the entropy of migration, a_0 the lattice parameter, and $v_{\alpha 0}$ the attempt frequency; k has its usual meaning as Boltzmann's constant, and g_α is a factor that takes into account the geometry of the crystal structure and the atomistic details of the diffusion process. The subscript α refers to the defect species controlling the diffusion process; i.e., in the case of the direct interstitial mechanism it indicates the chemical nature, geometrical configuration, electrical charge state, etc., of the interstitials involved, whereas in the case of indirect diffusion it characterizes the intrinsic defects acting as diffusion vehicles. In the latter case, we write

β instead of α if we wish to indicate that these intrinsic defects are monovacancies or monointerstitials.

The tracer self-diffusion coefficient, i.e., the diffusivity of radioactive self-atoms under thermal-equilibrium conditions, is given by

$$D^T = \sum_{\beta=I,V} f_\beta D_\beta^{SD} = \sum_{\beta=I,V} f_\beta D_\beta C_\beta^{eq} \tag{5}$$

where

$$C_\beta^{eq} = \exp(S_\beta^F/k)\exp(-H_\beta^F/kT) \tag{6}$$

are the concentrations[2] of self-interstitials ($\beta = $ I) and monovacancies ($\beta = $ V) in thermal equilibrium. In Eq. (5), contributions by clusters of I or V are neglected. The f_β denote correlation factors, $D_\beta^{SD} \equiv D_\beta C_\beta^{eq}$ contributions to the uncorrelated self-diffusion coefficient $\sum_{\beta=I,V} D_\beta^{SD}$, and S_β^F and H_β^F entropies and enthalpies of formation, respectively.

Insertion of Eqs. (3), (4), and (6) into Eq. (5) yields

$$D^T = \sum_{\beta=I,V} D_\beta^T = \sum_{\beta=I,V} f_\beta g_\beta a_0^2 v_{\beta 0} \exp(-G_\beta^{SD}/kT)$$
$$= \sum_{\beta=I,V} D_{\beta 0}^T \exp(-H_\beta^{SD}/kT) \tag{7}$$

with the preexponential factors

$$D_{\beta 0}^T = f_\beta g_\beta a_0^2 v_{\beta 0} \exp(S_\beta^{SD}/k) \tag{8}$$

the Gibbs free energy of self-diffusion

$$G_\beta^{SD} = H_\beta^{SD} - TS_\beta^{SD} \tag{9}$$

the self-diffusion enthalpy

$$H_\beta^{SD} = H_\beta^F + H_\beta^M \tag{10}$$

and the self-diffusion entropy

$$S_\beta^{SD} = S_\beta^F + S_\beta^M \tag{11}$$

The diffusion coefficient D^s of foreign substitutional atoms in thermal equilibrium may be derived from Eqs. (5) or (7) by inserting factors h_β under the summation signs. These factors account for the interaction between the intrinsic thermal equilibrium defects and the substitutional atoms. They depend on temperature and the atomic fraction of the substitutional atoms, unless this is small compared to unity.

Particle irradiation, illumination, charge-carrier injection, quenching from high temperatures, and other means of external stimulation lead to

---

[2] Usually concentrations are given in atomic fractions and denoted by C_α. In particular cases, in which the numbers of defects per unit volume are meant, we use c_α.

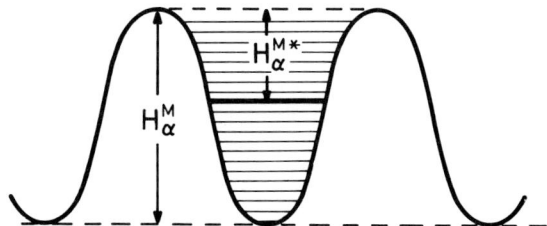

Fig. 4. Thermally assisted energy-release mechanism via the vibrational state marked by the heavy horizontal line.

deviations from the standard situation in which D_α possesses the Arrhenius-type form [Eqs. (3) and (4)] and/or to deviations of the concentrations C_β of intrinsic defects from their thermal-equilibrium values [Eq. (6)]. In the following discussion we confine ourselves to brief outlines of some examples of this so-called *enhanced diffusion* (which includes retarded diffusion as negatively enhanced diffusion). For a more detailed and complete discussion of this field we refer the reader to reviews by Bourgoin and Corbett (1978) and Frank *et al.* (1980).

As an example of *mobility-enhanced diffusion*, which is defined as a diffusion process in which D_α deviates from Eqs. (3) and (4) due to an external stimulation, we consider the charge-carrier, injection-enhanced diffusion of aluminium interstitials Al_i in silicon (Troxell *et al.*, 1979). Here the energy released from the electronic reservoir by carrier capture and recombination at the first donor level (corresponding to the charge-state transition Al_i^0–Al_i^+) is used to transfer Al_i into an excited vibrational state (schematically marked by the heavy horizontal line in Fig. 4). Thus under carrier injection the "normal" migration enthalpy of $H_\alpha^M = 1.2$ eV is lowered to $H_\alpha^{M*} = 0.27$ eV.

In the case of indirect diffusion, *concentration-enhanced* diffusion constitutes an alternative to mobility-enhanced diffusion. This arises from externally induced deviations of the atomic fractions of vacancies and/or self-interstitials from their equilibrium values C_β^{eq} [cf., e.g., Eq. (5) for D^T]. Oxidation-influenced diffusion of group III or group V elements in silicon (Section V) and the so-called anomalous diffusion phenomena in silicon (Section VIII) form typical examples of concentration-enhanced diffusion.

III. Self-Diffusion and Related Phenomena

A. General Remarks

Compared with metals, self-diffusion in semiconductors is a very slow process. For the elemental semiconductors this is illustrated in Fig. 5, in which the self-diffusivities of the cubic semiconductors Si and Ge and of the trigonal semiconductors Te and Se are compared with those of typical metals such as Cu, Ag, and Au on a temperature scale normalized to the melting temperature T_m. Figure 5 reveals the following differences between metals and semiconductors, already emphasized by Seeger and Chik (1968):

(i) Near the melting temperatures the self-diffusion in semiconductors is several orders of magnitude slower than in typical metals.

(ii) At lower normalized temperatures the ratio of the self-diffusivities of metals and semiconductors becomes even larger.

Generally speaking, the origin of these differences (and of others to be discussed later, e.g., the dependence of the self-diffusion of semiconductors on the concentration of electrically active dopants) lies in the homopolar bonding of the semiconductors considered here.

The basic question concerning the self-diffusion of elemental semiconductors is the same as in the case of metals: Does the self-diffusion occur

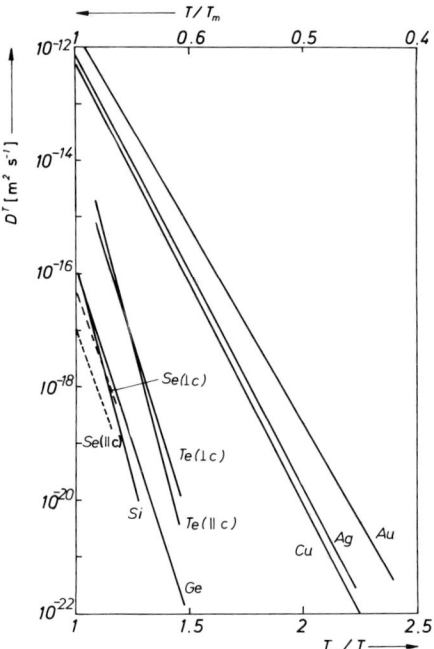

Fig. 5. Comparison between the self-diffusivities of the cubic semiconductors Ge (Fig. 6) and Si (Fig. 7), the trigonal semiconductors Te (Ghoshtagore, 1967; Werner, Mehrer, and Siethoff, 1983) and Se (Brätter and Gobrecht, 1970), and the typical metals Cu, Ag, Au.

directly (through nearest-neighbor exchange or a ring mechanism) or indirectly (via the vacancy or the interstitialcy mechanism)? The usual way to answer this question is to determine the diffusivity and the equilibrium concentrations of intrinsic point defects separately as functions of temperature and to investigate whether they are able to account for the measured self-diffusion [i.e., whether they obey a relationship of the type of Eq. (5)]. Up to now it has not been possible to obtain reliable information on equilibrium point defects in semiconductors by any of the techniques available for the study of high-temperature defects in metals (thermal expansion, specific heat, positron annihilation, or quenching from high temperatures). Therefore, for Ge and Si the above question had to be answered by employing less-direct approaches, particularly by carefully analyzing data on the diffusion of foreign atoms in these materials. In this way it was possible to separate the contributions $D_V C_V^{eq}$ and $D_I C_I^{eq}$ to the tracer self-diffusion coefficient D^T [Eq. (5)] but not to split $D_\beta C_\beta^{eq}$ ($\beta = I, V$) into the diffusivities D_β and the equilibrium concentrations C_β^{eq}.

It appears highly likely that the conventional techniques for studying intrinsic defects in thermal equilibrium fail in the case of elemental semiconductors, because in these materials the concentrations of equilibrium defects are extremely small, presumably $\lesssim 10^{-6}$ at T_m (Seeger and Chik, 1968). This is indicated by the relatively small self-diffusivities of the elemental semiconductors (Fig. 5), which must be attributed, at least to some extent, to large enthalpies of formation of vacancies and self-interstitials [see Eqs. (5) and (6)]. The alternative, that the small self-diffusivities might result from the possibility that self-diffusion in elemental semiconductors takes place by a direct diffusion mechanism, will be ruled out in this chapter by a number of arguments—at least for Ge and Si. On one hand, the small equilibrium concentrations of intrinsic defects lead to a considerable simplification of the discussion of the self-diffusion mechanisms in elemental semiconductors, since contributions by multiple defects (e.g., divacancies) may be neglected. On the other hand, the smallness of the self-diffusivities creates a number of experimental problems. Fortunately, these have been overcome to a considerable extent. This will become clear in Section III.B, in which techniques available for the study of self-diffusion in Ge and Si will be described.

B. Experimental Techniques

Let us first consider the conventional and well-established techniques of determining the tracer self-diffusion coefficient D^T based on studying the redistribution of radioactive or stable tracers initially deposited on the specimen surface by means of serial sectioning methods. In the case of radioactive isotopes, the redistribution may be investigated with radiation detection methods; for stable isotopes, secondary ion mass spectroscopy (SIMS) may be used.

For brittle materials, such as Si and Ge, the mechanical sectioning techniques are confined to lapping or grinding. These techniques require that the mean penetration distance $(D^T t)^{1/2}$ of the tracer atoms during the time t of a diffusion anneal exceed 10^{-6} m. For semiconductors this is a serious limitation, particularly for Si due to the short half-life (2.6 hr) of ^{31}Si, the only readily obtainable radioactive isotope of silicon. These were the main reasons why, until a few years ago, tracer measurements in semiconductors were limited to narrow temperature ranges in the vicinity of the melting temperature.

During the last decade several microsectioning techniques, permitting serial sectioning on a submicron level, have been developed. Sputter sectioning with Ar ions generated in a duoplasmatron-type ion source (Mehrer et al., 1978) has been successfully applied in radiotracer experiments on elemental and compound semiconductors. Sputtering in an ion microprobe and detection of stable tracer isotopes by means of mass spectroscopy of the secondary ions emitted during the sputtering process (SIMS) is also a well established technique for semiconductors. A more detailed discussion of microsectioning techniques is given by Rothman in Chapter 1. In favorable cases, these techniques have permitted the determination of tracer diffusivities down to values as low as 10^{-22} m^2 s^{-1} (Section III.C).

So-called ion-beam techniques, in which an ion beam extracted from an accelerator is used to create a concentration profile of limited depth, become increasingly important. In self-diffusion studies only the (p, γ)-resonance technique has been applied to monitor depth profiles of the tracer atom ^{30}Si. This method employs the reaction ^{30}Si$(p, \gamma)^{31}$Si, which has a narrow resonance of only 68 eV width at a proton energy of 620 keV. The basic result of such an experiment is the γ-ray yield of the resonance reaction as a function of the proton energy. The yield curves can be converted into ^{30}Si depth profiles if the specific energy loss of the protons is known. A deconvolution of the raw profiles is necessary in order to correct for the energy straggling of the proton beam and for the energy width of both accelerator and reaction.

C. SELF-DIFFUSION IN INTRINSIC GERMANIUM AND SILICON

1. *Germanium*

In intrinsic germanium the temperature dependence of the tracer self-diffusion coefficient of the radioactive isotope ^{71}Ge has been measured by several groups (Letaw et al., 1956; Valenta and Ramasastry, 1957; Widmer and Gunther-Mohr, 1961; Campbell, 1975; Vogel et al., 1983; Werner, 1984) by means of different techniques (Fig. 6). With the exception of the latest experiments, precision grinding techniques were used to remove sections with thicknesses of the order of 1 μm from the diffusion zone of the annealed specimens. As a consequence, the temperature range covered by the earlier experiments is rather limited. By means of a sputtering technique for serial sectioning, Vogel et al. (1983) and Werner (1984) have been able to extend the range of self-diffusion studies in Ge to diffusivities as low as 10^{-22} m² s⁻¹.

The overall agreement between Ge self-diffusion data of different authors is good. In the region of overlap a small difference between the data of Vogel et al. (1983) and those of the earlier workers may be seen. We tend

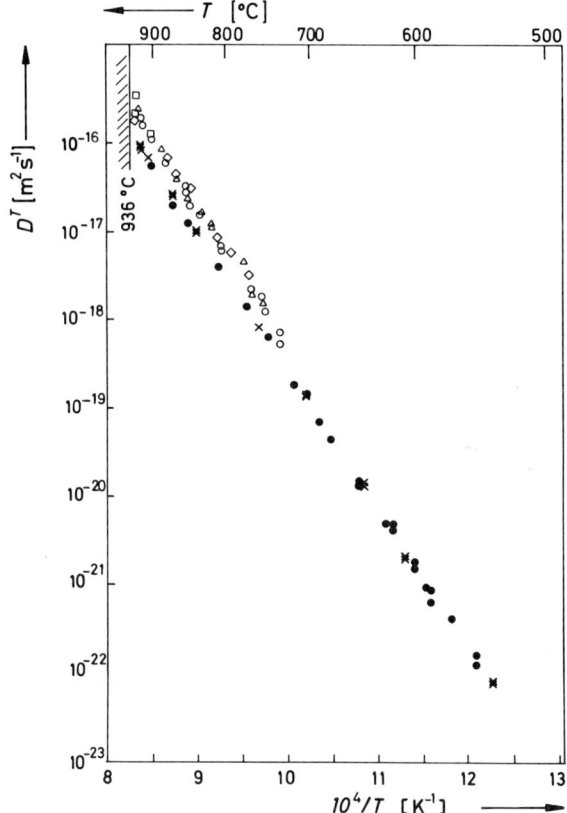

Fig. 6. Tracer self-diffusion coefficient of Ge as a function of temperature. Data from [◇] Letaw et al. (1956); [△] Valenta and Ramasastry (1957); [○] Widmer and Gunther-Mohr (1961); [□] Campbell (1975); [●] Vogel et al. (1983); [×] Werner (1984).

TABLE I

SELF-DIFFUSION DATA FOR GERMANIUM AND SILICON

Element	D_0^T (10^{-4} m^2 s^{-1})	H^{SD} (eV)	Temperature range (K)	Technique	References
Ge	7.8	2.95	1039–1201	Sectioning by grinding	Letaw et al. (1956)
	32	3.1	1023–1143	Sectioning by grinding	Valenta and Ramasastry (1957)
	44	3.12	1004–1188	Steigmann's method	Widmer and Gunther-Mohr (1961)
	10.8	2.99		Gruzin's method	
	24.8	3.14	822–1164	Sectioning by sputtering	Vogel et al. (1983)
	13.6	3.09	808–1177	Sectioning by sputtering	Werner (1984)
Si	1800	4.77	1473–1673	Hand lapping	Peart (1966)
	1200	4.72	1451–1573	Chemical sectioning, n activation of ^{30}Si	Ghoshtagore (1966)
	9000	5.13	1373–1573	Electrochemical sectioning	Masters and Fairfield (1966); Fairfield and Masters (1967)
	1460	5.02	1320–1660	Sectioning by sputtering	Mayer et al. (1977)
	8	4.1	1173–1373	(p, γ) resonance of ^{30}Si	Hirvonen and Anttila (1979)
	154	4.65	1128–1448	SIMS, ^{30}Si	Kalinowski and Seguin (1979, 1980)
	20	4.4	1103–1473	(p, γ) resonance of ^{30}Si	Demond et al. (1983)

to attribute this to problems in determining small diffusion coefficients during the earlier work. Widmer and Gunther-Mohr (1961) used Gruzin's or Steigmann's methods (Gruzin, 1952; Steigmann et al., 1939), both of which are known to be less reliable than the layer-counting method since these methods require a precise knowledge of the absorption coefficient of the radiation involved. In the work of Valenta and Ramasastry (1957), the condition $\delta \ll \sqrt{D^T t}$ (δ = thickness of the deposited tracer layer) was not always fulfilled. Since, nevertheless, these authors used the thin-film solution of the diffusion equation to deduce tracer diffusion coefficients, the obtained values are likely to be somewhat larger than the true D^T values.

As may be seen in Fig. 6, the temperature dependence of the D^T data of Ge is well described by an Arrhenius law. (The preexponential factors D_0^T and the self-diffusion enthalpies H^{SD} obtained from the measurements of different authors are compiled in Table I.) Seeger and Chik (1968) argued that this result may be accounted for in terms of an indirect self-diffusion mechanism involving one type of intrinsic defect [see Eq. (7)]. Guided by further observations, they suggested that it is the vacancy mechanism (Section II) that controls self-diffusion in Ge. The experimental results on Ge to be presented in Sections III.E–H, IV, VII.A, and IX.C.2 are in accordance with this interpretation.

Table I shows that the preexponential factor D_0^T of Ge is considerably larger than the D_0^T values typical for metals (10^{-6} m² s⁻¹ ≲ D_0^T ≲ 10^{-4} m² s⁻¹). Arguing that for an ordinary vacancy mechanism the product $f_V g_V a_0^2 v_{V0}$ in Eq. (8) for $D_{V0}^T (\equiv D_0^T)$ should be of the same order of magnitude for Ge and metals, Seeger and Chik (1968) interpreted the large D_0^T value of Ge in terms of a large self-diffusion entropy of the vacancy in Ge, $S_V^{SD} \approx 10\,k$. They suggested that this large S_V^{SD} value arises from a spreading out of the vacancy over several atomic volumes.

2. Silicon

A collection of self-diffusion data on intrinsic silicon is given in Fig. 7. The D_0^T and H^{SD} values from these studies are listed in Table I. According to the techniques used, the measurements fall into three categories.

(i) In three experiments the short-lived radiotracer ^{31}Si was diffused and studied by means of either a hand-lapping technique for sectioning (Peart, 1966), electrochemical sectioning (Fairfield and Masters, 1967), or sputter sectioning (Mayer et al., 1977).

(ii) In order to avoid the difficulties associated with the short-lived nuclide ^{31}Si, Ghoshtagore (1966) diffused the stable nuclide ^{30}Si into samples of natural Si which, however, contain ^{30}Si with an abundance of

3.09%. After the diffusion anneals, ^{30}Si was neutron-activated to ^{31}Si, and a chemical sectioning procedure was applied. As already pointed out by Hu (1973), it is questionable whether the natural ^{30}Si background was subtracted in a proper way. Kalinowski and Seguin (1979) also diffused ^{30}Si but determined the diffusion profiles by sputtering and by monitoring the signal of ^{30}Si by means of SIMS.

(iii) Hirvonen and Anttila (1979) were the first to demonstrate that the (p, γ)-resonance technique is applicable to self-diffusion in Si. They implanted $(2-4) \times 10^{20}$ m^{-2} ^{30}Si ions into float-zoned Si wafers and monitored the broadening of the implantation profiles during subsequent diffusion anneals by means of the reaction ^{30}Si$(p, \gamma)^{31}$P, using a proton beam. Demond et al. (1983) extended these experiments over a wider range of temperatures. They found D^T values about three times smaller than those obtained by Hirvonen and Anttila (1979). This difference may arise from different amounts of

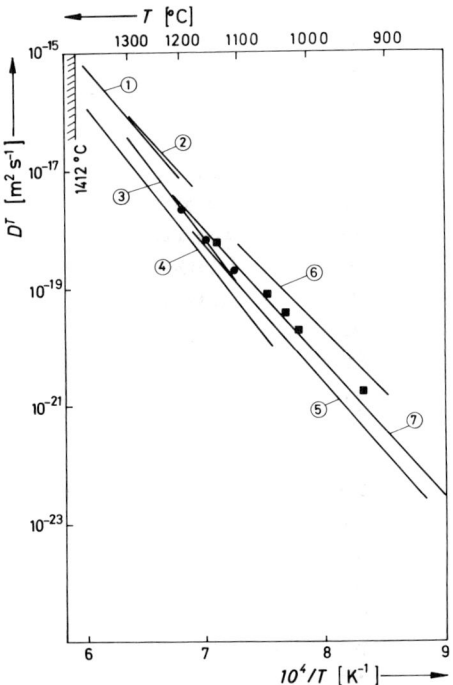

Fig. 7. Tracer self-diffusion coefficient of Si as a function of temperature. Data from [1] Peart (1966); [2] Ghoshtagore (1966); [3] Fairfield and Masters (1967); [4] Mayer et al. (1977); [5] Kalinowski and Seguin (1980); [6] Hirvonen and Antilla (1979); [7] Demond et al. (1983); [■, ●] from stacking-fault shrinkage.

^{30}Si-implantation-induced radiation damage not recovered prior to the diffusion anneals. This potential influence of residual radiation damage on the diffusion is a disadvantage of the (p, γ)-resonance technique.

Figure 7 and Table I show that the results on Si self-diffusion obtained by different groups are less consistent than those on Ge self-diffusion (Fig. 6) and by far less consistent than those on self-diffusion in metals (see, e.g., Peterson, 1978). Diffusion coefficients of metals have been reproduced in different laboratories within a few percent. In the case of silicon, however, discrepancies of several hundred percent are not unusual. This is very surprising, particularly since in the more recent diffusion experiments virtually dislocation-free Si single crystals of extremely high purity were used. The reasons for these discrepancies are not clear. They certainly deserve further attention.

Oxygen is one of the main impurities in both Czochralski-grown and float-zone Si crystals. It is well known that oxidation at the surface or oxide formation or dissolution in the bulk causes deviations of the concentrations of intrinsic point defects from their equilibrium values. This leads to enhanced diffusion of group III and group V elements in Si (Section V). We suppose that one cause of the discrepancies among the various Si self-diffusion data might be variation of the oxygen content to which, unfortunately, attention has not yet been paid in diffusion experiments.

In addition to the outstanding fact that in Si (and other elemental semiconductors) self-diffusion is considerably slower than in metals [items (i) and (ii) of Section III.A], the following features characteristic of Si self-diffusion deserve to be mentioned here, though their final discussion has to be postponed to later sections. Table I shows that attempts to describe the temperature dependence of D^T of Si by an Arrhenius expression yield H^{SD} values decreasing from about 5 eV near T_m to about 4 eV at about 1200 K and D_0^T values decreasing from $1-10^{-1}$ to $\lesssim 10^{-4}$ m^2 s^{-1} in this temperature regime. These observations indicate a change in the self-diffusion mechanism and/or in the self-diffusion parameters H^{SD} and S^{SD} as a function of temperature. Seeger and Chik (1968) and Seeger et al. (1977) suggested that both effects contribute to the nonlinearity of the Arrhenius plot. More specifically, they proposed that in Si at low temperatures self-diffusion mainly occurs via vacancies, whereas at high temperatures it is dominated by the interstitialcy mechanism. The enhancement of D_0^T with increasing temperature—leading to a $D_0^T(T_m)$ value that exceeds even that of Ge—was explained by these authors in terms of an increase of S^{SD} arising from a spreading out of the Si self-interstitial over several atomic volumes. The recent developments reported in the remaining sections of this chapter confirm this picture, which is in accordance with the evolving opinion that both vacancies and self-interstitials contribute to self-diffusion in Si (Prussin, 1972; Hu, 1974, 1977, 1981; Sirtl, 1977; Leroy, 1979; de Kock and van de Wiggert, 1980; Fair, 1981a,b; Lin et al., 1981a; Mizuo and Higuchi, 1981, 1982a,b,c,d; Antoniadis, 1982; Antoniadis and Moskowitz, 1982a,b; Tan and Ginsberg, 1983). The earlier assumption that Si self-diffusion occurs exclusively via vacancies (Masters, 1971; Petroff and de Kock, 1975; Shaw, 1975; Fair, 1977; van Vechten, 1978; Bourgoin and Lannoo, 1980; Kitagawa et al., 1982a, 1983) is no longer tenable.

D. Diffusion of Germanium Tracers in Silicon

Because at high temperatures Si and Ge form continuous solid solutions, and because the atomic sizes and the outer-shell electronic structures (sp^3 orbitals with p core) are similar, we may expect that a close relationship exists between the diffusion of Ge tracers in Si and Si self-diffusion. From the point of view of tracer experiments, the SiGe system has the advantage that

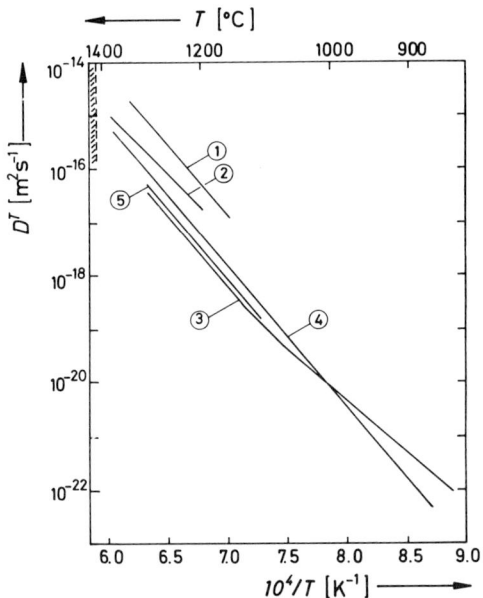

Fig. 8. Diffusion coefficient of Ge tracers in Si as a function of temperature. Data from [1] Petrov et al. (1975); [2] McVay and DuCharme (1973); [3] Hettich et al. (1979); [4] Dorner et al. (1982); [5] Ogino et al. (1982).

Ge possesses several radioisotopes suitable for diffusion measurements as well as a series of stable isotopes (^{70}Ge, ^{72}Ge, ^{73}Ge, ^{74}Ge, and ^{76}Ge) that may be distinguished by means of mass spectroscopy.

Data on the diffusion of Ge in Si are compiled in Fig. 8. Description of these data in terms of an Arrhenius expression yields the preexponential factors $D_0^{s'}$ and the activation enthalpies $H^{s'}$ given in Table II.

In three experiments the radiotracer method was employed either in combination with mechanical sectioning (Petrov et al., 1957; McVay and DuCharme, 1973, 1974, 1975) or sputter sectioning (Hettich et al., 1979). The SIMS method was applied by Dorner et al. (1982) and by Ogino et al. (1982). As in the case of Si self-diffusion, significant discrepancies are observed between the results obtained by various groups of authors (Fig. 8).

Among radiotracer studies, those of Hettich et al. (1979) cover the widest temperature range. In this work two temperature regimes have been distinguished (Table II): (i) The high-temperature regime in which an activation enthalpy and a preexponential factor were found similar to those in the high-temperature regime of Si self-diffusion, which Seeger and Chik (1968) attributed to the interstitialcy mechanism, and (ii) the low-temperature regime with a lower activation enthalpy and a lower preexponential factor, which come close to the corresponding values in the low-temperature regime of Si self-diffusion assigned to the vacancy mechanism (Seeger and Chik, 1968). The data in the high-temperature regime are in excellent agreement with those of Ogino et al. (1982), which, unfortunately, are confined to high temperatures. The measurements of Dorner et al. (1982), which do extend to lower temperatures, show no indication of a separate low-temperature regime. Again, it may be speculated whether these differences are due to different oxygen contents.

TABLE II

DATA ON GERMANIUM DIFFUSION IN SILICON

$D_0^{s'}$ (10^{-4} m^2 s^{-1})	$H^{s'}$ (eV)	Temperature range (K)	Technique	References
6.3×10^5	5.28	1423–1623	Sectioning by grinding	Petrov et al. (1957)
1500	4.7	1473–1653	Sectioning by grinding	McVay and DuCharme (1973, 1974, 1975)
2500	4.97	1303–1575	Sectioning by sputtering	Hettich et al. (1979)
0.35	3.92	1128–1273	Sectioning by sputtering	Hettich et al. (1979)
1.03×10^5	5.33	1149–1661	SIMS	Dorner et al. (1982)
7550	5.08	1373–1573	SIMS	Ogino et al. (1982)

TABLE III
Doping Effects on Ge and Si Self-Diffusion and on Ge Diffusion in Si

System	Type and dopant	Dopant concentration, c_s (cm^{-3})	Temperature range (K)	$D^{T,s'}(C_s)/D^{T,s'}(0)$	References
^{71}Ge in Ge	n:As	6×10^{18}	1023–1156	$\sim 2 - \sim 1$	Valenta and Ramasastry (1957)
	p:Ga	5×10^{19}	1023–1156	$\sim 0.6 - \sim 1$	
		10^{20}	1114–1156	~ 0.4	
^{71}Ge in Ge	n:Sb	1.5×10^{18}	894	1.19	Vogel et al. (1983)
	p:Ga	1.7×10^{18}	894	0.92	
^{71}Ge in Ge	n:Sb	3×10^{18}	973	1.19	Werner (1984)
	p:Ga	3×10^{18}	973	0.84	
	p:Ga	2×10^{19}	973	0.40	
^{31}Si in Si	n:P, As	$8 \times 10^{19} - 1.88 \times 10^{20}$	1363–1470	1.35–3.15	Fairfield and Masters (1967)
	p:B	$8 \times 10^{19} - 2.2 \times 10^{20}$		1–1.75	
^{31}Si in Si	p:B	$6 \times 10^{18} - 1.2 \times 10^{19}$	1318–1515	1.4–1.2	Hettich et al. (1979)
^{71}Ge in Si	n:P	1.1×10^{20}	1480–1650	$\sim 10 - \sim 2.5$	McVay and DuCharme (1975)
	p:B	4.5×10^{19}		$\sim 8 - \sim 1.6$	
^{71}Ge in Si	n:As	$\sim 4 \times 10^{19}$	1163–1493	1.8–2.5	Hettich et al. (1979)
			1128–1183	0.7–0.9	
	p:B	$6 \times 10^{18} - 1.2 \times 10^{19}$	1200–1542	1.6–2.6	

E. Effects of Doping

In metallic systems the effect of doping with foreign atoms on the tracer self-diffusion coefficient of the host atoms is well understood (for a review see Le Claire, 1978). Small concentrations C_s of substitutional impurities lead to an enhancement

$$D^T(C_s)/D^T(0) = 1 + bC_s \qquad (12)$$

that is linear in C_s. Typical values of the enhancement factor b lie between -20 and $+100$ ("enhancement" includes retardation as negative enhancement). Table III shows that the doping effects of group III and group V elements on the self-diffusion in Si and Ge and on the diffusion of Ge in Si are much larger than the corresponding effects in metals. We shall see that the doping effects in semiconductors are related to the fact that the dopants may act as donors or acceptors and that doping causes a change of the Fermi level, which in turn affects the diffusivity.

Diffusion enhancement or retardation via a Fermi level change is possible since defects in semiconductors may be electrically charged, so that, in general, both their Gibbs free energy of formation (Seeger and Chik, 1968) and migration (Frank et al., 1980) depend on the position of the Fermi level. It turns out that it is the influence of the Fermi level on the *formation* energy of the intrinsic point defects present in thermal equilibrium which controls the doping dependence of self-diffusion in Si and Ge and of Ge diffusion in Si.

In order to account for the acceptor nature of the vacancy predicted by James and Lark-Horovitz (1951), Seeger and Chik (1968) assumed that a vacancy in Si or Ge possesses a shallow acceptor level in the lower half of the band gap (Fig. 9a). Though more recent experiments have revealed a more complex level structure of the vacancy in Si (Watkins et al., 1979;

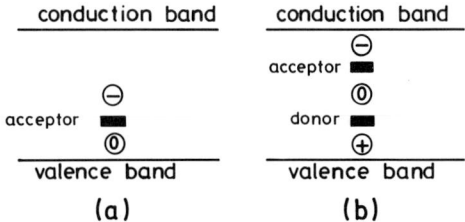

Fig. 9. Energy levels of (a) vacancies and (b) self-interstitials used in the interpretation of the influence of doping on self-diffusion (Section III.E) and on the impurity diffusion of group III and group V elements (Section VII). The defects are electrically neutral, positively charged, or negatively charged if the Fermi level lies in the part of the band gap marked by ⓪, ⊕, or ⊖, respectively.

Watkins, 1981), the model developed by Seeger and Chik (1968) allows correctly for those features that are essential for the description of the doping dependence. According to this model, by accepting an electron the formation enthalpy of the vacancy decreases as the distance by which the Fermi level lies above the acceptor level increases (Fig. 9a). As a consequence, doping with n-type impurities is expected to raise the equilibrium concentration of vacancies and hence to enhance indirect diffusion via the vacancy mechanism, whereas p doping should have the opposite effect. Table III shows that this doping dependence is observed for Ge self-diffusion in the entire temperature range investigated and for Ge impurity diffusion in Si below about 1200 K. This is in accordance with the view, mentioned in Sections III.C and III.D, that in these cases diffusion is dominated by the vacancy mechanism.

In the cases in which the interstitialcy mechanism dominates [self-diffusion (Section III.C.2) and Ge diffusion (Section III.D) in Si at high temperatures], both n and p doping lead to a diffusion enhancement. Seeger and Chik (1968) have shown that this finding may be understood by assuming that the Si self-interstitial possesses a donor level in the lower half and an acceptor level in the upper half of the band gap (Fig. 9b; Blount, 1959; Watkins et al., 1971) and by applying a line of reasoning analogous to that followed in order to explain the doping dependence of the diffusion via vacancies (see above).

Various authors (Shaw, 1975; van Vechten and Thurmond, 1976; Fair, 1977) have attempted to account for the doping dependence of the Si self-diffusion coefficient above about 1270 K in terms of a monovacancy mechanism. It is clear that formal agreement of the vacancy diffusion models with the doping dependence found by experiment can be enforced if the vacancy is arbitrarily endowed with essentially the same electronic levels that we attribute to the self-interstitial. Hence, experiments giving more specific information on the nature of the intrinsic defects involved in self-diffusion, e.g., those to be discussed in Section IX, are of great importance.

F. Isotope Effects

Measurements of the dependence of the diffusion coefficient on the isotope mass of the diffusing element (isotope effect) may provide information on the diffusion mechanism and, in the case of indirect foreign-atom diffusion, on the jump frequencies of intrinsic defects near foreign atoms (Peterson, 1975). This is due to the fact that the isotope effect, the strength of which is defined as

$$E = \frac{D^1/D^2 - 1}{(m_2/m_1)^{1/2} - 1} \tag{13}$$

is related via

$$E = f\,\Delta K \tag{14}$$

to the correlation factor f, which contains this valuable information. (D^l is the diffusivity of the isotope l of mass m_l, and f equals f_β with β either I or V [Section II.B].) In Eq. (14), which follows from Vineyard's (1957) many-body treatment of the atomic jump process (Mullen, 1961; Le Claire, 1966), ΔK is the fraction of the kinetic energy of the diffusing atom in the saddle point that is associated with the motion in the jump direction; i.e., by definition, ΔK is restricted to $0 < \Delta K \leq 1$. Limitations of the validity of Eq. (14) have been discussed by Peterson (1975) and Mehrer et al. (1976).

A prerequisite for isotope-effect measurements is that at least one suitable pair of isotopes exists. This condition is fulfilled for Ge self-diffusion and Ge diffusion in Si (Section III.D), and, indeed, in these cases isotope effects have been measured (Table IV). In Ge self-diffusion experiments, Campbell (1975) found $f\Delta K$ values between 0.26 and 0.30. Since for the vacancy self-diffusion mechanism in the diamond lattice $f = 0.5$ (Compaan and Haven, 1956), the interpretation of Ge self-diffusion in terms of the vacancy mechanism (Section III.C.1) leads us to $0.5 \lesssim \Delta K \lesssim 0.6$. This value is small compared to $\Delta K \approx 1$ characteristic of self-diffusion via monovacancies in close-packed metals (Peterson, 1975; Mehrer, 1978). Since the deviation of ΔK from unity increases with the number of atoms with which the diffusing atom shares its kinetic energy in the saddle point, we arrive at the satisfactory picture that in close-packed metals the vacancy is rather localized and unrelaxed, whereas in Ge it is more relaxed and/or extended. This is in accordance with the suggestion of Seeger and Chik (1968) that self-diffusion in Ge occurs via spread out vacancies (Section III.C.1).

TABLE IV

ISOTOPE EFFECT DATA ON DIFFUSION IN GE AND SI

System	Method	Temperature (K)	E	References
^{77}Ge/^{71}Ge in Ge	Radiotracer	1173	0.26 ± 0.015	Campbell (1975)
		1198	0.30 ± 0.06	
^{70}Ge/^{72}Ge/^{73}Ge ^{74}Ge/^{76}Ge in Si	SIMS	1498	0.20 ± 0.03	Södervall et al. (1982)
			0.29 ± 0.27	
		1422	0.34 ± 0.24	
		1323	0.50 ± 0.24	
			0.34 ± 0.10	
		1262	0.30 ± 0.13	

Södervall et al. (1982) measured the isotope effect of Ge impurity diffusion in Si at four temperatures. Their data appear to indicate that E increases up to 1323 K and decreases above this temperature, which might be interpreted in terms of a transition from a low- to a high-temperature process, as suggested in Section III.D. Due to the limited accuracy of the data (Table IV) only an average value of E over the entire temperature range investigated may be calculated. This value, $\bar{E} = 0.33 \pm 0.09$, is not suitable to draw any conclusions concerning the nature, relaxation, or spreading out of the intrinsic defects involved.

G. Pressure Effects

Further information on the nature and the properties of the defects involved in a diffusion process may be obtained from investigation of the effect of hydrostatic pressure on the diffusion coefficient. Making use of the thermodynamic relationship

$$V = (\partial G/\partial p)_T \tag{15}$$

relating the volume (V) to the pressure (p) derivative of the Gibbs free energy (G) at constant temperature, the activation volume of self-diffusion V^{SD} may be calculated from Eq. (7). If self-diffusion takes place via a single diffusion mechanism, i.e., either the vacancy or the interstitialcy mechanism [allowing us to drop the subscript β in Eq. (7)], we find

$$V^{SD} = [-\partial \ln D^T/\partial p + \partial \ln(gfv_0)/\partial p]kT \tag{16}$$

The first term on the right-hand side may be obtained from measurements of D^T as a function of the hydrostatic pressure. The second term represents only a small correction not exceeding a few percent of V^{SD} (Mehrer and Seeger, 1972). The subdivision of the Gibbs free energy of self-diffusion G^{SD} into the Gibbs free energy of formation $G^F (\equiv H^F - TS^F)$ and migration $G^M (\equiv H^M - TS^M)$ [see Eqs. (9–11)] leads to a splitting of the self-diffusion volume according to

$$V^{SD} = V^F + V^M \tag{17}$$

where V^F and V^M are the formation and migration volumes of the intrinsic defects controlling self-diffusion. However, the subdivision in Eq. (17) cannot be achieved on the basis of studies of pressure effects on self-diffusion alone.

Activation volumes V^{SD} have been measured on various noble metals and found to range from 0.6Ω to 0.9Ω (Ω = atomic volume), as reported in review articles by Peterson (1978) and Mehrer (1978) and as confirmed in recent experiments on Ag (Rein and Mehrer, 1982) and Au (Werner and Mehrer, 1983). It is well established that in these metals self-diffusion takes place mainly via monovacancies and that close to T_m minor contributions by divacancies exist. The most thoroughly investigated bcc metal is sodium (Mundy, 1971). The fact that V^{SD} in Na (0.4Ω to 0.55Ω) is smaller than in fcc metals presumably indicates that the defects acting as diffusion vehicles in Na are relaxed more strongly.

In our laboratory, measurements of the pressure dependence of self-diffusion in germanium have been performed (Werner et al., 1983; Werner, 1984). The activation volume varies from 0.24 to 0.41 Ω as the temperature increases from 876 to 1086 K. To our knowledge this is the first determination of V^{SD} in a semiconductor. These values are small, similar to Na. They are compatible with the concept that self-diffusion in Ge occurs via spread-out monovacancies (Section III.C.1). These values may very likely exclude extended self-interstitials as the defects controlling self-diffusion in Ge, since for self-interstitial mechanisms V^{SD} should be close to zero or even negative (Hu, 1973).

H. Self-Diffusion in Si–Ge Alloys

As discussed in Section III.C, Seeger and Chik (1968) suggested that in Ge self-diffusion occurs via vacancies, whereas in Si a transition takes place from the interstitialcy mechanism at high temperatures to the vacancy mechanism at low temperatures. At present, poor consistency among the various Si self-diffusion data (Fig. 7) makes it difficult to determine the temperature of this transition with satisfactory accuracy. However, in an indirect way (to be discussed in Section IX.C.1.b) this transition has been found to occur at about 1270 K. This estimate is confirmed by the fact that 1270 K lies right between 1330 K, where a knee in the Arrhenius plot of the data on the Ge tracer diffusion in intrinsic Si by Hettich et al. (1979) indicates a transition from the high- to low-temperature mechanism (Section III.D), and 1200 K, where such a transition is reflected by the doping dependence of the Ge impurity diffusion in Si (Section III.E).

The circumstance that at the temperatures at which diffusion experiments are performed Si and Ge form continuous solid solutions offers the possibility to test whether the concept just described is confirmed by diffusion studies on Si–Ge alloys. We expect (i) that above about 1270 K a transition from diffusion via extended self-interstitials to diffusion via extended vacancies takes place when the Ge concentration is enhanced, (ii) that the temperature of this transition increases with increasing Ge concentration, reaching T_m somewhere on the Ge-rich side, and (iii) that below about 1270 K the vacancy mechanism dominates, irrespective of the composition of the alloy.

The only systematic investigations of self-diffusion in Si–Ge alloys have been performed by McVay and DuCharme (1973, 1974, 1975) on polycrystalline material. Since these authors used the standard tracer method and conventional mechanical sectioning, their studies were limited to fairly high diffusivities ranging from 10^{-17} to 10^{-15} m^2 s^{-1}, i.e., to a temperature regime that does not permit one to check predictions (ii) and (iii) mentioned above. However, prediction (i) has been fully confirmed: In alloys with Ge concentrations above 35 at. % the activation enthalpy of diffusion turned out to be almost independent of the Ge content, namely about 3 eV, which is the value of the self-diffusion enthalpy in pure Ge. In Si-rich alloys, H^{SD} increases with decreasing Ge content to a value near 5 eV, which is characteristic of self-diffusion and Ge impurity diffusion in pure Si (Tables I and II). The preexponential factor of the self-diffusion coefficient was found to lie between 4×10^{-1} and 3×10^{-3} m^2 s^{-1} in alloys with more than 35 at. % Ge and between 10^{-2} and 10^{-1} m^2 s^{-1} in Si-rich alloys. The latter range of values is that of the preexponential factors found for Ge impurity diffusion and self-diffusion in pure Si at high temperatures (Tables I and II).

IV. Survey of Foreign-Atom Diffusion

From the technological point of view, an understanding of the diffusion of foreign atoms in Si and Ge is highly important because of two diametrically opposed reasons: On one hand, the diffusion of group III and group V elements plays a vital role in the doping of the base materials of electronic devices as well as in the various thermal annealing treatments during device fabrication. On the other hand, diffusion controls the incorporation of unwanted impurity atoms, e.g., of the metal atoms Fe, Ni, and Cu. Other metal atoms, e.g., Au, are used for tuning the minority-carrier lifetime.

From the scientific point of view, outstanding features of the diffusion of foreign atoms in Ge and Si are the grouping of the diffusivities of the group III and group V elements in a range slightly higher than the self-diffusion coefficients and the fact that the diffusivities of a number of foreign metal atoms are higher by several orders of magnitude (Figs. 10 and 11). These observations suggest the following interpretation, which is meant as a guideline for the remaining sections of this chapter.

(i) Group III and group V elements are exclusively incorporated on regular lattice sites and diffuse via the same mechanisms as self atoms. (Note that the same conclusion has been drawn for the group IV element Ge in Si [Section III.D].) A more detailed discussion of the diffusion mechanisms of group III and group V elements, which will consider the differences between these groups, will be given in Sections V–VIII.

2. DIFFUSION IN SILICON AND GERMANIUM

(ii) The fast diffusion of metal atoms is due to the fact that in thermal equilibrium a nonnegligible fraction C_i^{eq}/C_s^{eq} of these atoms are dissolved as interstitial atoms which undergo rapid diffusion [$C_i^{eq}(C_s^{eq})$ = solubility limit of interstitial (substitutional) atoms].

The differences between the diffusivities of various kinds of fast diffusors may be understood if, in addition to the parameter C_i^{eq}/C_s^{eq}, the migration enthalpies H_i^M of the interstitial configurations are considered. Extending the definition of "fast diffusors" from metal atoms to other species with diffusivities higher than those of the group III and group V solutes (Figs. 10 and 11) and assuming that in all these cases the reason for the relatively high mobility is a nonnegligible C_i^{eq}/C_s^{eq} value, the H_i^M values of the fast diffusors

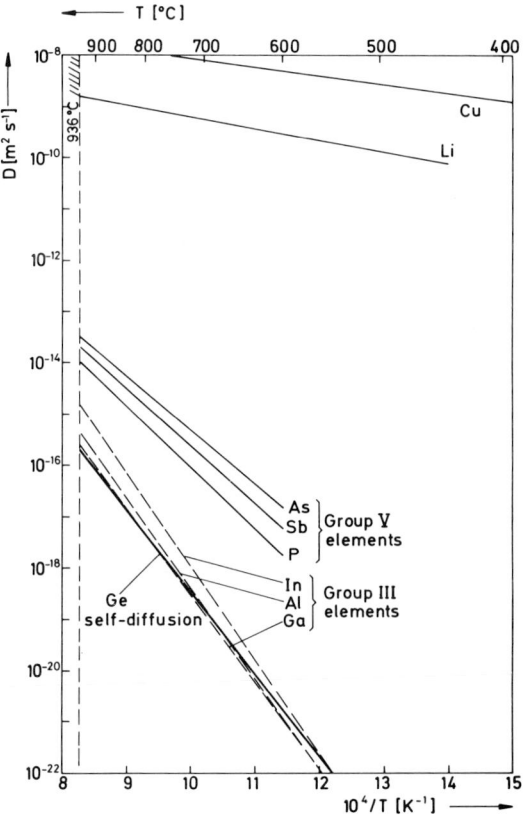

Fig. 10. Survey of the tracer diffusivities of foreign atoms in germanium [Cu (see Seeger and Chik, 1968), Li (Fuller and Ditzenberger, 1953), As, Sb, and Bi (Dunlap, 1954), In, Al, and Ga (Dorner, 1980)]. For comparison, Ge tracer self-diffusion data (Vogel et al., 1983) are included.

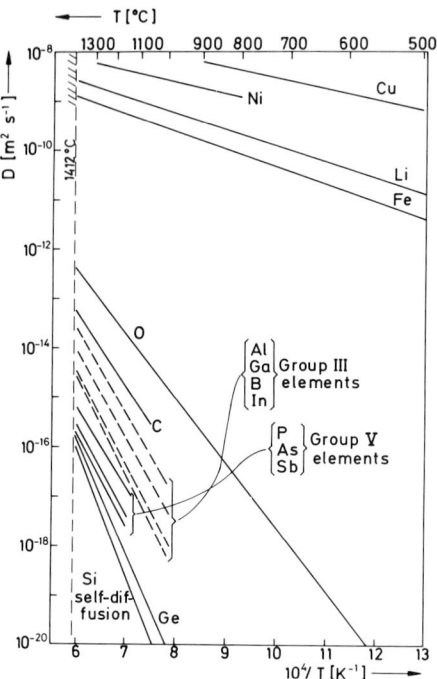

Fig. 11. Survey of the diffusivities of foreign atoms in silicon [Cu (Hall and Racette, 1964), Ni (Bakhadyrknanov et al., 1980), Li (Pell, 1960), Fe (see Weber, 1983), O (see Gösele and Tan, 1982), C (Newman and Wakefield, 1961a), Al, Ga, and In (Fuller and Ditzenberger, 1956), B (Hill, 1980), P (Ghoshtagore, 1971a), As and Sb (Ghoshtagore, 1971b), Ge (Hettich et al., 1979)]. For comparison, Si self-diffusion data (Mayer et al., 1977) are included.

in silicon (on which our attention will be mainly focused) naturally group themselves into "small" ($\lesssim 1$ eV) and "large" ($\gtrsim 2$ eV) ones (Table V). Within a given H_i^M group (i.e. within the same line of Table V) the diffusivities are presumed to increase mainly as a result of increasing values of C_i^{eq}/C_s^{eq} (Table V and Fig. 11). The gap between the H_i^M values of elements in different lines of Table V may be tentatively related to a difference in the bonding of the interstitials to the Si matrix. This will be illustrated in the following discussion.

In the case of the fast diffusing metal atoms, the outer electron shell either is complete or may be completed by the capture of an electron or a hole from the matrix, which lowers the free energy of the system considerably (deep-level impurities). As a consequence, the coupling of these interstitials to the Si lattice is weak, and therefore their migration enthalpy is "small." In contrast, under high-temperature equilibrium conditions Si self-intersti-

TABLE V

RANGES OF INTERSTITIAL FRACTIONS
AND INTERSTITIAL MIGRATION ENTHALPIES IN SILICON

Interstitial migration enthalpy (H_i^M [eV])	Interstitial fraction (C_i^{eq}/C_s^{eq})				
	$\lesssim 10^{-6}$	$\ll 1$	≈ 1	$\gg 1$	$\approx \infty$
$\lesssim 1$	—	Au, C(?)	—	Fe, Ni, Cu	—
$\gtrsim 2$	Si	—	C(?)	—	0

tials are extended over several atomic volumes, i.e., they are expected to be strongly coupled to the matrix and, accordingly, their migration enthalpy should be "large," namely about 2 eV at T_m (Sections III.C.2 and VI; Frank, 1981). Presumably for this reason and because of the smallness of C_I^{eq}/C_s^{eq} ($\equiv C_I^{eq}$) for Si self-interstitials [$C_I^{eq}(T_m) \lesssim 10^{-6}$ (Seeger et al., 1977)], self-diffusion in Si—a major part of which originates from the diffusion of self-interstitials (Section IX.C)—is "slow".

Compared to self-diffusion and to the diffusion of group III and group V elements, the diffusivity of carbon in Si,

$$D = 1.9 \times 10^{-4} \exp[-(3.1 \pm 0.2)eV/kT] \quad m^2 s^{-1} \qquad (18)$$

(Newman and Wakefield, 1961a,b), is fast, though it does not reach the high values of the diffusivities of some of the transition metals. Therefore, we propose that a nonnegligible fraction of C diffuses interstitially. On the other hand, a major portion, viz.

$$C_s^{eq} = 10^2 \exp(-2.3eV/kT) \qquad (19)$$

(Bean and Newman, 1971), must be substitutionally dissolved. This is indicated by a decrease of the lattice parameter of Si proportional to the content of carbon atoms (Baker et al., 1968), which are much smaller than Si atoms. The experimental data available at present do not enable us to decide between $C_i^{eq}/C_s^{eq} \ll 1$ or ≈ 1. In the former case, H_i^M should be "small," i.e., the diffusion of C in Si should resemble the diffusion of Au in Si (Section IX.C). If the latter possibility is realized, H_i^M should be close to the activation enthalpy in Eq. (18), namely about 3 eV. Such a "large" H_i^M value would reflect a strong coupling of the carbon interstitials to the Si matrix, which might be due to their isoelectric nature. It then remains open whether the strong coupling is due to a spreading out of the carbon interstitials similar to that of Si self-interstitials or whether for C in Si the mixed-dumbbell shape found in EPR studies on low-temperature irradiated Si (Watkins and

Brower, 1976) remains the stable interstitial configuration up to the high temperatures at which diffusion experiments are performed.

Using the calibration factor of Graff et al. (1973), from infrared absorption measurements (Craven, 1981) the solubility limit of oxygen in Si has been found to be

$$C_i^{eq} = 3 \times 10^{-2} \exp(-1.03 eV/kT) \tag{20}$$

The observation that the Si lattice is expanded proportional to the content of dissolved oxygen (Takano and Maki, 1973) is in accordance with the bond-centered interstitial configuration of O shown in Fig. 12 (Kaiser, 1957; Haas, 1960; Corbett et al., 1964). While, above room temperature, rotation of the kinked Si–O–Si bridge around the preferential $\langle 111 \rangle$ direction occurs rapidly, the elementary step of migration of this O interstitial involves a change into another $\langle 111 \rangle$ direction and thus requires the breaking of an Si–O bond. Therefore, the migration enthalpy H_i^M is expected to be "large." This is corroborated by several investigations of the diffusion of O in Si ranging from 540 K to T_m. By combining the data of Corbett et al. (1964), Mikkelson (1982), and Stavola et al. (1983), one finds for the oxygen diffusivity[3]

$$D = 1.7 \times 10^{-5} \exp(-2.54 eV/kT) \text{ m}^2 \text{ s}^{-1} \tag{21}$$

The fact that the reaction between interstitial O atoms and vacancies leads to A centers, i.e., interstitial-oxygen–vacancy pairs (Watkins et al., 1959), shows the strong tendency of O to avoid substitutional solution in Si. We conclude that oxygen in silicon is exclusively interstitially dissolved and that therefore the activation enthalpy in Eq. (21) may be identified with H_i^M.

Finally, we should like to focus attention on a further interesting feature related to Table V. The interstitials with $C_i^{eq}/C_s^{eq} \lesssim 10^{-6}$, i.e., the Si self-interstitials, diffuse via the interstitialcy mechanism, whereas oxygen in Si, for which $C_i^{eq}/C_s^{eq} \approx \infty$, migrates via the direct interstitial mechanism. The interstitialcy mechanism and direct interstitial mechanism may be considered as the two diametrically opposed limiting cases of the more general substitutional–interstitial interchange diffusion mechanisms (Section II.A). Indeed, as will be shown in Section IX.C, Au and Ni in Si undergo substitutional–interstitial diffusion, in agreement with the fact that for these impurities C_i^{eq}/C_s^{eq} is neither extremely small nor extremely large (Table V).

---

[3] The value of 2.7×10^{10} m² s⁻¹ measured at 720 K under different experimental conditions (Gaworzewski and Ritter, 1981), which is by about four orders of magnitude larger than that calculated from (21), was ascribed to the diffusion of O_2 molecules in Si by Gösele and Tan (1982), which presumably play a role in the formation of so-called thermal donors in Si (Patel, 1981).

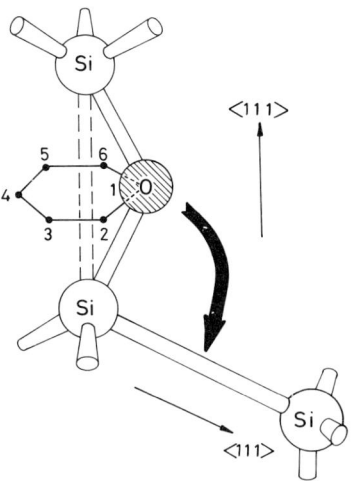

Fig. 12. Bond-centered configuration of the oxygen interstitial in the silicon lattice. The points labeled 1–6 mark equivalent positions of the O atom.

V. Oxidation-Influenced Diffusion of Group III and Group V Elements in Silicon

A. Introductory Remarks and Basic Concepts

In this section we shall deal with changes of the diffusivities of group III and group V elements in silicon that result from changes of the concentrations of self-interstitials and/or vacancies acting as diffusion vehicles. In Section II this effect was introduced as concentration-enhanced or concentration-retarded diffusion. More specifically, we shall restrict ourselves to cases in which the changes of the concentrations of intrinsic point defects are caused by surface oxidation. We shall see that the combination of results on the climbing of dislocation loops during surface oxidation with observations on oxidation-induced changes of the diffusivities of group III and group V dopants yields information on the diffusion mechanisms of these dopants and on the intrinsic point defects in silicon.

External oxidation of silicon leads to the formation and growth of an amorphous SiO_2 film on the oxidized surface. Such SiO_2 films are widely used in electronic devices because of their insulating properties. It was soon found that oxidation enhances the diffusion of phosphorus (Nicholas, 1966), boron (Willis, 1969; Bean and Gleim, 1969), and, to a lesser extent, of arsenic (Antoniadis et al., 1978a). In addition, surface oxidation may lead to the nucleation and growth of interstitial-type dislocation loops on {111} planes, which contain stacking faults and are therefore commonly termed *oxidation-induced stacking faults* or OSF (Booker and Tunstall, 1966; Jaccodine and Drum, 1966; Hu, 1974, 1981; Leroy, 1979; Antoniadis, 1982; Fair, 1981b).

Hu (1974) was the first who pointed out that the phenomenon of oxidation-enhanced diffusion (OED) is closely related to the growth of OSF. This may be deduced from the observation that under the same oxidation conditions both the diffusion enhancement of boron and the growth rate of OSF increase when the orientation of the oxidized surface is varied from {111} via {110} to {100}. The reason for the occurrence of both phenomena has turned out to be an oxidation-induced change in the concentration of intrinsic point defects. Hu (1974) argued that enhanced diffusion requires a supersaturation of the intrinsic defects which promote dopant diffusion. The growth of interstitial-type OSF showed that it had to be a supersaturation of self-interstitials, since a supersaturation of vacancies would lead to a shrinkage of these OSF. Therefore, Hu (1974) concluded that during surface oxidation Si self-interstitials are injected into the crystalline silicon. These excess self-interstitials enhance the diffusion of phosphorus or boron, which therefore must diffuse at least partly via an interstitialcy mechanism.

An experiment which led to a deeper insight into the diffusion mechanisms of group III and group V dopants in silicon was performed by Mizuo and Higuchi (1981). These authors found that, under the same oxidation conditions which lead to enhanced diffusion of phosphorus, the diffusion of antimony is retarded. Applying the arguments of Hu (1974) we may conclude that the observed oxidation-retarded diffusion of antimony indicates an undersaturation of the intrinsic point defect promoting antimony diffusion. The fact that under the same oxidation conditions interstitial-type OSF grow requires that it must be an undersaturation of vacancies, since a self-interstitial undersaturation would lead to a shrinkage of these OSF. Therefore, it follows that under thermal-equilibrium conditions as well as under oxidation conditions antimony predominantly diffuses via a vacancy mechanism.

The conclusion that the same type of surface oxidation simultaneously leads to a supersaturation of self-interstitials ($C_I > C_I^{eq}$) and to an undersaturation of vacancies ($C_V < C_V^{eq}$) may be rationalized by assuming that both vacancies and self-interstitials are present in silicon both in thermal equilibrium and under oxidation conditions and that local dynamical equilibrium

$$C_I C_V = C_I^{eq} C_V^{eq} \tag{22}$$

is maintained via the reaction

$$I + V \rightleftharpoons 0 \tag{23}$$

2. DIFFUSION IN SILICON AND GERMANIUM

where 0 denotes the perfect Si crystal. We emphasize that the equilibrium according to Eq. (22) can *exclusively* be established via reaction (23), i.e., both recombination and spontaneous thermally activated generation of Frenkel pairs in the bulk must take place.[4] Already Prussin (1972) and later Sirtl (1977) have considered this possibility.

A significant feature of the dynamical equilibrium [Eq. (22)] between vacancies and self-interstitials is that a supersaturation of self-interstitials cannot exist without an undersaturation of vacancies, and vice versa. The investigations of oxidation-influenced diffusion discussed above appear to reflect this peculiarity and to indicate that under thermal-equilibrium conditions the dopant diffusivity D^s is composed of a vacancy component D^s_V and a self-interstitial component D^s_I according to (Hu, 1974)

$$D^s = D^s_I + D^s_V \qquad (24)$$

The relative contributions

$$\phi_I = D^s_I/D^s \qquad (25)$$

and

$$\phi_V = 1 - \phi_I = D^s_V/D^s \qquad (26)$$

are different for different dopants, as may be seen most clearly from the fact that identical oxidation conditions enhance the diffusivity of B but retard that of Sb.

B. Influence of Surface Oxidation on the Concentrations of Intrinsic Point Defects

Under most oxidation conditions, oxygen from the gas phase diffuses in the form of O_2 molecules via holes in the SiO_2 network covering the silicon. The oxygen molecules reach the SiO_2–Si interface, react there with the silicon, and form new SiO_2 material (Rosencher *et al.*, 1979). This reaction is associated with a large volume increase ($\simeq 125\%$), which at sufficiently high temperatures is almost entirely accommodated by viscoelastic flow of the SiO_2 film (Tiller, 1980; Tan and Gösele, 1981). The remaining volume increase may be accommodated by reactions similar to those Saunders (1976) and Harris (1978) have summarized in the case of the oxidation of metals:

(i) The injection of Si self-interstitials from the SiO_2–Si interface into the silicon bulk leads to a supersaturation of self-interstitials.

---

[4] Reaction (23) may take place indirectly by the cooperation of other reactions involving I and V in a suitable way [see, e.g., item (i) in Section IX.B.3].

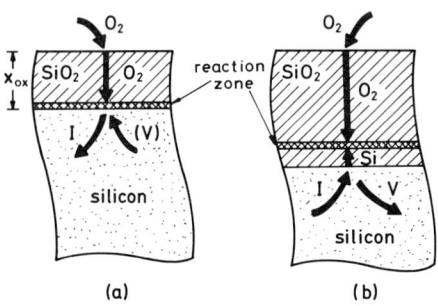

Fig. 13. Schematic illustration of the injection or absorption of intrinsic point defects induced by surface oxidation of silicon [according to Francis and Dobson (1979) and Tan and Gösele (1982a)]. (a) Thin oxide layer and/or moderate temperature, (b) thick oxide layer and/or high temperature.

(ii) Vacancies present in the silicon bulk under thermal-equilibrium conditions are absorbed at the SiO_2–Si interface and provide part of the required free volume. This reaction leads to an undersaturation of vacancies.

The two processes are schematically shown in Fig. 13a. When after a sufficiently long time dynamical equilibrium between vacancies and self-interstitials [Eq. (22)] has been established it can no longer be decided whether self-interstitial injection or vacancy absorption has been the dominant process. However, from experiments on oxidation-influenced diffusion at short times (<1 hr at 1373 K) one finds that self-interstitial injection is more important (Antoniadis and Moskowitz, 1982a; for a detailed discussion of these experiments see Section VI).

The oxidation process just described ceases to work at very high temperatures and/or for very thick oxide layers. Francis and Dobson (1979), Tan and Gösele (1982a), and Tan and Ginsberg (1983) presume that in these cases Si diffuses from the SiO_2–Si interface into the SiO_2 and reacts there with the oxygen diffusing in the opposite direction. The diffusion of Si from the interface into the SiO_2 film leads to a vacancy injection into the silicon or to an absorption of Si self-interstitials at the interface as indicated in Fig. 13b. Thus, irrespective of which of the two possibilities is realized, at high temperatures and/or after long oxidation periods surface oxidation produces a vacancy supersaturation coexisting with a self-interstitial undersaturation [Eq. (22)].

C. Qualitative Features of Oxidation-Influenced Diffusion

In the following section we assume that both vacancies and self-interstitials are present in Si under thermal-equilibrium conditions and that both contribute to the diffusivities of group III or group V dopants according to Eq. (24).[5] During surface oxidation the self-interstitial and vacancy concentrations deviate from their equilibrium values, so that the dopant diffusivities under oxidizing conditions are given by (Antoniadis, 1982)

$$D_{ox}^s = D_I^s C_I / C_I^{eq} + D_V^s C_V / C_V^{eq} \qquad (27)$$

Using the supersaturation ratios for self-interstitials,

$$s_I = (C_I - C_I^{eq})/C_I^{eq} \qquad (28)$$

and vacancies,

$$s_V = (C_V - C_V^{eq})/C_V^{eq} \qquad (29)$$

we obtain from Eqs. (24) and (27) the expression

$$\Delta_{ox}^s = \phi_I s_I + \phi_V s_V \qquad (30)$$

for the normalized oxidation-induced diffusivity change

$$\Delta_{ox}^s \equiv (D_{ox}^s - D^s)/D^s \qquad (31)$$

If the dynamical equilibrium [Eq. (22)] is established, s_I and s_V are interrelated via

$$s_V = -s_I/(1 + s_I) \qquad (32)$$

and Eq. (30) may be rewritten as

$$\Delta_{ox}^s = (2\phi_I + \phi_I s_I - 1)s_I/(1 + s_I) \qquad (33)$$

(Tan and Gösele, 1982a). In Fig. 14 Δ_{ox}^s versus s_I curves are shown as calculated for three ϕ_I values representing typical cases of dopant diffusion under thermal-equilibrium conditions: (i) predominantly via the interstitialcy mechanism ($\phi_I > 0.5$), (ii) predominantly via the vacancy mechanism ($\phi_I < 0.5$), and (iii) to the same extent via vacancies and self-interstitials ($\phi_I = \phi_V = 0.5$). In the case $\phi_I = \phi_V = 0.5$, both self-interstitial supersaturation ($s_I > 0$) and vacancy supersaturation ($s_I < 0$) lead to enhanced diffusion. For $\phi_I \gtrless 0.5$, depending on the regime of s_I, oxidation-enhanced diffusion (OED) or oxidation-retarded diffusion (ORD) may occur.

As mentioned in Section V.B, under most oxidation conditions employed in the processing of electronic devices a self-interstitial supersaturation ($s_I > 0$) is generated. Thus the part $s_I < 0$ of Fig. 14 representing a vacancy supersaturation is significant only at high temperatures and/or after long oxidation times (Francis and Dobson, 1979; Hill, 1981; Mizuo and Higuchi,

---

[5] Tan et al. (1983) have shown that a consistent interpretation of the data on oxidation-influenced diffusion is not possible if it is assumed that only one intrinsic point-defect species is present in Si under thermal-equilibrium conditions.

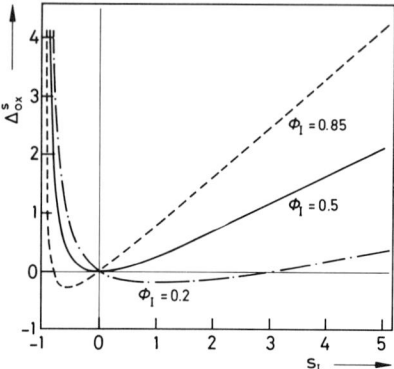

Fig. 14. Normalized oxidation-induced diffusivity change Δ_{ox}^s as a function of the self-interstitial supersaturation s_I, according to Eq. (33). The parameter ϕ_I denotes the fraction contributed by the interstitialcy mechanism to the diffusivity of the group III or V element considered.

1982c,d; Tan and Gösele, 1982a; Tan and Ginsberg, 1983) as well as in experiments performed in oxidizing atmospheres containing a sufficiently high percentage of chlorine compounds (Hill, 1981; Tan and Gösele, 1982b).

D. ANALYSIS OF EXPERIMENTS ON OXIDATION-ENHANCED AND OXIDATION-RETARDED DIFFUSION

Information on the diffusion mechanisms of various group III and group V elements in silicon, particularly on the ϕ_I values of these dopants, will now be extracted from OED and ORD experiments. For this purpose it is necessary to know the self-interstitial supersaturation ratio s_I for the oxidation conditions under which such experiments have been performed [see Eq. (33)]. In recent years it has turned out that a practicable way to determine s_I is the study of the growth or shrinkage kinetics of OSF; those that nucleate at the SiO_2–Si interface (surface OSF) are almost semicircular. Their radii, typically 5–100 μm, can be determined by etching and subsequent measurement of the etched intersection of the partial dislocation with the surface by means of optical microscopy. Smaller OSF, including bulk OSF, may be observed by electron microscopy.

The rate at which the radius r_{SF} of a circular stacking fault changes in the presence of mobile vacancies and self-interstitials due to climb processes is approximately given by

$$dr_{SF}/dt = -\alpha_{eff}[(D_I C_I^{eq} + D_V C_V^{eq})(\gamma/kT) - D_I C_I^{eq} s_I + D_V C_V^{eq} s_V]\bar{A}/\Omega \quad (34)$$

(Gösele and Frank, 1981; Tan and Gösele, 1982b; Gösele and Tan, 1983a).

Here γ ($= 0.026$ eV/atom) denotes the extrinsic stacking fault energy per atom (Alexander et al., 1980), $\bar{A}$ ($= 6.38 \times 10^{-20}$ m^2) the stacking fault area per atom, and Ω ($= 2.3 \times 10^{-29}$ m^3) the atomic volume. The dimensionless quantity α_{eff} accounts for the (elastic or electrostatic) interaction between the diffusing point defects and the partial dislocations bordering the OSF and for a potential reaction barrier against climb processes (Hu, 1981; Antoniadis, 1982). If there are neither reaction barriers nor long-range interactions, α_{eff} reduces to the purely geometrical factor

$$\alpha_d = 2\pi/\ln(8r_{\text{SF}}/r_0) \tag{35}$$

characteristic of diffusion-controlled processes (Flynn, 1972; Seeger and Gösele, 1977). The quantity r_0 denotes the distance from dislocation lines at which spontaneous capture of intrinsic point defects occurs. In the derivation of Eq. (34) it has been assumed that $\gamma/kT \ll 1$ is fulfilled (e.g., $\gamma/kT \approx 0.2$ at 1300 K) and that the line tension of the partial dislocation surrounding the stacking fault may be neglected in comparison with the stacking fault energy. The latter prerequisite is fulfilled for $r_{\text{SF}} \gtrsim 1$ μm. Although Eq. (34) has been derived for circular bulk OSF, it is reasonably good for describing the growth/shrinkage kinetics of surface OSF (Leroy, 1982).

On our way to determine s_{I} with the aid of Eq. (34), we first consider the behavior of OSF in an inert atmosphere that does not influence the point-defect equilibrium concentrations ($s_{\text{I}} = 0$, $s_{\text{V}} = 0$). In this case Eq. (34) reduces to

$$(dr_{\text{SF}}/dt)_{\text{in}} = -\alpha_{\text{eff}}(D_{\text{I}}C_{\text{I}}^{\text{eq}} + D_{\text{V}}C_{\text{V}}^{\text{eq}})(\gamma/kT)\bar{A}/\Omega \tag{36}$$

which predicts a constant shrinkage rate for preexisting OSF, as has been observed by means of optical microscopy (Hashimoto et al., 1977; Sugita et al., 1977; Wu and Washburn, 1977a,b; Shimizu et al., 1978; Claeys et al., 1979) as well as by electron microscopy (Sanders and Dobson, 1969; Lambert and Dobson, 1981). If the shrinkage is diffusion-controlled, α_{eff} is given by Eq. (35) and turns out to be about 0.5. Using this value and averaging over the shrinkage data available from optical-microscopy studies, Gösele and Frank (1981) deduced values of $D_{\text{I}}C_{\text{I}}^{\text{eq}} + D_{\text{V}}C_{\text{V}}^{\text{eq}}$ at various temperatures. From these values, D^{T} has been calculated with the aid of Eq. (5) by using $f_{\text{I}} = f_{\text{V}} = 0.5$. In Fig. 7, the D^{T} values so obtained are included as full circles. Values for D^{T} which have been determined in a corresponding way from electron microscopy data on the OSF shrinkage rate (Lambert and Dobson, 1981) are shown as full squares in Fig. 7. In both cases, the agreement with directly measured values of the tracer self-diffusion coefficient is satisfactory. This circumstance lends confidence to the subsequent attempts to extract information on the diffusion of group III and group V elements in Si from OED and ORD experiments.

Under oxidation conditions the growth rate $(dr_{SF}/dt)_{ox}$ of OSF is given by Eq. (34), with $s_I \equiv -s_V/(1 + s_V)$ [Eq. (32)], provided dynamical equilibrium between self-interstitials and vacancies is established. Quantitative information on s_I may be obtained by combination of the shrinkage rate $(dr_{SF}/dt)_{in}$ in an inert atmosphere [Eq. (36)] with the growth rate $(dr_{SF}/dt)_{ox}$ at the same temperature according to

$$s_I(T, t) \approx [1 - (dr_{SF}/dt)_{ox}/(dr_{SF}/dt)_{in}](\gamma/kT) \tag{37}$$

(Antoniadis, 1982). Equation (37) is a good approximation for $D_V C_V^{eq} < D_I C_I^{eq}$ [i.e., above about 1270 K (Section IX.C.1)] and/or $|s_I| \ll 1$.

In the case of thick oxides—the growth rate of which is controlled by the diffusion of oxygen through the oxide layer and thus decreases according to

$$dx_{ox}/dt \propto t^{-1/2} \tag{38}$$

when the oxide thickness x_{ox} increases in the course of the oxidation time t (Deal and Groove, 1965)—Eq. (37) yields for dry oxidation of a {100} Si surface at temperatures in the vicinity of 1400 K

$$s_I \approx 6.6 \times 10^{-9} t^{-1/4} \exp(2.52 eV/kT) \quad s^{1/4} \tag{39}$$

(Antoniadis, 1982; Tan and Gösele, 1982a). For {111} surfaces, the oxidation of which has been considered in detail by Tan and Ginsberg (1983) and Tan et al. (1983), the right-hand side of Eq. (39) has to be multiplied by a factor of 0.6 to 0.7 (Leroy, 1979). For wet oxidation, multiplication factors larger than unity have to be applied.

For thin oxides (i.e., low temperatures and/or short oxidation times) experimental data on $(dr_{SF}/dt)_{ox}$ do not exist. However, the relationship

$$s_I \propto (dx_{ox}/dt)^{1/2} \tag{40}$$

though not yet well founded by theory (Hu, 1974, 1981; Lin et al., 1981b; Fair, 1981b; Tan and Gösele, 1981; Antoniadis, 1982), appears to be valid for both thick and thin oxides and may therefore be used to calculate s_I from the growth rates of thin oxides, as given by Antoniadis and Moskowitz (1982a). For thin oxides, the growth is reaction-controlled and dx_{ox}/dt and s_I are time-independent, in contrast to what is true for thick oxides [see Eqs. (38) and (39)].

Since, in general, the oxidation-induced self-interstitial supersaturation s_I is time-dependent, the measured values of the normalized diffusivity change [see Eq. (33)] are average values of the time-dependent quantity $\Delta_{ox}^s(t)$,

$$\overline{\Delta_{ox}^s} = \frac{1}{t} \int_0^t \Delta_{ox}^s(t') \, dt' \tag{41}$$

Indeed, in the case of OED, $\overline{\Delta_{ox}^s}$ decreases slightly with oxidation time according to

$$\overline{\Delta_{ox}^s} \propto t^{-m} \qquad (42)$$

where m scatters between 0.5 and 0.1 (Taniguchi et al., 1980; Antoniadis et al., 1978b; Lin et al., 1981b; Ishikawa et al., 1982; Antoniadis and Moskowitz, 1982a). It turns out that Eq. (33) is a good approximation for $\overline{\Delta_{ox}^s}$ if s_I is replaced by its time average, $\bar{s}_I$. In the case of dry oxidation of $\{100\}$ surfaces this average is

$$\bar{s}_I = 8.8 \times 10^{-9} t^{-1/4} \exp(2.53eV/kT) \quad s^{1/4} \qquad (43)$$

Under oxidation conditions, for which $s_I > 0$ is fulfilled and Eq. (43) is applicable, all group III elements investigated so far (B, In, Ga, Al) show oxidation-enhanced diffusion (Mizuo and Higuchi, 1981, 1982a,b,c,d; Antoniadis and Moskowitz, 1982a,b). In Table VI we present ϕ_I values for B, Ga, and Al at 1373 K as determined by Gösele and Tan (1983a,b) from experimental data (Mizuo and Higuchi, 1981, 1982c,d) with the aid of Eqs. (33) and (43). At this temperature, all group III dopants appear to diffuse predominantly via the interstitialcy mechanism under thermal equilibrium conditions as indicated by $\phi_I > 0.5$. For group V dopants, the situation is

TABLE VI

Fractions (ϕ_I) Contributed at 1373 K by the Interstitialcy Mechanism to Diffusion Coefficients of Group III and Group V Dopants in Silicon[a] and Ratios (r_s/r_{Si}) of the Atomic Radii of the Dopants to the Atomic Radius of Si

Dopant	ϕ_I	r_s/r_{Si}	References
Group III			
B	0.8–1.0	0.75	Mizuo and Higuchi (1981)
Ga	0.6–0.7	1.08	Mizuo and Higuchi (1982c)
Al	0.5–0.7	1.08	Mizuo and Higuchi (1982d)
Group V			
P	0.5–1.0	1.94	Mizuo and Higuchi (1981); Antoniadis (1982); Ishikawa et al. (1982)
As	0.2–0.5	1.01	Mizuo and Higuchi (1981); Antoniadis and Moskowitz (1982a); Ishikawa et al. (1982)
Sb	0.02	1.16	Mizuo and Higuchi (1981); Antoniadis and Moskowitz (1982a)

[a] From Gösele and Tan, 1983a.

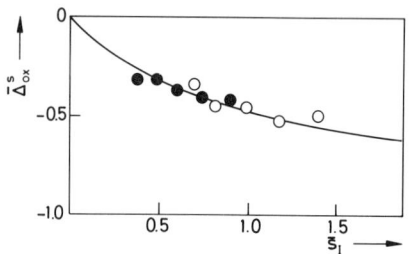

Fig. 15. Experimental data on the oxidation-retarded diffusion of Sb in Si at 1373 K, $\bar{\Delta}^s_{ox}$, as a function of $\bar{s}_I$. Data from [○] Mizuo and Higuchi (1981), [●] Tan and Ginsberg (1983). The solid line represents a fit of Eq. (33) for $\phi_I = 0.02$.

not so straightforward. The experimental data on oxidation-enhanced diffusion of P (Mizuo and Higuchi, 1981; Antoniadis, 1982; Ishikawa et al., 1982) scatter considerably and indicate ϕ_I values between 0.5 and 1.0 at 1373 K, whereas for As, ϕ_I values between 0.2 and 0.5 are obtained (Mizuo and Higuchi, 1981; Antoniadis and Moskowitz, 1982a; Ishikawa et al., 1982). Under the same oxidation conditions under which P shows oxidation-enhanced diffusion, Sb undergoes oxidation-retarded diffusion. The results of Mizuo and Higuchi (1981) and of Tan and Ginsberg (1983) on oxidation-retarded diffusion of Sb at 1373 K are presented in Fig. 15, together with a fit of Eq. (33) with $\bar{s}_I$ from (43) and $\phi_I = 1 - \phi_V \approx 0.02$ (Tan and Gösele, 1982a). A similar value for Sb at 1272 K, $\phi_I \approx 0.015$, has been determined by Antoniadis and Moskowitz (1982a).

From Table VI and Fig. 16 one realizes that at 1373 K the fractional interstitialcy component of the diffusivity of group V dopants, ϕ_I, strongly

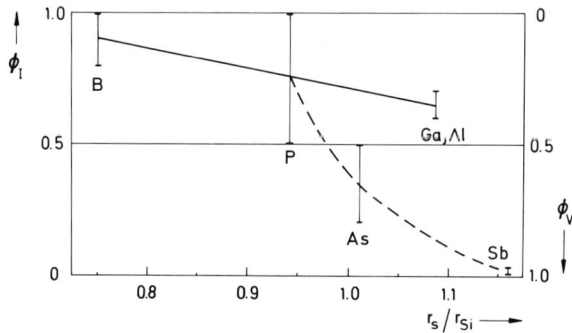

Fig. 16. Fractions ϕ_I contributed by the interstitialcy mechanism to the diffusivities of group III and group V elements versus their atomic radii r_s in units of the atomic radius r_{Si} of silicon.

decreases with increasing atomic radius r_s of the dopant. This and the fact that the relatively large Sb atoms diffuse almost exclusively via vacancies presumably reflect the elastic attraction between these dopants and the vacancies. In contrast to Sb, the group III dopants Al and Ga, although also relatively large, diffuse predominantly via the interstitialcy mechanism. This observation indicates that the actual value ϕ_I of a dopant is not only influenced by its size but also by its electric charge state (Mizuo and Higuchi, 1982c,d). The further discussion of these aspects is postponed to Section VII.

There are only a few reliable data on oxidation-influenced diffusion at other temperatures. At 1273 K Antoniadis and Moskowitz (1982a,b) found $\phi_I \approx 0.4$ for P, $\phi_I \approx 0.3$ for B, and $\phi_I \approx 0.35$ for In. From the diffusion data of Nabeta *et al.* (1976), Francis and Dobson (1979), Mizuo and Higuchi (1982c,d), and Hill (1981), Gösele and Tan (1983b) calculated for P $\phi_I > 0.75$ at 1423 K and $\phi_I > 0.95$ at 1540 K, for Ga and Al $\phi_I > 0.85$ at 1423 K, and for B $\phi_I > 0.83$ at 1473 K. In spite of the large scatter of the data, there is a clear trend that, at least for B, Ga, Al, and P, ϕ_I increases with increasing temperature [see also Matsumoto *et al.* (1983)]. As suggested by Seeger *et al.* (1979), this temperature dependence of ϕ_I is presumably a consequence of the fact that Si self-diffusion is dominated by the interstitialcy mechanism ($D_I C_I^{eq} > D_V C_V^{eq}$) or the vacancy mechanism ($D_I C_I^{eq} < D_V C_V^{eq}$) above or below about 1270 K, respectively (Sections III and IX).

E. Diffusion of Intrinsic Point Defects through Silicon Wafers

So far it has been assumed that the oxidation-induced point defects are homogeneously distributed in the region in which the influence of oxidation on dopant diffusion is measured. This is justified since usually this region is adjacent to the SiO_2–Si interface, where the excess point defects are generated. The situation is quite different if the influence of oxidation on dopant diffusion is measured far away from the SiO_2–Si interface. Under these circumstances the time dependence of the transport of oxidation-induced point defects into the silicon bulk via diffusion processes may play a role. Experiments in which this is the case have been performed at 1373 K by Mizuo and Higuchi (1982b, 1983). These authors investigated the influence of back side oxidation of float-zone[6] silicon wafers on the diffusion of B, P, or Sb implanted from the front side. The front side was covered with an

---

[6] Czochralski silicon wafers cannot be used for such experiments, since the oxide precipitates and the bulk stacking faults in such wafers act as sinks for oxidation-induced point defects and limit their diffusion lengths to 10–30 μm (Taniguchi *et al.*, 1980; Hu, 1981; Mizuo and Higuchi, 1982a).

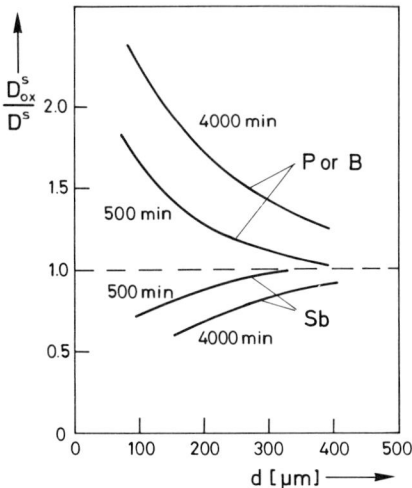

Fig. 17. Diffusion enhancement of P and B and diffusion retardation of Sb at the frontside of Si wafers oxidized at the backside [according to Mizuo and Higuchi (1982b, 1983)]; D_{ox}^s/D^s is given as a function of the wafer thickness d for 500- and 4000-min oxidations.

Si_3N_4 layer in order to prevent oxidation. In Fig. 17, D_{ox}^s/D^s values measured in these experiments are plotted versus the wafer thickness d for different times of oxidation. One finds that the time the oxidation-induced deviations from the point-defect equilibrium concentrations need to reach the front side increases with increasing wafer thickness. For a given value of d the same time is required to achieve a certain diffusion enhancement of B or P, or the corresponding diffusion retardation of Sb. Hence the oxidation-induced self-interstitial supersaturation $s_I > 0$, and the corresponding vacancy undersaturation $s_V \equiv -s_I/(1 + s_I) < 0$, reach the front side at the same time. We therefore conclude that, for the oxidation times employed, the local dynamical equilibrium [Eq. (22)] is established and that the oxidation-induced changes in C_I and C_V propagate with the same "effective diffusivity"

$$\tilde{D} \approx (D_I C_I^{eq} + D_V C_V^{eq})/(C_I^{eq} + C_V^{eq}) \qquad (44)$$

(Gösele and Tan, 1983b). The experiments of Mizuo and Higuchi (1982b, 1983) show that at 1373 K, $\tilde{D} \gtrsim 3 \times 10^{-13}$ m² s⁻¹. Since at this temperature $D_I C_I^{eq}$ and $D_V C_V^{eq}$ are of the same order of magnitude (Section IX.C.1), $\tilde{D}$ may be approximately identified with D_I or D_V, depending on whether $C_I^{eq} > C_V^{eq}$ or $C_V^{eq} > C_I^{eq}$ is valid. At present, it cannot be decided which of these two cases is realized. If such a decision becomes possible, the experiments discussed in this section will provide an efficient tool to measure D_I or D_V in a fairly direct way.

F. CONCLUSIONS

From a combined analysis of experiments on oxidation-enhanced/retarded diffusion of group III and group V dopants and of experiments on the growth and shrinkage of oxidation-induced stacking faults, the following conclusions concerning point defects in silicon may be drawn:

(i) At high temperatures both vacancies and self-interstitials are present in thermal equilibrium as well as during surface oxidation.

(ii) There exists a dynamical equilibrium between vacancies and self-interstitials. Its establishing requires fairly long times (e.g., more than 1 hr at 1373 K) and involves the recombination and spontaneous thermally activated generation of Frenkel pairs in the bulk.

(iii) Both vacancies and self-interstitials contribute to the diffusion of group III and group V dopants under thermal-equilibrium conditions. The relative contributions depend on the electric charge state and the atomic radius of the dopant as well as on temperature. The interstitialcy component increases with increasing temperature.

VI. A Barrier against Vacancy–Interstitial Recombination

Hu (1977) has suggested that in Si an energy barrier (hereafter more precisely referred to as an *enthalpy barrier*) of "a few to several eV" impedes the vacancy–interstitial recombination, so that, to some extent, vacancies and self-interstitials may behave independently of each other. In seeming contrast to this suggestion, the experiments on oxidation-retarded diffusion of Sb at 1373 K (Mizuo and Higuchi, 1981, 1982c) reported in Section V.D show that for oxidation times longer than 3 hr local dynamical equilibrium according to Eq. (22) is established, which requires spontaneous thermally activated creation of Frenkel pairs and vacancy–interstitial recombination. By performing the same type of experiments as Mizuo and Higuchi (1981) but for oxidation times ranging from 5 to 60 min, Antoniadis and Moskowitz (1982a) tested whether at 1373 K dynamical equilibrium has already been established after these shorter times. They found a small diffusion enhancement ($\bar{\Delta}^s_{ox} \approx 0.1$) of Sb after oxidation for 5 min, which gives way to a diffusion retardation after oxidation for 10 min. For the largest oxidation time applied by Antoniadis and Moskowitz (1982a), viz. about 60 min, the diffusion retardation approaches the value expected for dynamical equilibrium.

The observations described previously may be rationalized as follows: During oxidation self-interstitials are injected from the SiO_2–Si interface into the silicon, but it takes about 1 hr until vacancy–interstitial recombination has proceeded to such an extent that dynamical equilibrium is reached.

As a consequence, at the very beginning there is no vacancy undersaturation, so that the oxidation-induced interstitial supersaturation enhances the diffusion of Sb by the small interstitialcy diffusion component of $\phi_I \approx 0.02$.

We may use Waite's theory of diffusion-controlled reactions (Waite, 1957) to obtain from

$$\tau_{\text{dyn}} \lesssim \Omega/4\pi D^{\text{SD}} r_{\text{IV}} \tag{45}$$

an order-of-magnitude estimate for the time τ_{dyn} which would be required to establish the dynamical equilibrium [Eq. (22)] in the Antoniadis–Moskowitz (1982a) experiments if no recombination barrier were present (r_{IV} = recombination radius). In deriving Eq. (45), it has been assumed that the self-diffusion coefficient D^{SD} is controlled either predominantly by self-interstitials or, to comparable extents, by self-interstitials and vacancies (Section IX.C). With $D^{\text{SD}} \approx 10^{-19}$ m^2 s^{-1} at 1373 K (Kalinowski and Seguin, 1979) and $r_{\text{IV}} = 5 \times 10^{-10}$ m, Eq. (45) yields $\tau_{\text{dyn}} < 0.05$ s, which is about 10^5 times shorter than observed (Antoniadis and Moskowitz, 1982a). We therefore conclude that *vacancy–interstitial recombination is not controlled by diffusion but by overcoming a recombination barrier* that exceeds the Gibbs free energy of diffusion by ΔG. In this concept, the factor of 10^5 by which the experimental τ_{dyn} value differs from that estimated with the aid of Eq. (45) arises from the Boltzmann factor $\exp(\Delta G/kT)$, accounting for the enhancement of the recombination barrier. Following Hu's suggestion (Hu, 1977), Antoniadis and Moskowitz (1982a) interpreted their observation in terms of an enthalpy barrier corresponding to $\Delta H \approx 1.4$ eV, assuming that in the decomposition

$$\Delta G = \Delta H - T \Delta S \tag{46}$$

the entropy contribution is negligibly small. By contrast, Gösele et al. (1982) proposed that the main part of ΔG originates from the term $-T \Delta S$ where ΔS is negative, i.e., from an entropy barrier against vacancy–interstitial recombination.

The microscopic model standing behind the concept of an entropy barrier is the following. In order to account for the extraordinarily high preexponential factor in the self-diffusion coefficient, Seeger and Chik (1968) proposed that at high temperatures the self-interstitials in Si are smeared out over several atomic volumes (Section III.C). The high-temperature configuration of the vacancies may be extended, too, as in the case of germanium (Seeger and Chik, 1968). As a consequence, the recombination of a vacancy–interstitial pair requires a simultaneous contraction of both defects to about one atomic volume at about the same location. Since, in the sense of statistical mechanics, extended defect configurations comprise a much larger number of microstates than "point" defects, this contraction corresponds to a decrease of entropy, i.e., an entropy barrier against recombination $\Delta S < 0$ does exist. Assuming $|\Delta H| \ll |T \Delta S|$, this barrier may be estimated from $\exp(\Delta G/kT) = 10^5$ as $\Delta S = -11.5\ k$ at 1373 K. This value is very reasonable, as may be realized by comparison with the self-diffusion entropy of about $11\ k$ at 1373 K (Frank, 1981), which reflects the degree of the extension of the self-interstitials at this temperature.

Gösele et al. (1982) pointed out that the idea of the enthalpy nature of the barrier against vacancy–interstitial recombination in Si at high temperatures put forward by Antoniadis and Moskowitz (1982a) appears to be incompatible with experiments at lower temperatures. Already at room temperature, vacancies and interstitials induced by electron irradiation recombine readily [see, e.g., Matthews and Ashby (1973)]. If the high-temperature recombination barrier were an enthalpy barrier with $\Delta H = 1.4$ eV, a vacancy–interstitial encounter at room temperature would lead to recombination with a probability of 10^{-24}, i.e., at room temperature and below vacancy–interstitial recombination could not take place.

Within the concept of an entropy barrier the difficulty that at low temperatures vacancy–interstitial recombination should be forbidden does not exist. According to Seeger and his associates (Seeger and Chik, 1968; Seeger et al., 1977, 1979), the spreading out of Si self-interstitials is a high-temperature phenomenon. Internal friction measurements indicate that at temperatures below about 500 K various dumbbell configurations of the self-interstitial exist whose preferential axes depend on the electric charge state (Tan et al., 1973). It has been argued by Frank (1981) that at low temperatures the electrons forming the bonds between the two atoms of a dumbbell configuration are well localized in the center of gravity of the defect. When temperature increases, these electrons occupy excited states and may thus become more and more delocalized. As a consequence of this spreading out of electrons, the dumbbell relaxes and spreads out, too. Jackson proposed that the self-interstitial may first be extended over a ring of five atoms (see Kimerling, 1979). Such a configuration may be regarded as a intermediate state between the low-temperature dumbbells and the extended high-temperature configurations. According to this picture the shrinkage of the spreading out of self-interstitials with decreasing temperature leads to a gradual disappearance of the entropy barrier against vacancy–interstitial recombination with decreasing temperature. This explains why in radiation-damage experiments at low temperatures vacancies and self-interstitials recombine easily.

We emphasize that, independently of the detailed nature of the recombination barrier, the time τ_{dyn} required for establishing dynamical equilibrium in oxidation experiments, in which the deviation from the point-defect equilibrium concentrations is usually less than a factor of ten, strongly increases with decreasing temperature. Compared to $\tau_{dyn} \approx 1$ h at 1373 K, τ_{dyn} is expected to be of the order of days at 1170 K, so that from this temperature downwards vacancies and self-interstitials behave independently of each other during oxidation experiments lasting a few hours. Of course, this is not true if the point-defect concentrations C_I and C_V are orders of magnitude higher than C_I^{eq} and C_V^{eq}, respectively, as during concentration-enhanced diffusion under irradiation (see, e.g., Masters and Gorey, 1978).

VII. Diffusion of Group III and Group V Elements and Its Dependence on Doping

A. Germanium

In accordance with the idea that self-diffusion in Ge occurs via monovacancies (Section III.C.1) and that these introduce an acceptor level in the lower half of the energy gap, the Ge self-diffusion coefficient is enhanced by n doping and decreased by p doping (Section III.E). Thus the fact that the diffusivities of group III and group V elements in Ge show the same doping dependence as the Ge self-diffusion coefficient (see, e.g., Seeger and Chik, 1968) is not surprising but may be taken as an indication that these solutes diffuse in Ge via the vacancy mechanism, too. In fact, this view is confirmed further by the observations (Fig. 10) that the diffusivities of the group III elements are very similar to the self-diffusion coefficient (exceeding it by not more than an order of magnitude at T_m) whereas the diffusivities of the group V elements are considerably larger (by two orders of magnitude or more) than the self-diffusion coefficient: In the case of the group V donors, the probability to find acceptor-type vacancies on nearest-neighbor sites and thus to perform diffusional jumps is enhanced by Coulomb attraction, whereas for the group III acceptors this enhancement effect does not exist.

B. Silicon

Self-diffusion in Si is dominated by the vacancy or the interstitialcy mechanism below or above about 1270 K, respectively (Seeger and Chik, 1968; Sections III.C.2 and IX.C.1). The dependence of the diffusion coefficients for Si self-diffusion and for Ge tracer diffusion in Si on the doping with group III or group V elements found by experiment has been shown to be explicable within this picture, provided that the electronic properties of vacancies or self-interstitials may be described by an acceptor level in the lower half of the band gap or by Blount's donor–acceptor-pair model, respectively (Section III.E).

Impurity diffusion of group III and group V elements in Si, including its doping dependence, may be understood within the framework of the above model for Si self-diffusion if the electrostatic and the elastic interactions between the impurities and the intrinsic point defects are taken into account. Before this will be demonstrated in what follows, we recall our conclusion drawn from experiments on oxidation-influenced diffusion (Section V.D) that with increasing temperature the relative contribution by vacancies to the diffusion of both group III and group V elements decreases in favor of the diffusion via self-interstitials. This is in accordance with the corresponding trend in Si self-diffusion. Since most experiments yielding information on the diffusion of group III or group V elements have been performed in the temperature regime in which $D_I C_I^{eq} > D_V C_V^{eq}$, the interpretation given below is primarily valid under this prerequisite.

First we should like to give an explanation of the oxidation-influenced diffusion experiments at 1373 K (Fig. 16). In the case of the group III elements B, Ga, and Al, the preference of interstitialcy diffusion due to $D_I C_I^{eq} > D_V C_V^{eq}$ is enhanced by the Coulomb attraction between these negatively charged shallow-acceptor impurities and the self-interstitials, which in Si containing group III elements (p-Si) act as positively charged donors (Fig. 9b). As a result, all these group III elements diffuse preferentially via self-interstitials ($\phi_I > 0.5$). The increase of ϕ_V when going from B to Ga or Al is attributed to the fact that the elastic attraction between these impurities and vacancies increases with the size of the solute atoms. However, even for Ga and Al, the sizes of which exceed that of the Si atoms, ϕ_V remains below 0.5, since the Coulomb interaction between vacancies and group III elements is repulsive. This is different for the group V elements (positively charged shallow donors) which attract electrostatically both vacancies, which in Si containing group V element (n-Si) are negatively charged acceptors (Fig. 9a), and self-interstitials, which in n-Si also act as negatively charged acceptors (Fig. 9b). Therefore, for the relatively large group V atoms As and Sb (large elastic attraction to vacancies) ϕ_V exceeds 0.5, whereas for the smaller P atoms (small elastic attraction to vacancies) $D_I C_I^{eq} > D_V C_V^{eq}$ leads to a preference for the interstitialcy mechanism.

The doping dependence of the diffusivities of group III or group V elements may be understood along the lines of reasoning applied in the above interpretation of oxidation-influenced diffusion experiments. The observations to be explained are the increase or decrease of the diffusivities of the group III elements by p doping or n doping, respectively, and the opposite doping behavior of the group V elements (Seeger and Chik, 1968; Shaw, 1975; Fair, 1981a). Under the p-biased conditions prevailing during diffusion studies of group III elements, the doping dependence is controlled by the donor level of the self-interstitials (Fig. 9b) responsible for the diffusion of the group III elements (see above). Therefore, for group III elements, one expects the doping dependence observed. For those group V elements which diffuse preferentially via vacancies, e.g., As or Sb (see above), the doping dependence observed is in accordance with the acceptor nature of the vacancies (Fig. 9a). An interesting case is P, which, according to oxidation-influenced diffusion studies (Fig. 16), diffuses mainly via self-interstitials though it belongs to group V. Under normal doping conditions (i.e., except for very strong p doping) the doping dependence is governed by the acceptor level of the self-interstitials and is thus the same as for diffusion via vacancies. However, at very high levels of B doping the diffusion of P was found to be enhanced (Drake and Willoughby, 1969). We presume that in this case of extreme p-type doping the P diffusion is determined by the donor level of the self-interstitials.

A comparison of the diffusivities of the group III and group V elements with the self-diffusion coefficients shows that in Si the sequence of the diffusivities of acceptor and donor elements differs from that in Ge, which has been discussed in Section VII.A: In Si the diffusivities of the group III elements exceed the self-diffusion coefficient to a larger extent than the diffusivities of the group V elements, among which P diffuses fastest (Fig. 11). These observations fit well into the picture that in Si—in the temperature regime considered ($\gtrsim 1300$ K)—$D_I C_I^{eq}$ is larger than $D_V C_V^{eq}$, that the diffusion of the group III elements occurs via self-interstitials and is enhanced over $D_I C_I^{eq}$ by the interstitial–solute Coulomb attraction, and that, except for P, the group V elements prefer vacancies as diffusion vehicles.

VIII. Anomalous Diffusion Phenomena

A. Phenomenological Description

Diffusion of group III and group V elements in Si shows several unexpected features which have been termed "anomalous." They have been thoroughly discussed by Hu (1973), Willoughby (1977, 1978, 1981), and Seeger et al. (1979). According to Willoughby (1977) and Gösele and Strunk (1979), the three most prominent anomalous diffusion phenomena, which are apparently related to each other, may be characterized as follows:

  (i) The emitter-push effect is the extremely rapid diffusion of the base dopants (e.g., B or Ga) near the emitter region of double diffused n–p–n structures resulting in an enhanced movement of the base–collector boundary below the (usually strongly phosphorus-doped) emitter.

(ii) Movement of marked layers denotes the anomalously rapid diffusion of dopants (e.g., P or B) in marked or buried layers that is observed if the phosphorus concentration at the surface is high.

(iii) A kink-and-tail structure of P diffusion profiles occurs if such profiles are generated by high surface concentrations of P. In the tail the diffusion is much faster than expected from isoconcentration studies.

The following features common to all these anomalous diffusion phenomena[7] have been established:

(a) The diffusivity enhancements are caused by supersaturations of intrinsic point defects promoting the diffusion of the dopants. The enhancements are much higher than in oxidation experiments.

(b) The point-defect supersaturations are related to high surface concentrations of phosphorus or, in some cases, of boron (Lawrence, 1966; Claeys et al., 1978a) or arsenic (Shibayama et al., 1976).

(c) Mechanisms involving dislocations in an essential way, in particular dislocation climb, are not the origin of the supersaturations.

Until recently it was widely believed that the point defects present in supersaturation are vacancies, and elaborate models calculating the vacancy supersaturations have been put forward, e.g., by Fair and Tsai (1977), Yoshida (1979, 1980, 1983), and Mathiot and Pfister (1982a,b, 1983). In the past few years the nature of the excess point defects has been determined experimentally by several independent groups (Claeys et al., 1978b; Armigliato et al., 1977; Strunk et al., 1979; Jaccodine, 1983). The basic idea underlying all these experiments is the same as that standing behind experiments in which oxidation-enhanced diffusion and growth or shrinkage of oxidation-induced stacking faults have been combined: The observed diffusion enhancement requires a *supersaturation* of point defects and cannot be explained in terms of an undersaturation. Hence, isolated dislocations lying in the region of enhanced diffusion, climb by absorption and not by emission of point defects. The direction of climb of a given dislocation is therefore uniquely related to the nature of the point defects involved (i.e., opposite for vacancies or self-interstitials).

---

[7] It should be noted that, in contrast to the phenomena (i) to (iii) to be considered here, the diffusion retardation sometimes observed for arsenic emitters is caused by electric field effects (Willoughby, 1977, 1978, 1981; Mallam et al., 1981).

B. THE DISLOCATION CLIMB EXPERIMENTS OF CLAEYS et al.

Claeys et al. (1978b) investigated the growth or shrinkage kinetics of interstitial-type *bulk* stacking faults in Si under the influence of in-diffusing phosphorus in different atmospheres. The use of bulk OSF lying outside the region of high phosphorus concentration is essential in these experiments, since the shrinkage kinetics of *surface* OSF is strongly dependent on the doping level. If a high concentration of phosphorus, boron, or arsenic is diffused into silicon in an inert atmosphere, surface OSF shrink away much faster than in intrinsic materials (Hashimoto et al., 1976, 1977; Fair and Carim, 1982). In general, this enhanced shrinkage occurs only within and not outside the high-concentration diffusion zone of about 3 μm width (Hashimoto et al., 1976, 1977). Since it has been established that the phosphorus-induced point-defect supersaturation extends into the silicon more than 30 μm beneath this diffusion zone (Willoughby, 1977; Lecrosnier et al., 1979; Mizuo and Higuchi, 1982b), we conclude that the faster shrinkage of surface OSF reflects a doping effect and is not directly related to anomalous diffusion phenomena induced by point-defect supersaturations.

Claeys et al. (1978b) performed their experiments at 1423 K in both an oxidizing and an inert atmosphere in order to ensure that the effects observed were not due to self-interstitials produced by oxidation. An example of their

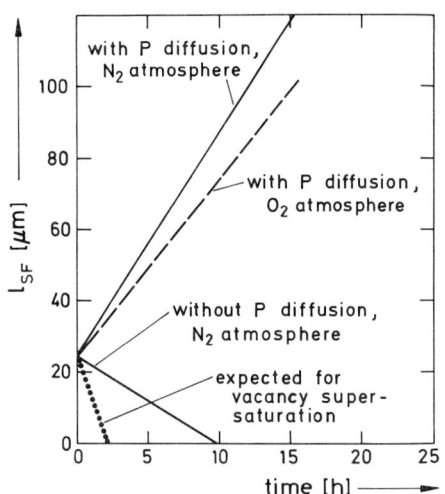

Fig. 18. Change of the lengths l_{SF} of oxidation-induced interstitial-type bulk stacking-faults as a function of time during in-diffusion of P at 1423 K in various atmospheres. [From Claeys et al. (1978b).]

results is shown in Fig. 18. In a nitrogen atmosphere the in-diffusion of phosphorus leads to a growth of bulk OSF, which shrink away in a nitrogen atmosphere without P diffusion. An oxygen atmosphere during the in-diffusion of P yields a slightly lower growth rate. The growth of interstitial-type OSF demonstrates that self-interstitials are present in supersaturation. A vacancy supersaturation would lead to a enhanced shrinkage rate, as also indicated in Fig. 18. Since in an inert atmosphere neither the shrinkage nor the growth rate depends on time, it is possible to estimate that the supersaturation ratio C_I/C_I^{eq} is less than two. Anomalous diffusion studies at lower temperatures, in which much higher supersaturations are built up, have been performed by Strunk et al. (1979) [Section VIII.C].

C. The Dislocation Climb Experiments of Strunk et al.

Strunk et al. (1979) investigated bipolar transistors with consecutively diffused boron base and phosphorus emitter in a high-voltage electron microscope. These transistors showed the emitter-push effect, i.e., enhanced diffusion of the base dopant (Fig. 19a). In several transistors isolated helical dislocations were detected in the emitter–base region (Fig. 19b). They had been formed by the climb of screw dislocations during the final emitter diffusion at 1223 K. An example is shown in Fig. 20. By contrast analysis of their dipole segments these helices were shown to be of *insterstitial* type.

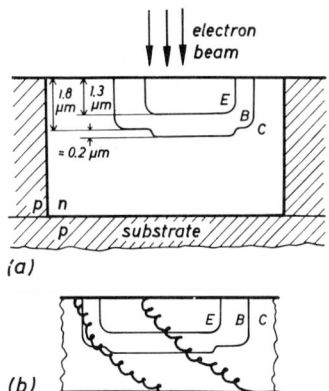

Fig. 19. Schematic illustration of the emitter-push effect in a bipolar transistor: (a) Specimen, about 2.5 μm thick, containing the emitter E, the base B, and part of the collector C, as used in the high-voltage electron-microscopy investigations of Strunk et al. (1979). (b) Helical dislocations formed by climb of the edge components of dislocations originally being almost purely screw-type.

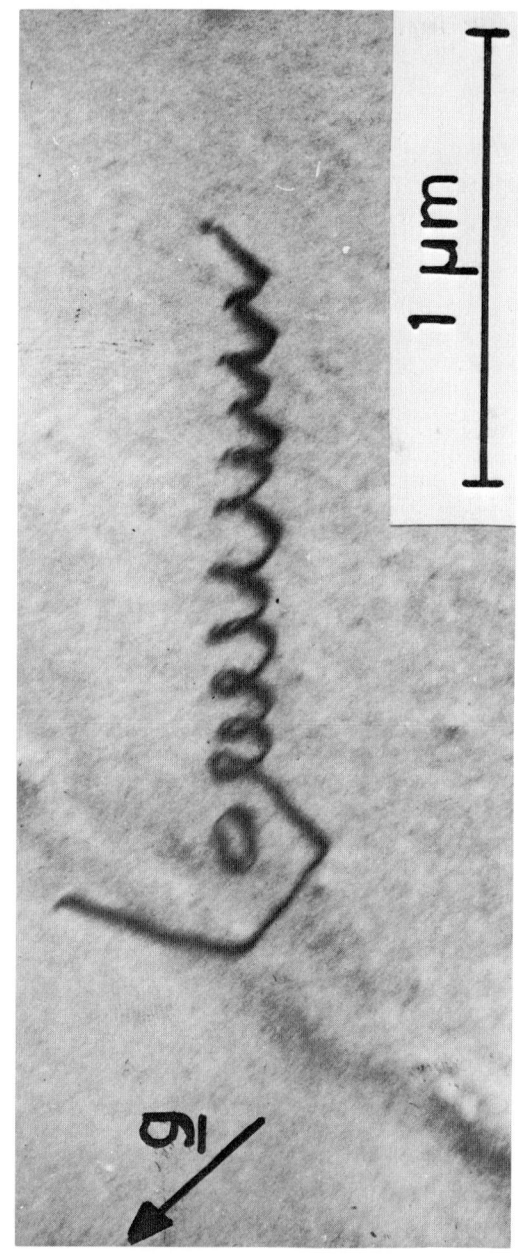

Fig. 20. High-voltage electron micrograph (**g** = diffraction vector) of a helical dislocation (compare Fig. 19b) running through emitter, base, and collector of a bipolar transistor [according to Strunk et al. (1979)].

The supersaturation ratio C_I/C_I^{eq} may be estimated from the number of turns per unit length, n_L, of the helices via (Weertman, 1957)

$$\ln(C_I/C_I^{eq}) = 2\pi n_L \mu b^4/kT, \qquad (47)$$

where b is the length of the Burgers vector and μ the shear modulus. The n_L value observed (≈ 9 turns/μm) corresponds to $C_I/C_I^{eq} \approx 140$ (Gösele and Frank, 1981). Since, under comparable conditions, surface oxidation would produce a supersaturation ratio below 10, this effect may be excluded as a significant contribution to the interstitial supersaturation in the experiments of Strunk et al. (1979). However, a minor influence by oxidation exists, as indicated by a comparison of studies of the emitter-push effect in oxygen with that in an inert atmosphere (Nicholas, 1966; Hill quoted by Willoughby, 1977; Hill, 1980).

D. Mechanisms of the Generation of Interstitial Supersaturations

Various mechanisms have been suggested for the generation of interstitial supersaturations by in-diffusing phosphorus as found by Claeys et al. (1978b), Strunk et al. (1979), and in related experiments (Grienauer and Mayer, 1975; Armigliato et al., 1977; Tseng et al., 1978; Jaccodine, 1983). Strunk et al. (1979) and Seeger et al. (1979) speculated that the interstitial generation might be associated with the stress produced by the undersized P atoms. Gösele and Strunk (1979) discussed the possibility that an interstitial supersaturation is produced in front of the in-diffusing phosphorus by some kind of "Kirkendall effect," i.e., a gradient in the P concentration may produce a supersaturation of self-interstitials, since these may act as vehicles for both self-diffusion and P diffusion and thus may couple the two diffusion processes. A semiquantitative treatment of this effect was given by Schaake (1984). However, it appears more likely that the breaking up of self-interstitial–phosphorus complexes plays a role in the formation of self-interstitials during the in-diffusion of P (Gösele and Strunk, 1979). This mechanism resembles the production of a vacancy supersaturation, as suggested and mathematically formulated by Fair and Tsai (1977) and Yoshida (1979, 1980). Nobili et al. (1982) observed that small precipitates of silicon phosphide are present in the high-concentration regions of phosphorous profiles. They claim that these precipitates are responsible for the fact that in such cases the total P concentration was found to be higher than the concentration of electrically active phosphorus as well as for the generation of a supersaturation of self-interstitials. A similar conclusion had been drawn earlier by Jaccodine (1968), who argued that phosphorus precipitation is accompanied by a volume expansion and that part of the resulting stresses may be relieved by the absorption of vacancies. Since a vacancy undersaturation cannot lead to enhanced diffusion, we rather conclude that the stress built up by P precipitation is relieved by the emission of self-interstitials, which may cause at least part of the interstitial supersaturation observed.

We emphasize that although anomalous diffusion of B, P, and Ga is caused by a supersaturation of self-interstitials, under thermal equilibrium conditions these dopants do not exclusively diffuse via the interstitialcy mechanism (Section V.D). Moreover, it is possible that a quantitative description of the phenomena of anomalous diffusion requires that a contribution by vacancies is taken into account (Hu et al., 1983).

IX. Substitutional–Interstitial Interchange Diffusion and Application to Gold and Nickel in Silicon and to Copper in Germanium

A. INTRODUCTORY REMARKS

As discussed in Section IV, the high diffusivities of the 3d transition metals in Ge (Fig. 10) or Si (Fig. 11) have two causes: (a) A nonnegligible fraction C_i^{eq}/C_s^{eq} of these impurities occupy interstitial sites, and (b) the migration enthalpies controlling the hopping of these impurities from interstice to interstice are particularly small due to the weak coupling of these interstitials to the host lattice.

Concerning Fe in Si, the view prevailing at present is that this element dissolves exclusively interstitially (Weber, 1983), though this picture cannot be reconciled satisfactorily with earlier results on the diffusion of Fe in Si, as discussed elsewhere (Seeger et al., 1979). For Co (Bergholz, 1983; Scheibe and Schröter, 1983) and Ni (Section IX.C.2) in Si it has been established by Mössbauer effect measurements and by electrical resistivity studies of the diffusion kinetics, respectively, that C_s^{eq} differs from zero, although C_i^{eq}/C_s^{eq} is considerably larger than unity. For Cu in Ge, van den Maesen and Brenkman (1955) estimated C_i^{eq}/C_s^{eq} by comparing electrical resistivity measurements with studies on radioactive Cu tracers and found $C_i^{eq}/C_s^{eq} \approx 1$ at 1075 K, but $C_i^{eq}/C_s^{eq} \ll 1$ at 875 K.

Regarding its diffusivity, Au in Si occupies an intermediate position between the slow by diffusing group III and group V elements on the one hand and the fast diffusing 3d metals on the other hand (Section IV). This is a consequence of the fact that, although the overwhelming fraction of the Au atoms are dissolved substitutionally, the long-range transport of Au is almost exclusively carried by the minute portion of the highly mobile Au atoms in interstices.

As demonstrated by the preceding examples, transition metals show the tendency to dissolve in Si or Ge both substitutionally and interstitialy, although the ratio C_i^{eq}/C_s^{eq} depends sensitively on the solute, the solvent, and on temperature. In general, the diffusion of such impurities involves an interchange between substitutional and interstitial sites, which—according to Section II.A—may take place via the kick-out mechanism [Eq. (1)], the dissociative mechanism [Eq. (2)], or a combination of these two mechanisms. In Section IX.B an outline of the theory and the major predictions of these mechanisms will be presented. This will be followed by a comparison with experiments (Section IX.C) mainly on the diffusion of Au in Si (Section IX.C.1). The cases of Ni in Si and of Cu in Ge will be briefly treated in Section IX.C.2.

B. Theory of Substitutional–Interstitial Interchange Diffusion

1. *The Kick-Out Mechanism*

a. *Basic Equations.* Kick-out diffusion involves the interchange of the diffusing foreign atoms (A) between substitutional (A_s) and interstitial (A_i) sites with the aid of self-interstitials (I) according to Eq. (1). Following Gösele et al. (1980), this type of diffusion may be described by the following equations:

$$\frac{\partial C_I}{\partial t} = D_I \frac{\partial^2 C_I}{\partial x^2} + \frac{\partial C_s}{\partial t} - K_I\left(\frac{C_I}{C_I^{eq}} - 1\right) \quad (48)$$

$$\frac{\partial C_i}{\partial t} = D_i \frac{\partial^2 C_i}{\partial x^2} - \frac{\partial C_s}{\partial t} \quad (49)$$

$$\frac{C_i}{C_I C_s} = \frac{C_i^{eq}}{C_I^{eq} C_s^{eq}} \quad (50)$$

Here x is the space coordinate in the diffusion direction considered,[8] t the diffusion time, C_α the concentration, and D_α the diffusivity [Eqs. (3) and (4)] of the defects of type α, where $\alpha =$ I, i, or s stands for self-interstitials A_i, or A_s, respectively. The equilibrium value of C_α is denoted by C_α^{eq}. The quantity K_I means the strength of sinks or sources for self-interstitials. If these sinks/sources are dislocations of density N_1, as will be assumed in what follows, K_I may be expressed as

$$K_I = \gamma_1 N_1 D_I C_I^{eq} \quad (51)$$

where the constant γ_1 characterizing the dislocation arrangement is of the order of magnitude of unity (Meyberg et al., 1983). By the law of mass action,

---

[8] Generalization of Eqs. (48–50) to three space dimensions is straightforward.

Eq. (50), it is assumed that local equilibrium between A_i, A_s, and I is established everywhere at any time via the kick-out reaction [Eq. (1)]. A numerical treatment of kick-out diffusion by Kitagawa et al. (1982b) has shown that in practice this requirement is not truly a restriction.

The coupled (nonlinear) Eqs. (48)–(50) for the unknown functions $C_I(x, t)$, $C_s(x, t)$, and $C_i(x, t)$ cannot be solved in full generality. Therefore, in the following subsections we shall consider a number of special cases of practical importance. Alternatively the assumptions $C_I = C_I^{eq}$ or $C_i = C_i^{eq}$ will be made. A detailed discussion of the criteria for the applicability of these assumptions has been given by Kitagawa et al. (1981).

b. *High Dislocation Density.* The high-dislocation-density case is characterized by the assumption that the sink density is high enough for the self-interstitial equilibrium concentration to be established instantaneously everywhere ($C_I = C_I^{eq}$). In this case Eqs. (49) and (50) yield for C_s a normal diffusion equation with the C_s-independent effective diffusion coefficient

$$D_{eff}^* = C_i^{eq} D_i / (C_I^{eq} + C_s^{eq}) \qquad (52)$$

The same result is obtained for the dissociative mechanism if the corresponding assumption $C_V = C_V^{eq}$ is made (Frank and Turnbull, 1956; Wilcox and LaChapelle, 1964; Gösele et al., 1980). Hence, in experiments involving high dislocation densities, the kick-out and the dissociative mechanism cannot be distinguished. That is, for diffusion into a semi-infinite solid, at the surface of which C_s is kept constant, both mechanisms predict the well-known erfc-type diffusion profiles typical of a concentration-independent diffusion coefficient (Carslaw and Jaeger, 1959).

c. *Dislocation Density Not High.* By definition, the dislocation density is considered *not* to be high if $C_I = C_I^{eq}$ is *not* fulfilled everywhere. Then Eq. (1) and the establishment of the A_i equilibrium concentration are the fastest reactions, so that—in addition to the simplified description of the kick-out reaction [Eq. (1)] by the mass-action law [Eq. (50)]—as a further simplification Eq. (49) may be replaced by $C_i = C_i^{eq}$. With this assumption, Eqs. (48) and (50) lead to the following differential equation for $C_s(x, t)$:

$$\frac{\partial C_s}{\partial t} = \frac{C_s^2}{C_s^2 + C_s^{eq} C_I^{eq}} \left[\frac{\partial}{\partial x} \left(\frac{C_s^{eq} C_I^{eq} D_I}{C_s^2} \frac{\partial C_s}{\partial x} \right) + K_I \left(\frac{C_s^{eq}}{C_s} - 1 \right) \right] \qquad (53)$$

As explicitly shown in the case of in-diffusion by Gösele et al. (1980), the use of the mass-action law [Eq. (50)] imposes the restriction

$$C_s \gtrsim (C_I^{eq} C_s^{eq})^{1/2} \qquad (54)$$

so that—with virtually no additional loss of generality—Eq. (53) may be

2. DIFFUSION IN SILICON AND GERMANIUM

replaced by

$$\frac{\partial C_s}{\partial t} = \frac{\partial}{\partial x}\left(D^{\text{I}}_{\text{eff}} \frac{\partial C_s}{\partial x}\right) + K_1\left(\frac{C_s^{\text{eq}}}{C_s} - 1\right) \qquad (55)$$

with

$$D^{\text{I}}_{\text{eff}} = (C_s^{\text{eq}} C_{\text{I}}^{\text{eq}} D_{\text{I}})/C_s^2 \qquad (56)$$

Note that $D^{\text{I}}_{\text{eff}}$ is strongly C_s-dependent. It is this concentration dependence of the effective diffusion coefficient $D^{\text{I}}_{\text{eff}}$ that gives rise to the distinct features of the kick-out diffusion. These have been discussed in a series of papers (Gösele et al., 1980; Seeger, 1980; Frank, 1981; Frank et al., 1981; Gösele and Frank, 1981; Gösele et al., 1981a,b; Seeger and Frank, 1982) and will be summarized in the remainder of Section IX.B.1.c.

For many purposes, it is possible and convenient to choose $t \geq 0$ as the regime of time in which Eqs. (55) and (56) are physically meaningful by using an initial condition of the form

$$C_s(0 < x < d, t = 0) = C_s^0 \qquad (57)$$

and an appropriate value of C_s^0. In the case of diffusion into a semi-infinite solid ($d = \infty$; Section IX.B.1.c.i) or into a wafer (d = wafer thickness; Section IX.B.1.c.ii) suitable choices of C_s^0 are

$$C_s^0 = (C_{\text{I}}^{\text{eq}} C_s^{\text{eq}})^{1/2} \qquad (58)$$

and, if C_s^0 is small compared to C_s^{eq},

$$C_s^0 = 0 \qquad (59)$$

(Gösele et al., 1980; Seeger, 1980). In the case of precipitation (Section IX.B.1.c.iii) C_s^0 is equal to the initial supersaturation of A_s atoms, i.e.,

$$C_s^0 > C_s^{\text{eq}} \qquad (60)$$

(i) *Diffusion into a Semi-Infinite Solid (Thick Specimen).* The diffusion of foreign atoms from the surface $x = 0$ of a thick specimen into the bulk via the kick-out mechanism is mathematically described by Eqs. (55) and (56), the boundary condition

$$C_s(x = 0, t) = C_s^{\text{eq}} \qquad (61)$$

and the initial condition (57). For a dislocation-free crystal ($K_1 = 0$) and $C_s^0 \neq 0$, Seeger (1980) derived an analytical solution of this problem in a parametric form. Here we only mention the approximate solution for $C_s^0 \ll C_s^{\text{eq}}$,

$$C_s = \frac{C_s^{\text{eq}}}{1 + |a_0|\eta} \qquad (62)$$

where

$$\eta^2 = \frac{C_s^{eq} x^2}{D_I C_I^{eq} t} \tag{63}$$

and

$$a_0 \exp(a_0^2) = -\frac{C_s^{eq}}{2\pi^{1/2} C_s^0} \tag{64}$$

The fact that the profiles characteristic of kick-out diffusion into a thick, dislocation-free specimen differ radically from the erfc-type profiles expected for dissociative diffusion (Section IX.B.2.b) plays a decisive role in revealing the main diffusion mechanism of Au in Si (Section IX.C.1.a).

(ii) *Diffusion into a Wafer.* In order to take advantage of the wafer symmetry, the wafer center is chosen as $x = 0$. Then $(\partial C_s/\partial x)(x = 0, t) = 0$, and the wafer surfaces are located at $x = \pm d/2$.

In the case of a *dislocation-free wafer* K_I and, hence, the second term on the right-hand side of Eq. (55) are equal to zero. Solutions of the resulting simplified differential equation have been determined for the boundary conditions

$$C_s(x = \pm d/2, t) = +\infty \tag{65}$$

It is obvious that these solutions fail when C_s comes close to C_s^{eq}, i.e., at long diffusion times and in the vicinity of the wafer surfaces.

Gösele *et al.* (1980) have shown that for $C_s^0 = 0$ solutions of the form

$$C_s(x, t) = X(x)(D_I t)^{1/2} \tag{66}$$

$[X(x) = $ function of x only$]$ fulfill Eqs. (55), (56), and (65) if $X(x)$ satisfies the ordinary differential equation

$$2 C_I^{eq} C_s^{eq} (d^2/dx^2)(X^{-1}) + X = 0 \tag{67}$$

The solution of Eq. (67), which satisfies the boundary condition (65), may be written in the implicit form

$$\pm \frac{(\pi C_I^{eq} C_s^{eq})^{1/2}}{X(0)} \mathrm{erf}\left[\left(\ln \frac{X}{X(0)}\right)^{1/2}\right] = x \tag{68}$$

with

$$X(0) = 2(\pi C_I^{eq} C_s^{eq})^{1/2}/d \tag{69}$$

Hence, the A_s concentration in the center of a dislocation-free wafer, C_s^m, is given by

$$C_s^m(t) = X(0)(D_I t)^{1/2} = (2/d)(\pi C_s^{eq} C_I^{eq} D_I t)^{1/2} \tag{70}$$

2. DIFFUSION IN SILICON AND GERMANIUM

For $C_s^0 \neq 0$ but otherwise equal assumptions, Meyberg (1981) derived the expression

$$C_s^m(t) = C_s^0 \{(C_s^{eq} C_I^{eq})^{1/2} \operatorname{erf}[C_s^0 d/4(C_s^{eq} C_I^{eq} D_I t)^{1/2}]\}^{-1} \tag{71}$$

which reduces to Eq. (70) for $C_s^0 \to 0$.

For a *dislocated wafer* Seeger and Frank (1982) have treated the case of $C_s \ll C_s^{eq}$ in which both terms on the right-hand side of Eq. (55) are important and $C_s^{eq}/C_s - 1$ may be approximated by C_s^{eq}/C_s. By separation of variables according to

$$C_s = (2 K_1 C_s^{eq} t)^{1/2} / Y(\xi) \tag{72}$$

with

$$\xi = \left(\frac{K_1}{D_I C_I^{eq}}\right)^{1/2} x = (\gamma_I N_I)^{1/2} x \tag{73}$$

they obtained

$$d^2 Y/d\xi^2 - Y + Y^{-1} = 0 \tag{74}$$

For approximate solutions of Eq. (74) fulfilling the boundary condition (65), the reader is referred to the original paper (Seeger and Frank, 1982). Here we restrict ourselves to two limiting expressions for $Y(0)$ permitting to calculate $C_s^m(t)$ by inserting $Y(0)$ in Eq. (72):

$$Y^2(0) = 1 - 6 \exp\left[-\left(\frac{\gamma_I N_I}{2}\right)^{1/2} d\right] + \cdots \tag{75}$$

for $[(\gamma_I N_I)/4]^{1/2} d \gg 1$,

$$\left(\frac{\gamma_I N_I}{4}\right)^{1/2} d = \left(\frac{\pi}{2}\right)^{1/2} \frac{Y(0)}{[1 - Y^2(0)]^{1/2}} \left(1 - \frac{\tfrac{1}{4} Y^2(0)}{1 - Y^2(0)} + \cdots\right) \tag{76}$$

for $[(\gamma_I N_I)/4]^{1/2} d \ll 1$.

We now review the case of $C_s^m(t)$ for $[(\gamma_I N_I)/4]^{1/2} d \gg 1$. While the expression for $C_s^m(t)$ obtainable from Eqs. (72) and (75) is restricted to $C_s^m \ll C_s^{eq}$, this treatment also covers the case $C_s^m \leq C_s^{eq}$.

For $[(\gamma_I N_I)/4]^{1/2} d \gg 1$, in the center of the wafer the first term on the right-hand side of Eq. (55) may be omitted (Gösele et al., 1980). The resulting ordinary differential equation

$$\frac{dC_s^m}{dt} = K_1 \left(\frac{C_s^{eq}}{C_s} - 1\right) \tag{77}$$

has the implicit solution

$$\ln\left(\frac{1 - C_s^m/C_s^{eq}}{1 - C_s^0/C_s^{eq}}\right) + \frac{C_s^m - C_s^0}{C_s^{eq}} + \frac{K_1 t}{C_s^{eq}} = 0 \tag{78}$$

For $C_s^m \ll C_s^{eq}$, Eq. (78) reduces to

$$C_s^m = (2K_I C_s^{eq} t + C_s^0)^{1/2} \tag{79}$$

which for $C_s^0 = 0$ becomes identical with

$$C_s^m = (2K_I C_s^{eq} t)^{1/2} \tag{80}$$

following from Eqs. (72) and (75) in the limit $N_I^{1/2} d \to \infty$.[9]

When considering the approach to saturation, we may put $C_s^m = C_s^{eq}$ in Eq. (78), except in the logarithmic term. This gives us

$$C_s^m = C_s^{eq} + (C_s^0 - C_s^{eq}) \exp\left(-\frac{K_I t + C_s^{eq} - C_s^0}{C_s^{eq}}\right) \tag{81}$$

(iii) Precipitation. By quenching of an A_s-saturated specimen, a supersaturation of A_s atoms, $C_s > C_s^{eq}$, may be frozen in. Then a driving force for the precipitation of A_s exists. Provided nucleation is *not* rate-controlling, the precipitation kinetics in the bulk of dislocated crystals may be described by the ordinary differential equation which obtains from Eq. (55) by neglecting the first term on the right-hand side. Since this equation is mathematically identical with Eq. (77), its solution $C_s(t)$ is given by Eq. (78) with C_s instead of C_s^m. Whereas at long times $C_s(t)$ approaches the form of Eq. (81), its short-time approximation (Gösele *et al.*, 1981a),

$$C_s/C_s^0 = 1 - (K_I t / C_s^0) \tag{82}$$

differs from Eq. (79), which is valid for $C_s \ll C_s^{eq}$, since during precipitation C_s is always larger than C_s^{eq}. According to Eq. (82), C_s/C_s^0 depends on the initial supersaturation C_s^0. This is not true for the dissociative mechanism, as may be seen from Eq. (92) [see Section IX.B.2.d].

2. The Dissociative Mechanism

a. Basic Equations. Dissociative diffusion involves the interchange of the diffusing foreign atoms between substitutional and interstitial sites with the aid of vacancies according to Eq. (2). It may be described by the following equations (Frank and Turnbull, 1956; Wilcox and LaChapelle, 1964; Gösele *et al.*, 1980):

$$\frac{\partial C_V}{\partial t} = D_V \frac{\partial^2 C_V}{\partial x^2} - \frac{\partial C_s}{\partial t} + K_V\left(1 - \frac{C_V}{C_V^{eq}}\right) \tag{83}$$

$$\frac{\partial C_i}{\partial t} = D_i \frac{\partial^2 C_i}{\partial x^2} - \frac{\partial C_s}{\partial t} \tag{84}$$

$$\frac{C_s}{C_V C_i} = \frac{C_s^{eq}}{C_V^{eq} C_i^{eq}} \tag{85}$$

2. DIFFUSION IN SILICON AND GERMANIUM

The newly introduced quantities with the subscript V refer to vacancies. Their meanings are analogous to those of the corresponding quantities for self-interstitials carrying the subscript I.

b. *Highly Dislocated and Dislocation-Free Specimens.* For *highly dislocated specimens*, Eq. (83) may be replaced by $C_V = C_V^{eq}$, which together with Eqs. (84) and (85) leads to the normal diffusion equation for C_s,

$$\frac{\partial C_s}{\partial t} = D_{\text{eff}} \frac{\partial^2 C_s}{\partial x^2} \tag{86}$$

with the same C_s-independent effective coefficient $D_{\text{eff}} \equiv D_{\text{eff}}^*$ [Eq. (52)] as in the case of kick-out diffusion in highly dislocated specimens. For the diffusion into a highly dislocated semi-infinite solid at $x \geq 0$ with an inexhaustible source of A_s atoms at its surface $x = 0$ [i.e., $C_s(x = 0, t) = C_s^{eq}$], the well-known solution of Eq. (86) for $C_s^0 = 0$ is

$$C_s = C_s^{eq} \operatorname{erfc}\{x/[2(D_{\text{eff}}t)^{1/2}]\} \tag{87}$$

with $D_{\text{eff}} \equiv D_{\text{eff}}^*$. Concerning the solution of Eq. (86) for the diffusion into a wafer, the reader is referred to the standard textbooks on diffusion or conduction of heat (e.g., Carslaw and Jaeger, 1959; Crank, 1975).

For *dislocation-free specimens*, $K_V = 0$ and $C_i = C_i^{eq}$. Together with Eqs. (83) and (85), this leads to the normal diffusion equation [Eq. (86)], where the C_s-independent effective diffusion coefficient D_{eff} has now to be identified with

$$D_{\text{eff}}^V = \frac{C_V^{eq} D_V}{C_V^{eq} + C_s^{eq}} \approx \frac{C_V^{eq}}{C_s^{eq}} D_V \tag{88}$$

The approximation made in Eq. (88) and in what follows is not an essential restriction, since—at least in the case of in-diffusion—$C_V^{eq} \gg C_s^{eq}$ is incompatible with the use of the mass-action law [Eq. (85)].

c. *Low and Intermediate Dislocation Density.* In this case ($K_V \neq 0$, $C_i = C_i^{eq}$), elimination of C_V from Eq. (83) with the aid of the mass-action law [Eq. (85)] yields

$$\frac{\partial C_s}{\partial t} = \frac{\partial}{\partial x}\left(D_{\text{eff}}^V \frac{\partial C_s}{\partial x}\right) + K_V\left(1 - \frac{C_s}{C_s^{eq}}\right) \tag{89}$$

Sturge (1959) has treated the diffusion into both a semi-infinite solid and into a wafer in the case of *low dislocation density*, in which both terms

---

[9] According to Gösele et al. (1981b), a convenient expression interpolating between the cases of a dislocation-free wafer [Eq. (70)] and a dislocated wafer with $N_I^{1/2}d \gg 1$ [Eq. (80)] is $C_s^m \approx [2D_I C_I^{eq} C_s^{eq} t(N_I + 2\pi d^{-2})]^{1/2}$.

on the right-hand side of Eq. (89) are important. His result on the wafer shows that the time dependence of C_s^m is weaker than $\propto t$ but, except for $C_s^m \approx C_s^{eq}$, stronger than $\propto t^{1/2}$, as predicted by the kick-out mechanism [cf. Eq. (70); Eqs. (72), (75), and (76) or Eq. (80)].

For a wafer with *intermediate dislocation density* the first term on the right-hand side of Eq. (89) may be omitted. Integration of the resulting ordinary differential equation gives us

$$C_s^m = C_s^{eq} + (C_s^0 - C_s^{eq})\exp[-(K_V t / C_s^{eq})] \tag{90}$$

In a regime of intermediate diffusion times, Eq. (90) may be approximated by

$$C_s^m = K_V t \tag{91}$$

d. *Precipitation.* After replacing C_s^m by C_s, Eq. (90) may be used to describe precipitation in a crystal with intermediate dislocation density. For short times, the approximation $C_s^0 \gg C_s^{eq}$ may be made, yielding

$$\frac{C_s}{C_s^0} = 1 - \frac{K_V t}{C_s^{eq}} \tag{92}$$

3. Combination of the Kick-Out and the Dissociative Mechanism

If foreign atoms are capable of interchanging between substitutional and interstitial sites via both the kick-out and the dissociative mechanism, the total rate at which C_s changes locally is given by

$$\frac{\partial C_s}{\partial t} = \left(\frac{\partial C_s}{\partial t}\right)_I + \left(\frac{\partial C_s}{\partial t}\right)_V \tag{93}$$

Here $(\partial C_s/\partial t)_I$ and $(\partial C_s/\partial t)_V$ are the contributions by the kick-out reaction [Eq. (1)] and the dissociative reaction [Eq. (2)], respectively. Excluding the case of high dislocation density by putting $C_i = C_i^{eq}$, confining ourselves to $C_s > (C_I^{eq} C_s^{eq})^{1/2}$ and $C_s^{eq} > C_V^{eq}$ for the reasons given above, and assuming the validity of the mass-action laws [Eqs. (50) and (85)], after replacing $\partial C_s/\partial t$ by $(\partial C_s/\partial t)_I$ in Eq. (48) and by $(\partial C_s/\partial t)_V$ in Eq. (83), we find from Eqs. (48), (83), and (93)

$$\frac{\partial C_s}{\partial t} = \frac{\partial}{\partial x}\left[(D_{\text{eff}}^I + D_{\text{eff}}^V)\frac{\partial C_s}{\partial x}\right] + K_I\left(\frac{C_s^{eq}}{C_s} - 1\right) + K_V\left(1 - \frac{C_s}{C_s^{eq}}\right) \tag{94}$$

with

$$D_{\text{eff}}^I + D_{\text{eff}}^V = \frac{D_I C_I^{eq}}{C_s^{eq}}\left[\left(\frac{C_s^{eq}}{C_s}\right)^2 + \frac{D_V C_V^{eq}}{D_I C_I^{eq}}\right] \tag{95}$$

Let us focus attention on a number of interesting features of the combined diffusion according to Eq. (94):

(i) If the mass-action laws (50) and (85) for the kick-out reaction

(1) and the dissociative reaction (2) are fulfilled simultaneously,

$$C_I C_V = C_I^{eq} C_V^{eq} \tag{22}$$

holds automatically. This means that the equilibrium between self-interstitials and vacancies can be established indirectly via reactions (1) and (2) even if the direct reaction

$$I + V \rightleftharpoons 0 \tag{23}$$

is hampered, e.g., by a barrier against self-interstitial–vacancy recombination as in the case of Si at high temperatures (Section VI) or by a high activation barrier against spontaneous formation of Frenkel pairs in the bulk.

(ii) For observing combined kick-out and dissociative diffusion according to Eqs. (94) and (95), it is sufficient that, in addition to $D_I C_I^{eq} \neq 0$ and $D_V C_V^{eq} \neq 0$, two of the three equations (22), (50), and (85) are fulfilled. As a consequence, the kick-out mechanism or the dissociative mechanism may *seemingly* operate though reactions (1) or (2) do not take place, respectively.

(iii) If $D_I C_I^{eq} = 0$ and/or if both reactions (1) and (23) are suppressed, neither a true nor a seeming contribution by the kick-out mechanism can be observed.

(iv) If $D_V C_V^{eq} = 0$ and/or if both reactions (2) and (23) are suppressed, neither a true nor a seeming contribution by the dissociative mechanism can be observed.

(v) If both the kick-out and the dissociative mechanism seemingly or truly operate and if $D_I C_I^{eq}$ and $D_V C_V^{eq}$ or K_I and K_V are of the same order of magnitude, then in in-diffusion experiments kick-out diffusion determines the features of diffusion, whereas in precipitation experiments the features of dissociative diffusion show up. The reason is that in Eq. (94) under these circumstances for $C_s/C_s^{eq} < 1$ the term with D_{eff}^I [cf. Eq. (56)] or with K_I becomes larger than the term with D_{eff}^V [cf. Eq. (88)] or with K_V, whereas for $C_s/C_s^{eq} > 1$ the situation is reverse.

C. Comparison with Experiments

1. *Gold in Silicon*

a. *Predominance of the Kick-Out Mechanism in In-Diffusion Experiments.* Among the early investigations of the diffusion of Au in Si (Dash, 1960; Sprokel and Fairfield, 1965; Bullis, 1966; Huntley and Willoughby, 1970, 1973a,b; Lambert, 1971, 1972; Yoshida, 1973; Kästner and Hesse, 1974; Scheibe and Schröter, 1977) the observation by Wilcox and LaChapelle (1964) and Wilcox et al. (1964) that between 1073 K and 1373 K the diffusion profiles in thick, dislocation-free specimens differ considerably from the erfc shape expected in the case of a concentration-independent diffusion

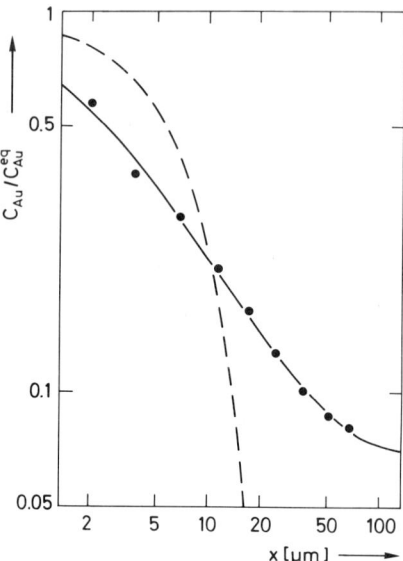

Fig. 21. Double-logarithmic plot of the Au penetration profile in a thick {111} Si specimen [(6–9) × 10^3 Ω cm, phosphorus-doped] after a 1-hr anneal at 1173 K [according to Stolwijk *et al.* (1983)]. The solid line represents a fit of the kick-out model (Seeger, 1980), the dashed line the attempt of a fit to an erfc dependence as predicted by the dissociative model.

coefficient was a surprising result that could not be rationalized at that time. Seeger (1980) and Gösele *et al.* (1981b) have shown that these profiles may be understood in terms of kick-out diffusion into a dislocation-free semi-infinite solid (Section IX.B.1.c.i). Recent high-accuracy measurements by Stolwijk *et al.* (1983), using neutron-activation analysis[10] and mechanical sectioning, confirmed the non-erfc nature of such profiles as well as the interpretation given by Seeger (1980) (Fig. 21).

Investigating the diffusion of Au into thin, dislocation-free Si slices, Stolwijk *et al.* (1983) found U-shaped diffusion profiles (Fig. 22) in accordance with observations by Hill *et al.* (1982), who used the spreading-resistance

---

[10] Though techniques making use of radioactive Au isotopes measure the total Au concentration C_{Au}, throughout Section IX.C.1 we put $C_s = C_{Au}$. Since in this way a consistent interpretation is achieved, we conclude that for Au in Si under high-temperature equilibrium conditions $C_s^{eq} \gg C_i^{eq}$ is valid. Under low-temperature non-equilibrium conditions the situation is more sophisticated: Whereas radioactive methods measure C_{Au}, it is unclear which Au configurations (Au_s, small complexes of Au_s and/or Au_i, etc.) contribute to electrical measurements. Thus the difficulties in interpreting data on Au in Si obtained by low-temperature electrical measurements (e.g., Lang *et al.*, 1980) become understandable.

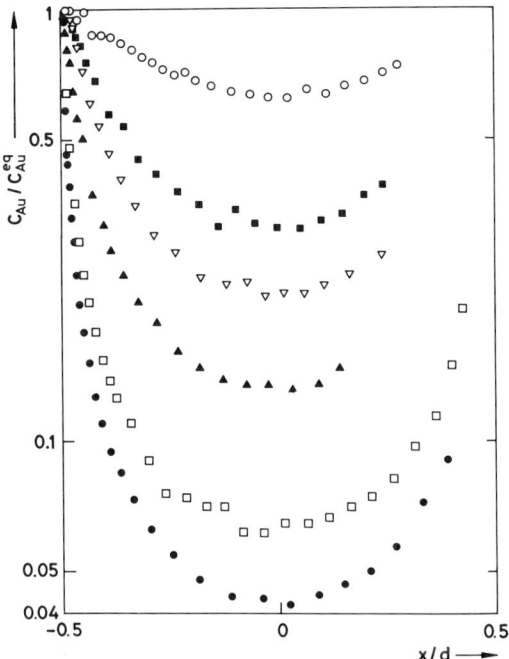

Fig. 22. Penetration profiles of Au in thin {100} Si wafers (8 Ω cm, boron-doped) for different annealing times at 1273 K; (●) 0.467 hr, (□) 1.03 hr, (▲) 4.27 hr, (▽) 4.27 hr, (■) 26.8 hr, (○) 100.6 hr [according to Stolwijk et al. (1983)]. The Au concentration C_{Au} is given in units of the solubility C_{Au}^{eq}, the penetration depth x as a fraction of the wafer thickness d. For one of the 4.27-hr anneals (▽) $d \approx 300$ μm, otherwise $d \approx 500$ μm.

technique.[11] These profiles are in qualitative agreement with both the kick-out and the dissociative mechanism: After the solubility of Au_i has been established almost instantaneously through the entire sample, the transformation of Au_i to Au_s occurs preferentially near the sample surfaces, which act as sinks for self-interstitials [kick-out mechanism (1)] or as sources for vacancies [dissociative mechanism (2)]. A distinction between the two mechanisms must therefore come from a quantitative analysis: As shown in Fig. 23 for the 1.03-hr diffusion anneal at 1237 K of Fig. 22, Stolwijk et al. (1983) have demonstrated that their U profiles may be well described by the

---

[11] The curved central parts of the U profiles of Stolwijk et al. (1983) indicate the absence of dislocations in these specimens. In previous studies on dislocated wafers (Huntley and Willoughby, 1970, 1973a,b; Kästner and Hesse, 1974) U profiles with flat central parts were observed.

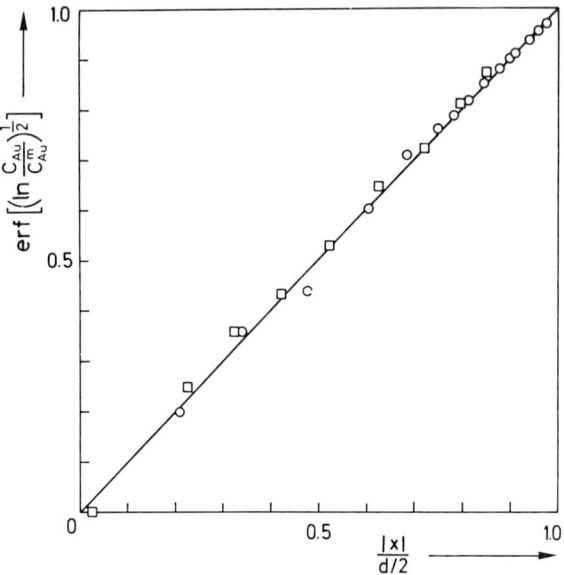

Fig. 23. Test of Eq. (96), predicted by the kick-out model, by comparison to the diffusion profile measured after a 1.03-hr anneal at 1273 K (see Fig. 22); (○) $x < 0$, (□) $x > 0$.

relationship

$$\mathrm{erf}\left[\ln(C_s/C_s^m)^{1/2}\right] = (2|x|)/d \tag{96}$$

predicted by Eqs. (66), (68), and (69) for kick-out diffusion into a thin, dislocation-free wafer. The dissociative diffusion model fails to account quantitatively for these data.

A quite specific prediction of the kick-out model is that during diffusion

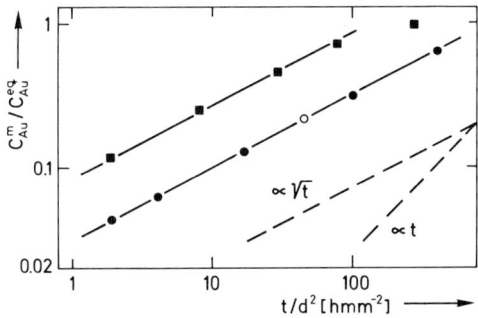

Fig. 24. Increase of the Au concentration C_{Au}^m in the center of a Si wafer as a function of t/d^2 [(●) 1273 K, $d \approx 500$ μm; (○) 1273 K, $d \approx 300$ μm; (■) 1371 K, $d \approx 500$ μm], according to Stolwijk et al. (1983).

into a wafer the Au_s concentration in the center increases proportionally to $t^{1/2}$, except for $C_s^m \approx C_s^{eq}$ and in highly dislocated wafers (Section IX.B.1.c.ii). Such a dependence of C_s^m has indeed been found by many authors (Sprokel and Fairfield, 1965; Lambert, 1971, 1972; Huntley and Willoughby, 1973a,b). Figure 24 shows the results obtained on dislocation-free wafers by Stolwijk et al. (1983). It is seen that not only the $t^{1/2}$ law is confirmed, but also the d^{-1} dependence of C_s^m on the wafer thickness d predicted by Eq. (70) for kick-out diffusion into a dislocation-free wafer.

Since the quantity

$$D^* \equiv (C_I^{eq} D_I)/C_s^{eq} \tag{97}$$

enters the theoretical expressions for kick-out diffusion (Section IX.B), it is possible to determine the contribution

$$D_I^{SD} \equiv C_I^{eq} D_I = C_s^{eq} D^* \tag{98}$$

of self-interstitials to the uncorrelated self-diffusion coefficient in Si,

$$D^{SD} = D_I^{SD} + D_V^{SD} = C_I^{eq} D_I + C_V^{eq} D_V \tag{99}$$

by combining data on the diffusion (D^*) and the substitutional solubility (C_s^{eq}) of Au in Si. Proceeding in this way, Stolwijk et al. (1983) computed the $D_I^{SD}/2$ values[12] represented by full symbols in Fig. 25 from their solubility (Fig. 26) and diffusion data of Au in dislocation-free specimens. More specifically, the full circle and the full squares were derived from diffusion profiles in thick specimens (Section IX.B.1.c.i) and from the increase of the Au concentration in the centers of wafers [Eq. (70)], respectively. The values so obtained may be described by

$$D_I^{SD} = 9.14 \times 10^{-2} \exp(-4.84 eV/kT) \quad m^2 \, s^{-1} \tag{100}$$

which corresponds to the straight line in Fig. 25. The open circles and crosses represent values of the tracer self-diffusion coefficient D^T measured by Mayer et al. (1977) and Kalinowski and Seguin (1980), respectively. The good agreement between the values of $D_I^{SD}/2$ and D^T shows that, in the temperature regime considered, the interstitialcy mechanism (Section II.A) yields a major contribution to Si self-diffusion. Therefore, we conclude that above 1130 K D_I^{SD} is larger than, or at least of the same order of magnitude as, D_V^{SD}.

b. *Contribution by the Dissociative Mechanism.* As reported in Section V, the injection of self-interstitials into Si by surface oxidation at 1373 K enhances the diffusion of P and B but retards the diffusion of Sb. From these

---

[12] $D_I^{SD}/2$ represents the contribution of self-interstitials to the tracer self-diffusion coefficient under the assumption that the correlation factor f_i is equal to 1/2.

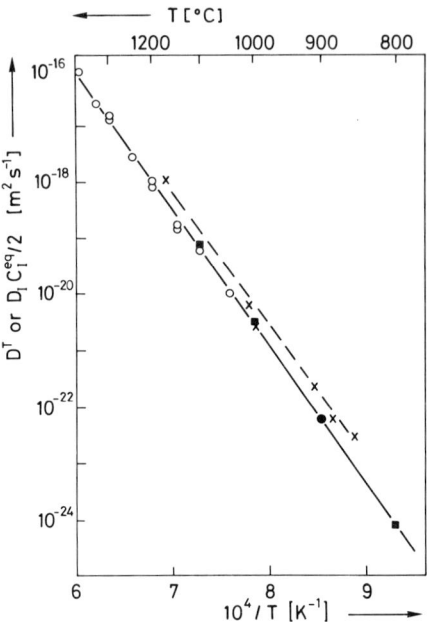

Fig. 25. Tracer self-diffusion coefficient D^T in silicon and interstitialcy component $\frac{1}{2}D_I C_I^{eq}$ of D^T as functions of the reciprocal temperature $1/T$. The $\frac{1}{2}D_I C_I^{eq}$ values have been calculated from the solubility (Fig. 26) and the diffusion data of Au in Si (Stolwijk et al., 1983) with the aid of Eq. (98). Data from [■, ●] Stolwijk et al. (1983), [○] Mayer et al. (1977), [×] Kalinowski and Seguin (1980).

observations it has been concluded that in Si both self-interstitials and vacancies are present in thermal equilibrium and that at 1373 K P and B diffuse predominantly via the interstitialcy mechanism, whereas Sb prefers to diffuse via the vacancy mechanism. Thus $D_V C_V^{eq}$ is large enough to be detected by suitable experiments. Hence, the good agreement of the diffusion data on Au in Si with the predictions of the kick-out model (Section IX.B.1.c) may be due either to the predominance of kick-out diffusion over dissociative diffusion in in-diffusion experiments [item (v) in Section IX.B.3] or to a blocking of the two reactions (2) and (23) [item (iv) in Section IX.B.3]. This question has been investigated by Morehead et al. (1982) by numerically fitting Eqs. (94) and (95) with $K_I = K_V = 0$ and the boundary conditions $C_s(x = \pm\frac{d}{2}, t) = C_s^{eq}$ to Au diffusion profiles measured on dislocation-free Si wafers at 1273 or 1373 K by Stolwijk et al. (1983). The fits to the 1273 K profiles are shown in Figs. 27a–c. The profile computed for $D_V^{SD}/D_I^{SD} = 0$ agrees well with the experimental data (full circles) measured after an annealing time t of 0.5 hr (Fig. 27a) but clearly misses the measuring points at

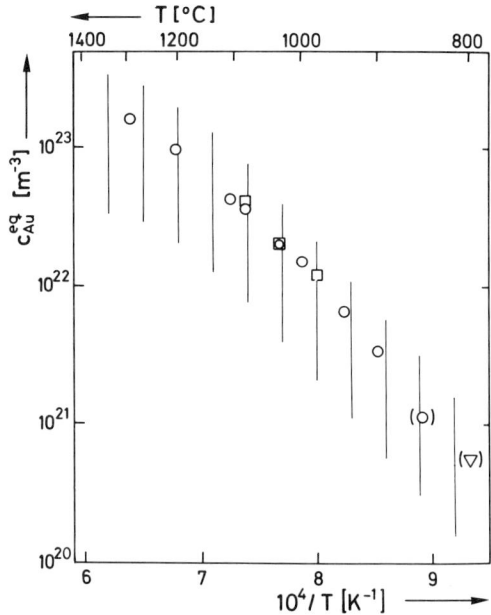

Fig. 26. Solubility C_{Au}^{eq} of Au in Si as a function of the reciprocal temperature $1/T$ [(○) 6–9 × 10^3 Ω cm, phosphorus-doped; (□) 0.3–0.4 Ω cm, boron-doped; (▽) 800–3000 Ω cm, boron-doped], according to Stolwijk et al. (1983). The hatched area indicates the scattering of the solubility data found in the literature. The data points in parentheses had to be corrected, since during the corresponding diffusion anneals the solubility limit had not been reached.

$t = 4.3$ hr (Fig. 27b) and even more so at $t = 100$ hr (Fig. 27c). Irrespective of the time of annealing, the profiles computed for $D_V^{SD}/D_I^{SD} = 1$ excellently fit the experimental data, whereas those for $D_I^{SD}/D_V^{SD} = 0$ radically fail to do so. From these findings, Morehead et al. (1983) concluded that at 1273 K, $D_I^{SD} \approx D_V^{SD}$, but the limited accuracy of the experimental data at this temperature did not allow a more precise statement to be made.

We want to focus attention on a paradox that may be seen from Figs. 27a–c. The influence of the admixture of dissociative diffusion (= difference between the profiles for $D_V^{SD}/D_I^{SD} = 1$ and $D_V^{SD}/D_I^{SD} = 0$) increases with the annealing time in the sense that, at a fixed distance from the wafer surfaces, C_s approaches C_s^{eq} faster than in the case of pure kick-out diffusion. As a consequence, the range of validity of the short-time approximation $C_s^m \propto t^{1/2}$ predicted by the kick-out mechanism is extended by the admixture of dissociative diffusion.

The global picture of self- and Au diffusion in Si evolving is as follows (Gösele and Tan, 1983b). Above about 1270 K, self-diffusion is dominated

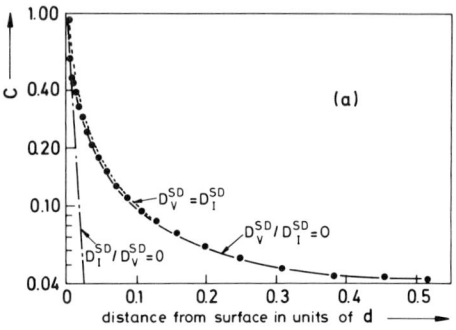
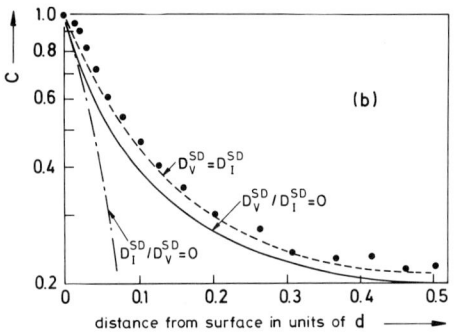
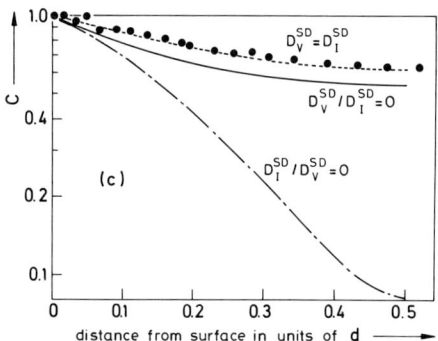

Fig. 27. Diffusion profiles of Au in Si wafers measured at 1273 K. (a) Annealing time $t = 0.5$ hr, wafer thickness $d = 0.5$ mm; (b) $t = 4.3$ hr, $d = 0.3$ mm; (c) $t = 100$ hr, $d = 0.5$ mm. C is the Au concentration in units of the solubility limit; the distance from the surface is given in units of d. The curves represent numerical fits for various ratios of D_I^{SD}/D_V^{SD} by Morehead *et al.* (1983), who assumed that the kick-out and the dissociative mechanism operate simultaneously.

2. DIFFUSION IN SILICON AND GERMANIUM

by the interstitialcy mechanism and below this temperature by the vacancy mechanism. Concerning Au diffusion, however, the predominance of the kick-out mechanism over the dissociative mechanism—as reflected by the shape of Au diffusion profiles in thick, dislocation-free specimens (Section IX.C.1.a)—reaches down to at least 1070 K due to the proportionality of $D_{\text{eff}}^{\text{I}}$ to C_s^{-2} [item (v) in Section IX.B.3]. Since below about 1170 K reaction (23) is considerably hampered by an entropy barrier against self-interstitial–vacancy recombination (Section VI), the predominance of the kick-out mechanism between 1170 and 1070 K may be taken as an indication that the kick-out reaction [Eq. (1)] is operating and is not simulated by a simultaneous operation of reactions (2) and (23) [item (ii) in Section IX.B.3]. The occurrence of erfc-shaped profiles in thick, dislocation-free samples at 970 K (Wilcox and LaChapelle, 1964; Wilcox et al., 1964) shows that at this temperature the dissociative reaction [Eq. (2)] still operates, whereas the kick-out reaction [Eq. (1)] is frozen in [item (iii) in Section IX.B.3]. The alternative explanation, that the ratio $D_{\text{I}}^{\text{SD}}/D_{\text{V}}^{\text{SD}}$ has become so small that a contribution by kick-out diffusion can no longer be detected, may be ruled out, since for the C_s^0/C_s^{eq} values of about 10^{-2} used in these experiments $D_{\text{eff}}^{\text{I}}$ should dominate over $D_{\text{eff}}^{\text{V}}$ during a considerable interval of C_s/C_s^{eq}.

It may be useful to compile the quantitative information on Au atoms and intrinsic point defects that have been obtained from diffusion and solubility measurements of Au in Si for the temperature interval 1070–1370 K. In Section IX.C.1.a, we have presented an expression for the contribution D_{I}^{SD} of self-interstitials to the Si self-diffusion coefficient [Eq. (100)] as well as the substitutional solubility C_s^{eq} of Au in Si. Between 1070–1370 K, C_s^{eq} may be described by

$$C_s^{\text{eq}} = 18 \exp(-1.98\,eV/kT) \tag{101}$$

According to Gösele et al. (1980) and Section IX.B.1.c, the results reported in Section IX.C.1.a show that

$$C_{\text{I}}^{\text{eq}} \ll C_s^{\text{eq}} \tag{102}$$

holds, so that from Eqs. (100)–(102)

$$D_{\text{I}} \gg 5.1 \times 10^{-3} \exp(-2.86\,eV/kT) \quad \text{m}^2\,\text{s}^{-1} \tag{103}$$

may be deduced. Furthermore, the internal consistency of the picture developed above justifies our assumptions

$$C_i^{\text{eq}} \ll C_s^{\text{eq}} \tag{104}$$

and

$$D_i \gg D_{\text{I}} \tag{105}$$

We recall that the validity of the inequality (105) was assumed when putting $C_i = C_i^{eq}$ (Section IX.B.1.c), since this implies that Au_i is the fastest diffusing defect involved in the kick-out diffusion. Concerning vacancies in Si, on one hand the analysis by Morehead *et al.* (1983) gives us $D_V C_V^{eq} \approx D_I C_I^{eq}$ at 1270 K, whereas on the other hand at 970 K one finds $D_V C_V^{eq} = 8.8 \times 10^{-26}$ m² s⁻¹ by combining $C_s^{eq} = 1.05 \times 10^{-6}$, obtained by extrapolating Eq. (101) to 970 K, and $D_{eff}^V = 8.4 \times 10^{-18}$ m² s⁻¹, taken from the erfc-shaped Au profiles observed at this temperature. This leads us to the estimate

$$D_V^{SD} = D_V C_V^{eq} \approx 0.57 \exp(-4.03 eV/kT) \quad m^2 \, s^{-1} \quad (106)$$

which in Fig. 28 is compared to D_I^{SD} according to Eq. (100).

In spite of the wealth of information summarized above, our knowledge on the equilibrium concentrations of intrinsic point defects in Si is still disappointingly poor. On vacancies we only have the information, obtained

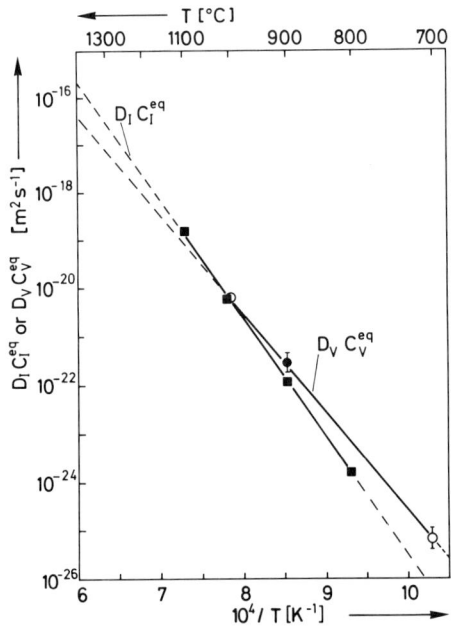

Fig. 28. Comparison of the contributions to the uncorrelated self-diffusion coefficient in Si by the interstitialcy mechanism, $D_I C_I^{eq}$, and by the vacancy mechanism, $D_V C_V^{eq}$. These contributions have been calculated from diffusion data of Au or Ni in Si. For details see text. Data from [■] Stolwijk *et al.* (1983), [○] Morehead *et al.* (1983), [ϕ] Wilcox *et al.* (1964), [♦] Kitagawa *et al.* (1982a).

2. DIFFUSION IN SILICON AND GERMANIUM

quite indirectly from the success in applying the approximation (88) in the analysis of Morehead *et al.* (1983), that

$$C_V^{eq} \ll C_s^{eq} \tag{107}$$

must be fulfilled. Concerning C_I^{eq}, in addition to (102) holding for 1070 K $\lesssim$ $T \lesssim 1370$ K,

$$2 \times 10^{-8} \lesssim C_I^{eq}(T_m) \lesssim 2 \times 10^{-6} \tag{108}$$

has been estimated from electron-microscope investigations of A-type swirl defects (Seeger *et al.*, 1977).

2. Nickel in Silicon and Copper in Germanium

Kitagawa *et al.* studied both in-diffusion and precipitation of Ni in Si at 1173 K (1982a, 1983) and of Cu in Ge between 873 and 1078 K (1982c) by means of four-point electrical resistivity measurements. In agreement with earlier experiments reviewed by Seeger and Chik (1968), they found that in both systems the dissociative mechanism operates, i.e., in-diffusion obeys Eqs. (90) and (91) and precipitation is independent of C_I^0 according to Eq. (92).

In a collaboration between researchers at the Max Planck Institute for Metals Research and the Lawrence Berkeley Laboratory the in-diffusion and solubility of Cu in ultrapure, undoped, dislocation-free Ge were investigated between 850 and 1200 K by means of the spreading-resistance technique. A major (unpublished) result of these studies was that erfc-type diffusion profiles were observed, as predicted by Eq. (87). From the effective diffusion coefficients [Eq. (88)] and from the solubility of Cu in Ge the values of $f_V C_V^{eq} D_V$ were calculated and found to be in good agreement with the values of the Ge tracer self-diffusion coefficient given in Fig. 6. This supports the idea that self-diffusion in Ge takes place via vacancies.

Regarding Ni in Si, it is expected that dissociative diffusion dominates already at higher temperatures than in the case of Au in Si for the following reasons: (a) For Ni the substitutional solubility C_s^{eq} is 100 to 1000 times smaller (Yoshida and Saito, 1967), and thus D_{eff}^V [Eq. (88)] is correspondingly larger than for Au (Stolwijk *et al.*, 1983); (b) A simulation of kick-out diffusion by the cooperation of reactions (2) and (23) is not possible, since below 1170 K during the short diffusion time required for the approach of C_s to C_s^{eq} via dissociative diffusion, reaction (23) is blocked (Section VI). Even at very high temperatures, the diffusion of Ni in Si will presumably not show the features of kick-out diffusion (Section IX.B.1.c), since for Ni in Si the case $C_s^{eq} \ll C_I^{eq}$ may be realized, i.e., C_s will have reached C_s^{eq} before Eqs. (55) and (56) become valid (Gösele *et al.*, 1980).

By evaluating the diffusion experiments of Kitagawa *et al.* (1982a, 1983) on Ni in Si in terms of pure dissociative diffusion, we find a $D_V C_V^{eq}$ value at 1173 K which lies on the Arrhenius plot for $D_V C_V^{eq}$ derived from diffusion data on Au in Si (Fig. 28). This supports the above interpretation of the diffusion data on Ni in silicon.

X. Concluding Remarks

After a quite sophisticated picture of the diffusion in silicon and germanium which has evolved in the preceding sections, the reader may appreciate the following summary of what the authors believe to be the highlights of this chapter.

(i) Self-diffusion in Ge may be explained in terms of the vacancy mechanism.

(ii) Both the interstitialcy and the vacancy mechanism contribute to self-diffusion in Si. The former mechanism dominates above about 1270 K, the latter below this temperature.

(iii) The combination of investigations of oxidation-enhanced diffusion of group III and group V elements with observations of the oxidation-induced growth of interstitial-type stacking faults leads to the conclusion that in Si both vacancies and self-interstitials contribute to the self-diffusion and to the diffusion of foreign substitutional atoms.

(iv) The most important step in understanding Si self-diffusion was the discovery that it is the kick-out mechanism which dominates the diffusion of Au in Si. An analysis of data on the diffusion of Au in Si allowing for both the kick-out and the dissociative mechanism has yielded quantitative information on the contributions of vacancies, $D_V C_V^{eq}$, and of self-interstitials, $D_I C_I^{eq}$, to self-diffusion.

(v) A reliable subdivision of $D_V C_V^{eq}$ and $D_I C_I^{eq}$ into the separate diffusivities and equilibrium concentrations of vacancies and self-interstitials remains as a challenging task for future research.

Acknowledgments

The authors are very grateful to the large number of researchers who have contributed to this work in various ways. Particular thanks are due to G. Hettich, J. Hölzl, B. O. Kolbesen, S. Mizuo, F. Morehead, B. Kühn, N. A. Stolwijk, H. Strunk, T. Y. Tan, G. Vogel, D. Weiler, and M. Werner. Part of the work on the diffusion of gold in silicon reported in Section IX was supported by the Bundesministerium für Forschung und Technologie, Federal Republic of Germany, within the framework of research project NT 2582/0.

References

Alexander, H., Eppenstein, H., Gottschalk, H., and Wendler, S. (1980). *J. Microsc. (Oxford)* **118**, 1.
Antoniadis, D. A. (1982). *J. Electrochem. Soc.* **129**, 1093.
Antoniadis, D. A., and Moskowitz, I. (1982a). *J. Appl. Phys.* **53**, 6788.
Antoniadis, D. A., and Moskowitz, I. (1982b). *J. Appl. Phys.* **53**, 9214.
Antoniadis, D. A., Lin, A. M., and Dutton, R. W. (1978a). *Appl. Phys. Lett.* **33**, 1030.
Antoniadis, D. A., Gonzales, A. G., and Dutton, R. W. (1978b). *J. Electrochem. Soc.* **125**, 813.
Armigliato, A., Servidori, M., Solmi, S., and Vecchi, I. (1977). *J. Appl. Phys.* **48**, 1806.
Baker, J. A., Tucker, T. N., Moyer, N. E., and Buschert, R. C. (1968). *J. Appl. Phys.* **39**, 4365.
Bakhadyrkhanov, M. K., Zainabinov, S., and Khamidov, A. (1980). *Sov. Phys. Semicond.* **14**, 243.
Bean, A. R., and Newman, R. C. (1971). *J. Phys. Chem. Solids* **32**, 1211.
Bean, K. E., and Gleim, P. S. (1969). *Proc. IEEE* **57**, 1469.
Bergholz, W. (1983). *Physica (Amsterdam)* **116B**, 312.
Blount, E. I. (1959). *J. Appl. Phys.* **30**, 1218.
Booker, G. R., and Tunstall, N. J. (1966). *Philos. Mag.* **13**, 71.
Bourgoin, J. C., and Corbett, J. W. (1978). *Radiat. Eff.* **36**, 157.
Bourgoin, J. C., and Lannoo, M. (1980). *Radiat. Eff.* **46**, 157.
Brätter, P., and Gobrecht, H. (1970). *Phys. Status Solidi* **37**, 869.
Bullis, W. M. (1966). *Solid-State Electron.* **9**, 143.
Campbell, D. R. (1975). *Phys. Rev. B* **12**, 2318.
Carslaw, H. S., and Jaeger, J. C. (1959). "Conduction of Heat in Solids." Oxford Univ. Press (Clarendon), London and New York.
Claeys, C. L., Declerck, G. J., and van Overstraeten, R. J. (1978a). *Rev. Phys. Appl.* **13**, 797.
Claeys, C. L., Declerck, G. J., and van Overstraeten, R. J. (1978b). *In* "Semiconductor Characterization Techniques" (P. A. Barnes and G. A. Rozgonyi, eds.), p. 366. The Electrochem. Soc., Princeton, New Jersey.
Claeys, L., Declerck, G. J., and van Overstraeten, R. J. (1979). *Appl. Phys. Lett.* **35**, 797.
Compaan, K., and Haven, Y. (1956). *Trans. Faraday Soc.* **52**, 786.
Corbett, J. W., McDonald, R. S., and Watkins, G. D. (1964). *J. Phys. Chem. Solids* **25**, 873.
Crank, J. (1975). "Mathematics of Diffusion." Oxford Univ. Press (Clarendon), London and New York.
Craven, R. A. (1981). *In* "Semiconductor Silicon 1981" (H. R. Huff, R. J. Kriegler, and Y. Takeishi, eds.), p. 254. The Electrochem. Soc., Pennington, New Jersey.
Dash, W. C. (1960). *J. Appl. Phys.* **31**, 2275.
Deal, B. E., and Groove, A. S. (1965). *J. Appl. Phys.* **16**, 3770.
Dederichs, P. H., Lehmann, C., Schober, H. R., Scholz, A., and Zeller, R. (1978). *J. Nucl. Mater.* **69 & 70**, 176.
De Kock, A. J. R., and van de Wiggert, W. M. (1980). *J. Cryst. Growth* **49**, 718.
Demond, F. J., Kalbitzer, S., Mannsperger, H., and Damjantschitsch, H. (1983). *Phys. Lett.* **93A**, 503.
Dorner, P. (1980). Ph.D. Thesis. University of Stuttgart, Stuttgart, Federal Republic of Germany.
Dorner, P., Gust, W., Lodding, A., Odelius, H., Predel, B., and Roll, U. (1983). *In* "DIMETA 82—Diffusion in Metals and Alloys" (F. J. Kedves and D. L. Beke, eds.). Monograph Series 7, Trans Tech Publications, p. 488.
Drake, W. H., and Willoughby, A. F. W. (1969). *Proc. Thomas Graham Mem. Symp. Diff. Proc., University of Strathclyde, England.* Gordon & Breach, New York.

Dunlap, W. C., Jr. (1954). *Phys. Rev.* **94**, 1531.
Fair, R. B. (1977). *In* "Semiconductor Silicon 1977" (H. R. Huff and E. Sirtl, eds.), p. 968. The Electrochem. Soc., Princeton, New Jersey.
Fair, R. B. (1981a). *In* "Impurity Doping Processes in Silicon" (F. F. Y. Wang, ed.), p. 315. North-Holland, New York.
Fair, R. B. (1981b). *J. Electrochem. Soc.* **128**, 1360.
Fair, R. B., and Carim, A. (1982). *J. Electrochem. Soc.* **129**, 2319.
Fair, R. B., and Tsai, J. C. C. (1977). *J. Electrochem. Soc.* **124**, 1107.
Fairfield, J. M., and Masters, B. J. (1967). *J. Appl. Phys.* **38**, 3148.
Flynn, C. P. (1972). "Point Defects and Diffusion," p. 486. Oxford Univ. Press (Clarendon), London and New York.
Francis, R., and Dobson, P. S. (1979). *J. Appl. Phys.* **50**, 280.
Frank, F. C., and Turnbull, D. (1956). *Phys. Rev.* **104**, 617.
Frank, W. (1981). *Adv. Solid State Phys.* **21**, 221.
Frank, W., Gösele, U., and Seeger, A. (1980). *In* "Radiation Physics of Semiconductors and Related Materials" (G. P. Kekelidze and V. I. Shakhovtsov, eds.), p. 110. Tbilisi State Univ. Press, Tbilisi, USSR.
Frank, W., Seeger, A., and Gösele, U. (1981). *In* "Defects in Semiconductors" (J. Narayan and T. Y. Tan, eds.), p. 31. North-Holland, New York.
Fuller, C. S., and Ditzenberger, J. A., (1953). *Phys. Rev.* **91**, 193.
Fuller, C. S., and Ditzenberger, J. A. (1956). *J. Appl. Phys.* **27**, 544.
Gaworzewski, P., and Ritter, G. (1981). *Phys. Status Solidi A* **67**, 511.
Ghoshtagore, R. N. (1966). *Phys. Rev. Lett.* **16**, 890.
Ghoshtagore, R. N. (1967). *Phys. Rev.* **155**, 598.
Ghoshtagore, R. N. (1971a). *Phys. Rev. B* **3**, 397.
Ghoshtagore, R. N. (1971b). *Phys. Rev. B* **3**, 389.
Glazman, V. B., and Myaken'kaya, G. S. (1977). *Izv. Akad. Nauk Kuz. SSR, Ser. Fiz. Mat.* **15**, 28.
Gösele, U., and Frank, W. (1981). *In* "Defects in Semiconductors" (J. Narayan and T. Y. Tan, eds.), p. 55. North-Holland, New York.
Gösele, U., and Morehead, F. (1981). *J. Appl. Phys.* **52**, 4617.
Gösele, U., and Strunk, H. (1979). *Appl. Phys.* **20**, 265.
Gösele, U., and Tan, T. Y. (1982). *Appl. Phys. A.* **28**, 79.
Gösele, U., and Tan, T. Y. (1983a). *In* "Aggregation Phenomena of Point Defects in Silicon" (E. Sirtl, ed.), p. 17. The Electrochem. Soc., Pennington, New Jersey.
Gösele, U., and Tan, T. Y. (1983b). *In* "Defects in Semiconductors" (J. W. Corbett and S. Mahajan, eds.), p. 45. North-Holland, New York.
Gösele, U., Frank, W., and Seeger, A. (1980). *Appl. Phys.* **23**, 361.
Gösele, U., Morehead, F., Föll, H., Frank, W., and Strunk, H. (1981a). *In* "Semiconductor Silicon 1981" (H. R. Huff, R. J. Kriegler, and Y. Takeishi, eds.), p. 766. The Electrochem. Soc., Pennington, New Jersey.
Gösele, U., Morehead, F., Frank, W., and Seeger, A. (1981b). *Appl. Phys. Lett.* **38**, 157.
Gösele, U., Frank, W., and Seeger, A. (1982). *Solid State Commun.* **45**, 31.
Graff, K., Gallrath, E., Ades, S., Goldbach, G., and Tölg, G. (1973). *Solid-State Electron.* **16**, 887.
Grienauer, H. S., and Mayer, K. R. (1975). *Inst. Phys. Conf. Ser.* **23**, 550.
Gruzin, P. L. (1952). *Dokl. Akad. Nauk SSSR* **86**, 289.
Haas, C. (1960). *J. Phys. Chem. Solids* **15**, 108.
Hall, R. N., and Racette, J. H. (1964). *J. Appl. Phys.* **35**, 379.

Harris, J. E. (1978). *Acta Metall.* **26**, 1033.
Hashimoto, H., Shibayama, H., Masaki, H., and Ishikawa, H. (1976). *J. Electrochem. Soc.* **123**, 1899.
Hashimoto, H., Shibayama, H., and Ishikawa, H. (1977). *Fujitsu Sci. Tech. J.* (March), 73.
Hettich, G., Mehrer, H., and Maier, K. (1979). *Inst. Phys. Conf. Ser.* **46**, 500.
Hill, C. (1980). Summer Course on Device Impact of New Microfabrication Technologies, Heverlee, Belgium, June 1980.
Hill, C. (1981). In "Semiconductor Silicon 1981" (H. R. Huff, R. J. Kriegler, and Y. Takeishi, eds.), p. 988. The Electrochem. Soc., Pennington, New Jersey.
Hill, M., Lietz, M., and Sittig, T. (1982). *J. Electrochem. Soc.* **129**, 1579.
Hirvonen, J., and Anttila, A. (1979). *Appl. Phys. Lett.* **35**, 703.
Hu, S. M. (1973). In "Atomic Diffusion in Semiconductors" (D. Shaw, ed.), p. 217. Plenum, New York.
Hu, S. M. (1974). *J. Appl. Phys.* **45**, 1567.
Hu, S. M. (1977). *J. Vac. Sci. Technol.* **14**, 17.
Hu, S. M. (1981). In "Defects in Semiconductors" (J. Naryan and T. Y. Tan, eds.), p. 333. North-Holland, New York.
Hu, S. M., Fahey, P., and Dutton, R. W. (1983). *J. Appl. Phys.* **54**, 6912.
Huntley, F. A., and Willoughby, A. F. W. (1970). *Solid-State Electron.* **13**, 1231.
Huntley, F. A., and Willoughby, A. F. W. (1973a). *J. Electrochem. Soc.* **120**, 414.
Huntley, F. A., and Willoughby, A. F. W. (1973b). *Philos. Mag.* **28**, 1319.
Ishikawa, Y., Sakino, Y., Tanaka, H., Matsumoto, S., and Niimi, T. (1982). *J. Electrochem. Soc.* **129**, 644.
Jaccodine, R. J. (1968). *J. Appl. Phys.* **39**, 3105.
Jaccodine, R. J. (1983). In "Defects in Semiconductors" (J. W. Corbett and S. Mahajan, eds.), p. 101. North-Holland, New York.
Jaccodine, R. J., and Drum, C. M. (1966). *Appl. Phys. Lett.* **8**, 29.
James, H. M., and Lark-Horovitz, K. (1951). *Z. Phys. Chem. (Leipzig)* **198**, 107.
Kaiser, W. (1957). *Phys. Rev.* **105**, 1751.
Kalinowski, L., and Seguin, R. (1979). *Appl. Phys. Lett.* **35**, 211.
Kalinowski, L., and Seguin, R. (1980). *Appl. Phys. Lett.* **36**, 171.
Kästner, S., and Hesse, J. (1974). *Phys. Status Solidi A* **25**, 261.
Kendall, D. L., and de Vries, D. (1969). In "Semiconductor Silicon 1969" (R. R. Haberecht and E. L. Kern, eds.), p. 358. The Electrochem Soc., Princeton, New Jersey.
Kimerling, L. C. (1979). *Inst. Phys. Conf. Ser.* **46**, 56.
Kitagawa, H., Hashimoto, K., and Yoshida, M. (1981). *Jpn. J. Appl. Phys.* **20**, 2033.
Kitagawa, H., Hashimoto, K., and Yoshida, M. (1982a). *Jpn. J. Appl. Phys.* **21**, 276.
Kitagawa, H., Hashimoto, K., and Yoshida, M. (1982b). *Jpn. J. Appl. Phys.* **21**, 446.
Kitagawa, H., Hashimoto, K., and Yoshida, M. (1982c). *Jpn. J. Appl. Phys.* **21**, 990.
Kitagawa, H., Hashimoto, K., and Yoshida, M. (1983). *Physica (Amsterdam)* **116B**, 323.
Lambert, J. A., and Dobson, P. S. (1981). *Philos. Mag. [Part] A* **44**, 1031.
Lambert, J. L. (1971). *Phys. Status Solidi A* **4**, K33.
Lambert, J. L. (1972). *Wiss. Ber. AEG-Telefunken* **45**, 153.
Lang, D. V., Grimmeiss, H. G., Meijer, E., and Jaros, M. (1980). *Phys. Rev. B* **22**, 3917.
Lawrence, J. E. (1966). *J. Appl. Phys.* **37**, 4106.
Le Claire, A. D. (1966). *Philos. Mag.* **14**, 1271.
Le Claire, A. D. (1978). *J. Nucl. Mater.* **69 & 70**, 70.
Lecrosnier, D., Gauneau, M., Paugam, J., Pelous, G., Richou, F., and Henoc, P. (1979). *Appl. Phys. Lett.* **34**, 224.

Leroy, B. (1979). *J. Appl. Phys.* **50**, 7996.
Leroy, B. (1982). *J. Appl. Phys* **53**, 4779.
Letaw, H., Jr., Portnoy, W. M., and Slifkin, L. (1956). *Phys. Rev.* **102**, 636.
Lin, A. M., Antoniadis, D. A., and Dutton, R. W. (1981a). *J. Electrochem. Soc.* **128**, 1131.
Lin, A. M., Dutton, R. W., Antoniadis, D. A., and Tiller, W. A. (1981b). *J. Electrochem. Soc.* **128**, 1121.
Mallam, N., Jones, C. L., and Willoughby, A. F. W. (1981). *In* "Semiconductor Silicon 1981" (H. R. Huff, R. J. Kriegler, and Y. Takeishi, eds.), p. 979. The Electrochem. Soc., Princeton, New Jersey.
Masters, B. J. (1971). *Solid State Commun.* **9**, 283.
Masters, B. J., and Fairfield, J. M. (1966). *Appl. Phys. Lett.* **8**, 280.
Masters, B. J., and Gorey, E. F. (1978). *J. Appl. Phys.* **49**, 2717.
Mathiot, D., and Pfister, J. C. (1982a). *J. Appl. Phys.* **53**, 3053.
Mathiot, D., and Pfister, J. C. (1982b). *J. Phys. Lett.* **43**, 453.
Mathiot, D., and Pfister, J. C. (1983). *Physica B (Amsterdam)* **116**, 95.
Matthews, M. D., and Ashby, S. J. (1973). *Philos. Mag.* **27**, 1313.
Mayer, H. J., Mehrer, H., and Maier, K. (1977). *Inst. Phys. Conf. Ser.* **31**, 186.
McVay, G. L., and DuCharme, A. R. (1973). *J. Appl. Phys.* **44**, 1409.
McVay, G. L., and DuCharme, A. R. (1974). *Phys. Rev. B* **9**, 627.
McVay, G. L., and DuCharme, A. R. (1975). *Inst. Phys. Conf. Ser.* **23**, 91.
Mehrer, H. (1978). *J. Nucl. Mater.* **69** & **70**, 38.
Mehrer, H., and Seeger, A. (1972). *Cryst. Lattice Defects* **3**, 1.
Mehrer, H., Seeger, A., and Steiner, E. (1976). *Phys. Status Solidi B* **73**, 131.
Mehrer, H., Maier, K., Hettich, G., Mayer, H. J., and Rein, G. (1978). *J. Nucl. Mater.* **69** & **70**, 545.
Meyberg, W. (1981). Unpublished.
Meyberg, W., Frank, W., Seeger, A., Peretti, H. A., and Mondino, M. A. (1983). *Cryst. Lattice Defects Amorphous Mater.* **10**, 1.
Mikkelsen, J. C. (1982). *Appl. Phys. Lett.* **40**, 336.
Mizuo, S., and Higuchi, H. (1981). *Jpn. J. Appl. Phys.* **20**, 739.
Mizuo, S., and Higuchi, H. (1982a). *Jpn. J. Appl. Phys.* **21**, 281.
Mizuo, S., and Higuchi, H. (1982b). *J. Electrochem. Soc.* **129**, 2292.
Mizuo, S., and Higuchi, H. (1982c). *Denki Kagaku* **50**, 338.
Mizuo, S., and Higuchi, H. (1982d). *Jpn. J. Appl. Phys.* **21**, 56.
Mizuo, S., and Higuchi, H. (1983). *J. Electrochem. Soc.* **130**, 1942.
Morehead, F., Stolwijk, N. A., Meyberg, W., and Gösele, U. (1983). *Appl. Phys. Lett.* **42**, 690.
Mullen, J. G. (1961). *Phys. Rev.* **121**, 1649.
Mundy, J. N. (1971). *Phys. Rev. B* **3**, 2431.
Nabeta, Y., Uno, T., Kubo, S., and Tsukamoto, H. (1976). *J. Electrochem. Soc.* **123**, 1416.
Newman, R. C., and Wakefield, J. (1961a). *In* "Metallurgy of Semiconductor Materials" (J. B. Schroeder, ed.), Vol. 15, p. 201. Wiley (Interscience), New York.
Newman, R. C., and Wakefield, J. (1961b). *J. Phys. Chem. Solids* **19**, 230.
Nicholas, K. H. (1966). *Solid-State Electron.* **9**, 35.
Nobili, D., Armigliato, A., Finetti, M., and Solmi, S. (1982). *J. Appl. Phys.* **53**, 1484.
Ogino, M., Oano, Y., and Watanabe, M. (1982). *Phys. Status Solidi A* **72**, 535.
Patel, J. R. (1981). *In* "Semiconductor Silicon 1981" (H. R. Huff, R. J. Kriegler, and Y. Takeishi, eds.), p. 189. The Electrochem. Soc., Pennington, New Jersey.
Peart, R. F. (1966). *Phys. Status Solidi* **15**, K119.
Pell, E. M. (1960). *Phys. Rev.* **119**, 1014 and 1222.

Peterson, N. L. (1975). In "Diffusion in Solids–Recent Developments" (A. S. Nowick and J. J. Burton, eds.), p. 115. Academic Press, New York.
Peterson, N. L. (1978). J. Nucl. Mater. **69 & 70**, 3.
Petrov, D. A., Shaskov, Yu. M., and Akimchenko, I. P. (1957). Vopr. Met. Fiz. Pomprov. Akad. Nauk SSSR, Tr. Vtorogo Soveshch., 130.
Petroff, P. M., and de Kock, A. J. R. (1975). J. Cryst. Growth **30**, 117.
Prussin, S. (1972). J. Appl. Phys. **43**, 2850.
Rein, G., and Mehrer, H. (1982). Philos. Mag. A **45**, 467.
Rosencher, E., Straboni, A., Rigo, S., and Amsel, G. (1979). Appl. Phys. Lett. **34**, 254.
Sanders, I. R., and Dobson, P. S. (1969). Philos. Mag. **20**, 881.
Saunders, S. R. J. (1976). Sci. Prog. (London) **63**, 163.
Schaake, H. F. (1984). J. Appl. Phys. **55**, 1208.
Scheibe, E., and Schröter, W. (1977). Inst. Phys. Conf. Ser. **31**, 272.
Scheibe, E., and Schröter, W. (1983). Physica (Amsterdam) **116B**, 318.
Seeger, A. (1980). Phys. Status Solidi A **61**, 521.
Seeger, A., and Chik, K. P. (1968). Phys. Status Solidi **29**, 455.
Seeger, A., and Frank, W. (1982). Appl. Phys. A **27**, 171.
Seeger, A., and Gösele, U. (1977). Phys. Lett. **81A**, 423.
Seeger, A., Föll, H., and Frank, W. (1977). Inst. Phys. Conf. Ser. **31**, 12.
Seeger, A., Frank, W., and Gösele, U. (1979). Inst. Phys. Conf. Ser. **46**, 148.
Sharma, B. L. (1970). "Diffusion in Semiconductors." Trans Tech, Clausthal-Zellerfeld, West Germany.
Shaw, D. (1975). Phys. Status Solidi B **72**, 11 (1975).
Shibayama, H., Masaki, H., Ishikawa, H., and Hashimoto, H. (1976). J. Electrochem. Soc. **123**, 743.
Shimizu, H., Yoshinaka, A., and Sugita, S. (1978). Jpn. J. Appl. Phys. **17**, 747.
Sirtl, E. (1977). In "Semiconductor Silicon 1977" (H. R. Huff and E. Sirtl, eds.), p. 4. The Electrochem. Soc., Princeton, New Jersey.
Södervall, U., Roll, U., Predel, B., Odelius, H., Loading, A., and Gust, W. (1983). In "DIMETA 82–Diffusion in Metals and Alloys" (F. J. Kedves and D. L. Beke, eds.). Diffusion and Defect Monograph Series 7, Trans Tech Publications, p. 492.
Sprokel, G. J., and Fairfield, J. M. (1965). J. Electrochem. Soc. **112**, 200.
Stavola, M., Patel, J. R., Kimerling, L. C., and Freeland, P. E. (1983). Appl. Phys. Lett. **42**, 73.
Steigmann, J., Shockley, W., and Nix, F. C. (1939). Phys. Rev. **56**, 13.
Stolwijk, N. A., Schuster, B., Hölzl, J., Mehrer, H., and Frank, W. (1983). Physica (Amsterdam) **116B**, 335.
Strunk, H., Gösele, U., and Kolbesen, B. O. (1979). Appl. Phys. Lett. **34**, 530.
Sturge, M. D. (1959). Proc. Phys. Soc. London **73**, 297.
Sugita, Y., Shimizu, H., Yoshinaka, A., and Aoshima, T. (1977). J. Vac. Sci. Technol. **14**, 44.
Tanako, Y., and Maki, M. (1973). In "Semiconductor Silicon 1973" (H. R. Huff and R. R. Burgess, eds.), p. 469. The Electrochem. Soc., Princeton, New Jersey.
Tan, S. I., Berry, B. S., and Frank, W. (1973). In "Ion Implantation in Semiconductors and Other Materials" (B. L. Crowder, ed.), p. 19. Plenum, New York.
Tan, T. Y., and Ginsberg, B. J. (1983). Appl. Phys. Lett. **42**, 448.
Tan, T. Y., and Gösele, U. (1981). Appl. Phys. Lett. **39**, 86.
Tan, T. Y., and Gösele, U. (1982a). Appl. Phys. Lett. **40**, 616.
Tan, T. Y., and Gösele, U. (1982b). J. Appl. Phys. **53**, 4767.
Tan, T. Y., Gösele, U., and Morehead, F. (1983). Appl. Phys. A **31**, 97.
Taniguchi, K., Karosawa, K., and Kashiwagi, M. (1980). J. Electrochem. Soc. **127**, 2243.

Tiller, W. A. (1980). *J. Electrochem. Soc.* **127,** 625.
Troxell, J. R., Chatterjee, A. P., Watkins, G. D., and Kimerling, L. C. (1979). *Phys. Rev. B* **19,** 5336.
Tseng, W. F., Lau, S. S., and Mayer, J. W. (1978). *Phys. Lett.* **68A,** 93.
Valenta, M. W., and Ramasastry, C. (1957). *Phys. Rev.* **106,** 73.
van den Maesen, F., and Brenkman, J. A. (1955). *J. Electrochem. Soc.* **10,** 229.
van Ommen, A. H. (1983). *J. Appl. Phys.* **54,** 5055.
van Vechten, J. A. (1978). *Phys. Rev. B* **17,** 3197.
van Vechten, J. A., and Thurmond, C. D. (1976). *Phys. Rev. B* **14,** 3551.
Vineyard, G. H. (1957). *J. Phys. Chem. Solids* **3,** 121.
Vogel, G., Hettich, G., and Mehrer, H. (1983). *J. Phys. C* **16,** 6197.
Vorob'ev, V. H., Murav'ev, V. A., and Panteleev, V. A. (1981). *Sov. Phys. Solid State* **23,** 2055.
Waite, T. R. (1957). *Phys. Rev.* **107,** 463.
Watkins, G. D. (1981). *In* "Defects in Semiconductors" (J. Narayan and T. Y. Tan, eds.), p. 21. North-Holland, New York.
Watkins, G. D., and Brower, K. L. (1976). *Phys. Rev. Lett.* **36,** 1329.
Watkins, G. D., and Corbett, J. W. (1964). *Phys. Rev. A* **134,** 1359.
Watkins, G. D., Corbett, J. W., and Walker, R. M. (1959). *J. Appl. Phys.* **30,** 1198.
Watkins, G. D., Messmer, R. P., Weigel, C., Peak, D., and Corbett, J. W. (1971). *Phys. Rev. Lett.* **27,** 1573.
Watkins, G. D., Troxell, J. D., and Chatterjee, A. P. (1979). *Inst. Phys. Conf. Ser.* **46,** 16.
Weber, E. R. (1983). *Appl. Phys. A* **30,** 1.
Weertman, J. (1957). *Phys. Rev.* **107,** 1259.
Weiler, D., and Mehrer, H. (1983). *Philos. Mag. A* **49,** 309.
Werner, M. (1984). Ph.D. thesis, University of Stuttgart, Stuttgart, Federal Republic of Germany.
Werner, M., and Mehrer, H. (1983). *In* "DIMETA 82—Diffusion in Metals and Alloys" (J. F. Kedves and D. L. Beke, eds.), Diffusion and Defect Monograph Series 7, Trans Tech Publications, p. 393.
Werner, M., Mehrer, H., and Hochheimer, H. D. (1983). *Verh. Dtsch. Phys. Ges. (VI)* **18,** 761.
Werner, M., Mehrer, H., Siethoff, H. (1983). *J. Phys. C* **16,** 6185.
Widmer, H., and Gunther-Mohr, G. R. (1961). *Helv. Phys. Acta* **34,** 635.
Wilcox, W. R., and LaChapelle, T. J. (1964). *J. Appl. Phys.* **35,** 240.
Wilcox, W. R., LaChapelle, T. J., and Forbes, D. H. (1964). *J. Electrochem. Soc.* **111,** 1377.
Willoughby, A. F. W. (1977). *J. Phys. D* **10,** 455.
Willoughby, A. F. W. (1978). *Rep. Prog. Phys.* **41,** 1665.
Willoughby, A. F. W. (1981). *In* "Impurity Doping Processes in Silicon" (F. F. Y. Wang, ed.), p. 1. North-Holland, New York.
Wills, G. N. (1969). *Solid-State Electron.* **12,** 133.
Wu, W.-K., and Washburn, J. (1977a). *J. Appl. Phys.* **48,** 3742.
Wu, W.-K., and Washburn, J. (1977b). *J. Appl. Phys.* **48,** 3747.
Yoshida, M. (1973). *Jpn. J. Appl. Phys.* **12,** 1956.
Yoshida, M. (1979). *Jpn. J. Appl. Phys.* **18,** 479.
Yoshida, M. (1980). *Jpn. J. Appl. Phys.* **19,** 2427.
Yoshida, M. (1983). *Jpn. J. Appl. Phys.* **22,** 1404.
Yoshida, M., and Saito, K. (1967). *Jpn. J. Appl. Phys.* **6,** 573.

3

Atom Transport in Oxides of the Fluorite Structure

A. S. NOWICK

HENRY KRUMB SCHOOL OF MINES
COLUMBIA UNIVERSITY
NEW YORK, NEW YORK

I.	Introduction	143
II.	Diffusion Studies	145
	A. Cationic Diffusion and Ordering	146
	B. Oxygen Diffusion	149
III.	Conductivity and Relaxation	152
	A. Theory of Ionic Conductivity	152
	B. Theory of Dielectric and Anelastic Relaxation	158
	C. Studies of Conductivity and Relaxation	161
IV.	Comparison of Ionic Conductivity and Oxygen Diffusion	183
	References	185

I. Introduction

The purpose of this chapter is to examine how defect properties and defect interactions control mass transport in a well-studied and relatively simple family of ionic systems, viz., oxides (MO_2) that possess the fluorite structure. This structure is a relatively open one; it shows exceptional tolerance for high levels of disorder, which may be introduced either by dopants or by reduction or oxidation (i.e., departures from stoichiometry). These oxides also have a wide variety of actual and potential technological applications: ZrO_2, ThO_2, and CeO_2 are of interest as electrolytes in high-temperature fuel cells, water electrolyzers, oxygen sensors, and automobile exhaust emission controls, while UO_2 is important as a nuclear reactor fuel. It will be shown that the state of our knowledge of atomic transport in these systems depends not only on diffusion studies, but also greatly on conductivity and relaxation studies, which usually are carried out at lower temperatures than diffusion. We will concentrate on investigations that reveal the principles and will not attempt to present a comprehensive review of available data. Such reviews are already available (Etsell and Flengas, 1970; Freer, 1980; Matzke, 1981).

We shall begin by reviewing the crystal structure and basic defect structure of these oxides. The unit cell of the fluorite structure may be regarded as made up of eight small cubes containing anions at their corners, with every alternate cube having a cation at its center. Figure 1 (regarded for the moment without defects present) shows one-half the unit cell. Of the binary oxides, ThO_2, CeO_2, PrO_2, UO_2, and PuO_2 possess this structure in the pure state, but PrO_2 is difficult to obtain in a composition close to stoichiometry. Because these oxides have very high melting points (2000–3000°C), it is very difficult to study them in an intrinsic–stoichiometric range in order to determine the intrinsic defect. Nevertheless, by analogy with the well-studied fluorides of the fluorite structure (e.g., CaF_2, SrF_2, etc.), where it is well established that the anion Frenkel defect is the intrinsic defect (Lidiard, 1975), there is good reason to expect the same for the oxides. In fact, computer simulation-type calculations for UO_2 (Catlow and Lidiard, 1974; Catlow, 1977) firmly support this conclusion.[1] Thus the basic intrinsic defects are, in Kröger–Vink notation, $O_i'' + V_O^{..}$.

Because nominally "pure" samples tend to be dominated by unknown impurities, the oxides are better characterized with regard to defects when they are either deliberately doped or grossly nonstoichiometric. Thus lower-valent cation dopant ions, designated D^{2+} or D^{3+}, should introduce oxygen-ion vacancies V_O. Numerous studies of x ray lattice parameters in combination with macroscopic density measurements show that this expectation is indeed correct (Etsell and Flengas, 1970). Similarly, departure from stoichiometry to the hypostoichiometric side (i.e., MO_{2-x}) produces V_O defects, while on the hyperstoichiometric (MO_{2+x}) side, O_i defects are anticipated. The latter has been confirmed by density measurements (see Belle, 1961) and by neutron diffraction studies on UO_{2+x} (Willis, 1963, 1978). These results are fully consistent with the anion Frenkel as the intrinsic defect.

For the doped oxides, as the doping level becomes sufficiently high (≥ 15 mole % dopant) various types of ordered structures may be produced (Carter and Roth, 1968; Rossell, 1976, 1981; Rossell and Scott, 1977). This will be discussed later in greater detail.

Finally, the oxides ZrO_2 and HfO_2, which do not possess the fluorite structure at ordinary temperatures in the pure stoichiometric state, take on

---

[1] There will be several references to computer simulation calculations in this chapter. These calculations use the generalization Mott–Littleton procedures that have been incorporated into the HADES and related computer programs. A detailed discussion of the techniques and their applications is given by Catlow and Mackrodt (1982).

3. ATOM TRANSPORT IN OXIDES OF THE FLUORITE STRUCTURE

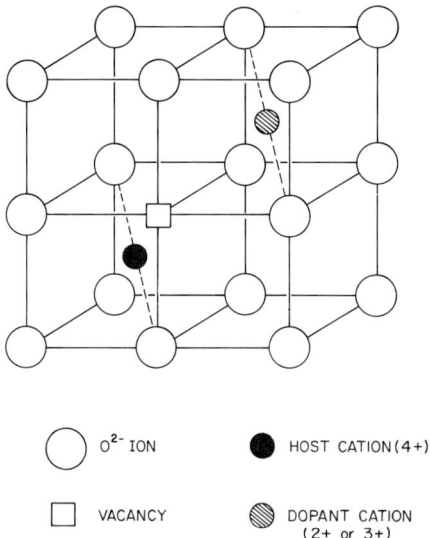

Fig. 1. Diagram of one-half the unit cell of a fluorite-type oxide showing a dopant cation and an oxygen ion vacancy as a nearest-neighbor.

this structure when doped with lower-valent cations to the extent of $\gtrsim 8$ mole %. Particularly well known are the solid solutions $Zr_{1-x}Ca_xO_{2-x}$, called calcia-stabilized zirconia (CSZ), and the corresponding solution yttria-stabilized zirconia (YSZ).

II. Diffusion Studies

In discussing diffusion studies we shall confine ourselves almost exclusively to tracer diffusion, where, assuming a single dominant defect mechanism, the tracer diffusion coefficient D^* takes the form

$$D^* = \tfrac{1}{6}\Gamma^*\lambda^{*2}f_d^* \qquad (1)$$

Here Γ^* is the jump frequency of tracer atoms, λ^* their jump distance, and f_d^* the tracer correlation factor for diffusion via defect d (Manning, 1968; Le Claire, 1970). For self diffusion at low defect concentrations f_d^* is a pure number; For example, for anion diffusion by a vacancy mechanism $f = 0.653$. The jump frequency is given by

$$\Gamma^* = c_d z w_d \qquad (2)$$

where c_d is the mole fraction of defects (probability of a defect on an adjacent site), z the number of adjacent defect sites, and w_d the jump frequency of the defect between two specified sites, given by

$$w_d = v_0 \exp(-g_d^m/kT) \tag{3}$$

Here v_0 is the attempt frequency and g_d^m the Gibbs free energy of activation for motion of the defect

$$g_d^m = h_d^m - Ts_d^m \tag{4}$$

where h_d^m and s_d^m are the corresponding enthalpy and entropy terms.

The above equations show that the temperature dependence of D^* involves not only the activation enthalpy for motion h_d^m, but also terms coming from a dependence of c_d on temperature. In a high-temperature intrinsic stoichiometric range, c_d should depend exponentially on temperature, while in other regions it may be fixed at a value determined by doping or by nonstoichiometry.

In order to evaluate c_d for various situations, it is necessary to make use of the appropriate mass action relations for defect equilibrium. As already mentioned, the predominant defect is the anion Frenkel (denoted Fa), and its corresponding mass action equation is

$$c_{V_O} c_{O_i} = K_{Fa} \propto \exp(-h_{Fa}/kT) \tag{5}$$

where h_{Fa} is the enthalpy of formation. Similarly, for the Schottky equilibrium, we have

$$c_{V_M} c_{V_O}^2 = K_S \propto \exp(-h_S/kT) \tag{6}$$

and for the cation Frenkel (Fc)

$$c_{V_M} c_{M_i} = K_{Fc} \propto \exp(-h_{Fc}/kT) \tag{7}$$

A. Cationic Diffusion and Ordering

It is consistently found that for the fluorite oxides the cationic diffusivity D_M^* is many orders of magnitude lower than the oxygen diffusivity D_O^* (Etsell and Flengas, 1970; Freer, 1980). In fact, because of the low values of D_M^* at all except very high temperatures and the general use of sintered polycrystalline samples, cationic diffusion measurements offer numerous problems (e.g., short-circuiting paths, evaporation loss of tracers) to the extent that Matzke (1981) has concluded that most such data must be rejected, or taken as only an upper limit. The most reliable work on UO_2 gives 5.6 eV and for ThO_2 6.5–7.0 eV, while recent work on Er_2O_3-stabilized HfO_2 of different compositions gives 6–7 eV for the activation enthalpy h_M^* for diffusion of cation tracers (Matzke, 1981, 1983; Tesch et al., 1982). Such values are extremely high compared with the usual activation enthalpies for diffusion, and certainly relative to those for oxygen diffusion.

To obtain further information it is helpful to study D_M^* as a function of departure from stoichiometry x in $MO_{2\pm x}$. Such a study is particularly possible for UO_2, which departs from stoichiometry in both the hypo and hyper directions. Uranium dioxide (UO_2) and $U(Pu)O_2$ have also been much studied because of their interest as nuclear fuels. Figure 2 shows the results of such a study, in which D_{Pu}^* decreases with decreasing x, for $x > 0$ reaching a minimum at $x = -0.02$ (O/M ratio of 1.98) and then increasing again. They can be interpreted in terms of mass action considerations, together with the expectation that V_M is more mobile than M_i. Thus for $x > 0$, O_i defects are introduced in proportion to the departure from stoichiometry. From Eqs. (5) and (6), enhancement of c_{O_i} results in an increase in c_{V_M}, so the increase in D_M^* with increasing x is readily understood. In the case of hypostoichiometry, V_O defects predominate, giving rise to a suppression of c_{V_M} and, by Eq. (7), an enhancement of c_{M_i}. Because the mobility of cation vacancies is greater than that of the interstitials, the former continue to predominate until as x decreases the product $c_{V_M} w_{V_M}$ becomes equal to $c_{M_i} w_{M_i}$, at which point D_M^* goes through a minimum at x_{min}. The corresponding activation enthalpy h_M^* should then be $h_S - 2h_{Fa} + h_{V_M}^M$ for the region $x > 0$, $h_S - h_{Fa} + h_{V_M}^m$ for $x = 0$, $h_S + h_{V_M}^m$ for $x_{min} < x < 0$, and, finally, $h_{Fc} - h_S + h_{M_i}^m$ for $x < x_{min}$. The assumption that $w_{V_M} \gg w_{M_i}$ is

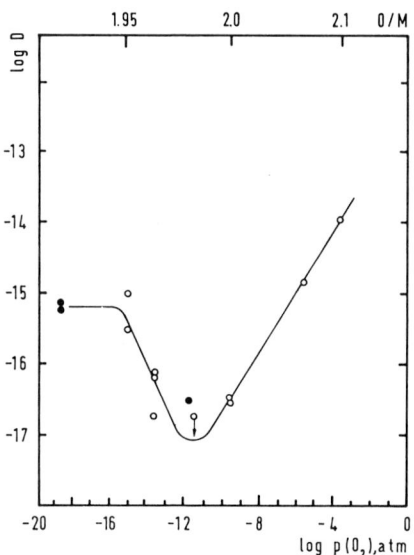

Fig. 2. Dependence of the tracer diffusion coefficient D_{Pu}^* on oxygen-to-metal (O/M) ratio [and also on $p(O_2)$] in $(U_{0.8}Pu_{0.2})O_{2\pm x}$ at 1500°C (○) CO/CO_2; (●) H_2. [From Matzke (1981).]

further supported by computer simulation calculations (Catlow, 1977). The above interpretation also leads to the conclusion [from Eqs. (5) and (6)] that D_U^* should vary as x^2 in the region UO_{2+x}, which was confirmed by Matzke (1976). It also leads to the conclusion that h_{Fa} is one-half the difference between the activation enthalpies in the near hypo- and the hyperstoichiometric regions. This gives (Matzke, 1981) $h_{Fa} \cong 2.5$ eV. Catlow (1977) predicts, however, a value ~ 5 eV for h_{Fa}.

To reconcile the data with his calculations, Catlow has suggested that the minimum of D_M^* is really *at* stoichiometry (due to difficulties in measuring the exact stoichiometric point)[2] and that the change from vacancy to interstitialcy mechanism occurs in passing through $x = 0$. Accordingly, the difference in activation enthalpies between $x > 0$ and the region of the minimum is h_{Fa} rather than $\frac{1}{2}h_{Fa}$, in better agreement with his calculations. This controversy in the interpretation of the minimum in D_M^* still needs to be resolved by further work. We shall see, however, that further information on h_{Fa} is obtainable from studies of oxygen diffusion (Section II.B).

Tesch *et al.* (1982) carried out careful tracer diffusion measurements of both Er and Hf in Er_2O_3-stabilized HfO_2 over the range 10–40 mole % Er_2O_3. In this work, grain boundary diffusion effects were separated from bulk diffusion. For bulk diffusion, values of $h_M^* \sim 6$–7 eV were obtained both for Hf and Er, and in neither case was there a strong concentration dependence of D_M^*. These results suggest that in this range of composition, >10 mole %, no longer are there simple mass-action type dependences of D_M^* on defect concentrations. Instead, the authors consider that extensive clustering of defects into microdomains must be operative. Nevertheless, the presence of high c_{V_O} must lead to a strong suppression of c_{V_M}, by Eq. (6), and, therefore, to abnormally low values of the preexponential D_{M_0} in the Arrhenius expression

$$D_M^* = D_{M_0} \exp(-h_M^*/kT) \qquad (8)$$

Instead, anomalously large values are obtained for D_{M_0}, typically 10^4–10^6 cm^2 s^{-1}. (The usual values obtained by substituting for Γ^* into Eq. (1) are ~ 0.01–0.1 cm^2 s^{-1}). The question of why these high preexponential values appear is not dealt with by Tesch *et al.* (1982).

Still another manifestation of cation migration is in the process of ordering. As mentioned earlier, ordered structures are often observed in heavily doped fluorite oxides. Since oxygen ions are much more mobile than cations, the motion of the latter must determine the ordering rate. Some authors have followed the process of ordering in CSZ for $T > 1000°C$ through the

---

[2] Matzke (1983a) believes, however, that the occurrence of such an error in the stoichiometry is out of the question.

decrease in the electrical conductivity with time (e.g., Tien and Subbarao, 1963; Subbarao and Sutter, 1964). The kinetics do not have the simple form of, say, an exponential decay in time. Nevertheless, the principal changes occur in $\sim 10^4$ sec at 1000°C, or in $\sim 10^3$ s at 1200°C. Assuming that, on average, one cation jumps in this period, and employing Eq. (3) with the usual preexponential ($v_0 \exp s_d^m/k \sim 10^{14}$ s^{-1}) we obtain an activation enthalpy $\sim 5 \pm 0.5$ eV, in reasonable agreement with self-diffusion values.

Because the ordering kinetics are so slow, it is difficult to attain true equilibrium of these oxides in temperature ranges below ~ 1200°C. Therefore, it is reasonable to regard the cation sublattice as essentially frozen-in below 1000°C, in a partially ordered state. For the more dilute solid solutions, the assumption that cation dopants are almost randomly distributed may then be expected to be a good approximation. We will see later that it is.

B. Oxygen Diffusion

There is no long-lived radioactive isotope of oxygen, but, nevertheless, self diffusion can be carried out using the stable isotopes ^{17}O and ^{18}O. The methods may involve (a) measurement of the depletion of isotope from an enriched gas, (b) isotope exchange followed by mass-spectrographic determination of isotope content at various depths in the sample, or (c) nuclear reaction depth profiling (Marin and Contamin, 1969). Data on oxygen diffusion exist for UO_2, PuO_2, $(U, Pu)O_2$, CeO_2, ThO_2, and stabilized zirconia, and have been well reviewed (Matzke, 1981; Freer, 1980; Murch, 1983). Some of these systems have been studied in the pure and stoichiometric state, some nonstoichiometric ($MO_{2\pm x}$), and some doped with lower-valent cations. The most widely studied material has been UO_2 in stoichiometric as well as hyper- and hypostoichiometric states. We will first deal with this material. For stoichiometric UO_2, Matzke (1981) gives as the best D_O^* value

$$D_O^* = 0.26 \exp(-2.58 eV/kT) \text{ cm}^2 \text{ s}^{-1}$$

If this truly represents the intrinsic range, and the anion Frenkel is the dominant intrinsic defect, then $h_O^* = \frac{1}{2}h_{Fa} + h^m$, where h^m is the motion energy of the faster migrating of V_O and O_i, which are generated in equal numbers. Based on analogy with the fluorides as well as computer calculations, it is concluded that V_O migrates faster.

Murch and Thorn (1978) have calculated D_O^* values for various x in $UO_{2\pm x}$ using the statistical mechanical formalism of Thorn and Winslow (1966), which includes contributions from both vacancy and interstitialcy mechanisms. They chose values $h_{Fa} = 5$ eV, $h_{V_O}^m = 0.25$ eV, and $h_{O_i}^m = 1$ eV. Their calculated results for the hyperstoichiometric region are shown in

Fig. 3 along with most of the available data. Figure 3 shows reasonable agreement in both stoichiometric and nonstoichiometric ranges, but also shows the large extent of scatter among various investigators. An interesting result of the calculations is that between the two ranges a transition region appears, where the slope becomes even steeper than the stoichiometric slope. This is due to the switch from interstitial dominance in the hyperstoichiometric range to vacancy dominance in the stoichiometric range. The effect has not yet been observed experimentally for UO_2, but such behavior occurs in AgCl doped with divalent cations (Corish and Jacobs, 1972).

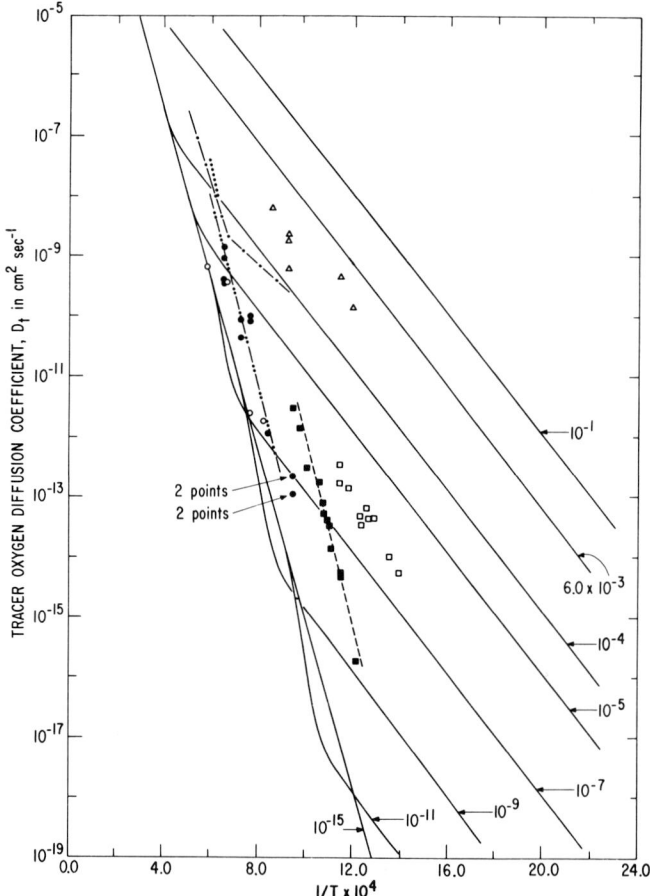

Fig. 3. The calculated temperature dependence of the oxygen tracer diffusion coefficient in UO_{2+x} for various values of x (as labeled on the curves) and comparison with experimental data from several sources. [From Murch and Thorn (1978).]

Computer simulation calculations by Catlow (1977) for UO_2 give $h_{Fa} = 5$ eV and $h_{V_O} \sim 0.25$ eV. These are the same as the values chosen by Murch and Thorn (1978), and they are also consistent with the intrinsic value $h_O^* = 2.58$ eV given above. However, Matzke (1981) argued that all other fluorite oxides of similar melting points have substantially higher values of $h_{V_O}^m$, in the range 0.6–1.2 eV (as we shall see below), and therefore questioned a value as low as 0.25 eV for UO_2. In fact, he suggested a higher motion energy and a value $h_{Fa} \sim 3$ eV, as also obtained from his interpretation of the cationic diffusion (Fig. 2). A study of oxygen diffusion in the hypostoichiometric range (UO_{2-x}) by Kim and Olander (1981) gave a value of 0.51 eV, presumably representing $h_{V_O}^m$. This then suggests that $h_{Fa} \cong 4.2$ eV, a compromise between Matzke's and Catlow's values.

One complication in all of the above is the role of clusters, as opposed to elementary point defects. On the interstitial side, there is the well-established 2:2:2 cluster (Willis, 1963, 1964, 1978; Catlow, 1973), which is an interstitial dimer stabilized by additional large displacements. It is difficult to consider the effects of such clusters on diffusion, though such calculations have been attempted (de Bruin and Murch, 1973). In hyperstoichiometric UO_2, the movement of clusters is probably also associated with hole hopping (U^{5+} or U^{6+} ions).

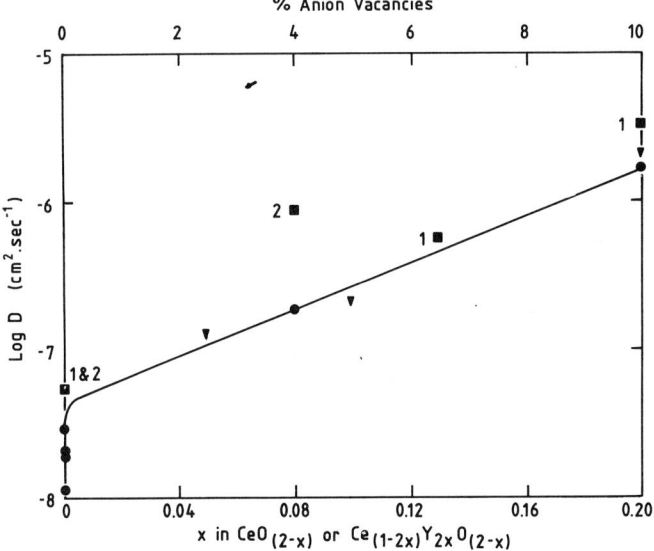

Fig. 4. Variation of oxygen diffusion coefficient with x both in reduced ceria, CeO_{2-x}, and in yttria-doped ceria $Ce_{1-2x}Y_{2x}O_{2-x}$ at 1100°C: (●) ceria, single crystal; (■) ceria, polycrystalline (two preparations); (▼) ceria-yttria, polycrystalline. [From Steele and Floyd (1971).]

For fluorite oxides other than UO_2, it is doubtful if true stoichiometric behavior can be observed, because at high temperatures the materials depart from stoichiometry. The best information is probably that available from distinctly nonstoichiometric or doped materials.

In the case of CeO_2, a study of oxygen diffusion was made by Steele and Floyd (1971) (see also Floyd, 1973), both for CeO_{2-x} and $Ce_{1-2x}Y_{2x}O_{2-x}$. Figure 4 shows that there is remarkably good agreement when both sets of data are plotted versus x (or versus the percent of anion vacancies). For the 5% Y_2O_3 ($x = 0.048$) material, they obtain an activation enthalpy of $h_O^* = 0.83$ eV and preexponential of 1.5×10^{-4} cm^2 s^{-1}. Similar values are obtained for the other Y^{3+}-doped samples. On the other hand, for CeO_{2-x} much lower values of h_O^* are reported, especially for $x = 0.2$, where h_O^* and the preexponential are both extremely low. On this basis, the agreement in Fig. 4 at 1100°C between pure ceria and ceria-yttria would have to be regarded as fortuitous. Examination of the Arrhenius plots (Floyd, 1973) for large x shows that the conclusions about their slopes rest on very little data with large scatter. Therefore, it seems best to keep open the question of activation enthalpies of the reduced ceria until further work is done. In a later section (Section IV), we shall compare Floyd's diffusion results with conductivity data on $Ce(Y)O_{2-x}$.

Several investigations on CSZ, viz., $Zr_{0.85}Ca_{0.15}O_{1.85}$, show that $h_O^* = 1.2–1.35$ eV (Kingery et al., 1959; Simpson and Carter, 1966). Presumably, then, for zirconia as well as for ceria, $h_{V_O}^m$ falls in the range of 1 eV. It is this relatively low $h_{V_O}^m$ that is responsible for the fact that these materials are fast ionic conductors at elevated temperatures.

III. Conductivity and Relaxation

In view of the limitations of oxygen diffusivity measurements, we now turn to the use of ionic conductivity and of dielectric and anelastic relaxation to obtain further information about defects and oxygen transport in these oxides. Ionic conductivity studies must be confined to fluorite oxides doped with aliovalent impurities (D^{2+} or D^{3+}) so as to introduce oxygen-ion vacancies by charge compensation. [Undoped materials generally show a dominant electronic conductivity if they become even slightly nonstoichiometric (Tuller, 1981).] We begin by reviewing the necessary theoretical background.

A. Theory of Ionic Conductivity

The ionic conductivity σ is, in general, given by a sum of contributions from various ionic carriers i having number density n_i, charge q_i, and mobility (drift velocity per unit electric field) μ_i, such that

$$\sigma = \sum_i n_i q_i \mu_i \tag{9}$$

Usually, one carrier predominates, in our case the oxygen-ion vacancy (henceforth denoted by V rather than V_O), so that

$$\sigma = c_V N_M q_V \mu_V \tag{10}$$

where c_V is the mole fraction of these vacancies and N_M the number of cation sites per unit volume. The mobility μ_V is given by

$$\mu_V = q_V \lambda_V^2 \Gamma_V f_I / 6kT \tag{11}$$

Here λ_V is the jump distance of the vacancy [which is the same as λ^* of the tracer, Eq. (1)], and Γ_V its jump frequency given by

$$\Gamma_V = zw_V \tag{12}$$

in which z is the anion coordination number of the vacancy (here $z = 6$) and w_V the specific jump frequency given by Eq. (3). Equation (11), without the factor f_I, is well known (Lidiard, 1957). The quantity f_I is the "conductivity correlation factor," first introduced by Sato and Kikuchi (1971) and recently reviewed in detail by Murch (1982). This factor deviates from unity only at high defect concentrations, where local order is present. It originates in the fact that, following a given jump which lowers the local order, subsequent jumps will preferentially act to reverse that jump. We define a diffusivity obtained from conductivity by

$$D_\sigma \equiv kT\sigma/N_M q_V^2 \tag{13}$$

Comparing this with Eq. (1) for tracer diffusion, incorporating Eqs. (2), (11), and (12), yields

$$D^*/D_\sigma = f_V^*/f_I \equiv H_R \tag{14}$$

where H_R is called the Haven ratio. At low defect concentrations, where $f_I = 1$, D^* and D_σ differ only by the correlation factor f_V^*, which equals 0.653. Thus conductivity measurements can be used to obtain diffusion information, and over a much wider temperature range than direct diffusion measurements. There is one important precaution, however. If there exists another mechanism that contributes to either D^* or σ and not to the other, the above equations will not be valid. Thus neutral defect pairs may contribute to D^* and not to σ, or electronic conduction may contribute to σ and not to D^*. In order to employ Eqs. (13) or (14) one must be sure that such complexities are not present.

In studying conductivity, a useful expression is

$$\sigma T = (N_M q_V^2 f_1 \lambda_V^2 / 6k) c_V \Gamma_V \tag{15}$$

Then, inserting $\lambda_V = a$ and $N_M = \frac{1}{2}a^{-3}$ [where a is half the lattice parameter of the fluorite unit cell, $\Gamma_V = 6w_V$ with w_V given by Eq. (3), $f_1 = 1$, and $q_V = 2e$ (where e is the electronic charge)], we obtain

$$\sigma T = (2e^2/ka) v_0 c_V \exp(-q_V^m/kT) \tag{16}$$

Equation (16) is the basis of the "conductivity plot" of $\ln \sigma T$ versus $1/T$. This plot often takes the empirical (Arrhenius) form

$$\sigma T = A \exp(-h_\sigma/kT) \tag{17}$$

where A is the preexponential and h_σ the conductivity activation enthalpy. Comparing Eqs. (16) and (17) shows that h_σ depends on the variation of c_V with temperature. Generally, we define three ranges (Dreyfus and Nowick, 1962): At highest temperatures region I, the *intrinsic* range, is obtained. Here $c_V \propto \exp(-h_{Fa}/2kT)$. Because of the high melting point and the presence of impurities, this region is not observed for the doped oxides. At intermediate temperatures one obtains region II, the *extrinsic dissociated* range, in which all oxygen vacancies which compensate the dopant are available for conduction. Here c_V is a constant, equal to the dopant concentration c_0. Accordingly, h_σ is given by

$$h_\sigma^{II} = h_V^m \tag{18}$$

and the preexponential A by

$$A^{II} = (2e^2 v_0 c_0/ka) \exp(s_V^m/k) \tag{19}$$

Finally, at lower temperatures impurity–vacancy association takes place, and so the concentration of free vacancies is determined by the association equilibrium. In stage III, the *extrinsic associated* range, association is nearly complete and an extra enthalpy term enters into h_σ from the association reaction. The three ranges are shown schematically in a conductivity plot in Fig. 5.

The behavior in stage III is different for the two cases of divalent and trivalent dopant ions. For dopant D^{2+}, which in the lattice is effectively doubly negatively charged (D_M''), one obtains association with the positively charged vacancy to form neutral pairs:

$$D_M'' + V_O^{\cdot\cdot} \rightleftarrows (D_M V_O)^x \tag{20}$$

On the other hand, for D^{3+} dopants, only one vacancy is generated for two

3. ATOM TRANSPORT IN OXIDES OF THE FLUORITE STRUCTURE

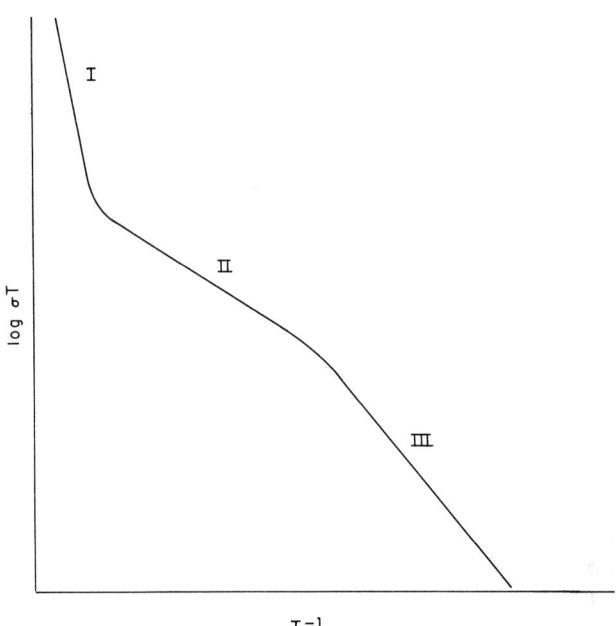

Fig. 5. Schematic conductivity plot (log σT versus $1/T$) showing stages I (intrinsic), II (extrinsic, dissociated), and III (extrinsic, associated).

dopant ions (D'_M) Because the atomic mobility of the dopant cations is frozen in at relatively high temperatures (see Section II.A), we expect to find, basically, a random distribution of D^{3+} ions, particularly for dilute systems. Association, under the action of Coulomb forces, can then give rise only to incompletely compensated (DV_0) pairs, i.e., each pair carries a net charge which is compensated by an unassociated D'_M located elsewhere. For both types of dopants, the association reaction leads to the mass-action equation

$$(c_D c_V)/c_P = K_A = \tfrac{1}{4}\exp(-g_A/kT) \tag{21}$$

where c_D is the concentration of unassociated dopants, c_P that of pairs, K_A the association equilibrium constant, and g_A the free energy of association, given, in the usual way, by

$$g_A = h_A - Ts_A \tag{22}$$

To Eq. (21) must be added equations of dopant and charge conservation. These take different forms for the cases of divalent and trivalent dopants as follows, where the D^{2+} case appears on the left and the D^{3+} case appears on the right:

	D^{2+} case	D^{3+} case	
	$c_D + c_P = c_0$	$c_D + c_P = 2c_0$	(23)
	$c_V = c_D$	$2c_V + c_P = c_D$	(24)

where c_0 is the total mole fraction of DO or D_2O_3, respectively. From Eqs. (21), (23), and (24), we may eliminate c_P and c_D to obtain an equation for c_V alone.

$$\frac{c_V^2}{(c_0 - c_V)} = K_a \qquad \frac{c_V(c_0 + c_V)}{(c_0 - c_V)} = K_A \qquad (25)$$

For stage III, almost complete association, $c_V/c_0 \ll 1$ and so

$$c_V = (K_A c_0)^{1/2} \qquad c_V = K_A \qquad (26)$$

Using the expression for K_A in Eq. (21) as well as Eq. (22) we obtain the following equations for h_σ and A in stage III:

$$h_\sigma^{III} = h_V^m + \tfrac{1}{2} h_A \qquad h_\sigma^{III} = h_V^m + h_A \qquad (27)$$

$$A^{III} = e^2 v_0' c_0^{1/2}/ka \qquad A^{III} = e^2 v_0'/2ka \qquad (28)$$

where

$$v_0' = v_0 \exp[(s_V^m + rs_A)/k] \qquad (29)$$

in which $r = \tfrac{1}{2}$ for D^{2+}, and $r = 1$ for D^{3+}. Note that for D^{2+} dopant, $c_V \propto c_0^{1/2}$ in stage III and, therefore, we expect that $\sigma \propto c_0^{1/2}$ in this range. For D^{3+} dopant, on the other hand, we obtain the surprising result that σ is independent of c_0 in stage III. It is also noteworthy that h_σ^{III} includes only half of the association enthalpy h_A for D^{2+}, but the full h_A for D^{3+}.

To evaluate the preexponentials A^{III}, all factors are known except v_0', particularly the entropy factor in Eq. (29). However, usually such factors are ~ 10, so that it is reasonable to take $v_0' \sim 3 \times 10^{13}$ s^{-1}. Then, using a typical half-lattice parameter $a = 2.7 Å$, we estimate

$$A^{III} \sim 2 \times 10^6 c_0^{1/2} \qquad A^{III} \sim 1 \times 10^6 K/\Omega \, cm \qquad (30)$$

for the two cases.

Actually, with decreasing temperature the transition from stages II to III is a very gradual one, and it is often desirable to fit the data using the complete formula [Eq. (25)] rather than simply laying down a ruler and attributing the slope to Eq. (27). For example, it is easy to show that for the conductivity to display almost no departure (say, $<20\%$) from the stage III formula for temperatures up to $1000/T = 1.0$, for either D^{2+} or D^{3+} doping, one requires that $h_A > 0.4$ eV for $c_0 = 0.01$ or >0.6 eV for $c_0 = 0.001$.

Thus only for relatively large association enthalpies can we avoid using the full equation [Eq. (25)] in the temperature range in which data are usually obtained.

Thus far we have only discussed doping with lower-valent cations, which are compensated by oxygen vacancies. No mention has been made of the possibility of doping with higher valent cations, e.g., D^{5+}, the reason being that there is almost no literature dealing with such cases (however, see Naik and Tien, 1979). Nevertheless, it is worth pointing out what can be learned from such studies. Doping with D^{5+} cations must inevitably lead to the introduction of oxygen-ion interstitials O_i'' as the compensating defect. For every molecule of D_2O_5 there would be two $D_M^{\cdot}$ and one O_i''. The situation should then be analogous to the D^{3+} case, with the formation of $(DO_i)'$ pairs by association, and an enthalpy in the stage III of

$$h_\sigma^{III} = h_i^m + h_A \tag{31}$$

where h_A is the association energy of the pair and h_i^m the motion energy of the oxygen interstitial. Based on our knowledge of fluorides that have the same structure, we can expect that $h_i^m > h_V^m$ or that $\mu_i \ll \mu_V$.

Any treatment based on the mass action principle, such as that given above, cannot be expected to apply beyond defect concentrations $\sim 1\%$, if that far. In addition, this treatment has limited itself to considering only the strong nearest-neighbor interactions involved in pair formation, but otherwise no interactions at all. Yet Coulomb interactions are long range and therefore should not be negligible even in dilute systems. Lidiard (1954, 1957) suggested how these might be taken into account by adopting the Debye–Hückel (DH) theory designed for liquid electrolytes. (See, for example, Fowler and Guggenheim, 1949). In Lidiard's approach, close pairs are still treated as a distinct defect species, but the defects at larger separations are treated by the DH approach. In this theory, Coulomb interactions lead to an average distribution of oppositely charged defects about any given charged defect, which is just sufficient to screen its Coulomb field. The key parameter is the DH screening length κ^{-1}, which is determined by the concentration of all the charged species. The qualitative effect is to lower defect formation or association energies by small amounts, so that the concentration of charged defects reaches values higher than that given by the simple theory. In fact, for two defects of opposite charges $\pm q$, which associate to form pairs, the association energy is lowered by

$$\Delta h = q^2 \kappa / 4\pi\varepsilon_0 \varepsilon (1 + \kappa d) \tag{32}$$

where d is the distance of closest approach of the defects, ε_0 the permittivity of vacuum and ε the dielectric constant. In addition, there is a small lowering of the mobility (a "mobility drag factor"), but this is less important.

The Lidiard DH model is only valid for very small concentrations of the charged species, only a few ppm at most, since it requires the condition $\kappa d \ll 1$. This is a major limitation of the theory.

Inevitably, the pairing model must break down at $\sim 1\%$ dopant, since interaction effects leading to the formation of clusters larger than single pairs then become important. There have been several sophisticated statistical mechanical approaches to ionic conductivity in systems with high concentrations of defects. These are generally based on lattice gas models and include the path probability method of Kikuchi and Sato (Kikuchi, 1966; Sato and Kikuchi, 1971; Kikuchi and Sato, 1972) and dynamical models of Beyeler *et al.* (1979) and of Bunde *et al.* (1982), the latter including long-range coulombic interactions. Unfortunately, results of such approaches have not yet reached the point where they are readily applicable to the analysis of experiments such as those to be described herein (Section III.C.3). Furthermore, these theories do not take into account specific crystallographic clusters that may have a high formation energy, such as the Willis 2:2:2 cluster and others of even greater complexity (Catlow, 1981).

B. Theory of Dielectric and Anelastic Relaxation

While ionic conductivity is a manifestation of free oxygen vacancies, the presence of associated pairs can be detected through both dielectric and anelastic relaxation, which, therefore, provide a tool that is complementary to conductivity measurements. An associated pair can act both as an electric or elastic dipole. In the absence of an external field, the vacancy may reside, with equal probability, on any one of the eight nearest neighbor positions to the cation shown in Fig. 1. Application of an electric field, say in the x direction, favors the four sites in which V_O lies in the $+x$ side of the dopant over the four sites on the $-x$ side. A readjustment of the occupancies of the eight sites then occurs, with relaxation time τ, leading to an extra polarization δP (or, correspondingly, to an increase in the electric displacement δD). This "electrically active" mode of relaxation is shown in Fig. 6 under the heading T_{1u}. This group theoretical symbol implies three-fold degeneracy, since the field direction could equally well have been along the y or z axes. Similarly, a stress field that has a shear component could also produce preferential redistribution among the eight sites, since the strain field about the defect causes it to act as an "elastic dipole" (Nowick and Heller, 1963). In the presence of a uniaxial stress along a $\langle 111 \rangle$ direction, or of a shear stress in the $\{100\}$ faces, a redistribution will occur in the manner shown in Fig. 6 under the heading T_{2g}. This "elastically active" mode is also three-fold degenerate. Also shown is a nondegenerate mode A_{2u} which, however, cannot be excited by either a uniform electric or elastic field, and a symmetric

3. ATOM TRANSPORT IN OXIDES OF THE FLUORITE STRUCTURE

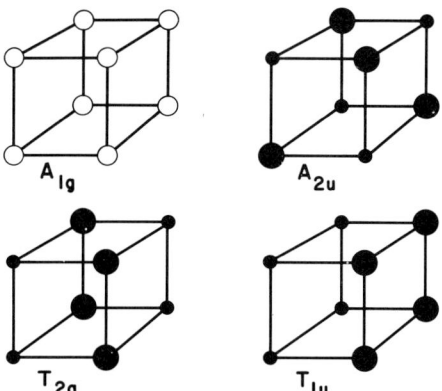

Fig. 6. Diagrams showing the various relaxational normal modes for the nn vacancy–impurity pairs in the fluorite lattice, where the impurity is at the center of the cube. Open circle means a random probability of occupancy by a vacancy, large filled circle a slightly greater-than-random probability, and small filled circle a slightly smaller-than-random probability. The T_{2g} mode type can be induced by stress and T_{1u} by an electric field.

mode labeled A_{1g}. The latter is active only under hydrostatic stress and can only occur if two different species of defects are present [e.g., nearest neighbor (nn) and next-nearest neighbor (nnn) pairs]; It is therefore inoperative in the present case.

It is relatively straightforward to calculate the relaxation time τ associated with the dielectric (T_{1u}) and anelastic (T_{2g}) modes in terms of the jump frequency w_b of the vacancy betweeen two adjacent bound sites. The result is (Nowick, 1972):

$$\tau^{-1}(T_{1u}) = 2w_b, \quad \text{dielectric,}$$
$$\tau^{-1}(T_{2g}) = 4w_b, \quad \text{anelastic}$$
(33)

Thus there is a factor of 2 in the ratio of τ values between these two modes. Note that w_b, which appears here, is not the same as w_V of the free vacancy, which appeared in Eq. (12), but represents the jump between two adjacent *bound* sites (nn → nn).

The magnitude of the relaxation is also obtainable from the general theory (Nowick and Heller, 1965). In the case of dielectric relaxation, the quantity $\delta\varepsilon = \varepsilon_R - \varepsilon_U$ [the difference between the dielectric constant ε_R at low enough frequencies ω, where the defect distribution is essentially always relaxed ($\omega\tau \ll 1$), and ε_U, the value at high enough frequencies that it is unrelaxed ($\omega\tau \gg 1$)] is given by

$$\delta\varepsilon = c_p\mu^2/3v_0kT \qquad (34)$$

Here v_0 is the molecular volume and μ the effective dipole moment, which can with reasonable accuracy be taken simply as charge times separation. In the case of anelastic relaxation, we similarly deal with $\delta s = s_R - s_U$, the difference between the relaxed and unrelaxed elastic compliance, which is given by

$$\delta s = \alpha v_0 c_P (\Delta \lambda)^2 / kT \tag{35}$$

Here α is a numerical constant which depends on the defect symmetry and type of stress employed. For a trigonal defect and a $\langle 111 \rangle$ oriented uniaxial stress $\alpha = 1/27$, while for the same defect and a $\langle 100 \rangle$ stress $\alpha = 0$, a result that is characteristic of the trigonal defect. The quantity $\Delta \lambda$ is the difference between principal values of components of the elastic dipole tensor, which characterizes the strain about the defect (Nowick and Heller, 1963). Note that both $\delta \varepsilon$ and δs are proportional to c_P, the concentration of pairs, as we might have expected.

When dielectric and anelastic measurements are made by ac methods (i.e., with alternating field applied) the response (electric displacement or strain, respectively), in general, lags behind the applied field by an angle δ (the "loss angle"). The measurement gives $\tan \delta$, which in both cases, for a simple defect, takes the form of a Debye peak,

$$\tan \delta = \Delta \omega \tau / (1 + \omega^2 \tau^2) \tag{36}$$

Here $\Delta = \delta \varepsilon / \varepsilon$ in the dielectric case, and $\delta s/s$ in the anelastic case, is called the relaxation strength. (It is assumed in the above that $\Delta \ll 1$). In addition to the ac methods, a highly sensitive dc method for dielectric relaxation, called thermally stimulated depolarization currents (TSDC, also known as ionic thermocurrent or ITC), has been developed (Bucci et al., 1966). In this method, the sample is first polarized, by applying an electric field at a temperature at which dipole reorientation can readily occur, and then cooled with the field on to a temperature low enough to freeze in a state of preferred orientation. The field is then removed and the sample is heated at a uniform rate during which time one measures the depolarization current $i(T)$, which takes the form of a peak as a function of temperature. A useful formula is that the temperature T_m of the peak is implicitly given by

$$kT_m^2 = bh_b^m \tau(T_m) \tag{37}$$

where b is the heating rate, $\tau(T_m)$ is the relaxation time at T_m, and h_b^m the activation enthalpy for the relaxation (i.e., for w_b).

These relaxation methods have the advantage that they are spectrographic in nature, i.e., that different defects give rise to peaks at different temperatures. By contrast, conductivity is an integrated property; it does not give such selective information.

C. Studies of Conductivity and Relaxation

1. Introduction

The study of ionic conductivity in the fluorite oxides has been taking place for a long time and there are several extensive reviews of the topic (Etsell and Flengas, 1970; Takahashi, 1972; Nowick, 1979; Kilner and Steele, 1981). A most striking general observation is that the isothermal conductivity shows a maximum as a function of dopant concentration for almost all such systems. In many cases, the conductivity near the maximum is sufficiently high at temperatures above 700–800°C that the material can be classified as a fast ion conductor. Such high-conductivity oxygen-ion conductors are of practical interest as already mentioned in Section I. The occurrence of a maximum in conductivity is not predicted by the theory of Section III.A. However, as already mentioned, we have no reason to expect that results derived by the use of the elementary mass action principle should be applicable above a defect concentration of $\sim 1\%$. Therefore, it is clear that we should distinguish between the *dilute* or *low-concentration range*, where mass-action equations are applicable, and the *high-concentration range*, where they are not. In this subsection, we make that distinction and discuss these two regions separately.

For the widely studied zirconia-based oxides (e.g., CSZ and YSZ), as well as the doped hafnia systems, there exists no dilute range, because it requires ~ 8 mole % of dopant to stabilize the cubic fluorite structure. Therefore, these materials will be dealt with entirely under the high concentration heading. For the study of the dilute range, ceria (CeO_2) and thoria (ThO_2) are the best host materials.

The key to the dilute range is the formation of nn $D_M V_O$ pairs for both D^{2+} and D^{3+} dopants. In this connection, we will show that relaxation experiments provide the best guide to the range of validity of the theory, which is based on such pairs and their dissociation. Both dielectric and anelastic relaxation peaks have been observed, which, at sufficient dilution, are close to Debye peaks [Eq. (36)]. We may then regard that the onset of broadening of these peaks means the onset of interactions of the type that render the simple theory invalid.

2. The Dilute Range

Since relaxation studies provide the most convenient basis for determining the dilute range, we begin with these. The presence of anelastic and dielectric relaxations due to $D_M V_O$ pairs was first observed by Wachtman (1963) for CaO-doped ThO_2. Later, Lay and Whitmore (1971) made similar observations in CaO-doped CeO_2. Peaks were found in both dielectric and anelastic relaxation, and it was verified that $\tau_{\text{diel}}/\tau_{\text{anel}} \cong 2$. Since this result characterizes the eight-position model of Figs. 1 and 6 [see Eq. (33)], it can be regarded as a strong verification of the model. Another prediction of the model is that anelastic relaxation takes place only for uniaxial stress in the $\langle 111 \rangle$, and not in the $\langle 100 \rangle$ direction (i.e., $\alpha = 0$ in Eq. (35) for the $\langle 100 \rangle$ direction). This prediction could not be checked for the present materials, since single crystals are not available. (Due to the extremely high melting points of these materials they have, thus far, been prepared only as sintered polycrystals.) It is interesting, however, that in the analogous system of NaF-doped CaF_2, which has the same structure and basic defect types, the absence of relaxation under $\langle 100 \rangle$ stress was verified (Johnson et al., 1969). In addition, application of Eq. (34) to the results for $Th(Ca)O_{2-x}$ and $Ce(Ca)O_{2-x}$, inserting for the concentration of dipoles the full Ca^{2+} concentration, gives reasonable values for the electric dipole strength μ. Therefore, there is little doubt that, at low temperatures, nearly complete association into nn $Ca_M V_O$ pairs takes place.

These studies have also yielded the activation enthalpy h_b^m for the relaxation process, i.e., for the jump frequency w_b in Eq. (33). Values of h_b^m and of the corresponding preexponential w_{b0} are given in Table I for the two cases discussed above, as well as for the case of $Ce(Y)O_{2-x}$, which will be discussed below. It is important to recognize that since h_b^m applies to the jump of a *bound* vacancy there is no reason for it to be identical to h_v^m, the activation enthalpy for motion of a free vacancy, which enters into the expression for the conductivity, Eq. (16); however, the two quantities should be comparable in magnitude.

The peak widths obtained for concentrations up to 1 mole % CaO correspond very nearly to that given by the Debye equation. Appreciable broadening develops, however, for concentrations beyond 1% (Lay and Whitmore, 1971; Nowick and Park, 1976; Wang, 1983), indicative of interaction effects and a departure from the simple association model.

For the case of trivalent cation dopants (D^{3+}), similar experiments are particularly important because of the need to establish the correctness of

TABLE I

KINETIC PARAMETERS FOR BOUND VACANCY JUMP FROM RELAXATION EXPERIMENTS

System	h_b^m (eV)	$w_{b0}(10^{13}\ s^{-1})$
$Th(Ca)O_{2-x}$	1.02	2.2
$Ce(Ca)O_{2-x}$	0.82	3.5
$Ce(Y)O_{2-x}$	0.67	6.0

our earlier assumptions that the D^{3+} ions are essentially randomly distributed and that, in the dilute range, vacancies form charged $(D_M V_O)^{\cdot}$ pairs instead of neutral triplets. Such nn pairs are still bound by electrostatic interaction and constitute trigonal defects which occupy the same eight positions as the neutral $(Ca_M V_O)^x$ pair. The anelastic experiments on Y_2O_3-doped CeO_2 carried out by Anderson and Nowick (1981) are illustrated in Fig. 7, while corresponding dielectric relaxation observed by Wang and Nowick (1983) is shown in Fig. 8. The anelastic results again show that for <1 mole % Y_2O_3 very nearly a Debye peak is obtained, but that for higher concentrations a substantial broadening toward higher temperatures occurs. Anderson and Nowick resolved the broadened peak into additional relaxations that were attributed to clusters larger than simple pairs. In the case of the dielectric work (Fig. 8), it is found that two distinct peaks are observed for Y_2O_3 concentrations $\leqq 1\%$. The lower peak is close to a Debye peak and has a fixed temperature (for a given frequency) independent of concentration. Also, this peak is of the correct magnitude for a pair-relaxation effect and its height varies linearly with the dopant concentration. All of these features support the pair-relaxation hypothesis. The second peak, at higher temperatures, has too large an integrated intensity to be a dipole peak; its origin will be discussed later. In addition to the dielectric measurements by the TSDC technique, Wang and Nowick (1983) also made ac measurements in

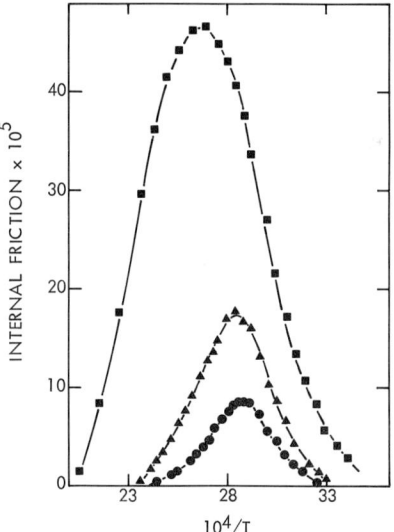

Fig. 7. Internal friction as a function of $1/T$ for CeO_2 containing (●) 0.5; (▲) 1.0, and (■) 2.5 mole % Y_2O_3. [From Anderson and Nowick (1981).]

the kHz range. By combining these results and comparing them with the anelastic measurements of Anderson, it was found that $\tau_{diel}/\tau_{anel} = 1.94$, in agreement with the theoretical ratio of 2.00 to within experimental error. The values of h_b^m and $\log w_{bo}$ for this case are entered in Table I. These results strongly support the expectation that, in the dilute solutions, charged $(Y_{Ce}V_O)^{\cdot}$ defects are responsible for the relaxations, leaving an equal number of Y'_{Ce} defects (i.e., Y^{3+} ions in a cubic environment) for charge compensation. While conductivity and relaxation experiments do not provide direct evidence for the D'_M cubic defects, such evidence is in fact provided in cases in which the D^{3+} ions are lanthanides. Then, electron-spin resonance measurements show the presence of a large fraction of D^{3+} in cubic environments (Abraham et al., 1965, 1966). This picture of an array of charged

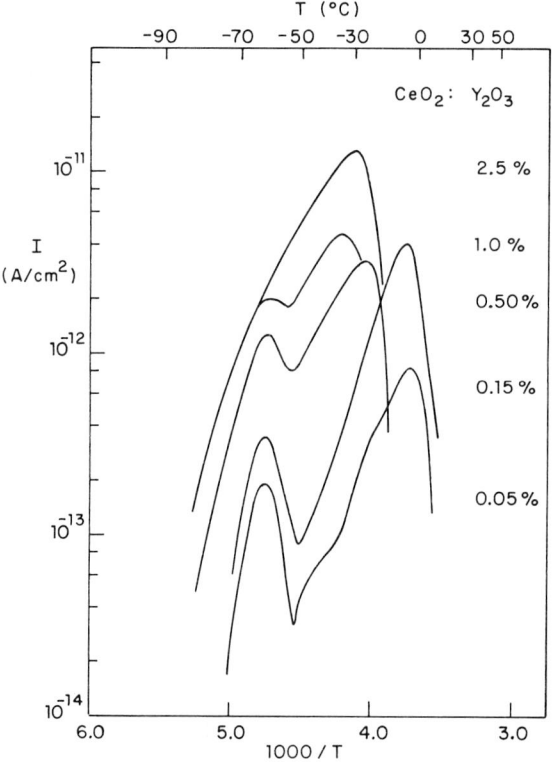

Fig. 8. Dielectric relaxation in the form of TSDC measurements (depolarization current versus $1/T$) for five dilute CeO_2:Y_2O_3 solid solutions (Y_2O_3 concentrations are given in mole %); $E = 965$ V/cm. [Reprinted with permission from *Journal of Physics and Chemistry of Solids* **44**, D. Y. Wang and A. S. Nowick, copyright 1983, Pergamon Press, Ltd.]

defects, $(D_M V_O)'$ and D'_M, in the D^{3+} case is strikingly different from that in the D^{2+} case, where at low temperatures only neutral $(D_M V_O)^x$ pairs are present.

We turn now to conductivity measurements in the dilute range. Unfortunately, measurements made in different laboratories often do not agree as well as one might wish. One of the reasons is the complication of interfacial phenomena. First, there is always the problem of electrode effects, since the migrating ions do not pass freely into and out of the electrodes. (See, for example, Kleitz and Dupuy, 1976). Second, in these polycrystalline ceramics there is often a "grain-boundary effect" that involves blocking of flow in the vicinity of grain boundaries (Wang and Nowick, 1980). In order properly to extract a true "bulk" (or "lattice") conductivity, ac impedance measurements are required as a function of frequency over a wide range. Such measurements were not normally done in early work, where it was customary to use a fixed frequency, typically 1 kHz. However, such methods have become more widely used in recent years, particularly since the work of Bauerle (1969). Figure 9 illustrates schematically complex-impedance analysis that permits the separation of the various processes, each represented by an equivalent circuit element consisting of parallel resistance and capacitance. If one plots the negative imaginary part (Z'') of the ac complex

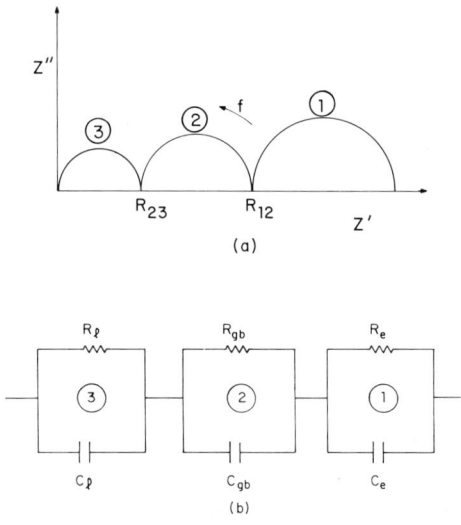

Fig. 9. (a) Schematic complex impedance plot showing three arcs. Arrow shows direction of increasing frequency; (b) The equivalent circuit that gives rise to the three arcs of the complex impedance plot: subscripts e, gb and ℓ refer to electrodes, grain boundary and lattice contributions, respectively.

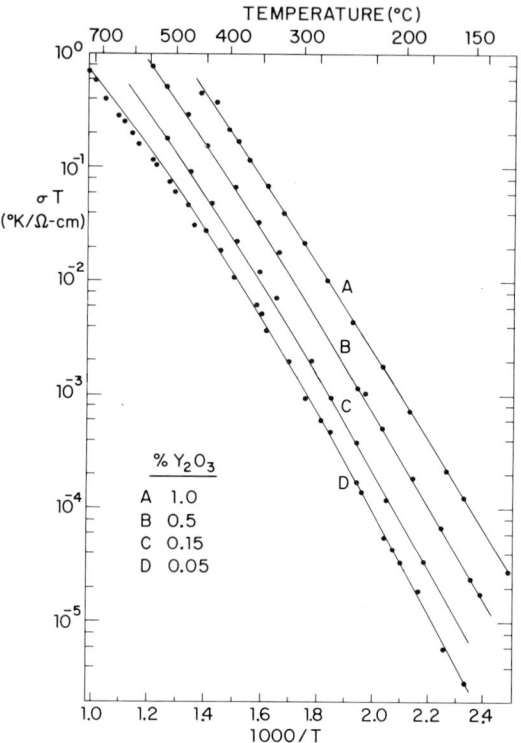

Fig. 10. Conductivity plots (log σT versus $1/T$) for four dilute $CeO_2 : Y_2O_3$ solid solutions: (A) 1.0, (B) 0.5, (C) 0.15, and (D) 0.05 % Y_2O_3. [From Wang et al. (1981).]

impedance against the real part (Z') for the various frequencies of measurement, a series of arcs are observed, each related to a different process.[3] The highest frequency arc, which is closest to the origin, is the one representing the bulk (lattice) behavior. Usually, only a portion of the three arcs is observed at any one temperature, and for a fixed experimental range of frequencies the observations fall further to the right in the impedance diagram as the temperature increases. This means that measurements made at a single fixed frequency cease to represent bulk behavior as the temperature increases. Except where otherwise stated, all conductivity results quoted here will be true bulk values extracted from data taken over a range of frequency by complex impedance analysis.

Since the conductivity of CeO_2 doped with trivalent cations, especially Y^{3+}, has been studied most thoroughly in the dilute range, we begin by

---

[3] Actually, electrode behavior is more complex than that represented by a single R-C element (Macdonald, 1976).

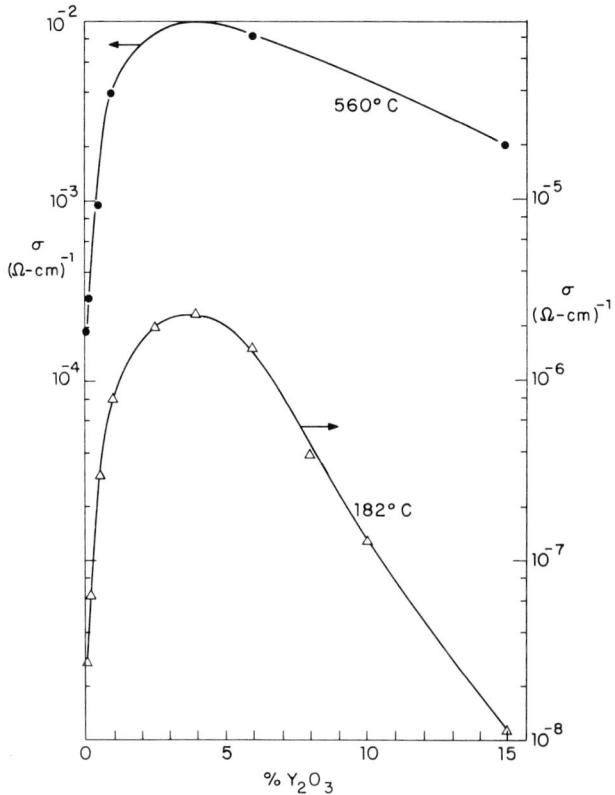

Fig. 11. Isothermal variation of conductivity σ with composition of $CeO_2:Y_2O_3$ solid solutions at 182° and 560°C. Composition is in mole % Y_2O_3. [From Wang et al. (1981).]

discussing those results. Having established that the principal defects at the lower temperatures are the $(D_M V_O)^{\cdot}$ pair and the isolated D'_M, it follows that the free vacancies required for conduction must come from the dissociation of this pair. In Section III.A we have shown that in the extrinsic associated stage (stage III) the conductivity σ should be independent of the total dopant concentration c_0. Conductivity data for $Ce(Y)O_{2-x}$ in the dilute range, 0.05–1.0 mole % Y_2O_3, are presented in Fig. 10 as Arrhenius plots. First of all, these results show that the prediction of independence of c_0 is not confirmed. Secondly, all four curves show departure from a straight line, the greatest curvature occurring for the lowest concentrations. This suggests that the transition between stages II and III is involved at all of these compositions. When $CeO_2-Y_2O_3$ solutions at higher concentrations are studied, it is found that the isothermal conductivity goes through a maximum as a function of c_0 (as has often been observed for other systems). Figure 11 shows that this maximum is sharper at low temperatures than it is at higher temperatures.

While the theoretical prediction that σ is independent of c_0 in the dilute range is not fulfilled, neither is the related expectation that the activation enthalpy h_σ should be independent of concentration. In Fig. 12 we repeat the data for the lower curve of Fig. 11 on a plot that gives the concentration on a logarithmic scale in order to stretch out the dilute range. At the same time, we have added the variation of h'_σ (the empirical activation enthalpy obtained in the lowest temperature region of Fig. 10). It is now clear that the maximum in σ at ~ 3 mole % Y_2O_3 goes hand in hand with a minimum in h'_σ, indicating that the variation of σ with c_0 may be mainly controlled by variation in activation enthalpy. In view of the curvature of these Arrhenius plots, h'_σ should not be regarded as h^{III}_σ. In fact, to extract the fundamental parameters, we have made use of the full mass-action equation [Eq. (25)] together with Eq. (16), but in doing so some additional assumptions are

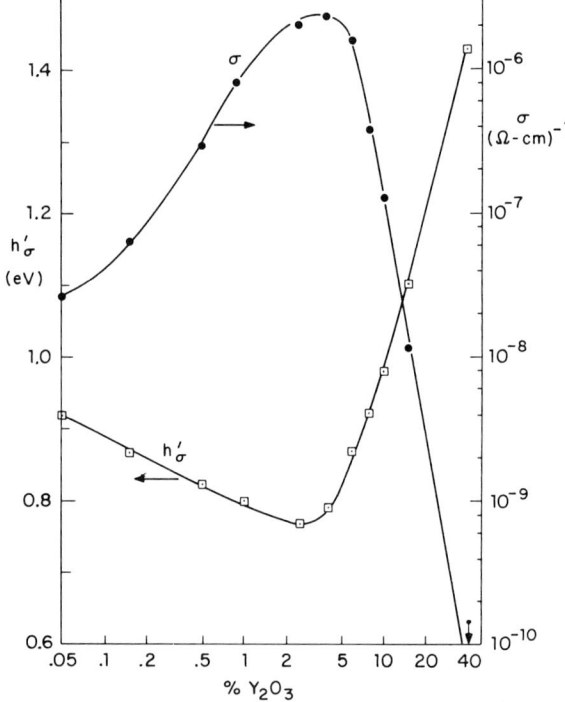

Fig. 12. Variation of the activation enthalpy at low temperatures h'_σ of $CeO_2:Y_2O_3$ solutions with composition, the latter plotted on a logarithmic scale. Also shown (right-hand scale) is the conductivity at 182°C as a function of composition. [From Wang et al. (1981).]

required. First, we take $h_v^m = 0.61$ eV for CeO_2, based on available evidence. (For example, the lowest slope for Gd^{3+}-doped ceria at elevated temperatures is 0.63 eV; clearly, h_v^m cannot exceed this value. Other evidence is given by Wang et al., 1981.) Second, we allow h_A to be function of c_0 but not of temperature, i.e., h_A is constant for each curve in Fig. 10. Third, we take $s_A = 0$, which is not unreasonable. With these assumptions, fitting the curves to Eqs. (25) and (16) gives the results in Table II for h_σ^{III}, preexponential A^{III}, and $h_A = h_\sigma - h_v^m$. The values obtained for h_σ^{III} fall ~ 0.03–0.05 eV higher than the empirical quantity h'_σ, showing that, even in the lowest temperature range of these measurements, stage III is not quite fully attained. The solid curves drawn in Fig. 10 are the fitted curves following this procedure. It is clear that the transition from stage II to III is a very gradual one as compared with the usual schematic drawings (e.g., Fig. 5). Table II shows that A^{III} is independent of c_0 to within experimental error, in agreement with the theoretical prediction. Further, the actual value of A^{III} is in excellent agreement with the calculated value [Eq. (30)], which is valid to within a factor ~ 3. We thus conclude that the only lack in the theory is its failure to consider that h_A may be a function of dopant concentration, for it is this dependence $h_A(c_0)$ that accounts completely for the increase in σ with c_0.

In order to understand $h_A(c_0)$, we must recall the unique aspect of the D^{3+} doped oxides in the dilute range, viz., that defect association involves incomplete compensation, i.e., that $(Y_{Ce}V_O)^\cdot$ and Y'_M defects are present in essentially equal concentrations. This means that in the associated range (stage III) the current carrying vacancies $V_O^{\cdot\cdot}$ migrate in an environment of a strongly fluctuating electrostatic field produced by this array of alternate charges. The strongest evidence for this model quoted thus far was the fact (from relaxation experiments) that the YV_O pair is a trigonal defect. There is, however, additional direct evidence, viz., the existence of a second dielectric relaxation peak (see Fig. 8), which is too large to be due to dipole reorientation. No such extra peak has been observed in Ca^{2+} doped samples (Nowick

TABLE II

PARAMETERS OBTAINED FOR STAGE III CONDUCTIVITY OF DILUTE $CeO_2:Y_2O_3$ SOLID SOLUTIONS

Percent Y_2O_3	h_σ^{III} (eV)	$\log_{10} A^{III}$ (K/Ω cm)	h_A (eV)
0.05	0.955	5.70	0.345
0.15	0.91	5.55	0.30
0.50	0.86	5.55	0.25
1.0	0.82	5.70	0.21

and Park, 1976; Wang, 1983). Further, it was shown that the relaxation time for this peak is controlled by the same activation energy as the conductivity (Wang and Nowick, 1983). All of this evidence leads to the conclusion that the peak originates in the relaxation of the array of charged defects by means of migration of V_O from one Y^{3+} to another. Specifically, to minimize electrostatic energy each $(YV_O)^{\cdot}$ pair tends to surround itself by Y' defects and vice versa, forming a sort of "superlattice" (except that the locations are random rather than regular). Application of an electric field will give rise to a relaxation of this array through the migration of oxygen vacancies. A simplified model based on "wrong pairs" in this superlattice (Wang and Nowick, 1983) shows that it gives the correct order of magnitude for the height of the peak and its dependence on concentration. We thus have further evidence for the existence of the charged defect array. It is interesting that such an array is unique to trivalent-doped fluorite oxides, since no such situation exists in other previously well-studied systems, such as alkali halides and alkaline earth fluorides (Lidiard, 1957, 1974; Franklin, 1972).

Therefore, it is reasonable to expect that the dependence of h_A on c_0 observed in the conductivity study is a consequence of electrostatic interactions in which a freed $V_O^{\cdot\cdot}$ migrating in the crystal spends more time in regions of favorable electrostatic interactions than in regions of unfavorable ones. Qualitatively, this effect on h_A due to electrical interactions also appears in the Debye–Hückel type modification of conductivity theory [Eq. (32)]; however, that approximation is valid only up to ~ 10 ppm of charged species, which is much less than those involved here (up to $\sim 1\%$). Accordingly, we must develop a different treatment to account for $h_A(c_0)$.

Let us first write

$$h_A = h_A^0 - \Delta h_{int} \qquad (38)$$

where h_A^0 is the association enthalpy at infinite dilution and Δh_{int} is the mean electrostatic interaction energy contributed by the surrounding ions to the association process. We assume that Δh_{int} varies as the mean spacing r_0 of Y^{3+} ions in the crystal and write the Coulombic expression

$$\Delta h_{int} = 2e^2/4\pi\varepsilon_0\varepsilon\beta r_0 \qquad (39)$$

where ε_0 is the permittivity of vacuum and $\varepsilon(\cong 22)$ is dielectric constant of the medium. The quantity β is a dimensionless parameter inserted to represent the averaging (i.e., βr_0 is a weighted average separation of the vacancy from opposite charges). Now, c_0 is given by

$$r_0^3 = a^3/c_0 \qquad (40)$$

Inserting Eq. (40) into (39) and evaluating numerical constants yields

$$\Delta h_{int} = 0.42 c_0^{1/3}/\beta \quad \text{eV} \qquad (41)$$

A plot of h_A versus $c_0^{1/3}$ (Fig. 13) shows that this relation is obeyed. The intercept of the plot gives $h_A^0 = 0.43$ eV, while the slope gives $\beta = 0.39$. From Fig. 13 we see that Δh_{int} is still quite significant (0.09 eV) for the 0.05% Y_2O_3 sample, i.e., that h_A has not yet leveled off even at this relatively low concentration.

When Eqs. (38) and (41) are inserted into Eqs. (17) and (27) we obtain, for the concentration dependence of the stage III conductivity in the dilute range,

$$\sigma \propto \exp(0.42 c_0^{1/3}/\beta kT) \tag{42}$$

A plot of $\log \sigma$ versus $c_0^{1/3}$ in Fig. 13 verifies this relation without the analysis involved in determining h_A. The value for 2.5% Y_2O_3 is included in Fig. 13 to show the sharp departure from Eq. (42) as the conductivity maximum is approached.) From the slope of the straight-line portion of the plot the value $\beta = 0.40$ is obtained, in agreement with the value obtained from the h_A plot.

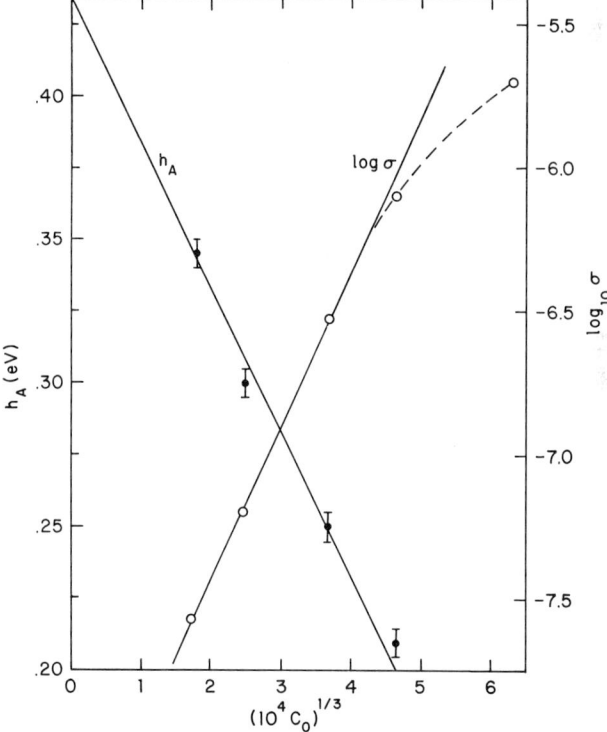

Fig. 13. Variation of association enthalpy h_A and of $\log \sigma$ (at 182°C) with the $\frac{1}{3}$ power of the yttria concentration c_0 over the dilute range. [From Wang *et al.* (1981).]

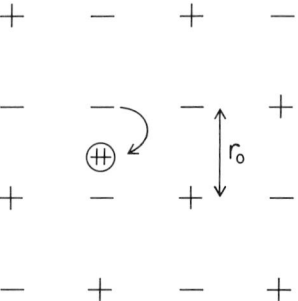

Fig. 14. Schematic representation of the "lattice" of Y ions in ceria. Symbol + represents the $(YV_O)^{\cdot}$ pair, $-$ represents isolated Y'_{Ce}, and the curved arrow shows an oxygen vacancy $V_O^{\cdot\cdot}$ breaking away from a $(YV_O)^{\cdot}$ pair.

Up to this point, we have arrived at Eq. (41) by little more than dimensional reasoning, the key parameter being β. However, a rough estimate of β can be made as follows. We replace the random distribution of Y^{3+} ions by a cubical array, letting the $(YV_O)^{\cdot}$ and Y' defects alternate to form an NaCl-type superlattice with spacing r_0, as shown in Fig. 14. Such an array represents the fully associated state. To simulate dissociation, we move a $V_O^{\cdot\cdot}$ to a position midway between the Y^{3+}, to which it was originally bound, and one of its nearest neighbors, as shown in Fig. 14. The energy change in this process gives (in cgs units)

$$h_A = h_A^0 + 2\alpha_M e^2/r_0 - 8e^2/r_0 \tag{43}$$

Here the first term h_A^0 includes the Y–V_O interaction at nn separation. The second term is the change in electrostatic energy of the lattice of Fig. 14, ignoring the $V_O^{\cdot\cdot}$ defect, and is equal to twice the Madelung energy per charge, where $\alpha_M = 1.748$ is the Madelung constant. The final term is the interaction of the doubly charged $V_O^{\cdot\cdot}$ with its two closest negative charges, the effects of all other charges cancelling by symmetry. Comparing with Eqs. (38) and (39), we see that Eq. (43) has the same form, with $\beta = (4 - \alpha_M)^{-1} = 0.44$. This is in good agreement with the experimental value of 0.40, considering the simplicity of the model.

It is concluded that ionic conductivity in the dilute range is reasonably well understood for the $Ce(Y)O_{2-x}$ system. The behavior described also occurs in related systems, as shown in Fig. 15, where we see that a similar decrease in h_σ occurs in the systems $Th(Y)O_{2-x}$ and $Ce(Gd)O_{2-x}$.

Following the detailed study of Y^{3+}-doped ceria by Wang et al. (1981) similar conductivity measurements were carried out for various D^{3+} dopants of different ionic radii, all at the 1 mole % D_2O_3 level (Gerhardt-Anderson and Nowick, 1981). The dopants studied were La^{3+}, Gd^{3+}, Y^{3+}, and Sc^{3+}, in order of decreasing radius. Results are shown in Fig. 16. Fitting was again

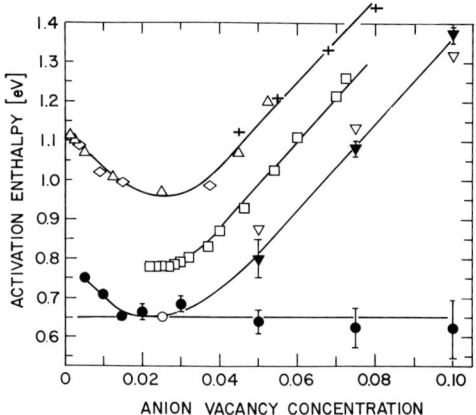

Fig. 15. Variation of activation enthalpy with dopant concentration for $M(D)O_{2-x}$, where $M(D)$ is (□) Zr(Y); (+) Hf(Y); △ and ◇ Th(Y); ▼ and ● Ce(Gd). [From Hohnke (1981).]

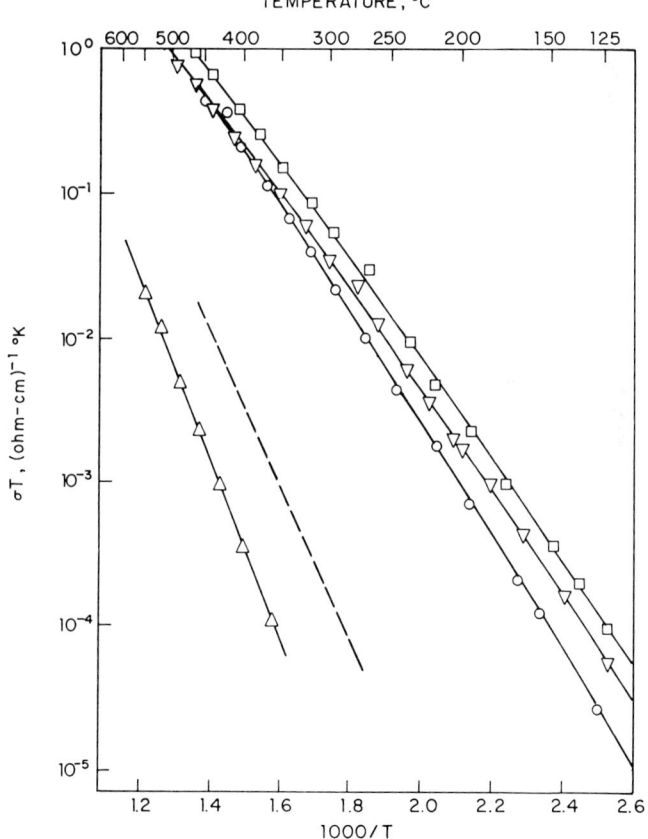

Fig. 16. Conductivity curves for 1 mole % solutions of four different trivalent dopants in ceria: (□) Gd_2O_3, (▽) La_2O_3, (○) Y_2O_3, (△) Sc_2O_3. The dashed curve is for ceria double doped with 1% $Sc_2O_3 + \frac{1}{2}\% Y_2O_3$. [From Gerhardt-Anderson and Nowick (1981).]

TABLE III

Association Enthalpy h_A Obtained for 1% and 6% D_2O_3 Solid Solutions for Four Dopants of Different Ionic Radii in CeO_2

Dopant	Size ratio (D^{3+}/Ce^{4+})	$c_0^{-1} \Delta a/a_0$ (%)	h_A (eV) From 1% D_2O_3	h_A (eV) From 6% D_2O_3
La^{3+}	1.21	5.5	0.14	0.18
Gd^{3+}	1.03	0.9	0.12	0.16
Y^{3+}	0.98	−0.9	0.21	0.26
Sc^{3+}	0.86	−3.7	0.67	—

carried out by using Eqs. (16) and (25), taking $h_V^m = 0.61$ eV, and with the preexponential and h_A as adjustable parameters. The values of h_A thus obtained are listed in Table III, where they are compared with the ratio of ionic size of dopant to that of Ce^{4+}, and also with the fractional change in lattice parameter per unit concentration: $c_0^{-1} \Delta a/a_0$. (The last column represents results for a higher concentration, which will be discussed later.) It is clear that there are large variations in h_A, depending on the dopant radius. The quantity h_A varies, of course, in an inverse manner to the conductivity. The smallest h_A and the highest σ occur for Gd^{3+}, whose radius is only slightly larger than that of Ce^{4+}, while h_A increases both for larger and smaller dopants. Butler *et al.* (1983) have carried out computer simulation calculations for these same dopants in CeO_2, which show generally good agreement with the experimental results (Fig. 17). They predict a minimum in h_A when the dopant and host radii are the same, a small rate of increase toward larger dopants, and a strikingly large increase for smaller dopants. Similar results were obtained by Mackrodt and Stewart (1979) for divalent dopants in MgO, and by Kilner and Brook (1982) for rare earth aluminates. These calculations show that elastic strain energy provides an important component of the dopant-vacancy association energy; this term is minimized when there is a small size mismatch between host and dopant.

The case of Sc^{3+}-doped CeO_2 is of special interest, because here h_A is so large as to give rise to a scavenging effect on vacancies V_0 introduced by other dopants, if the present model is correct. The scavenging effect is strikingly demonstrated in double-doping experiments in which Y_2O_3 was used to introduce vacancies in addition to those contributed by Sc_2O_3 (Gerhardt-Anderson and Nowick, 1981). For example, with $\frac{1}{2}$% Y_2O_3 and 1% Sc_2O_3 (the dashed curve of Fig. 16), the value of σ was about 30 times smaller than that for $\frac{1}{2}$% Y_2O_3 alone, and h_σ was correspondingly elevated. This result shows that the effects of the two dopants are not additive, but rather that the Sc^{3+} ions, which would be in cubic environment in the singly doped material,

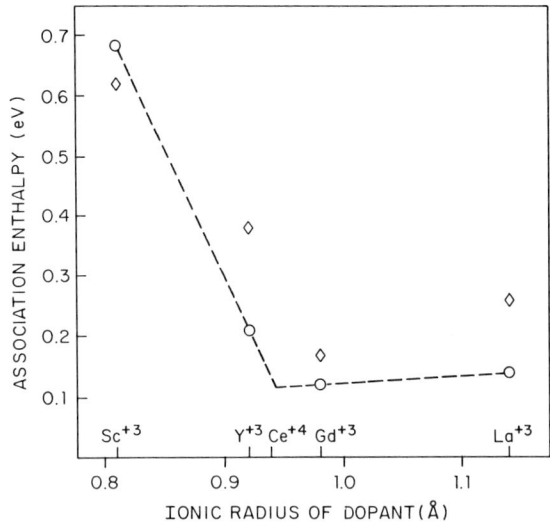

Fig. 17. Comparison of experimental vacancy-dopant association energy for cation dopants of different radii obtained from Fig. 16, with values calculated by computer simulation: (○) experimental; (◇) calculated. [From Butler et al. (1983).]

now associate with nearly all of the V_O contributed by the Y_2O_3 doping, leaving the Y^{3+} ions all largely unassociated. This result is completely consistent with the model we have developed earlier that, in D^{3+}-doped material, only half of the D^{3+} ions are associated with vacancies into charged $(D_{Ce}V_O)^{\cdot}$ pairs.

Until now, we have been discussing the conductivity behavior of trivalent-cation doped oxides, since this case has been studied in some detail in the dilute range. For divalent doping, primarily $Ce(Ca)O_{2-x}$ has been studied. For this material Nowick and Park (1976) reported a constant h_σ of 0.9 eV over a range of composition beginning at 1% CaO, and no maximum in σ versus x. However, the recent work of Hohnke and Hurley (1984), which begins with lower Ca concentrations, shows a decrease in h_σ from 1.2 eV at low concentrations to a minimum value ~ 0.9 eV in the range of 1 to 4% CaO. At the same time they show a shallow maximum in σ versus x at $x \sim 0.06$ for low temperatures (~ 500 K), which is no longer present at higher temperatures (1250 K).[4] The variation of h_σ does not appear to fit the $c_0^{1/3}$ law [Eq. (41)], which was valid for the Y_2O_3-doped case, while the preexponential A, rather than remaining constant, slowly increases with increasing c_0. Of course, we do not expect the theory developed for trivalent

---

[4] These results, however, are in disagreement with earlier work of Blumenthal et al. (1973), which show a maximum even at 1273 K.

systems, where the associated pairs are electrically charged, to apply in the D^{2+} case in which pairs are neutral. It is probably reasonable here to apply the Debye–Hückel approach, but then one cannot analyze an entire conductivity curve in terms of a constant h_A, since the degree of dissociation (and therefore the concentration of charge carriers) is changing rapidly over the entire temperature range of the measurements. This requires further examination.

A value of $h_\sigma = 1.2$ eV for $Ce(Ca)O_{2-x}$ at low concentrations, taken together with $h_V^m = 0.61$ gives, from Eq. (27), the value $h_A = 1.18$ eV. This is much higher than the h_A values obtained for D^{3+} ions (Table III) but is consistent with the higher effective charge of Ca''_{Ce}. In fact, a value $h_A = 1.08$ eV for the CaV_O pair has been obtained in the same set of computer simulation calculations from which results were presented for D^{3+} ions in Fig. 17 (Butler, 1982).

3. The Concentrated Range

The results already presented (see Fig. 12) show that for Y_2O_3-doped CeO_2 the behavior is very different in the dilute and concentrated ranges. It is well known that for most trivalent doped fluorite oxides, the conductivity goes through a maximum at $\sim 1.5-2\%$ anion vacancies (3–4 mole % D_2O_3). The study of the $CeO_2:Y_2O_3$ system has shown (Wang et al., 1981) that the Arrhenius plots become straight over the temperature range of measurement at concentrations above the maximum. At the same time, h_σ increases

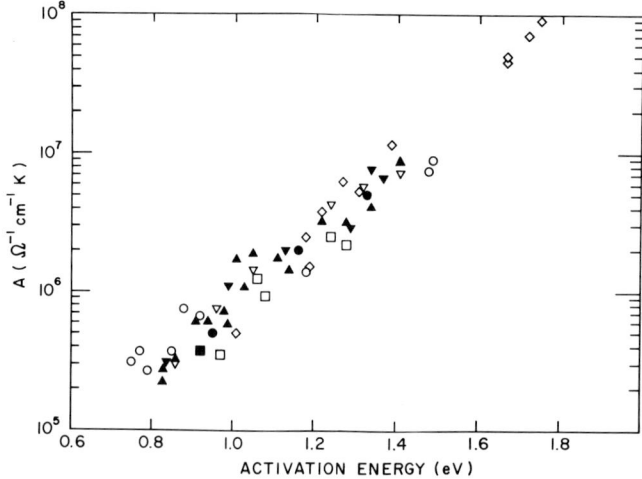

Fig. 18. Variation of the preexponential factor A with activation energy for conductivity, for a variety of dopants and dopant concentrations in zirconia. [From Hohnke (1979).]

3. ATOM TRANSPORT IN OXIDES OF THE FLUORITE STRUCTURE

sharply (Fig. 12) and A also increases from the value $\sim 4 \times 10^5$ at low concentrations to $\sim 10^7$ K/Ω cm at 15% Y_2O_3. Thus it is the increase in h_σ that is responsible for the strong decrease in σ beyond the maximum. However, as we noted earlier, the maximum is not as sharp at more elevated temperatures (Fig. 11).

Hohnke (1979, 1981) compiled conductivity data for many of the doped fluorite oxides and found that for most of them, in the high-concentration range $dh_\sigma/dx \cong 6$ eV. (This result may be seen in Fig. 15, but he also demonstrated it for several ZrO_2-based systems.) Furthermore, Hohnke correlates A and h_σ by noting that there is a linear variation of ln A with h_σ for many systems (Fig. 18) with values of A going up to 10^8 K/Ω cm and higher. Taken together, these two empirical correlations imply that A increases exponentially with concentration. To interpret these results, Hohnke started with mass-action theory and then introduced an *extra* free energy term, which includes the increasing enthalpy as well as an entropy term $\bar{S}(x)$. The latter is then responsible for the exponential increase in A with concentration. However, this approach is only empirical, since the extra free energy is not derived in terms of basic principles.

Interestingly enough, in spite of the complexity of the high-concentration range, the dopant-radius effect described earlier for the dilute range is still maintained. For example, the last column of Table III shows that the values of h_A maintain the same order for dopants in CeO_2 at 6 mole % D_2O_3 as at 1 mole %. (Unfortunately, Sc_2O_3 is not soluble to such a high concentration.) In other work, Kudo and Obayashi (1975) who studied $Ce_{0.7}D_{0.3}O_{1.85}$ (corresponding to 17.6 mole % D_2O_3), with Y^{3+} and various lanthanides as dopants, also showed that the highest conductivity occurs for D^{3+} ions with radii closest to that of Ce^{4+}. (Unfortunately, they carried out ac measurements at a fixed frequency, so that their h_σ values are probably not reliable.)

In other systems, Strickler and Carlson (1965) (also using a fixed ac frequency), studying $Zr(D)O_{2-x}$ for $x = 0.1$, found a strong size effect in measurements above 800°C. Since Zr^{4+} is a small ion, all of the D^{3+} ions were larger, except for Sc^{3+} which is comparable in size to Zr^{4+}. It was found that h_σ increases strongly with increasing radius of dopant. Thus in spite of the fact that we are no longer dealing with simple DV_O pairs at such high concentrations, the size effect criterion is still applicable, at least qualitatively.

We turn now to the theory of ionic conductivity in the concentrated range. In Section III.A we presented a mass-action type theory which, in modified form, was applicable to the dilute range. It is difficult to justify the application of such a theory to the concentrated range, even as a point of departure, for the following reasons:

(i) Mass action equations (utilizing concentrations rather than activities) are, quite generally, valid only for a dilute range.

(ii) For concentrations near the conductivity maximum and beyond, it is no longer meaningful to consider association of a single V_O with a single dopant ion since, even with a random distribution of dopant ions the probability that a vacancy has more than one D ion close to it rapidly increases. For example, each cation has 12 nearest and six next-nearest neighbors. If x is the fraction of dopant cations, the probability of finding an *isolated* dopant, defined as one with no other dopant in nn or nnn position, is $(1 - x)^{18}$. Table IV lists values of this probability, showing that it decreases rapidly for $x > 0.02$ (2% DO or 1% D_2O_3). For a 10% solid solution, the concept of an isolated dopant is clearly meaningless.

(iii) Experimentally, substantial broadening of anelastic relaxation was observed for concentrations greater than 1% D_2O_3 (Fig. 7), while the dielectric pair peak fused with the peak due to the charged array at these higher concentrations (Fig. 8).

Early approaches attempted to explain the conductivity maximum by introducing configurational constraints to the movement of vacancies due to their mutual repulsion. For example, it was proposed that no vacancy jump could occur to a site in which another V_O would be a nn (or even a nnn). This is the approach first used by O'Keefe (1970) and by Barker and Knop (1971), and indeed, it leads to prediction of a maximum in σ versus c_0. Such theories, which regard the maximum as a purely statistical effect, cannot provide the whole answer, however, in view of the more recent results already discussed that show that the maximum is primarily related to changes in h_σ. Subbarao and Ramakrishnan (1979) extended the statistical model by considering vacancy trapping at cation dopants as well as V_O–V_O repulsions. They assume that only free vacancies (with no D ions as nn) contribute to the conductivity. The maximum in σ calculated in this way still does not drop rapidly enough on the high-concentration side.

TABLE IV

PROBABILITY OF FINDING AN ISOLATED DOPANT
(NO OTHER DOPANT IN NN AND NNN SITES)
FOR VARIOUS CATIONIC FRACTIONS, x

x	%DO	%D_2O_3	$(1 - x)^{18}$
0.001	0.1	0.05	0.98
0.003	0.3	0.15	0.95
0.01	1	0.50	0.84
0.02	2	1.01	0.70
0.04	4	2.04	0.50
0.10	10	5.3	0.15
0.20	20	11.1	0.02

The problem with a theory which attributes conduction only to "free" vacancies is that, at high concentrations, there no longer exist connected paths of such vacancies. It is, therefore, necessary to consider contributions to the conductivity from vacancies that are adjacent to dopant ions. Such an approach is that of Nakamura and Wagner (1980), who postulate that conduction proceeds by a multimode mechanism through and between different degrees of $D-V_O$ associates. Three possible pathways through the crystal are considered, consisting of connected anion sites having different dopant coordination numbers. The primary mechanism is V_O transport through channels of one-fold bonded $D-V_O$ complexes. A V_O in one-fold coordination is considered as being in a free state, while two-fold and higher coordinated V_O are regarded as more tightly bound. A maximum in conductivity is obtained when all anion sites are effectively one-fold coordinated. At higher concentrations, the effective coordination number becomes greater than one and σ decreases (because the effective path for migration through one-fold complexes decreases). Applying the theory to CSZ, Nakamura and Wagner assume that the coordination number for Ca^{2+} in a nnn site is 1/3. (In this way they give some weight to nnn coordinations.) This ad hoc assumption is made so that the concentration at maximum σ will come out right. Finally, the mobilities needed for the two types of channels are evaluated from experimental data. Thus the theory is really semiempirical in nature, although it does succeed in fitting experimental data well.

Another approach is that of Catlow and Parker (1982), who start from fundamental principles and consider jumps between sites of different coordinations. They assume a frozen-in random distribution of dopants and classify the anion sites according to the number of dopants in surrounding shells. For simplicity, however, the calculation considers only the first shell and labels the anion sites $i = 0, 1, 2, 3, 4$, assigning a binding energy e_i to a vacancy in each site. The calculation of e_i is carried out by the HADES computer simulation. Vacancies are then distributed among these sites according to a Fermi–Dirac distribution function (Catlow, 1978). What is then needed is the rate of V_O jumps between sites, with activation energy e_{ij} (for an $i \to j$ jump); these are also calculated using the HADES code. Also needed are p_{ij}, the probability of an i site surrounded by j, and a correlation-type factor f_{ij} (the probability that jump $i \to j$ is not nullified subsequently by $j \to i$). All of these terms were calculated, but with simplifying assumptions. The result for $Ce(Y)O_{2-x}$ shows a maximum in σ at the right concentration but does not continue to fall steeply enough. It was felt that the ionic size factor had not been taken into account and, accordingly, an arbitrary scaling factor in the values of e_i and e_{ij} was introduced. This yielded a more satisfactory result.

This calculation by Catlow and Parker is a first attempt that clearly needs refinement, but it shows the qualitative effects very well. In particular, the calculation brings out the point that the pronounced drop in σ after the maximum is attributable to an increasing number of deep-trap sites at the higher concentrations, which effectively immobilize the vacancies. This approach includes such features as the change in effective migration energy h_V^m due to changes in coordination, and also correlation effects. That correlation effects are to be expected in the theory of conductivity at high concentrations, as well as in tracer diffusion, has recently been pointed out (Sato and Kikuchi, 1971; Murch, 1982). The assumption of random dopant distribution, however, may be the greatest problem with this calculation. While this assumption is probably valid in the dilute range, there is ample evidence from a large number of structural studies of these oxides for short-range order and for the existence of microdomains. This point is of sufficient importance that we will digress briefly to give some background. Extensive reviews have been given by Rossell (1976, 1981).

For certain compositions and choices of cations, long-range ordered structures of the doped oxides can be obtained. Usually, a requirement for cationic order to be present is an appreciable size difference between the radii of the host (M^{4+}) cation and the dopant (D^{2+} or D^{3+}). One such ordered structure is the pyrochlore structure of the type $Zr_2Ln_2O_7$, where Ln represents a lanthanide ion. This structure is produced from a $2 \times 2 \times 2$ set of fluorite subcells, by ordering of cations, and the ordered omission of one anion in every eight. It is only found for lanthanides whose ionic radius is larger than that of Gd (i.e., a radius ratio to Zr^{4+} of 1.2–1.6). The ordered phase also increases in stability with increasing size difference. Other ordered structures are of the form $M_3D_4O_{12}$ (e.g., $Zr_3Yb_4O_{12}$, $Zr_3Er_4O_{12}$), and for D^{2+} cases, M_4DO_9 (e.g., Zr_4CaO_9, Hf_4CaO_9) and $M_7D_2O_{16}$ (e.g., $Hf_7Ca_2O_{16}$). In all of these cases large differences in cationic radii are involved. In some solutions that have small size differences only anion order occurs (e.g., $Zr_3Sc_4O_{12}$ and $Hf_3Sc_4O_{12}$), while the cations remain randomly distributed. A basic feature of these ordered structures is that oxygen vacancies occur as pairs along the body diagonal of the MO_8 cube of the parent fluorite structure. The cation inside the cube then becomes six-fold coordinated, and the remaining anions are then displaced along $\langle 100 \rangle$ directions so as to move closer to the vertices of a regular octahedron. At the same time, the tetrahedron of cations about each V_O expands radially, along $\langle 111 \rangle$, by small distances. In the ordered structures, the smaller cation preferentially occupies these octahedral sites while the larger cation occupies the remaining sites that are either eight- or seven-fold coordinated.

Compositions that are not long-range ordered may still possess *microdomains*, which are small localized coherent regions, generally of different composition from the average value, and with a structure resembling that of a nearby ordered phase. The principal evidence for microdomains comes from high resolution electron microscopy, which permits one to obtain a diffraction pattern from a very small region (as small as a few angstroms across). It is anticipated that such microdomains will be present equally in all possible orientations to the fluorite lattice. Such a picture may be oversimplified, since this assumption does not explain quantitatively the diffuse x-ray scattering from stabilized zirconias (Morinaga et al., 1979, 1980). Nevertheless, it is difficult to dispute the observations of microdomains by electron microscopy techniques.

The effect of ordering in lowering the conductivity has already been mentioned (Section II.A). A dramatic decrease in conductivity with long-range ordering has been demonstrated for the case of $Zr_3Er_4O_{12}$, as shown in Fig. 19. Here it is seen that the conductivity of the ordered structure is about two orders of magnitude smaller than that of the disordered structure obtained by taking the material to $\sim 1500°C$ and cooling relatively rapidly. The qualitative reason for the low conductivity of the ordered structure is clear: what were oxygen vacancies in the original fluorite structure are no longer vacancies in the lattice of the ordered material, viz., they are now an inherent part of the structure and are not free to migrate.

A more complex situation exists in the case of the pyrochlores, as shown in a conductivity study by van Dijk et al. (1980). Here $Gd_2Zr_2O_7$ and

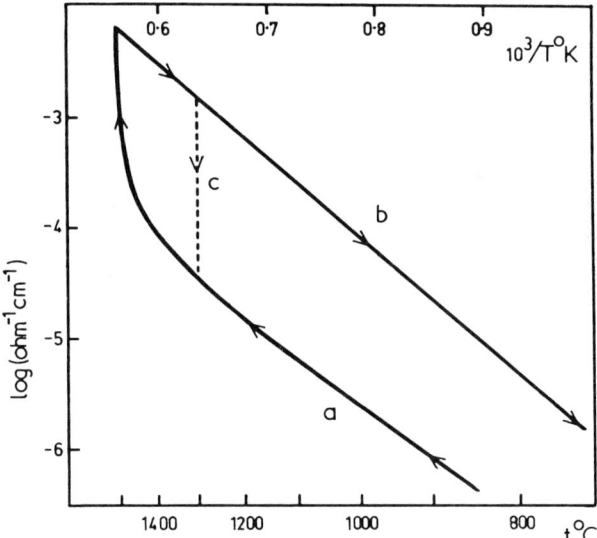

Fig. 19. Effect of ordering on the conductivity of $Zr_3Er_4O_{12}$. Curve (a) is for the ordered (well annealed) specimen, and curve (b) was obtained during relatively rapid cooling from 1500°C. Curve (c) shows annealing at a fixed temperature. [From Rossell (1981).]

$Nd_2Zr_2O_7$ were studied over a range of composition close to stoichiometry. It was found that both the preexponential A and the activation enthalpy h_σ decrease with increasing order. Since these two terms oppose each other in their contributions to σ [see Eq. (17)], the net result was a maximum σ at stoichiometry for $Gd_2Zr_2O_7$ and a minimum for $Nd_2Zr_2O_7$. In later work (Burgraaf et al., 1981), a low A was correlated with anionic order, while a low h_σ was related to cationic order, specifically, the creation of preferential diffusion paths in the structure.

These concepts of the role of order on conductivity can be important even in long-range disordered solid solutions through the presence of microdomains which possess some of the features of ordered structures. An interesting example is the case of $Zr_{1-2x}Gd_{2x}O_{2-x}$, discussed above in relation to the pyrochlore at $x = 0.25$. In this system it is found (van Dijk et al., 1980) that h_σ shows a peak in the fluorite range (at $x = 0.17$) and then sharply decreases, continuing without a break into the pyrochlore phase. This turnaround of h_σ, which is in striking contrast to the $Ce_{1-2x}Y_{2x}O_{2-x}$ system (see Fig. 12), can be attributed to the formation of microdomains of the pyrochlore structure within the fluorite phase. Work of van Dijk et al. (1983) verifies this suggestion. In this system the nearly ordered structure (exact pyrochlore composition) shows the highest conductivity and the lowest h_σ. More usually, ordering gives rise to a lowering of conductivity (as in Fig. 19), in which case it may be expected that the higher the fraction of microdomain regions, the lower the conductivity of the sample. These concepts are difficult to handle quantitatively without a more detailed knowledge of the structure. In fact, since a true equilibrium state is unattainable below $\sim 1200°C$ due to the slow cationic transport, virtually all oxides with high concentrations of dopants must be regarded as being in nonequilibrium states. It would then be necessary to document the thermal history of the sample and to understand fully the short-range order that is present, in order to interpret the conductivity. It appears, however, that the simple concept of a random cation sublattice, as used in the theory of Catlow and Parker (1982), is more reasonable for the systems $Ce(D)O_{2-x}$, where cation size differences are small and long-range ordered structures usually do not form, than systems based on zirconia. In the latter case, the cation size factor is usually large, particularly for the lanthanide dopants, and so local ordering is much more prevalent.

4. Other Related Oxide Structures

Most of the rare earths, as well as Sc and Y, are trivalent and form sesquioxides M_2O_3 of various structures. Among these are the hexagonal

A-type structure exhibited by La_2O_3 and Nd_2O_3 and the monoclinic B-type of Sm_2O_3 and Gd_2O_3. But the most common structure is the cubic C-type exhibited by Sc_2O_3 and also by Y_2O_3 and many of the rare earths (Sm_2O_3, Dy_2O_3, Eu_2O_3, Er_2O_3) at lower temperatures, with transformation into other forms at high temperatures. This C-type structure is closely related to the fluorite structure. It can be derived from the fluorite by ordered removal of one-fourth of the oxygen atoms which lie along the four $\langle 111 \rangle$ directions of the fluorite structure, in such a way that all cations become six-coordinated.

Most of these sesquioxides are mixed conductors with a substantial electronic component (Kofstad, 1972) and, therefore, it is not readily possible to study ionic transport by conductivity measurements.

One especially interesting example of a sesquioxide, however, is $\delta\text{-}Bi_2O_3$, the high-temperature form of this oxide that is stable above 729°C. It has a defect fluorite structure with 25% vacant anion sites and has been shown (Harwig and Gerards, 1978) to be the best oxygen-ion conductor known, with the exceptionally low activation enthalpy for conductivity of ~ 0.4 eV. Takahashi and coworkers (Takahashi et al., 1972, 1975a,b, 1976, 1977; Takahashi and Iwahara, 1973, 1978; Esaka et al., 1982) in an impressive series of papers, showed that the $\delta\text{-}Bi_2O_3$ phase can be stabilized to lower temperatures by various additions, typically in the range 15–25 mole %. Among the solutes that they used are divalent CaO and SrO, trivalent Gd_2O_3 and Y_2O_3, pentavalent Nb_2O_5, and hexavalent WO_3. In spite of the wide variety of these solutes, all seem to stabilize the $\delta\text{-}Bi_2O_3$ structure and to provide good oxygen-ion solid electrolytes, though not as good as the extrapolated values for pure $\delta\text{-}Bi_2O_3$. Following the work by the Japanese group, the study of these Bi_2O_3-based materials was taken up by other groups (Boivin and Thomas, 1981a,b; Conflant et al., 1976, 1980; Verkerk and Burggraaf, 1981; Verkerk et al., 1980, 1982; Burggraaf et al., 1981; Cahan et al., 1980). Some of these later studies show that there is much structural complexity in these oxides, and the understanding of their behavior has certainly not reached the level of that of the conventional fluorite oxides discussed herein. Limitations of space do not permit us to go into further details, but the interested reader may consult the above references for details.

IV. Comparison of Ionic Conductivity and Oxygen Diffusion

The earliest detailed comparison of ionic conductivity and oxygen diffusion measurements was made by Kingery et al. (1959) for $Zr_{0.85}Ca_{0.15}O_{1.85}$. They measured σ at a fixed frequency (10^3 Hz) and oxygen diffusion by the decrease in ^{18}O content of a fixed volume of gas in equilibrium with the solid. They used Eqs. (13) and (14) to convert σ to D, but with $H_R = 1$ instead of 0.65. The agreement that they reported between the two measurements was excellent (it comes out even better for $H_R = 0.65$!) Simpson and Carter (1966) repeated the oxygen diffusion measurements using a sectioning technique and a mass spectrometer to determine the diffusion profile, both for single crystals and polycrystals of CSZ. Their results showed higher D values and more scatter than that reported by Kingery. They also compared these results with the conductivity data of Carter and Roth (1968) to obtain a rough agreement, but indicate that the excellent agreement by Kingery et al. (1959) may have been somewhat fortuitous.

More recently, we have Steele and Floyd's (1971) diffusion measurements on $Ce(Y)_{2-x}$ and the conductivity measurements of Wang et al. (1981), both quoted earlier. Floyd, like Kingery et al., used the rate of depletion of ^{18}O from the gas surrounding the sample and obtained a monotonic increase in diffusivity with Y content (Fig. 4). On the other hand, the conductivity study shows a maximum at $\sim 3\%$ Y_2O_3, but measurements are only up to 300–500°C because of the onset of electrode effects at higher temperatures. Therefore, in attempting to compare these two types of measurements a long extrapolation is required. Nevertheless, extrapolation of conductivity data for 10 mole % Y_2O_3 gives reasonably good agreement with Floyd's diffusion data for the same composition. However, it seems doubtful that we can reconcile the monotonic diffusion behavior of Fig. 4 with the maxima in conductivity of Fig. 11, in spite of the fact that the conductivity maximum becomes milder with increasing temperature. This statement is based on the fact that to reconcile the two sets of data would require crossing over of the curves at elevated temperatures, resulting in a very low activation enthalpy for diffusion at the higher concentrations.

We conclude from the present review that, thus far, considerably more information about defect parameters and defect interactions in these oxides has been obtained from conductivity studies than from measurements of oxygen diffusion. It is interesting to note, however, that a major improvement in the technique of oxygen diffusion measurement has recently been made (Kilner et al., 1984). This method uses secondary ion mass spectrometry (SIMS) to measure the ^{18}O profile after diffusion. The principle of the method is that the sample is bombarded by a primary, mass analyzed, ion beam ($^{40}Ar^+$) of ~ 10 keV energy in an UHV analysis chamber. The secondary ions produced by sputtering are then analyzed either by a magnetic sector or a quadrupole mass analyzer. It turns out to be advantageous to analyze for the negative secondary oxygen ions produced by the bombardment. The problem of charging of insulating samples during the measurement is overcome by simultaneous bombardment with a low-energy electron beam.

By using this method the ^{18}O profile can be obtained with excellent precision over distances considerably less than 1 μm. Results obtained on other oxides indicate that it would be very profitable to redo many of the measurements of oxygen diffusion on the fluorite oxides by this much improved technique.

ACKNOWLEDGMENTS

I am grateful to the U.S. Department of Energy, and particularly to Dr. R. J. Gottschall, for continued support of our research on this subject. Drs. Hj. Matzke, D. K. Hohnke, M. F. Berard, J. A. Kilner, and G. E. Murch kindly contributed material, some not yet published, for inclusion. Finally, I wish to thank Dr. G. E. Murch and Prof. A. J. Burggraaf for many helpful comments on the manuscript.

REFERENCES

Abraham, M., Weeks, R. A., Clark, G. W., and Finch, C. B. (1965). *Phys. Rev. A* **137**, 138.
Abraham, M., Weeks, R. A., Clark, G. W., and Finch, C. B. (1966). *Phys. Rev.* **148**, 350.
Anderson, M. P., and Nowick, A. S. (1981). *J. Phys. (Paris)* **42**, C5-823.
Barker, W. W., and Knop, O. (1971). *Proc. Br. Ceram. Soc.* **19**, 15.
Bauerle, J. E. (1969). *J. Phys. Chem. Solids* **30**, 2657.
Belle, J., ed. (1961). "Uranium Dioxide: Properties and Nuclear Applications." U.S. Government Printing Office, Washington, D.C.
Beyeler, H. U., Brüesch, P., Pietronero, L., Schneider, W., Strässler, S., and Zeller, H. R. (1979). *In* "Physics of Superionic Conductors" (M. B. Salamon, ed.), Chap. 4. Springer-Verlag, Berlin.
Blumenthal, R. N., Brugner, F. S., and Garnier, J. E. (1973). *J. Electrochem. Soc.* **120**, 1230.
Boivin, J. C., and Thomas, D. J. (1981a). *Solid State Ionics* **3/4**, 457.
Boivin, J. C., and Thomas, D. J. (1981b). *Solid State Ionics* **5**, 523.
Bucci, C., Fieschi, R., and Guidi, G. (1966). *Phys. Rev.* **148**, 816.
Bunde, A., Chaturvedi, D. K., and Dieterich, W. (1982). *Z. Phys. B* **47**, 209.
Burggraaf, A. J., van Dijk, T., and Verkerk, M. J. (1981). *Solid State Ionics* **5**, 519.
Butler, V. (1982). Thesis. University College, London.
Butler, V., Catlow, C. R. A., Fender, B. E. F., and Harding, J. H. (1983). *Solid State Ionics* **8**, 109.
Cahan, M. T., van der Belt, T. G. M., de Wit, J. H. W., and Broers, G. H. J. (1980). *Solid State Ionics* **1**, 411.
Carter, R. E., and Roth, W. L. (1968). *In* "Electromotive Force Measurements in High Temperature Systems" (C. B. Alcock, ed.), pp. 125-144. Elsevier, Amsterdam.
Catlow, C. R. A. (1973). *J. Phys. C* **6**, L64.
Catlow, C. R. A. (1977). *Proc. R. Soc. London, Ser. A* **353**, 533.
Catlow, C. R. A. (1978). *Phys. Status Solidi A* **46**, 191.
Catlow, C. R. A. (1981). *In* "Nonstoichiometric Oxides" (O. T. Sorensen, ed.), Chap. 2. Academic Press, New York.
Catlow, C. R. A., and Lidiard, A. B. (1974). *In* "Thermodynamics of Nuclear Materials 1974," Vol. II, pp. 27-43. IAEA, Vienna.

Catlow, C. R. A., and Mackrodt, W. C., eds. (1982). "Computer Simulation of Solids." Lecture Notes in Physics, Vol. 166. Springer-Verlag, Berlin.
Catlow, C. R. A., and Parker, S. C. (1982). *In* "Computer Simulation of Solids" (C. R. A. Catlow and W. C. Mackrodt, eds.), Chap. 15. Springer-Verlag, Berlin.
Conflant, P., Boivan, J. C., and Thomas, D. J. (1976). *J. Solid State Chem.* **18**, 133.
Conflant, P., Boivan, J. C., and Thomas, D. J. (1980). *J. Solid State Chem.* **35**, 192.
Corish, J., and Jacobs, P. W. M. (1972). *J. Phys. Chem. Solids* **33**, 1799.
de Bruin, H. J., and Murch, G. E. (1973). *Philos. Mag.* **27**, 1475.
Dreyfus, R. W., and Nowick, A. S. (1962). *Phys. Rev.* **126**, 1367.
Esaka, T., Iwahara, H., and Kunieda, H. (1982). *J. Appl. Electrochem.* **12**, 235.
Etsell, T. H., and Flengas, S. N. (1970). *Chem. Rev.* **70**, 339.
Floyd, J. M. (1973). *Ind. J. Tech.* **11**, 589.
Fowler, R., and Guggenheim, E. A. (1949). "Statistical Thermodynamics," Chap. 9. Cambridge Univ. Press, Cambridge.
Franklin, A. D. (1972). *In* "Point Defects in Solids" (J. H. Crawford and L. M. Slifkin, eds.), Vol. 1, Chap. 1. Plenum, New York.
Freer, R. (1980). *J. Mater. Sci.* **15**, 803.
Gerhardt-Anderson, R., and Nowick, A. S. (1981). *Solid State Ionics* **5**, 547.
Harwig, H. A., and Gerards, A. G. (1978). *J. Solid State Chem.* **26**, 265.
Hohnke, D. K. (1979). *In* "Fast Ion Transport in Solids" (P. Vashishta, J. Mundy and G. Shenoy, eds.), pp. 669–672. North-Holland, Amsterdam.
Hohnke, D. K. (1981). *Solid State Ionics* **5**, 531.
Hohnke, D. K., and Hurley, M. D. (1984). To be published.
Johnson, H. B., Tolar, N., Miller, G. R., and Cutler, I. B. (1969). *J. Phys. Chem. Solids* **30**, 31.
Kikuchi, R. (1966). *Prog. Theor. Phys. Suppl.* 35, p. 1.
Kikuchi, R., and Sato, H. (1972). *J. Chem. Phys.* **57**, 4962.
Kilner, J. A., and Brook, R. J. (1982). *Solid State Ionics* **6**, 237.
Kilner, J. A., and Steele, B. C. H. (1981). *In* "Nonstoichiometric Oxides" (O. T. Sorensen, ed.), Chap. 5. Academic Press, New York.
Kilner, J. A., Ilkov, L., and Steele, B. C. H. (1984). *In* "Transport in Nonstoichiometric Compounds." *Solid State Ionics* in press.
Kim, K. C., and Olander, D. R. (1981). *J. Nucl. Mater.* **102**, 192.
Kingery, W. D., Pappis, J., Doty, M. E., and Hill, D. C. (1959). *J. Am. Ceram. Soc.* **42**, 393.
Kleitz, M., and Dupuy, J., eds. (1976). "Electrode Processes in Solid State Ionics." Reidel, Dortrecht.
Kofstad, P. (1972). "Nonstoichiometry, Diffusion and Electrical Conductivity in Binary Oxides," Chap. 12. Wiley, New York.
Kudo, T., and Obayashi, H. (1975). *J. Electrochem. Soc.* **122**, 142.
Lay, K. W., and Whitmore, D. H. (1971). *Phys. Status Solidi B* **43**, 175.
Le Claire, A. D. (1970). *In* "Physical Chemistry: An Advanced Treatise" Vol. X (H. Eyring, D. Henderson, and W. Jost, eds.), Chap. 5. Academic Press, New York.
Lidiard, A. B. (1954). *Phys. Rev.* **94**, 29.
Lidiard, A. B. (1957). *Handb. Physik.* **20**, 246.
Lidiard, A. B. (1974). *In* "Crystals with the Fluorite Structure" (W. Hayes, ed.), Chap. 3. Oxford Univ. Press (Clarendon), London and New York.
Macdonald, J. R. (1976). *In* "Electrode Processes in Solid State Ionics" (M. Kleitz and J. Dupuy, eds.), p. 149. Reidel, Dortrecht.
Mackrodt, W. C., and Stewart, R. F. (1979). *J. Phys. C* **12**, 5015.
Manning, J. R. (1968). "Diffusion Kinetics for Atoms in Crystals." Van Nostrand, Princeton, New Jersey.

Marin, J. R., and Contamin, P. (1969). *J. Nucl. Mater.* **30**, 16.
Matzke, Hj. (1976). *In* "Plutonium 1975 and Other Actinides" (H. Blank and R. Lindner, eds.), pp. 801–831. North-Holland, Amsterdam.
Matzke, Hj. (1981). *In* "Nonstoichiometric Oxides" (O. T. Sorensen, ed.), Chap. 4. Academic Press, New York.
Matzke, Hj. (1983a). Private Communication.
Matzke, Hj. (1983b). *J. Nucl. Mater.* **114**, 121.
Morinaga, M., Cohen, J. B., and Faber, J. (1979). *Acta Crystallogr. A* **35**, 789.
Morinaga, M., Cohen, J. B., and Faber, J. (1980). *Acta Crystallogr. A* **36**, 520.
Murch, G. E. (1982). *Solid State Ionics* **7**, 177.
Murch, G. E. (1983). *Diffus. Defect Data* **32**, 9.
Murch, G. E., and Thorn, R. J. (1978). *J. Nucl. Mater.* **71**, 219.
Naik, I. K., and Tien, T. Y. (1979). *J. Electrochem. Soc.* **126**, 562.
Nakamura, A., and Wagner, J. B., Jr. (1980). *J. Electrochem. Soc.* **127**, 2325.
Nowick, A. S. (1972). *Adv. Phys.* **16**, 1.
Nowick, A. S. (1979). *Comm. Solid State Phys.* **9**, 85.
Nowick, A. S., and Dreyfus, R. W. (1962). *J. Appl. Phys.* **33**, 473.
Nowick, A. S., and Heller, W. R. (1963). *Adv. Phys.* **12**, 251.
Nowick, A. S., and Heller, W. R. (1965). *Adv. Phys.* **14**, 101.
Nowick, A. S., and Park, D. S. (1976). *In* "Superionic Conductors" (G. Mahan and W. Roth, eds.), pp. 395–412. Plenum, New York.
O'Keeffe, M. (1970). *In* "The Chemistry of Extended Defects in Non-Metallic Solids" (L. Eyring and M. O'Keeffe, eds.), pp. 609–628. North-Holland, Amsterdam.
Rossell, H. J. (1976). *J. Solid State Chem.* **19**, 103.
Rossell, H. J. (1981). *Adv. Ceramics* **3**, 47.
Rossell, H. J., and Scott, H. G. (1977). *J. Phys. (Paris)* **38**, C7-28.
Sato, H., and Kikuchi, R. (1971). *J. Chem. Phys.* **55**, 677.
Simpson, L. A., and Carter, R. E. (1966). *J. Am. Ceram. Soc.* **49**, 139.
Steele, B. C. H., and Floyd, J. M. (1971). *Proc. Br. Ceram. Soc.* **19**, 55.
Strickler, D. W., and Carlson, W. G. (1965). *J. Am. Ceram. Soc.* **48**, 286.
Subbarao, E. C., and Ramakrishnan, T. V. (1979). *In* "Fast Ion Transport in Solids" (P. Vashishta, J. Mundy, and G. Shenoy, eds.), pp. 653–656. North-Holland, Amsterdam.
Subbarao, E. C., and Sutter, P. H. (1964). *J. Phys. Chem. Solids* **25**, 148.
Takahashi, T. (1972). *In* "Physics of Electrolytes" (J. Hladik, ed.), Vol. 2, Chap. 24. Academic Press, New York.
Takahashi, T., and Iwahara, H. (1973). *J. Appl. Electrochem.* **3**, 65.
Takahashi, T., and Iwahara, H. (1978). *Mater. Res. Bull.* **13**, 1447.
Takahashi, T., Iwahara, H., and Nagai, Y. (1972). *J. Appl. Electrochem.* **2**, 97.
Tskahashi, T., Esaka, T., and Iwahara, H. (1975a). *J. Appl. Electrochem.* **5**, 197.
Takahashi, T., Iwahara, H., and Arao, T. (1975b). *J. Appl. Electrochem.* **5**, 187.
Takahashi, T., Esaka, T., and Iwahara, H. (1976). *J. Solid State Chem.* **16**, 317.
Takahashi, T., Iwahara, H., and Esaka, T. (1977). *J. Electrochem. Soc.* **124**, 1563.
Tesch, R. J., Wirkus, C. D., and Berard, M. F. (1982). *J. Am. Ceram. Soc.* **65**, 511.
Thorn, R. J., and Winslow, G. H. (1966). *J. Chem. Phys.* **44**, 2632.
Tien, T. Y., and Subbarao, E. C. (1963). *J. Chem. Phys.* **39**, 1041.
Tuller, H. L. (1981). *In* "Nonstoichiometric Oxides" (O. T. Sorensen, ed.), Chap. 6. Academic Press, New York.
van Dijk, T., de Vries, K. J., and Burggraaf, A. J. (1980). *Phys. Status Solidi A* **58**, 115.
van Dijk, M. P., de Vries, K. J., and Burggraaf, A. J. (1983). *Solid State Ionics* **9/10**, 913.
Verkerk, M. J., and Burggraaf, A. J. (1981). *J. Electrochem. Soc.* **128**, 75.

Verkerk, M. J., Keizer, K., and Burggraaf, A. J. (1980). *J. Appl. Electrochem.* **10,** 81.
Verkerk, M. J., van de Velde, G. M. H., and Burggraaf, A. J. (1982). *J. Phys. Chem. Solids* **43,** 1129.
Wachtman, J. B., Jr. (1963). *Phys. Rev.* **131,** 517.
Wang, D. Y. (1983). Unpublished work.
Wang, D. Y., and Nowick, A. S. (1980). *J. Solid State Chem.* **35,** 325.
Wang, D. Y., and Nowick, A. S. (1983). *J. Phys. Chem. Solids* **44,** 639.
Wang, D. Y., Park, D. S., Griffith, J., and Nowick, A. S. (1981). *Solid State Ionics* **2,** 95.
Willis, B. T. M. (1963). *Proc. Br. Ceram. Soc.* **1,** 9.
Willis, B. T. M. (1964). *J. Phys. (Paris)* **25,** 431.
Willis, B. T. M. (1978). *Acta Crystallogr. A* **34,** 88.

4

Tracer Diffusion in Concentrated Alloys

H. BAKKER

LABORATORY OF NATURAL SCIENCES
UNIVERSITY OF AMSTERDAM
AMSTERDAM, THE NETHERLANDS

I.	Introduction	189
II.	Theoretical Background	191
III.	Empirical Rules	193
IV.	Theoretical Considerations of the Kinetics of Diffusion in Random Alloys	197
V.	Tracer Diffusion Experiments in Primary (Terminal) Phases	200
	A. Binary Face-Centered Cubic Alloys	200
	B. Binary Body-Centered Cubic Alloys	208
VI.	Theoretical Considerations of Diffusion in Ordered Structures	213
	A. Introduction	213
	B. Models for the Elementary Atomic Jump	215
	C. The Six-Jump Cycle Mechanism	216
	D. Expression for the Self-Diffusion Coefficient in Ordered Alloys in Terms of Averaged Quantities	220
	E. Kikuchi and Sato's Path Probability Method	221
	F. Monte Carlo Diffusion Simulation in Ordered Alloys	226
	G. Random-Walk Approach to the Calculation of Correlation Factors	228
	H. Alloys with Short-Range Order Only	231
VII.	Tracer Diffusion Experiments in Intermediate Phases	235
	A. Intermediate Phases with the B2 (CsCl) Structure	235
	B. Intermediate Phases with the DO_3 (BiF_3 or Fe_3Si) Structure	248
	C. Intermediate Phases with the $L1_2$ (Cu_3Au) Structure	251
VIII.	Conclusions	253
	References	253

I. Introduction

Hardly any metal is used for practical purposes in its pure form. Usually, mixtures of two or more metals are produced to obtain desired physical and chemical properties. In these production processes diffusion plays an important role. It determines the rate at which a casting homogenizes, precipitates are formed, or surface segregation occurs. Thus for the understanding of these phenomena knowledge of the values of diffusion coefficients is important.

A special class of alloys is the intermetallic compounds. At present about 700 of these intermediate phases are known to exist. In general, investigations on these compounds are scarce. Yet some of them are already used for diverse practical purposes; For example, Nb_3Sn, V_3Ga, and related compounds are used as technical superconductors, and Ag_3Sn as a dental powder. Knowledge of their diffusion behavior is of interest for the production of these materials or for their use in practical applications. The technical superconductors, for example, are produced embedded in copper or a bronze by a solid-state reaction, where diffusion plays an obvious role. The amalgamation of dental powders occurs also by a solid-state chemical reaction. Therefore, it is not surprising that tracer diffusion in these powders has been studied (Okabe *et al.*, 1976; Shires *et al.*, 1977).

Apart from its technical importance, diffusion is also interesting from a fundamental point of view. The first task is to investigate the mechanism by which the atoms migrate through the material. It appears that substitutional diffusion in alloys occurs via lattice vacancies. Diffusion measurements, therefore, give information on the number and mobility of these lattice defects. On the other hand, the diffusivity also strongly depends on the state of atomic order of the alloy and, in turn, diffusion measurements will contribute to the knowledge of ordering and disordering phenomena. For intermediate phases in particular, large numbers of (structural) vacancies combined with disordering may occur. Knowledge of the vacancy content, the degree of disorder, and the rate at which thermodynamic equilibrium is established, if at all at a certain temperature, is essential for the understanding of the physical properties of these materials.

In this chapter a number of tracer diffusion experiments in concentrated binary alloys will be described. A concentrated alloy will be defined (arbitrarily) to contain at least 10 atomic % of any component. We will confine ourselves to homogeneous systems in which mainly self-diffusion will be considered. Moreover, the systems will have a metallic character; For example, work on group III–V compounds such as GaAs will not be included. Neither diffusion of the well-known interstitial impurities such as H, C, O and N, nor the effects of grain boundaries, dislocations, interphase boundaries, radiation damage, magnetic transitions, and driving forces will be reviewed. Apart from describing experimental results, we will pay attention to the various methods that have been developed to interpret diffusion measurements in these systems: empirical rules, analytical theories, and Monte Carlo computer simulation results.

The contents of the chapter are as follows. After some introductory remarks on diffusion in pure metals, which comprises Section II, Section III deals with empirical rules from which rough estimates of diffusion parameters in primary phases can be made. In Section IV, Manning's atomistic theory of the kinetics of diffusion in random alloys will be presented. Tracer diffusion experiments in primary (terminal) face-centered cubic (fcc) and body-centered cubic (bcc) phases will be reviewed in Section V. A number of investigations in these alloys has been carried out in order to test the validity of empirical rules and to obtain numerical values of empirical constants describing diffusion parameters as a function of the concentration of the solute. It is to be expected that Manning's random alloy theory can be applied to alloys where long-range order is absent and the degree of short-range order is low. Examples of such applications will be given. Finally, in this section, diffusion experiments in primary titanium alloys will be discussed in connection with the anomalous behavior of diffusion in pure titanium. In contrast to a primary phase, where the crystallographic structure of the majority component is maintained, and where physical properties usually vary gradually as a function of the concentration of the solute, the structure of an intermediate phase in the phase diagram may be quite different from that of both components. Such a phase may also display physical properties that do not resemble those of the constituting species. These systems usually crystallize in ordered structures that are sometimes called intermetallic compounds. Therefore, before proceeding to a discussion of experimental results of diffusion in these phases, in Section VI diffusion in ordered structures will be considered from a theoretical point of view. Finally, in Section VII a review of experimental diffusion studies in intermediate phases crystallizing in various crystallographic structures will be given.

A number of reviews of the same or related topics have been published before. Chapters on diffusion in alloys and intermetallic compounds are found in the treatise by Adda and Philibert (1966); Girifalco (1973) described diffusion in binary order–disorder alloys and Larikov *et al.* (1975) reviewed diffusion processes in ordered alloys. Diffusion data in copper and copper alloys have been collected and critically evaluated by Butrymowicz *et al.* (1977). Abstracts of papers on diffusion are being published regularly in *Diffusion and Defect Data* (1967 to date).

Without pursuing absolute completeness, our presentation of experimental and theoretical results will range mainly over work published in the past fifteen years.

II. Theoretical Background

Since the description of diffusion in alloys will appear to be an extension of that in pure metals, we will summarize briefly the essentials of self-diffusion in pure metals (Shewmon 1963; Girifalco, 1964a; Adda and Philibert, 1966).

Let us consider as an example a pure bcc metal and diffusion in the (100) direction. Then the tracer diffusion coefficient is given by

$$D = \langle X^2 \rangle / 2t \tag{1}$$

where X is the displacement of an atom in the (100) direction during a long time t. This displacement is made up of n elementary jump vectors to nearest-neighbor positions, all of length x, and with a projection x_i along the (100) direction:

$$X = \sum_{i=1}^{n} x_i \tag{2}$$

This yields

$$D = nx^2 f / 2t \tag{3}$$

with

$$f = 1 + \frac{2}{n} \sum_{i=1}^{n-1} \left(\frac{\langle x_i x_{i+1} \rangle}{x^2} + \frac{\langle x_i x_{i+2} \rangle}{x^2} + \cdots \right) \tag{4}$$

where f, called the correlation factor, accounts for the nonrandomness of the directions of jumps in a sequence. This correlation effect diminishes the efficiency of the atomic jumping for displacements over longer distances. When diffusion occurs by atom–vacancy exchanges, the atomic jump frequency is given by

$$n/t = zc_V w \tag{5}$$

Here c_V is the probability of finding any lattice site unoccupied, z is the number of nearest neighbors, and w the atom–vacancy exchange rate. This leads to a diffusion coefficient of the Arrhenius form

$$D = D_0 \exp(-Q/kT) \tag{6}$$

where k is the Boltzmann constant, T is absolute temperature, and the pre-exponential D_0 is

$$D_0 = fa^2 v \exp[(S_f + S_m)/k] \tag{7}$$

where a is the lattice parameter and v a vibrational frequency of the order of the Debye frequency. The activation entropies for the formation and migration of a vacancy are S_f and S_m, respectively. The activation energy Q is composed of corresponding enthalpies

$$Q = H_f + H_m \tag{8}$$

The correlation factor can be shown to equal (Manning, 1968)

$$f = H_0/(H_0 + 2w^*) \tag{9}$$

4. TRACER DIFFUSION IN CONCENTRATED ALLOYS

Here H_0 is the escape frequency of the vacancy from the tracer, which is a function of the host-atom–vacancy exchange rate w. The quantity w^* is the tracer–vacancy exchange rate. For self-diffusion, $w^* \simeq w$, so that we usually omit the asterisk. On the basis of Eq. (9), f becomes a numerical factor for self-diffusion in a pure metal. By using Eq. (9) it can be shown that the isotope effect parameter E, measurable by the diffusion of two different isotopes a and b of the same element, is

$$E \equiv [1 - (D_a/D_b)]/[1 - (m_b/m_a)^{1/2}] = f \, \Delta K \qquad (10)$$

where m_a and m_b are the (slightly) different masses of the isotopes. The so-called kinetic energy factor ΔK is the fraction of the kinetic energy of the jump that resides in the jumping atom (Peterson, 1975). Equation (10) makes f, in principle, accessible to experimental determination.

III. Empirical Rules

The validity of a number of empirical rules for diffusion in pure metals (Shewmon, 1963) has now been extended to alloys. A thorough reexamination of these rules and a test of their validity has been recently published (Brown and Ashby, 1980). These rules are presented below as Rules 1–4. The numerical values were taken from Brown and Ashby (1980).

Rule 1. The diffusion coefficient at the solidus temperature T_s is roughly constant for a given structure and bond type.

For the fcc structure $D(T_s) \simeq 5 \times 10^{-9}$ cm^2 s^{-1}; For the bcc structure (with the exception of the alkali metals, ε-Pu, δ-Ce, γ-La, and γ-Yb) $D(T_s) \simeq 3 \times 10^{-8}$ cm^2 s^{-1}. A variation of this rule is:

Rule 2. The diffusion coefficient at the melting point is constant over the phase diagram.

Rule 3. Q/kT_s is roughly constant for a given structure and bond type.

For the structures mentioned under Rule 1, the value for this quantity is

$$Q/kT_s \simeq 18 \qquad (11)$$

Rule 4. The preexponential is roughly constant for a given structure and bond type.

This rule is a consequence of Rules 1–3. For the fcc structure, $D_0 \simeq 0.3$ cm^2 s^{-1}; For the bcc structure, with the exceptions mentioned above, $D_0 \simeq 1.6$ cm^2 s^{-1}. From the evaluation of experimental material it turns out that the numerical values given in Rules 1–4 show large standard deviations. The standard deviation in the value given for Q/kT_s is typically ± 2. Individual values for $D(T_s)$ and D_0 may differ even by a factor of 10 from the values given in Rule 1 and Rule 4, respectively (Brown and Ashby, 1980).

The following rules apply especially to dilute alloys. They predict, essentially, the influence of concentration on the diffusion coefficients.

Rule 5.   The addition of a faster diffusing species as a finite impurity causes increases in the rates of diffusion of both constituents, whereas the opposite effect is found for the addition of slow-diffusing impurities (Lazarus, 1960).

Rule 6. (Le Claire's Rule).   *An increase in concentration of a solute increases the diffusion rate of that component at a given temperature if it also decreases the melting point of the solvent, or decreases the diffusion rate if it increases the melting point* (see Murdock and McHargue, 1968).

The validity of the Rule 6 has been extended also to concentrated alloys (Murdock and McHargue, 1968).

Rules 7 and 8 give a quantitative description of the concentration dependence of the diffusion coefficients in (concentrated) alloys.

Rule 7.   At a fixed temperature the concentration dependence of the diffusion coefficient of the majority component A in the alloy AB can be described by

$$D_{AB}^A = D_A^A \exp(bc_B) \tag{12}$$

Here c_B is the fraction of B atoms in the alloy and D_A^A is the self-diffusion coefficient in pure A. The variable b is generally temperature dependent (Hoffmann et al., 1955).

Rule 8.   At a certain fraction $a = T/T_{mAB}$ of the temperature T_{mAB} the logarithm of the diffusion coefficient of any component in the alloy is a linear function of composition:

$$\ln D_{AB}^Z(aT_{mAB}) = \ln D_A^Z(aT_{mA}) + K^Z c_B \tag{13}$$

Here $Z = A$ or B, the superscript denotes "diffusion of," the subscript "diffusion in." The melting point of pure A is T_{mA} and T_{mAB} is defined as the average of the solidus and liquidus temperature at the composition c_B (Kučera and Million, 1970a). In fact, this rule is a refinement and extension of Rule 7. In many cases K^Z was found to be a function of temperature (see Section V). Kučera and Million (1970a) argued that because of the Arrhenius behavior of the diffusion coefficients, K must be of the form

$$K^Z = K_0^Z + \frac{K_1^Z T_{mAB}}{T} \tag{14}$$

4. TRACER DIFFUSION IN CONCENTRATED ALLOYS

As a consequence of Eq. (14), the concentration dependence of Q and D_0 are

$$D_{0AB}^Z = D_{0A}^Z \exp(K_0^Z c_B) \tag{15}$$

$$Q_{AB}^Z/kT_{mAB} = Q_A^Z/kT_{mA} - K_1^Z c_B \tag{16}$$

Kučera and Million (1970a) also investigated systems exhibiting complete mutual solubility over the whole concentration range and forming (nearly) regular solutions. It is not difficult to demonstrate that for such systems Eq. (16) leads to

$$Q_{AB}^Z/T_{mAB} = c_A(Q_A^Z/T_{mA}) + c_B(Q_B^Z/T_{mB}) \tag{17}$$

This result can be considered as a generalization of Rule 3. Furthermore,

$$D_{0AB}^Z = (D_{0A}^Z)^{c_A} \cdot (D_{0B}^Z)^{c_B} \tag{18}$$

If all other factors except the entropies in the preexponentials are constant, Eq. (18) leads to

$$S_{AB}^Z = c_A S_A^Z + c_B S_B^Z \tag{19}$$

where S_A^A is the activation entropy for self-diffusion in pure A, etc. Before proceeding to the next rule, it must be remarked that Kučera and Million's use of T_{mAB} (the average of the solidus and liquidus temperature) can be criticized, because both the liquidus and solidus temperature are determined also by the free energy in the *liquid* state. It is not clear a priori what the free energy in the liquid state has to do with diffusion in the *solid* state (Radelaar, 1982).

In Rule 9 the deviations of the diffusion coefficients of both components from ideal behavior are related.

Rule 9. *If the deviation from ideal behavior of component* $Z(= A \ or \ B)$ *in the alloy* AB *is defined as*

$$\Delta \ln D_{AB}^Z \equiv \ln D_{AB}^Z - (c_A \ln D_A^Z + c_B \ln D_B^Z) \tag{20}$$

then this deviation is equal for both components at a given temperature and composition. In mathematical terms,

$$\Delta \ln D_{AB}^A = \Delta \ln D_{AB}^B \tag{21}$$

This is an empirical rule of Borovskiy *et al.* (1970). These authors demonstrated on the basis of calculations using nearest-neighbor pair interactions that the relation is satisfied automatically for regular solutions. Ugaste (1971) examined the implications of the above rule. He obtained a linear relationship between the difference in activation energy for the diffusion of both components and composition. His expression is not given here explicitly, because it immediately follows from Eq. (17) if one takes $T_{mAB} = T_{mA} = T_{mB}$. He also obtained an equation for the logarithm of the ratio of the pre-exponentials. By use of Eq. (18), this relation also follows directly.

The final rule, Rule 10, relates the activation energy to the preexponential.

Rule 10.   For a given alloy system $\ln D_0$ is a linear function of Q/kT_s.

This rule would be trivial if Rules 3 and 4 were exactly obeyed, because following these rules $\ln D_0$ and Q/kT_s are constants.

Calculations have been performed with the aim of giving some of the empirical rules cited above a theoretical base. We shall conclude this section with a brief discussion of this work. The calculations of Beke et al., (1979) were performed for random alloys. Essential to their treatment was the assumption of an interatomic pair potential of the form

$$\phi_{ij}(|\mathbf{r}_k - \mathbf{r}_l|) = \varepsilon_{ij} \phi \left(\frac{|\mathbf{r}_k - \mathbf{r}_l|}{a} \right) \tag{22}$$

The atom of species i ($=$ A or B) is situated at the position $\mathbf{r}_k$, the atom of species j ($=$ A or B) at $\mathbf{r}_l$, a is the lattice parameter, ε_{ij} are parameters with the dimension of an energy. These energy parameters were taken to be constant over the composition range studied. Also ε_{AA} and ε_{BB} were assumed to be representative of the energy parameters in pure A and B, respectively. First the potential Eq. (22) was used to calculate the crystal energy of the perfect, random binary alloy. As a consequence of the very form of the potential such a summation yields a function of the ε_{ij}s and c_B multiplied by a numerical factor. This factor results from the summation of ϕ over the crystal and is a universal constant for crystals of the same structure. In the same way, the energy of the crystal with a vacancy and of the crystal with an atom in the saddle-point position were obtained. Relaxations were taken into account. For numerical estimates of the quantities occurring in the expressions, use was made of the melting points of the pure components and the formation energy of the alloy. From these calculations Beke et al. (1979) obtained a linear relationship between the difference in activation energy of both components and concentration. For a number of systems they also reproduced Kučera and Million's (1970a) empirical rule and showed a relationship to exist between the empirical coefficients K_0^Z and K_1^Z. For details, the reader is referred to the original paper.

In conclusion, it must be emphasized that the ten empirical rules given in this section are only rules of thumb without absolute validity. Taken together, some of these rules are even contradictory. The fact that some of the rules can be reproduced theoretically is encouraging for their validity. However, we must bear in mind the limitations of the model and the approximations used in this theoretical treatment. Nevertheless, the above rules can

4. TRACER DIFFUSION IN CONCENTRATED ALLOYS

be used with caution in order to estimate diffusion parameters in cases, where experimental data are not available. In Section V a number of the rules will be applied to experimental results and values of the empirical constants K_0 and K_1 will be obtained.

IV. Theoretical Considerations of the Kinetics of Diffusion in Random Alloys

In the previous section a number of empirical rules were given by which values for the activation energies, preexponentials, and diffusion coefficients in alloys can be estimated. These rules have a phenomenological character. In this section, diffusion in alloys will be treated on an atomistic level. However, in order to perform calculations of the kinetics of diffusion in alloys in terms of atomic quantities, the alloy model to be used must first be specified. As a natural starting point for such calculations Manning (1968, 1971a, 1971b, 1971c) chose a simple alloy model, the random alloy model. Later on, more sophisticated models were elaborated by several authors (see Section VI). In this section, Manning's treatment will be reviewed.

A random alloy is defined as a homogeneous alloy, where no correlation between the occupation of various lattice sites exists. This means that for a binary alloy with components A and B the probability of finding an A atom on any lattice site is given just by the fraction c_A of A atoms. The same also holds for the B component. For such a random alloy, Manning postulated also a random distribution of vacancies. This means that the probability of finding a lattice site unoccupied is independent of the occupation of its nearest neighbor (nn) positions. Moreover, Manning assumed the vacancy fraction c_V to be small and diffusion to occur by nn vacancy–atom exchanges, w_A and w_B being the exchange rates for A and B atoms, respectively. In contrast to pure metals, the vacancy does not pursue a real random walk, through an alloy. The directions of subsequent vacancy jumps are not independent, but correlated. This introduces a vacancy correlation factor $f_V \neq 1$. Similar to the case of self-diffusion or impurity diffusion in pure metals, the atomic self-diffusion coefficients in alloys will be proportional to the probability c_V of finding an unoccupied nn position of the atom, to the atom–vacancy exchange rates w_A and w_B, respectively, and also to the correlation factors f_A and f_B, reflecting the cooperative character of diffusion in alloys. This results for bcc and fcc alloys in the expression (i = A or B)

$$D_i = f_i c_V w_i a^2 \qquad (23)$$

where a is the lattice parameter. In a treatment of the kinetics of diffusion in such alloys, the main difficulty is the evaluation of the correlation factors in terms of exchange frequencies and composition. Crucial in Manning's approach is the assumption that the correlation factors are of the same form as for self-diffusion and diffusion of an isolated impurity in pure metals [Eq. (9)],

$$f_i = H_a/(2w_i + H_a) \tag{24}$$

It is also assumed that in the alloy the frequency H_a with which the vacancy escapes from the atom under consideration does not depend on the species of atom that has made the jump. This is analogous to the case of diffusion of an isolated impurity in a pure crystal, where the expression for the escape frequency only contains host atom–vacancy exchange rates (Manning, 1968). Because of the correlated nature of the vacancy jumps, the escape frequency H_a will be proportional to the effective vacancy jump frequency $f_V w_V$, rather than to the vacancy jump frequency w_V only,

$$H_a = I_0 f_V w_V \tag{25}$$

The proportionality constant I_0 is chosen in such a way that, in the case of self-diffusion in a pure solid, the correlation factor f_0 for self-diffusion results. The problem now is to find an expression for $f_V w_V$ in terms of atomic correlation factors and exchange rates. When one realizes that among a large number n_V of vacancy jumps a number n_A will be due to exchanges with A atoms and a number n_B due to exchanges with B atoms, partial vacancy correlation factors f_V^A and f_V^B can be introduced as

$$f_V n_V = f_V^A n_A + f_V^B n_B \tag{26}$$

The relation of these partial vacancy correlation factors to the atomic correlation factors can be shown to be (Manning, 1971a)

$$f_V^i = f_i/f_0 \tag{27}$$

Since

$$n_A \propto z c_A w_A \tag{28}$$

where z is the number of nearest neighbors, the expression for $f_V w_V$ becomes

$$f_V w_V = f_0^{-1}(z c_A f_A w_A + z c_B f_B w_B) \tag{29}$$

Substituting this result into Eq. (24) using Eq. (25), both correlation factors can be obtained as a function of exchange frequencies and composition as was the aim of the calculation. However, after some algebra the result can also be brought in a more practicable form in terms of diffusion coefficients,

$$f_i = 1 - \frac{(1 - f_0)D_i}{c_A D_A + c_B D_B} \tag{30}$$

4. TRACER DIFFUSION IN CONCENTRATED ALLOYS

From the above equation, correlation factors can be deduced if D_A and D_B have been measured. And, from the temperature dependence of D_A and D_B, the temperature dependence of the correlation factors can be found. This latter function is important, because it contributes to the apparent activation energy of diffusion (Le Claire, 1962) as found from the slope of the Arrhenius plots. The real difference in activation energy for A and B jumps is derived from the temperature dependence of (see also Manning, 1971b)

$$\frac{w_A}{w_B} = \frac{(1-f_A)f_A^{-1}}{(1-f_B)f_B^{-1}} \tag{31}$$

which result is easily obtained from Eq. (30). From Eq. (31) it is clear that $w_A > w_B$ implies $f_A < f_B$, and so differences in the jump frequencies are decreased by correlation effects. From Eq. (30) it also follows that for the random alloy

$$c_A f_A + c_B f_B = f_0 \tag{32}$$

We will see in Section VII.A.1 that this equation is no longer obeyed for ordered alloys. Since negative values for the correlation factors are not allowed, at lower concentrations c_A not all ratios of the diffusion coefficients of A and B are permissible [Eq. (30)]. In Fig. 1 this forbidden region is shown for fcc alloys. If for an alloy system negative values are found in this region, the random alloy model is clearly not applicable to the alloy.

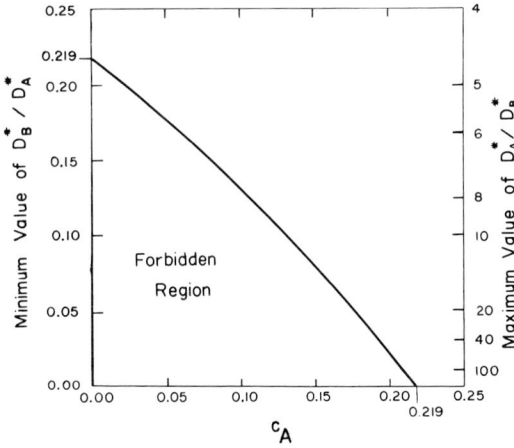

Fig. 1. Forbidden region for diffusion by vacancy mechanism in random binary fcc alloy. Values of D_A^*/D_B^* in the forbidden region are not allowed in the random alloy model. [From Manning (1971a).]

The validity of the results from Manning's calculations was confirmed by Monte Carlo computer simulation of diffusion in fcc, bcc and sc alloys by De Bruin *et al.* (1975, 1977) and by Murch and Rothman (1981). The Manning values were reproduced nearly exactly by these studies. More details on Monte Carlo diffusion simulation are given in Chapter 7.

It is to be expected that Manning's theory could be applied successfully to primary phases without long-range order and where the degree of short-range order is low. Examples of such applications will be given in Section V.

V. Tracer Diffusion Experiments in Primary (Terminal) Phases

A. BINARY FACE-CENTERED CUBIC ALLOYS

1. *The Application of Empirical Rules*

In a number of primary metallic fcc phases, diffusion experiments were undertaken to test the validity of empirical rules and to obtain a description of the diffusion parameters as a function of concentration. Some of these rules, such as Rule 8 and Rule 9, have been suggested to apply to regular solutions (Kučera and Million, 1970a; Borovskiy *et al.*, 1970; see also Beke *et al.*, 1979). Although some short-range order will always be present in an alloy, it will appear that, for example, the Kučera and Million empirical rule is obeyed for a considerable number of systems. Deviations from the empirical rules at certain compositions may indicate the appearance of a substantial degree of ordering.

The systems Ag–Au, Cu–Ni, Co–Ni, and Fe–Ni are examples in which the primary phase extends over the whole composition range. The Eqs. (17) and (18) from Section III were verified by Kučera and Million (1970a) for gold and silver diffusion in Ag–Au (Mallard *et al.*, 1963; Mead and Birchenall, 1957) and for copper and nickel diffusion in Cu–Ni (Monma *et al.*, 1964a). The results for Ag–Au are displayed in Fig. 2. These equations could also be applied to nickel diffusion in Co–Ni (Million and Kučera, 1971) and to nickel diffusion in Fe–Ni (Kučera and Stránský, 1982).

Kučera and Million (1970a) fitted lead and thallium diffusion measurements in Pb–Tl alloys with 0–40 at. % Tl (Resing and Nachtrieb, 1961) to the empirical Eqs. (15) and (16), and obtained numerical values for the empirical constants K_0 and K_1 (see Table I). Resing and Nachtrieb inferred from their measurements the existence of a superlattice near the composition $PbTl_7$. In Table II results are given for K_0 and K_1 for silver diffusion in Ag–Al and Ag–Pd (Kučera and Million, 1970a). The Kučera and Million empirical rule was also applied successfully to copper diffusion in copper-rich Cu–Al (up to 12 at. % Al), yielding a K_1^{Cu} value nearly

4. TRACER DIFFUSION IN CONCENTRATED ALLOYS

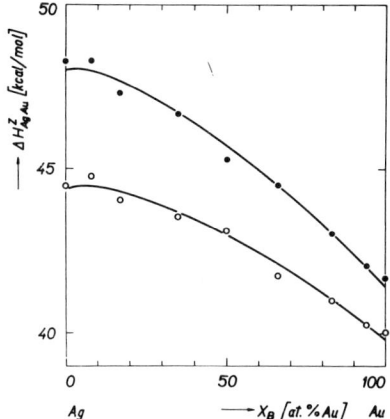

Fig. 2. Concentration dependence of silver (open circles) and gold (closed circles) activation energies in the Ag–Au system. The solid lines were calculated using Eq. (17). [From Kučera and Million (1970a); copyright American Society for Metals 1970).]

equal to zero (Kučera and Million, 1970b). Also, for Cu–Zn a linear dependence of $\ln D_{CuZn}^{Zn}$ on concentration was found at certain fractions of the temperature T_{mCuZn} [Eq. (13)]. These measurements were performed by Kučera et al. (1972). The concentration range studied was from 8–38 at. % Zn and the temperature interval from 390–810°C. As a consequence also the Eqs. (15) and (16) were obeyed. By inserting the T_{mCuZn} dependence on concentration the preexponential and the activation energy could be described directly as a function of concentration. Silver and cadmium self-diffusion measurements in Ag–Cd by Schoen (1958) were extended by

TABLE I

PRIMARY FCC PHASES EXTENDING OVER THE WHOLE COMPOSITION RANGE OBEYING EQS. (17) AND (18) FROM SECTION III

System (A–B)	Composition (at. % B)	Diffusing element	Reference
Ag–Au	0–100	Ag, Au	Kučera and Million (1970a)
Cu–Ni	0–100	Cu, Ni	Kučera and Million (1970a)
Fe–Ni	0–100	Ni	Kučera and Stransky (1982)
Co–Ni	0–100	Ni	Million and Kučera (1971)[a]

[a] Hässner and Lange (1965) found a deviation at the 95 at. % Co composition. Million and Kučera (1971) found a local maximum in D for cobalt diffusion near the composition $CoNi_3$. This was not corroborated by low temperature measurements (Million, 1972).

TABLE II
Values of K_0 and K_1 from Eqs. (15) and (16) for FCC Primary Phases

System (A–B)	Composition (at. % B)	Diffusing element	$K_0{}^a$	$K_1{}^a$	Reference	Deviations at
Ag–Al	0–15	Ag	8.9	−8.3	Kučera and Million (1970a)	—
Ag–Cd	0–30	Ag	−5.4	2.8	Kučera and Million (1970a)	31–37 at. % Cd
Ag–Pd	0–25	Ag	−16.1	16.1	Kučera and Million (1970a)	—
Al–Zn	0–53	Zn	0.03 ± 0.28	−1.84 ± 0.3	Gödeny et al. (1975) Cermak et al. (1980)	53–57.5 at. % Zn
Cu–Al	0–12	Cu	2.4	1.1	Kučera and Million (1970b)	—
Cu–Zn	8–38	Zn	−10.0 ± 1.4	—^{b}	Kučera et al. (1972)	—
Pb–Tl	0–40	Pb	−3.4	6.2	Kučera et al. (1970a)	PbTl$_7$
Pb–Tl	0–40	Tl	2.4	−5.3	Kučera et al. (1970a)	PbTl$_7$

^{a} c_B in Eqs. (15) and (16) in atomic fraction.
^{b} The concentration dependence of T_{mCuFn} was included in the analysis.

Gardner et al. (1968) to concentrations near the phase boundary, i.e., 31–37 at. % Cd. Activation energies for the diffusion of both constituents decreased linearly with cadmium concentration, but leveled out at the higher concentrations, where the measurements by Gardner et al. were made. The authors interpreted this effect as a result of a substantial degree of short-range order at these higher concentrations. The silver diffusion data in the range 0–30 at. % Cd could be fitted to the Eqs. (15) and (16) (Kučera and Million, 1970a). Kučera and Million (1975) reported results on the diffusion of platinum in ten Fe–Pt alloys, containing 15–60 at. % Pt at temperatures from 780–1420°C. All compositions showed an Arrhenius behavior within experimental error and yielded D_0 and Q values compatible with a single vacancy mechanism. The Q/T_m values exhibited a pronounced maximum at the equiatomic composition due to the existence of the ordered FePt structure.

Zinc diffusivity in Al–Zn alloys was studied by Hilliard et al. (1959) in the composition range up to 62.9 at. % Zn, by Gödény et al. (1975) *up to* 58.5 at. % Zn, and by Cermák et al. (1980) up to 57.5 at. % Zn. There is fair agreement between the results of these investigations for the 0–53 at. % Zn alloys. Although the Al–Zn system is far from being an ideal one, Gödény et al. found that the empirical Eqs. (15) and (16) were obeyed for the 0–30 at. % Zn compositions. By adding values for another seven compositions, Cermák et al. even demonstrated a validity in the range 0–53 at. % Zn. Another interesting result was that extrapolation of Q and $\ln D_0$ to 100% zinc yielded values that were compatible with measured self-diffusion coefficients in pure polycrystalline zinc. This gives support to the concept that the polycrystalline fcc and hcp structures are comparable from the point of view of diffusion. This had been suggested before by Lazarus (1965), and was confirmed later by Brown and Ashby (1980). Furthermore, Cermák et al. found increased Q/T_m and $\ln D_0$ values in solid solutions with ~ 53–57.5 at. % Zn. This effect was attributed to the occurrence of the β phase (Elliot, 1965). Chatterjee and Fabian (1970) also studied the Al–Zn system and suggested a substantial influence of grain boundary diffusion below 400°C (0–20 at. % Zn). Values for the activation energy and preexponential for the diffusion of zinc in monocrystalline aluminum were unusually large and moreover not compatible with results on the same material given by Peterson and Rothman (1970a). Besides, the number of measured values in Chatterjee and Fabian's experiment is quite small, thereby casting doubt on their general conclusions.

Hässner and Lange (1965) diffused both nickel and cobalt tracers into thirteen Co–Ni alloys in the composition range 0–100 at. % Ni at temperatures from 1100 to 1416°C. In this system the primary phase extends over all concentrations. The chosen temperatures were high enough to avoid the effects of the paramagnetic–ferromagnetic transition. The results were analyzed in terms of an Arrhenius behavior. The extrapolation of the diffusion coefficients to the solidus temperature gave values of $D(T_s) = 5-7 \times 10^{-9} \text{ cm}^2 \text{ s}^{-1}$, in excellent agreement with the values given in Rule 1 of Section III. Although there was a slight tendency for the cobalt diffusion coefficients to be somewhat higher than the nickel diffusion coefficients, both diffusion coefficients were nearly equal and so were the D_0 and Q values within experimental error. At all tempertures a deviation of the diffusion coefficients of both species from ideal behavior was observed, i.e., $\Delta \ln D$ from Eq. (20) was not equal to zero for all concentrations. This deviation increased to lower temperatures and had its maximum at the equiatomic composition. However the empirical rule Eq. (21) was always obeyed. Furthermore, a linear relation was found between $\ln D_0$ and Q/T_s (Rule 10). A local maximum occurred in the activation entropy and energy near the 95 at. % Co composition both for cobalt and for nickel diffusion. Million and Kučera (1969) reexamined the diffusion of cobalt in alloys with 20–100 at. % Co and at temperatures from 1050 to 1300°C. Measurements at nearby compositions showed fair agreement for Q and D_0 with the Hässner and Lange (1965) results. The investigation was undertaken especially to study the behavior at the composition $CoNi_3$, where strong short-range order is expected, because of the evidence of a structure with long-range order at lower temperatures (Hansen, 1958). At this composition a local maximum was found in $\ln D$ corresponding to a local minimum in the activation energy and preexponential. These effects were absent in a study of nickel diffusion in the same compositions by the same authors (Million and Kučera, 1971). This was possibly due to the asymmetry in composition of $CoNi_3$. In contrast to cobalt diffusion, the diffusion behavior of nickel in this system offered an excellent example of the Kučera and Million rules [Eqs. (17) and (18)]. The experimental values of D_0 and Q turned out to be definitely higher than those from Hässner and Lange (1965) over the whole composition range. The effect of ordering on the cobalt diffusion in 0–50 at. % Co Co–Ni alloys was examined at lower temperatures (580–980°C) by Million (1972). In spite of the evidence of superlattice formation in the vicinity of the composition $CoNi_3$ (Hansen, 1958), surprisingly no effect within experimental error was found in the vicinity of this composition. It is not clear how this can be reconciled with the high-temperature results.

It was mentioned above that the Co–Ni system is an example of the applicability of the empirical Rule 10, the linear relationship between $\ln D_0$ and Q/T_s. This rule was also demonstrated to apply to the systems Ag–Au, Cu–Ni, and Pb–Tl (Beke et al., 1979).

In Table I results are given for alloys for which the Kučera and Million empirical rules (Section III, Rule 8) were obeyed. Table I gives the systems in which the primary phase ranges over all concentrations and for which

Eqs. (17) and (18) were found to be valid. In Table II are presented the results on diffusion in alloys that were analyzed in terms of Eqs. (15) and (16). From the value of the constant K_1, the difference in Q/kT_m between the diffusion in the pure metal and in the alloy is directly derivable. This difference turns out to be not very dramatic in most cases. With the exception of silver diffusion in Ag-Pd the variation of Q/kT_{mAB} with concentration is within the standard deviation of ± 2 given for Rule 3. The variation of D_0 with concentration is derivable from the constant K_0. The evaluation of such a variation also gives an idea of the validity of Rule 4. For most systems the variation is within a factor of 10. For silver diffusion in Ag–Pd, and also for zinc diffusion in Cu–Zn, the variation is larger. Undoubtedly, the empirical rule studied most systematically is Kučera and Million's Rule 8. Such studies reveal which systems this rule is valid for and where deviations occur, but they are also useful for the evaluation of unknown diffusion coefficients by interpolation. However, such investigations do not give much information of what occurs on an atomistic level.

2. The Application of Manning's Random Alloy Model

One can also consider diffusion measurements in primary alloys from another point of view, by interpreting the results in terms of Manning's model (Manning, 1971a; Section IV), and estimating correlation factors and their contribution to the apparent activation energies. Then measurements of the diffusivities of both components are required. This approach has been demonstrated to be extremely fruitful for the interpretation of isotope effect investigations.

An excellent confirmation of Manning's random alloy model for α-CuZn (30.6 at. % Zn) at 895.8°C was given by Peterson and Rothman (1970b) in the measurement of the isotope effect parameters for both copper and zinc. The ratio of both isotope effect parameters turned out to be

$$E_{Zn}/E_{Cu} = (f \Delta K)_{Zn}/(f \Delta K)_{Cu} = 0.706 \pm 0.023$$

whereas on the basis of the ratio of both diffusion coefficients Eq. (30) predicts $f_{Zn}/f_{Cu} = 0.650 \pm 0.025$. If one assumes the ΔK factors for zinc and copper diffusion to be equal, the agreement between theory and experiment is quite satisfactory. Thus it appears that short-range order in this alloy has no dramatic effect on the correlation factors. As an additional result, a value of $\Delta K = 0.76 \pm 0.03$ was obtained. At about the same time a similar experiment in 50 at. % fcc Fe–Co was published by Fishman et al. (1970), yielding at 1060°C

$$E_{Fe}/E_{Co} = 0.92 \pm 0.10$$

while according to Manning at that temperature we have $f_{Fe}/f_{Co} = 0.95 \pm 0.05$. This again confirmed the Manning model. From the experiments a ΔK value of 0.88 ± 0.10 was estimated, in good agreement with ΔK values found for fcc systems (Peterson, 1975; Mehrer, 1978; Herzig et al., 1978). In the same alloy Fishman et al. (1970) also measured diffusion coefficients of both species as a function of temperature (1012–1164°C). Hirano and Cohen (1972) studied cobalt diffusion in Fe–49.6 at. % Co as well. Unfortunately, the value given in this paper for the activation energy is obviously in error, because it does not reproduce their measurements. Their diffusion coefficients appear to be somewhat smaller than those of Fishman et al., which, on the other hand, fit in well with those of Wanin and Kohn (1968). An analysis of the measurement of Fishman et al. on the basis of the Manning model yields $f_{Fe} \simeq 0.75$, $f_{Co} \simeq 0.81$, with only a small temperature dependence. By using straight line fits to Eq. (30) the apparent activation energies caused by the temperature dependence of the correlation factors can be obtained as

$$Q(f) \equiv -d(\ln f)/d(1/kT) \qquad (33)$$

The apparent activation energies are $Q(f_{Fe}) = 0.006$ eV and $Q(f_{Co}) = -0.005$ eV, which are an order of magnitude smaller than the experimental errors in the measured activation energies $Q_{Fe} = 2.97 \pm 0.09$ eV and $Q_{Co} = 3.01 \pm 0.09$ eV.

Measurements of iron and palladium diffusion in eleven compositions of Fe–Pd (0–100 at. % Fe) at temperatures from 1100 to 1250°C were performed by Fillon and Calais (1977). At fixed temperature the diffusion coefficients exhibited a maximum at about 45 at. % Fe. This corresponds to a minimum in the liquidus in the Fe–Pd phase diagram. Iron was always the more mobile component. There were no indications of deviations from an Arrhenius behavior. Activation energies for iron and palladium were always equal within experimental error (~ 2.7 eV), and also hardly any composition dependence was detectable. The authors calculated the correlation factors following Manning (1971a) for all compositions and temperatures. The correlation factors were always positive, and therefore consistent with the Manning model. Since $D_{Fe} > D_{Pd}$, the correlation factor for iron diffusion was always smaller than for palladium diffusion. For the concentrated alloys f_{Fe} increased with iron content from 0.72–0.78, whereas f_{Pd} exhibited a minimum value of 0.79 at 10 at. % Fe and a maximum of 0.85 at 80 at. % Fe. The temperature dependence was weak. From straight-line fits to Eq. (30) the apparent activation energies due to the temperature dependence of the correlation factors for the 50–50 at. % alloy are $Q(f_{Fe}) = 0.003$ eV and $Q(f_{Pd}) = -0.003$ eV, which are of no influence on the measured activation energies. The authors concluded that their results were consistent with a vacancy mechanism for diffusion. Askill (1971) reported measurements of the diffusion coefficients of both chromium and nickel in a Ni–Cr alloy, containing 62 at. % Ni, at temperatures from 890 to 1360°C. No curvature was found in the Arrhenius plots. The activation energies and preexponentials are given in Table III. The results are in excellent agreement with those of Gruzin and Fedorov (1955) for a 77 at. % Ni alloy, but at variance with measurements of Kalinovich et al. (1972, 1973) for a 67 at. % Ni alloy at temperatures from 950–1300°C (Table III). In the latter measurements no conventional sectioning technique was used, but a special technique was applied, because the main object of the study was a measurement of electrotransport. On the other hand, Monma et al. (1964b) obtained for the 67 at. % Ni alloy Q and D_0 values, neither compatible with the results of Askill, nor with those of Kalinovich et al. Thus poor agreement exists between various authors concerning diffusion in nickel-rich Ni–Cr alloys. Because of the similarity of Askill's and Gruzin and Fedorov's values, and the fact that a conventional sectioning technique was used, Askill's diffusion parameters are preferable. If we apply Manning's model to these measurements, we find that at the lower temperatures, where the diffusion coefficients are nearly equal, the correlation factors are similar to those of self-diffusion. The correlation factor for nickel increases to 0.86 at the highest temperature, that for chromium decreases to 0.65. By a straight-line fit, the activation energy residing in the temperature dependence of f_{Ni} is obtained as 0.04 eV and of f_{Cr} as -0.08 eV. The difference in activation energy of w_{Ni} and w_{Cr} is found from a straight-line fit to Eq. (31) to equal -0.54 eV, whereas the difference in activation energy of the diffusion coefficients is -0.42 eV. The ratios of the corresponding preexponentials are 0.006 and 0.02, respectively. From these numbers it is clear that differences in the preexponentials and activation energies in the vacancy–atom exchange rates are decreased in the diffusion coefficients by correlation effects.

TABLE III

PREEXPONENTIALS AND ACTIVATION ENERGIES FOR DIFFUSION OF CHROMIUM AND COBALT IN NI–CR ALLOYS

Concentration (at. % Cr)	D_{0Cr} (cm² s⁻¹)	D_{0Ni} (cm² s⁻¹)	Q_{Cr} (eV)	Q_{Ni} (eV)	Reference
62	+0.2 0.2 −0.1	+4 4 × 10⁻³ −2	2.54 ± 0.11	2.13 ± 0.11	Askill (1971)
67	2.38	7.21	2.32	2.45	Kalinovich et al. (1972, 1973)
67	3.2	—	3.01	—	Monma et al. (1964b)

None of the systems discussed in this section exhibits a diffusion behavior in conflict with the Manning model. Although some short-range order is present in any concentrated alloy, the applicability of Manning's theory was confirmed by isotope effect measurements in the systems α-CuZn and in equiatomic disordered Fe–Co. Also the other systems that were examined showed quite "normal" values for the correlation factors calculated on the basis of the diffusion coefficients of both components. However, this only means that the diffusion coefficients of the constituents are not very different and cannot be considered as a firm argument for the applicability of the Manning model. From the correlation factors and their temperature dependence it was possible to derive their contribution to the apparent activation energy for diffusion. These contributions were only small and even within the experimental errors for the systems examined here.

B. BINARY BODY-CENTERED CUBIC ALLOYS

In this section experimental results of diffusion in b.c.c. alloys will be reviewed. The systems Fe–Cr, Fe–V, and Fe–Si will be considered in the framework of the empirical rules. Other systems will be treated in the light of Manning's theory. From the point of view of diffusion a special class of bcc alloys is formed by the titanium alloys. Experiments in these alloys have been performed especially to get more insight into the anomalous diffusion behavior of pure titanium. Possible reasons for this behavior of titanium and some of its alloys will be discussed.

Kučera and Stránský (1982) applied the empirical Eqs. (15) and (16) to chromium diffusion in Cr–Fe alloys in the concentration range 0–100 at. % Fe. However, the standard deviations in the values of the empirical constants, which they obtained, are large and one could doubt whether on the basis of the experimental material used, a fit to the equations is justified. Vanadium diffusion in Fe–V alloys with 0–100 % V was studied in the temperature range 839–1320°C by Obrtlík and Kučera (1979). All Arrhenius plots were straight lines. The activation energies and frequency factors agree fairly well with results from other investigations (Stanley and Wert, 1961; Bowen and Leak, 1970). A local maximum in the activation energy and preexponential was found near the 25 at. % V composition, which was explained by the influence of the ordered Fe_3V structure. Apart from this maximum the $\ln D_0$ values remained constant in the concentration ranges 1.6–28 at. % V, whereas the activation energy itself, rather than Q/kT_{mFeV}, could be fitted to a linear function of the vanadium concentration. In the concentration range 47–100 at. % V an attempt was made to fit the experimental values of $\ln D_0$ and Q to a parabolic function of the vanadium concentration. All together it must be concluded that the application of the original Kučera and Million rule to the system Fe–V was not successful. On the other hand, Rule 10, the linear relationship between $\ln D_0$ and Q/T_m, was obeyed for the whole concentration range. Diffusion coefficients of iron in six monocrystalline specimens of Fe–Si (5.5–19.2 at. % Si; 740–1100°C) were reported by Million (1977). Plots of $\ln D$ versus $1/T$ gave straight lines. A comparison with measurements of Borg and Lai (1970) shows systematically higher D values in the latter experiment, probably due to the use of polycrystalline samples. It turned out that at a fraction of 0.8 of the temperature T_{mFeSi} $\ln D$ is a linear function of the silicon concentration, in agreement with the empirical equation [Eq. (13)]. Although this is not salient from the figure given in the paper, the author stated that for a fraction of 0.7 of T_{mFeSi} the diffusion coefficients at higher concentrations are lower than to be expected on the basis of Eq. (13), probably due to the influence of the ordered Fe_3Si structure. In conclusion, Million (1977) gave as an expression for D_{FeSi}^{Fe} as a function of concentration in the whole temperature interval

$$D_{FeSi}^{Fe} = D_{Fe}^{Fe} \exp(20 c_{Si})$$

Although the number of systems discussed above is only small, from these results it will be clear that the application of the Kučera and Million empirical rule to bcc alloys has so far been less successful than to fcc alloys.

An example of the application of Manning's model to a bcc primary phase is an investigation by Fishman et al. (1970) of the isotope effect of iron in disordered bcc Fe–Co, containing 50 at. % Co. Also the diffusion coefficients of both components were measured in the temperature range 795–944.5°C. The results are

$$E_{Fe} = 0.52 \pm 0.08$$

$$D_{0Fe} = 0.25 \pm 0.10 \text{ cm}^2 \text{ s}^{-1}, \quad Q_{Fe} = 2.39 \pm 0.04 \text{ eV}$$

$$D_{0Co} = 2.00 \pm 0.50 \text{ cm}^2 \text{ s}^{-1}, \quad Q_{Co} = 2.60 \pm 0.09 \text{ eV}$$

Measurements of cobalt diffusion in the temperature range 750–850°C in Fe–Co with 49.6 at. % Co by Hirano and Cohen (1972) yielded an activation energy (2.56 ± 0.11 eV) in good agreement with Fishman et al. (1970). The D_0 value 6.59 (+0.74; −0.66) × 10^{-2} cm^2 s^{-1} is substantially lower, and so are the values of the diffusion coefficients. On the basis of Manning's model Fishman et al. (1970) found an average value of f_{Fe} of 0.70 ± 0.05 so that a ΔK value of 0.73 ± 0.10 resulted. From isotope effect measurements of iron in pure bcc iron, Walter and Peterson (1969), deGonzales and deReca (1971), and Irmer and Feller-Kniepmeier (1972) concluded that the value of ΔK is about 0.5. On the other hand, Mehrer (1978) reanalyzed measurements in another bcc metal, namely, sodium, and proposed a value of 0.68 for ΔK for diffusion via single vacancies. Still, the ΔK value from the present experiment seems somewhat high for a bcc system, possibly indicating that an application of Manning's theory is somewhat less successful for this system. From an analysis in terms of Manning's theory the apparent activation energies due to the temperature dependence of the correlation factors are obtained as $Q(f_{Fe}) = 0.04$ eV and $Q(f_{Co}) = -0.04$ eV. These are within experimental error. Measurements in the Mo–W system (99.9–0.1 % W) were performed by Frantsevich et al. (1969). A special technique developed by the authors was applied. This technique was also applied to the Cr–Ni system, where results were obtained at variance with other investigations (see Section V.A.2). Therefore, the results presented below might be considered with some caution. From 15–85 % W the tungsten activation energy in Mo–W varied from 3.17–5.29 eV, and the molybdenum activation energy from 4.57–3.47 eV, and the corresponding values for D_0 were 1.4–24 and 265–8 × 10^{-3} cm^2 s^{-1}, respectively. For the 15 and 85% compositions, the correlation factor for the minority component obtained with the aid of Manning's model becomes negative (i.e., the forbidden region, similar to Fig. 1). This makes the interpretation of the results on other compositions within this model rather delicate. On the other hand, rather normal values are obtained, for example, for the 50% alloy, namely $f_W \simeq 0.89$ and $f_{Mo} \simeq 0.56$. Again it appears that "normal" values calculated for the correlation factors at a certain composition are not a guarantee of the randomness of the solution. Askill (1971) reported results on chromium diffusion in a Ni–Cr alloy with 19.1 at. % Ni. The Arrhenius plot showed a complex structure. Above 1415°C up to 1477°C an upward curvature was found, which was interpreted as a divacancy contribution to the diffusion process. Nonlinearity below 1100°C was explained by a change in phase to the two-phase region.

From the above discussion it appears that for the description and interpretation of diffusion in bcc alloys, empirical rules and Manning's model are less successful than for fcc alloys. However, the experimental material is rather scarce so a definite conclusion is premature. Really strong deviations from the empirical rules and the random alloy model have been found for some titanium alloys. This is not surprising, because the diffusion in pure titanium itself shows a number of anomalous features. Some examples of diffusion in concentrated titanium alloys will now be presented.

Titanium alloys have been explored mainly to get more insight into the anomalous behavior of diffusion in pure titanium. A feature of this behavior is a curved Arrhenius plot that can be analyzed in terms of two exponentials,

$$D(T) = D_{01} \exp(-Q_1/kT) + D_{02} \exp(-Q_2/kT)$$

The first, the high temperature term, exhibits D_{01} and Q_1 values that are normal in the sense of the empirical Rules 3 and 4 from Section III and are compatible with a monovacancy mechanism. The value of D_{02} is some orders of magnitude smaller than D_{01} and $Q_2 \simeq \frac{1}{2}Q_1$. Various mechanisms have been proposed to explain the low-temperature diffusion, including introduction of extrinsic vacancies by interstitial impurities, occurrence of a single highly relaxed vacancy-like defect, diffusion via dislocations created during the α–β phase transformation, softening of an elastic constant corresponding to the α–β transition, and formation of ω-phase embryos that enhance the diffusion (for a full discussion and references, see Pontau and Lazarus, 1979). Mainly dilute alloys ($c < 10$ at. %) have been investigated in an attempt to clarify the diffusion mechanism at lower temperatures. We will restrict ourselves to concentrated alloys of titanium only. Murdock and McHargue (1968) diffused both vanadium and titanium into eleven compositions of the Ti–V system in the concentration range 0–100 % Ti in steps of 10 wt % Ti. The temperature range was from about 750°C to about 150°C below the solidus. In this temperature region about fifteen measurements were performed for each composition and tracer. Arrhenius plots were curved, the curvature being the stronger with higher titanium concentration. On the titanium-rich side some of the empirical rules given in Section III were violated:

(a) At, for example, $0.9T_m$ the diffusion coefficients were not constant as a function of composition, but decreased sharply as more vanadium was added until they became constant from 70–100 % V (Rule 2).

(b) The addition of vanadium to titanium lowers the solidus temperature until a minimum is reached at 30% V. However, at a fixed temperature, both diffusion coefficients *decreased* over the composition range. In contrast, vanadium-rich alloys behaved in agreement with empirical Rule 6.

(c) Vanadium diffused faster in titanium than titanium itself. Yet the addition of vanadium *decreased* the titanium diffusion coefficient (Rule 5).

Accordingly, diffusion in vanadium-rich alloys was consistent with the empirical rules, whereas the titanium-rich alloys showed unusual features. Finally, by the uncritical application of the Manning model the present author found rather normal values for the correlation factors over the whole concentration range. This indicates once more that "normal" values for the correlation factors, obtained with the aid of Eq. (30), are not at all a guarantee of a "normal" diffusion behavior. Santos and Dyment (1975) studied titanium diffusion in Ti–Mn alloys with compositions 9.7, 13.3, 17.9, and 20.6 at.% Mn over large temperature intervals ($\simeq 500$°C) up to 50°C below the solidus. Though large numbers of measurements were performed over these wide temperature intervals (e.g., 18 measurements in the 13.3 at.% Mn alloy) no curvature of the Arrhenius plots was detected. A straight-line fit yielded an increase in activation energy from 1.77 ± 0.1 eV for the 9.7% alloy to 2.16 ± 0.08 eV for the 20.6% alloy. The preexponential D_0 varied from 1.9×10^{-2} to 5.47×10^{-1} cm^2 s^{-1}. In the same concentration range, the solidus temperature *decreases* rather sharply. Moreover, the addition of the fast diffusant manganese to titanium accelerated the titanium diffusion at higher temperature, but decreased the diffusion rate at lower temperatures. The diffusivity of manganese, measured at 898 and 1240°C, decreased also with increasing manganese content. The above-mentioned features are for the greater part at variance with empirical rules, just as for titanium-rich Ti–V alloys. In this system it is possible to retain the β phase at ambient temperature by quenching. So the effect of passing through the phase transition during the diffusion anneal can be studied. An interesting outcome of the experiment was that samples that underwent the transformation did not reveal any effect within experimental error. On the contrary, they showed a slight tendency to have the smaller diffusivity. This seems to exclude the possibility of low-temperature enchancement of diffusion by the formation of dislocations. Application of Manning's theory to the 9.7% alloy yielded a negative value of f_{Mn} at 1240°C. At 900°C f_{Mn} varied from 0.23–0.50 in the composition range from 9.7–20.6%, while $f_{Ti} \simeq 0.79$. The authors expressed doubt on the validity of the random alloy model for this alloy and argued that even a vacancy mechanism seems open to question. They did not exclude an interstitial type of diffusion of manganese or an interstitial–vacancy pair mechanism. As an argument, the fact was brought forward that the manganese diffusion rate in titanium is little affected by the phase transition, in contrast to the titanium self-diffusion rate. Pontau and Lazarus (1979) performed titanium and niobium diffusion measurements in Ti–Nb alloys up to 35.7 at. % Nb. Like vanadium and manganese, niobium in titanium acts as a β-phase stabilizer. Moreover, the system is able to form a metastable ω phase at approximately 20 at. % Nb. A straight-line fit gave Q_{Ti} values varying from 1.36 eV at 0 at. % Nb to 2.56 eV at 35.7 at. % Nb, and D_{0Ti} values from 4.5×10^{-4} to 2.5×10^{-1} cm^2 s^{-1}. The corresponding results for Nb were Q_{Nb} from 1.35 eV to 2.68 eV and D_{0Nb} from 2.9×10^{-4} to 3×10^{-1} cm^2 s^{-1}. The Arrhenius plots were moderately curved, but no significant enhancement or depression was found at lower temperatures, in comparison with pure titanium. Previous heat treatment of the samples did not show any measurable effect. Accordingly, this work did not provide any support for an influence of ω phase (19.6 at. % alloy) or dislocation enhancement (35.7 at. % alloy). In contrast to Ti–V and Ti–Mn, the activation energy increased with increasing solidus temperature and the addition of the slower diffusant niobium decreased the diffusion coefficients of both titanium and niobium. Therefore, empirical rules are not strongly violated. An analysis in terms of Manning's model yielded rather normal values for the correlation factors, compatible with a monovacancy mechanism. The experimental results were compared with previous experiments of Gibbs et al. (1963). For the two concentrated alloys (19.6 and 35.7 at. %) the agreement was reasonably good. The authors concluded that although an enhancement of diffusion by ω phase fluctuations or effects of the $\alpha-\beta$ transformation can not be ruled out, these are certainly not supported. The data were consistent with a normal relaxed vacancy–divacancy mechanism. Thus this alloy system seems less anomalous than Ti–V and Ti–Mn.

The results of diffusion in primary phases can be summarized as follows. For fcc alloys the situation is relatively clear. With some caution, values for the activation energies, preexponentials, and diffusion coefficients can be estimated with the aid of the empirical rules. The concentration dependence of these quantities can also be described by the empirical rules. Usually, deviations can be explained by the occurrence of ordering in the alloy. There are no measurements in conflict with Manning's model. Besides, the Manning theory is supported by isotope effect measurements in α-CuZn and equiatomic fcc Fe–Co. The situation is less clear for "normal" bcc alloys. However, the number of bcc systems that has been investigated is relatively small. Diffusion experiments in "anomalous" bcc titanium alloys reveal strongly deviating diffusion properties. The systems Ti–V and Ti–Mn appear to behave more anomalously than Ti–Nb. The reason for this is not clear.

VI. Theoretical Considerations of Diffusion in Ordered Structures

A. Introduction

It is generally accepted that vacancies play an important role in diffusion in ordered alloys. Therefore, most theoretical work starts from the concept of vacancies exchanging with atoms via nn jumps. With only a few exceptions, theories have been developed for ordered binary alloys crystallizing in the B2 (CsC.) structure. This structure (Fig. 3) consists of two interpenetrating simple cubic sublattices, which we shall call the α and β sublattices. In the following discussion we shall bear this structure in mind. The diffusion coefficients in such binary ordered alloys will depend on the number of vacancies present and also on their mobility. Intuitively, it is also obvious that the migration of both species will have a cooperative character. For example, it will be difficult for one component to migrate over long distances

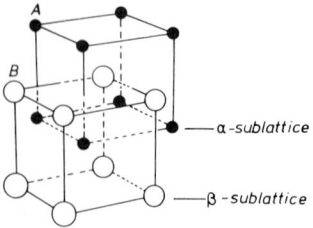

Fig. 3. The B2 (or CsCl) structure.

if the mobility of the other component is small. This effect will be primarily incorporated in the correlation factor for diffusion. In the following discussion we will neither review the extensive literature on vacancy formation in alloys, nor will we try to calculate activation energies for diffusion from first principles. We will concentrate mainly on the kinetics of diffusion and the description of the quantities occurring in the diffusion coefficient in terms of simple interaction energies between the constituting species. Before we proceed to an discussion of the models used for the description of the elementary atomic jump process, we will very briefly discuss some aspects of ordering.

Let us take as an example the intermediate phase β-CuZn. In this alloy interaction between the atoms favors the atoms to be surrounded by atoms of the different kind. At higher temperatures thermal agitations tend to make the distribution of atoms more random than at lower temperatures. However, always short-range order exists, i.e., the probability of finding a copper atom next to a zinc atom is greater than random. When the temperature is lowered, the distribution of atoms becomes more and more regular, until below the order–disorder temperature a structure with long-range order is formed, in which two sublattices are distinguishable (Fig. 3). The long-range order is complete if each atom is surrounded by atoms of the different kind as much as possible (see Sato, 1970). A popular lattice model used for calculations on these systems is the analog of the Ising model for magnetism. Only nearest-neighbor interactions are assumed. That is, for the B2 structure (Fig. 3) and an alloy AB, interaction energies $-V_{AA}$, $-V_{BB}$, and $-V_{AB}$ are defined between neighboring AA, BB, and AB atoms, respectively. The system exhibits a tendency to order if the ordering energy

$$v = V_{AB} - (V_{AA} + V_{BB})/2 \qquad (34)$$

is positive. The degree of order depends on this energy and on temperature. The degree of long-range order can be characterized by quantities like c_A^α, etc., being the fraction of α sites occupied by A atoms and the degree of short-range order by quantities like $p_{AB}^{\alpha\beta}$, etc.; for example, the probability of finding adjacent to an A atom on the α sublattice a B atom on the β sublattice.

It has so far not been possible to construct—even within this simple model—an exact analytical expression for the free energy of the system to be minimized in a calculation of the thermodynamic state of order at a certain temperature. However, it is feasible to handle the problem by using approximations of increasing levels of sophistication. A general formulation has been devised by Kikuchi (see, for example, Sato, 1970, or Kikuchi, 1977), the so-called cluster variation method. In this, atomic clusters of increasing size are used as the basic cluster for the construction of the free energy. Then one obtains approximations using one atom as a basic cluster (Bragg–Williams approximation), a pair of atoms (Bethe or pair approximation), etc. The problem can be solved "exactly" by computer simulation (Guttman, 1961; Murch, 1982a). In the following we will usually employ the nn interaction model to make calculations on diffusion properties. Occasionally, results will also be used which have been obtained in pair approximation.

B. Models for the Elementary Atomic Jump

In the previous section it was mentioned that it is usually assumed that diffusion in ordered alloys occurs by the exchange of atoms with nearest-neighboring vacancies. In this section we will briefly discuss the different models that have been devised to describe this elementary jump process. In one of the models the difference in configurational energy of the system before and after the jump is used as the activation energy for migration. This difference can be evaluated, for example, by the use of nn pair interactions as outlined in the previous section. However, in this model the existence of the saddle point configuration is disregarded, and such a model is not adequate to describe the kinetics of the diffusion process. On the other hand, it will give correct results if the aim is to construct the state of thermodynamic equilibrium at a certain temperature, for example, by computer simulation (for references see Girifalco, 1973). It is more realistic to define the activation energy of a jump as the difference between the energy of the saddle point configuration and the initial configuration. The best way to calculate this energy would be to apply reliable interatomic potentials, allowing the system to relax in both configurations, and make the relevant calculations. Wynblatt (1967), for example, used a Morse potential to calculate activation energies for various jump processes in completely ordered AgMg. However, apart from the lack of reliable potentials for most systems, such a procedure would be very consuming of computer time for the study of the kinetics of systems with more or less disorder. Therefore, simplifying assumptions have been made. Kikuchi and Sato (1969) used a pair interaction model in their study on bcc alloys. The activation energy in their model is the difference between the energy to break the seven nn bonds of the atom before the jump

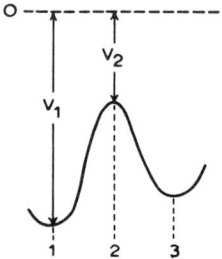

Fig. 4. Potential barrier for an atomic jump from position 1 to position 3. If an A atom that is going to jump has, e.g., 4A− and 3B− nearest neighbors in the position 1, then $v_1 = -4V_{AA} - 3V_{AB}$ and $v_2 = u_A$ in the Kikuchi and Sato model. [From Radelaar (1970).]

into the adjacent vacancy and the saddle point energy. This saddle point energy is assumed to depend on the species of the jumping atom only and not on the detailed saddle-point surroundings (Fig. 4). Nearest neighbor interactions in the saddle point can also be included as was done by several authors (for example, Girifalco, 1964b; Schoijet and Girifalco, 1968; Radelaar, 1970; Kinoshita et al., 1978). However, estimates of numerical values of these quantities are even more difficult than estimates of V_{AA}, etc.

C. The Six-Jump Cycle Mechanism

If one assumes diffusion in a completely ordered structure—for example the B2 structure—to take place by means of vacancies and nearest-neighbor jumps, it is obvious that a random vacancy path would cause an appreciable amount of disorder, locally increasing the energy of the crystal. To find a way out of this dilemma, Huntington (see Elcock and McCombie, 1958)

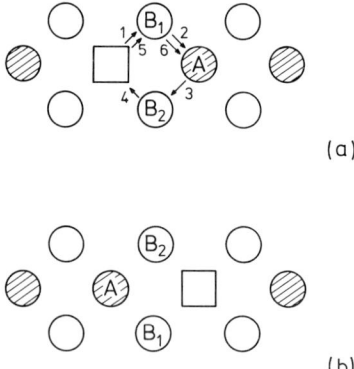

Fig. 5. Six-jump cycle in the ordered square lattice; (a) the arrows indicate the path of the vacancy, (b) displacements resulting from the complete cycle.

suggested that migration of atoms could occur via a special sequence of six vacancy jumps with only a temporary minimal increase of energy. Such a six-jump cycle is presented for the square two-dimensional lattice in Fig. 5a. The first three vacancy jumps (numbered 1–3) displace the three atoms, involved in the process to the wrong sublattices. Each of these steps leads to an increment of the crystal energy, which can be evaluated easily in terms of the nearest-neighbor interaction energies V_{AA}, etc. The last three jumps (numbered 4–6) restore complete order. The potential energy scheme of this jumping process is given in Fig. 6 (Wynblatt, 1967). It is clear that the efficiency of the first elementary jump frequencies will be low: The probability to reverse such jumps is large. The final result of the full cycle is shown in Fig. 5b. The A atom and the vacancy have exchanged positions, and so have the two B atoms. Thus one A atom and two B atoms have been displaced over a distance of $a\sqrt{2}$, if a is the lattice parameter. From Fig. 5 it is clear that for the square lattice the following relation holds for the ratio of both self-diffusion coefficients

$$\frac{D_A}{D_B} = \frac{c_V^\alpha w_6^\alpha + 2c_V^\beta w_6^\beta}{2c_V^\alpha w_6^\alpha + c_V^\beta w_6^\beta} \tag{35}$$

where c_V^α is the vacancy fraction on the α sublattice, etc., and w_6^α is the frequency with which a vacancy, starting on the α sublattice, completes a full cycle. Elcock (1959) demonstrated Eq. (35) to be also valid for the simple cubic alloy, if six-jump cycles with a minimal increase of energy are considered. For the B2 structure, Wynblatt (1967) showed that there are three

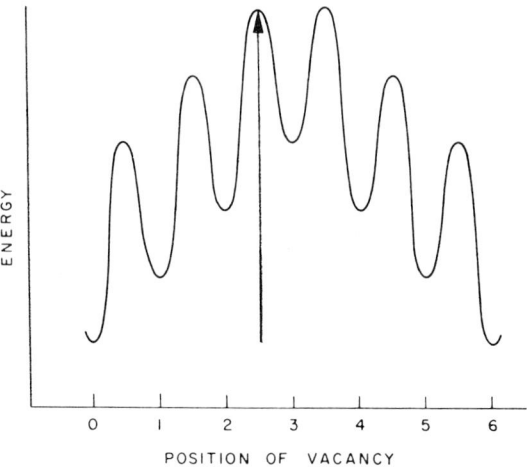

Fig. 6. Schematic of energy changes during six-jump vacancy cycle. [From Wynblatt (1967).]

possible forms of six-jump cycles, leading to 48 possible cycles, 12 of which displace the atoms over a distance of $a\sqrt{2}$ (a is the lattice parameter) and 36 causing a jump vector of length a (Arnhold, 1981). For this structure this leads also to Eq. (35) (Elcock and McCombie, 1958). Although other expressions for the B2 structure are found in literature, in our opinion Eq. (35) is the correct one. From Eq. (35) it can be deduced that the ratio of diffusion coefficients of both species is restricted to values between 0.5 and 2 (Gupta and Lieberman, 1971; Stolwijk et al., 1980; Frohberg, 1981).

In Elcock and McCombie's treatment correlation effects between successive cycles were not taken into consideration. Such a correlation factor could be defined in such a way that it would account for the efficiency of the six-jump cycle frequency for displacements over long distances. For the square lattice such a correlation factor is easily found from Fig. 5. Comparing the starting situation of Fig. 5a with the final configuration of Fig. 5b, it is clear that the migration of the A atom can be visualized as taking place by jumps with a virtual atom–vacancy exchange frequency w_6^z to second nn positions. Thus it is as if the migration of the A atom occurs over the α sublattice with a jump distance of length $a\sqrt{2}$. The α sublattice is also a square lattice. Thus the diffusion of this A atom by means of this vacancy is equivalent to self-diffusion in a square lattice, leading to a correlation factor of 0.47 belonging to the six jumps cycle frequency w_6^z. For the B2 structure the situation is more complicated, due to the different types of cycles that are possible in this structure. An important aspect of correlation effects lies in the interpretation of isotope effect measurements (Peterson, 1975). For simple situations, such as self-diffusion in pure metals via monovacancies or diffusion of an isolated impurity in a pure metal, the isotope effect parameter is equal to the correlation factor multiplied by the so-called kinetic energy factor [see Section II, Eq. (10)]. This relation can be found on the basis of the expression of the correlation factor in terms of the atom–vacancy exchange rates (Peterson, 1975). A proof of such a relation for a somewhat more complex diffusion mechanism, such as diffusion via divacancies in fcc metals, is already laborious (Bakker, 1971). From this it is clear that in more complicated situations such as diffusion via the six-jump cycle there is no justification for equating the isotope effect parameter simply to some correlation factor multiplied by a kinetic energy factor. Essential for the interpretation of isotope effect measurements is the knowledge of a detailed functional relationship between the diffusion coefficient and the various separate atom–vacancy exchange rates occurring in the six-jump cycle. Knowledge of the way the isotope mass enters into the six-jump cycle correlation factor and the six-jump cycle frequency, and so into the diffusion coefficient, is indispensable for a meaningful interpretation of the results of isotope effect measurements. So far no such analysis exists for the six-jump cycle, so that the interpretation of isotope effect measurements in case of a six-jump cycle mechanism is not possible.

In spite of this, some authors have attempted to interpret isotope effect measurements under the assumption that a six-jump cycle mechanism is operative. These interpretations either show a lack of any firm base or are incorrect. Fishman et al. (1970) presumed intuitively that the isotope effect parameter will be of the order of $f^6 \Delta K$ for the six-jump cycle in the B2 structure. Here f is the correlation factor for the six separate elementary jumps. They assumed the correlation factor f to be of the same magnitude as the correlation factor f_0 for self-diffusion in bcc metals and ΔK to be equal to about 0.5. In this way they postulated a low value of 0.07 for the isotope effect parameter. Such a treatment is obviously an oversimplification of the problem. In contrast, Frohberg (1971) and Arnhold (1981) suggested rather high values for the correlation factor for the six-jump cycle. By Monte Carlo computer simulation of the six-jump cycle, Frohberg claimed a value of 0.78 for f in the B2 structure and Arnhold values between 0.80 and 0.85. These values were based on the use of the equation

$$f = \langle R^2 \rangle / nr^2 \tag{36}$$

Arnhold (1981) explained how these results have to be interpreted. Since there are 48 possible six-jump cycles, from which 12 give rise to a displacement vector $a\sqrt{2}$ and 36 to a displacement a, one obtains for the mean squared displacement per cycle and per atom

$$\langle R^2 \rangle = (12/48)2a^2 + (36/48)a^2 \tag{37}$$

Since the length r of an elementary jump vector equals $\frac{1}{2}a\sqrt{3}$, and taking a number of two effective jumps per individual atom ($n = 2$) in the completed cycle, from Eq. (36) a value of f of 0.83 follows. However, this number is not representative for any correlation effect, because during the six-jump cycle the atoms will make many jumps to and fro before the cycle is completed. Thus there is no justification for equating n to 2 in Eq. (36).

From the above discussion it becomes clear that attempts to calculate correlation effects for the six-jump cycle have not been successful so far, and that the interpretation of isotope effect measurements, if diffusion occurs by six-jump cycles, is still an open question. Thus isotope effect measurements can not be used to evaluate the efficiency of the six-jump cycle mechanism. However, such an evaluation is possible by means of computer simulation. Then it turns out that the six-jump cycle is only one of the diffusion processes occurring (Arnhold, 1981). This will be discussed in more detail in Section VI.F.

D. EXPRESSION FOR THE SELF-DIFFUSION COEFFICIENT IN ORDERED ALLOYS IN TERMS OF AVERAGED QUANTITIES

A different approach for the description of diffusion in ordered alloys, more closely linked to diffusion in pure metals, starts from the concept of averaged quantities, like averaged atom–vacancy exchange rates, averaged probabilities of finding a vacancy adjacent to an atom, and averaged correlation factors. This approach was initiated by Kikuchi and Sato (1969). The theoretical background of such an approach and a discussion of Kikuchi and Sato's path probability method will be presented in the Section VI.E. In this section a simple scheme will be given, by which it is possible to reproduce one of Kikuchi and Sato results, namely the expression of the diffusion coefficients in terms of averaged quantities (Bakker et al., 1976). An advantage of this scheme is its simplicity and transparency. We will give examples of two crystallographic structures. The treatment applies to any degree of substitutional disorder and to any composition. Again we assume atomic migration to occur via nn atom–vacancy exchanges. Starting point is the equation for the diffusion constant (for component A) in the form [see Eqs. (3) and (4)]

$$D_A = n_A x^2 f_A / 2t \tag{38}$$

The first example is the B2 (L2$_0$ or CsCl) structure (Fig. 3) of stoichiometric formula AB. We define N_A^α as the number of A atoms on the α sublattice and N_A as the total number of A atoms, etc. The number of A atom jumps per second is the weighted average of the jump frequencies for A atoms on both sublattices,

$$n_A/t = (N_A^\alpha/N_A)\Gamma_A^\alpha + (N_A^\beta/N_A)\Gamma_A^\beta \tag{39}$$

where Γ_A^α is the jump frequency of A atoms on the α sublattice, etc. Assuming that jumps occur between both sublattices, thermodynamic equilibrium requires that the number of A atom jumps from the α to the β sublattice is equal to the number of opposite jumps, so that the degree of order remains unchanged

$$N_A^\alpha \Gamma_A^\alpha = N_A^\beta \Gamma_A^\beta \tag{40}$$

When diffusion takes place by means of vacancies, we write

$$\Gamma_A^\alpha = 8 p_{AV}^{\alpha\beta} w_A^\alpha \tag{41}$$

where $p_{AV}^{\alpha\beta}$ is the probability of finding a vacancy on the β sublattice next to an A atom on the α sublattice and w_A^α is the A atom–vacancy exchange rate for A atoms on the α sublattice. Applying Eqs. (38)–(41) we get

$$D_A = 2(N_A^\alpha/N_A) f_A p_{AV}^{\alpha\beta} w_A^\alpha a^2 = 2(N_A^\beta/N_A) f_A p_{AV}^{\beta\alpha} w_A^\beta a^2 \tag{42}$$

4. TRACER DIFFUSION IN CONCENTRATED ALLOYS

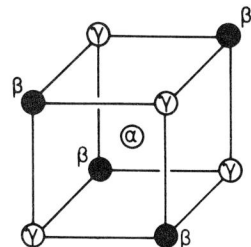

Fig. 7. One-eighth of the unit cell of the DO$_3$ (or Fe$_3$Si) structure. Sublattices are indicated by α, β, and γ. [From Wever and Frohberg (1974).]

in which x is set equal to $\frac{1}{2}a$, a being the lattice parameter. A similar equation holds for D_B.

A somewhat different expression is obtained by defining an average value p_{AV} as

$$p_{AV} \equiv (N_A^\alpha/N_A)p_{AV}^{\alpha\beta} + (N_A^\beta/N_A)p_{AV}^{\beta\alpha} \quad (43)$$

and an average jump frequency w_A as

$$1/w_A \equiv \tfrac{1}{2}[(1/w_A^\alpha) + (1/w_A^\beta)] \quad (44)$$

Then an equation is obtained similar to that of self-diffusion in a pure solid

$$D_A = f_A p_{AV} w_A a^2 \quad (45)$$

The same result was obtained by Kikuchi and Sato (1969) in a different way.

As a second example we will treat the DO3 (Fe$_3$Si) structure with stoichiometric formula A$_3$B. In this structure the B atoms form an fcc lattice (β sublattice), in which both the tetrahedral positions (α sublattice) and the octahedral positions (γ sublattice) are occupied by A atoms. One eighth of the unit cell is drawn in Fig. 7. Let us assume that A atoms jump between the α and γ sublattices, B atoms between the α and β sublattices. Then

$$n_A/t = (N_A^\alpha/N_A)\Gamma_A^\alpha + (N_A^\gamma/N_A)\Gamma_A^\gamma \quad (46)$$

$$N_A^\alpha \Gamma_A^\alpha = N_A^\gamma \Gamma_A^\gamma \quad (47)$$

$$\Gamma_A^\gamma = 8 p_{AV}^{\gamma\alpha} w_A^\gamma \quad \text{and} \quad \Gamma_A^\alpha = 4 p_{AV}^{\alpha\gamma} w_A^\alpha \quad (48)$$

$$D_A = (N_A^\alpha/N_A) f_A p_{AV}^{\alpha\gamma} w_A^\alpha a^2 = 2(N_A^\gamma/N_A) f_A p_{AV}^{\gamma\alpha} w_A^\gamma a^2 \quad (49)$$

Similar equations can be found for D_B by replacing A by B and γ by β. Other structures can be treated in a similar way.

E. Kikuchi and Sato's Path Probability Method

Until 1969, the only mechanism devised to describe the kinetics of diffusion in ordered systems with, e.g., the B2 structure was the six-jump cycle. In that year Kikuchi and Sato presented their path probability method based on the principles of time-dependent statistical mechanics. Using a model of nn interactions and applying the pair approximation, values for the various quantities occurring in the expression of the diffusion coefficient were obtained. The method has a wider applicability than the six-jump cycle; It can be used also for systems without complete order and for off-stoichiometric compounds. Kikuchi and Sato (1969) argued that in fact the applicability of the six-jump-cycle mechanism is limited. The assumption of diffusion via such a mechanism is restricted to stoichiometric compounds in which the degree of order is extremely high. They stated that the system as a whole must be considered in thermodynamic equilibrium and that from this point of view no necessity exists for restoring order during each unit process, because the disordering caused by one jump can be compensated by a reverse jump at another place of the system. Accordingly a general treatment must be based rather upon the application of the principles of statistical mechanics. Along these lines Kikuchi and Sato (1969, 1970, 1972) developed the path-probability method as the kinetic counterpart of Kikuchi's cluster variation method (see, for example, Kikuchi, 1977; Sato, 1970). Though the path-probability method has been applied also to other systems, in the present connection the calculations on (ordered) alloys with the bcc structure are most important. Since these computations are complicated, only a short outline of the method will be given here and the results will be summarized.

An ordered alloy AB with a low concentration of vacancies is considered in which some B atoms are replaced by tracer B* atoms in such a way that the distribution of A and (B + B*) atoms remains constant and in thermodynamic equilibrium. A steady state is attained in which the concentration gradient of B* atoms is maintained by placing the system in contact with two infinitely large particle reservoirs, acting as a source and a sink of B* particles. Migration takes place via vacancies and nn jumps. A model of nn interactions is applied, in which the interaction energies between atoms and vacancies are formally put equal to zero. The saddle point energy of a jumping atom only depends on the species of atom and not on the saddle point configuration as described in Section VI.B. Diffusion in the (100) direction is studied and the flux of B* particles is related to the concentration gradient. This flux is expressed in so-called path variables, i.e., the probability that a pair of certain constituents transforms into another pair during a time interval between t and $t + \Delta t$. Then the most probable expression of the path variables in terms of the state variables is obtained. (State variables c_A^α, $p_{AV}^{\alpha\beta}$, etc., are the quantities describing the equilibrium state of the system at a certain temperature.) Thus the next task is to evaluate the state variables as a function of temperature and interaction energies. Until 1969, the problem of ordering of an alloy *with vacancies* was unsolved. Kikuchi and Sato (1969) were also able to settle this problem and to obtain the state variables as a function of the interaction energies in pair approximation. Since there are some typographical errors in the expressions given in the paper of Kikuchi and Sato (1969), their results have been summarized elsewhere (Bakker, 1979). By computer simulation Stolwijk et al. (1980) showed that such a pair approximation applied to the ordering of a bcc alloy with vacancies gives satisfactory predictions for the state variables. There is only a deviation in the location of the ordering temperature. (In 1969, Schapink (1969) also published results on the ordering of alloys with vacancies, using a different method.) The results of the path probability method calculations, outlined above can be summarized as follows. First, Kikuchi and Sato showed that the diffusion coefficient, e.g., for the B component, can be partitioned into a number of averaged atomistic quantities,

$$D_B = f_B p_{BV} w_B a^2 \tag{50}$$

It was demonstrated in Section VI.D that this result also follows from a random-walk approach. Furthermore, the separate quantities occurring in the right-hand side of this equation were calculated in terms of the interaction energies. In the following these results will be discussed.

In Eq. (50) the correlation factor f_B accounts for the cooperative nature of migration of A and B atoms. Results for f_B are given by solid lines in Fig. 8a for the ordered alloy as a function of the reduced reciprocal temperature T_c/T (the ordering temperature T_c indicated by an arrow in the figures is the ordering temperature obtained by computer simulation). For the results for the disordered alloy the reader is referred to Fig. 13 in Section VI.H. It turns out that for a fixed concentration and a fixed temperature, f_B is determined by the parameter

$$U = (V_{BB} - V_{AA})/2v \tag{51}$$

where v is the ordering energy (Eq. 34). Here the saddle point energy for A and B jumps was taken to be equal and so were the attempt frequencies v_A and v_B. As a measure of p_{BV}, Kikuchi and Sato defined the so-called vacancy availability factor $\tilde{V}_B$ as the probability per unit vacancy fraction of finding a vacancy adjacent to a B atom. This quantity is plotted in Fig. 8b as solid lines. The averaged B atom–vacancy exchange rate w_B is composed of w_B^α and w_B^β following Eq. (44). The path probability method yields, for w_B^α as an example,

$$w_B^\alpha = \{p_{BA}^{\alpha\beta} \exp(-\beta V_{AB}) + p_{BB}^{\alpha\beta} \exp(-\beta V_{BB})\}^7 v_B \exp(-\beta u_B) \tag{52}$$

where v_B is a vibrational frequency, u_B the saddle point energy for a B atom, and $\beta = 1/kT$. The expression is easily interpretable. For a jump the B atom has to break 7 bonds with neighboring atoms. Let us, for example, assume the neighborhood to consist of 6 B atoms and 1 A atom, then the depth of the potential well (similar to Fig. 4) is $-6V_{BB} - V_{AB}$ and the probability of such a configuration in pair approximation is $7p_{BA}p_{BB}^6$. This probability is found as a prefactor of one of the eight terms, which leads to the expression in braces of Eq. (52). In this configuration the potential barrier to overcome is $6V_{BB} + V_{AB} + u_B$, and this quantity occurs in the exponential of the same

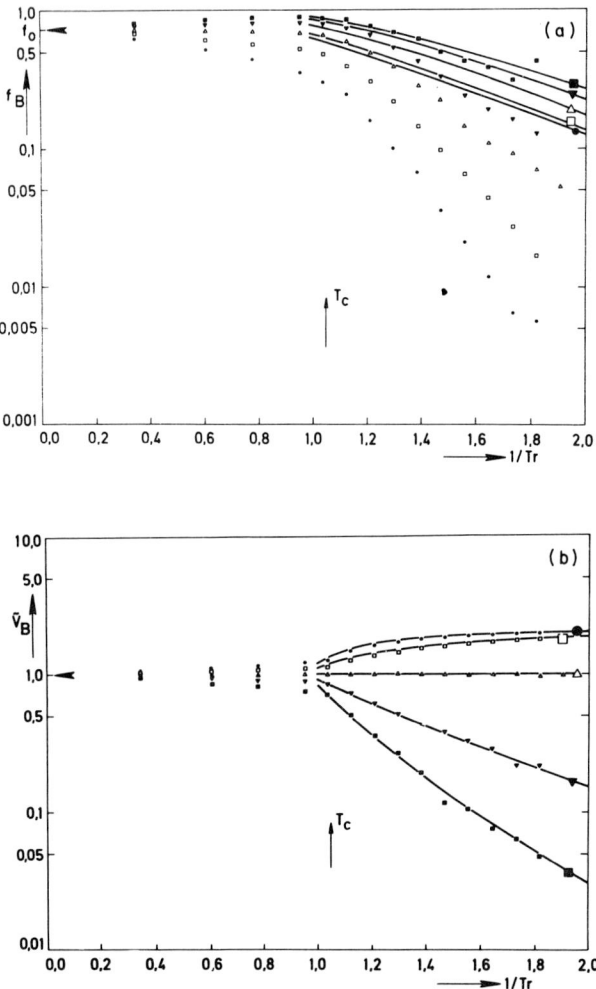

Fig. 8. The quantities occurring in the diffusion coefficient as a function of reciprocal reduced temperature for the B2 (and bcc) structure and equiatomic composition. (a) Correlation factor; (b) vacancy availability factor; (c) reduced vacancy–atom exchange rate; (d) reduced

4. TRACER DIFFUSION IN CONCENTRATED ALLOYS

term. Thus Eq. (52) represents a weighted average of all possible configurations. As a measure of w_B, Kikuchi and Sato defined $\tilde{W}_B$ as

$$\tilde{W}_B \equiv w_B \exp(7\beta V_{AB}). \tag{53}$$

This quantity is drawn as solid lines in Fig. 8c, where u_B was taken to be equal to zero and v_B to unity. Finally, in Fig. 8d the results are presented for the

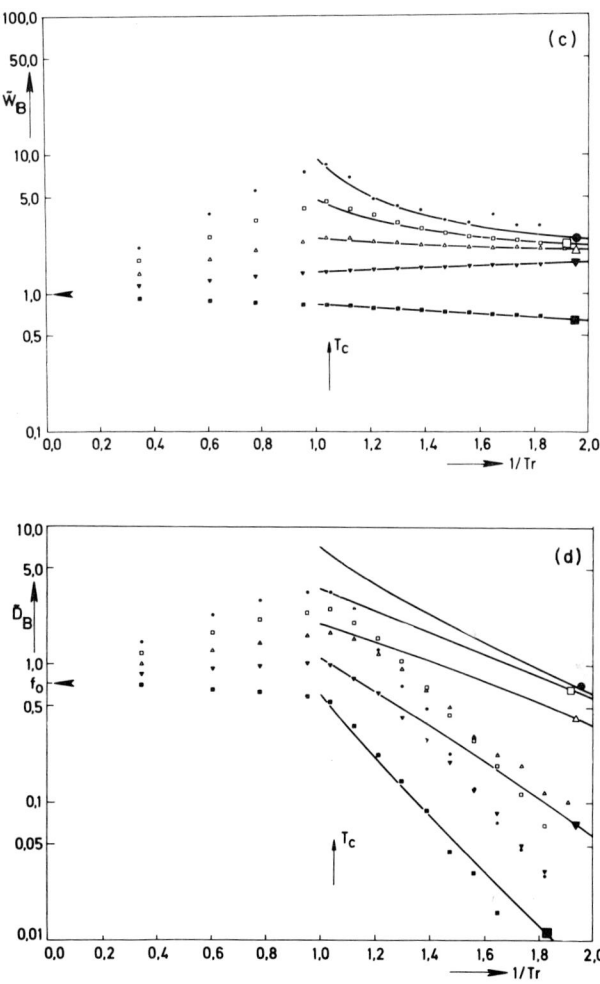

diffusion coefficient. Solid lines are results from Kikuchi and Sato (1972); symbols are computer simulation values of U (Bakker et al., 1976): (●) −1.2; (□) −0.6; (△) 0.0; (▼) 0.6; (■) 1.2.

diffusion coefficient itself in the form

$$\tilde{D}_B = f_B \tilde{V}_B \tilde{W}_B. \tag{54}$$

In Section VI.F we shall compare the results of Kikuchi and Sato (1969, 1970, 1972) with computer simulation results in order to find out whether the various quantities occurring in the right-hand side of Eq. (54), and so $\tilde{D}_B$ itself, are predicted correctly by this theory.

F. Monte Carlo Diffusion Simulation in Ordered Alloys

A review of the methods and applications of Monte Carlo computer simulation of diffusion is given by Murch in Chapter 7. Therefore, we will confine ourselves to some main points. Bakker *et al.* (1976) simulated diffusion in the B2 structure, above and below the ordering temperature, at several temperatures and in three compositions of 40% A–60% B, 50% A–50% B and 60% A–40% B. A nn interaction model was applied. For comparison, the same sets of energy parameters were chosen as those of Kikuchi and Sato (1972). For a given temperature and given values of the interaction energies, a lattice in thermodynamic equilibrium was constructed following Guttman (1961). Next, a vacancy was introduced, which was allowed to exchange with atoms by the Monte Carlo method. The energy of the atom in the saddle point was—for comparison in the same approximation as used

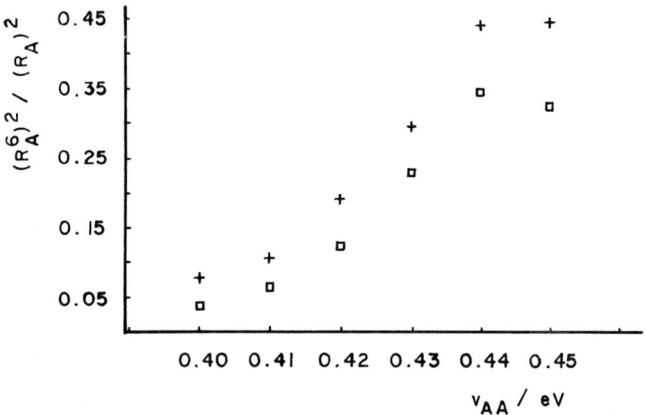

Fig. 9. The ratio of the mean squared displacement of A atoms in six-jump cycles and the total mean squared displacement of A atoms as a function of the interaction energy V_{AA}, while $V_{AA} + V_{BB} = 0.90$ eV: (□) six-jump cycles; (+) cycles where another atom was involved. [From Arnhold (1981).]

4. TRACER DIFFUSION IN CONCENTRATED ALLOYS

by Kikuchi and Sato—assumed to be independent of the detailed neighborhood of the saddle point position, but only dependent on the species of the atom. The migration energy in this model was given by the difference of the energy of the atom in its equilibrium position as a nn of the vacancy and its saddle-point energy. For example, when the seven neighboring atoms of an A atom that was going to jump consisted of four A atoms and three B atoms, the activation energy for this jump was $u_A + 4V_{AA} + 3V_{AB}$, where u_A denotes the saddle point energy (see Fig. 4). The "exact" results obtained from these computer experiments are given in parts a–d of Fig. 8 for the equiatomic composition and compared with the Kikuchi and Sato values. The definitions of the various quantities displayed in these figures were given in Section VI.E. It is seen that the agreement is good for the vacancy availability factor $\tilde{V}_B$ and for the average exchange rate $\tilde{W}_B$, but less satisfactory for the correlation factor and so for $\tilde{D}_B$. The reason for this disagreement for the correlation factor is the implicit assumption that if after an atom–vacancy exchange the vacancy dissociates from the tracer atom to a nnn position, it is lost for any further correlation. This gives a rough approximation for the value of the correlation factor.[1]

Starting from the same diffusion model, Arnhold and Heumann (1982; Arnhold, 1981) also studied diffusion in the B2 structure by simulation. For the same energy parameters the values obtained from their program were in good agreement with those of Bakker *et al.* (1976). However, the program was specially designed to study the efficiency of the six-jump cycle. All kinds of jump sequences were taken into account that eventually gave, as a net result, the six-jump cycle. First, in the stoichiometric compound the degree of long-range order was chosen to be $S = 0.96$. The ordering energy was equal to 0.05 eV and $V_{AA} + V_{BB} = 0.90$ eV. Subsequently, V_{AA} was varied. It was concluded that in the most favorable case ($V_{AA} = V_{BB}$) about 45% of the mean squared displacement was due to any possible six-jump cycle with the rest of the migration taking place via normal single vacancy transport (see Fig. 9). The contribution of the six-jump cycle decreased sharply with increasing difference between V_{AA} and V_{BB}. The efficiency of the six-jump-cycle mechanism in off-stoichiometric compounds turned out to be much smaller for the majority component. For a 46–54 at. % compound, again at most 45% of the mean squared displacement of the minority component occurred in six-jump cycles, and for the majority component only 10%. Murch (1982b) simulated diffusion in an ordered simple cubic alloy over the whole composition range and obtained values for exchange frequencies,

---

[1] In Sato and Kikuchi (1983) it is demonstrated that for the random alloy and the disordered alloy much better results are obtained by taking time averaging instead of ensemble averaging. Similar results are to be expected for the ordered alloy.

vacancy availability factors, and correlation factors as a function of composition. The same (Kikuchi and Sato) model for the elementary atomic jump was applied. Of special interest are the Arrhenius plots of the diffusion coefficients. It was shown that the plots at compositions that exhibit an ordered structure at lower temperatures bend downward on entry to the ordered region. The change of slope has a maximum at stoichiometry. Another remarkable result is the large contribution of the temperature dependence of the correlation factor to the apparent activation energy. This contribution amounts up to about 30%. Furthermore, the activation energy has its maximum precisely at stoichiometry, which was shown to be also due to the temperature dependence of the correlation factor.

G. Random-Walk Approach to the Calculation of Correlation Factors

As was demonstrated in Section VI.D, the diffusion coefficient in an ordered B2 structure is proportional to a number of physical quantities. In Section VI.F it was shown that, with a model of nn interactions, the vacancy availability factor and the average exchange frequency are predicted satisfactorily by the application of the pair approximation. This approximation appeared to be less successful in the case of the correlation factor. In an attempt to produce values for this quantity which are in better agreement with computer simulation results, Bakker (1979) used random walk theory and an extension of Manning's theory (1968, 1971) for random alloys. Let us consider diffusion in the (100) direction and let us use the expression Eq. (54) for the correlation factor. Let an A atom in a long sequence of jumps make n_A^α jumps from the α to the β sublattice and n_A^β jumps of the opposite type, then

$$f_A = \frac{n_A^\alpha}{n_A^\alpha + n_A^\beta}\left[1 + \frac{2}{n_A^\alpha}\sum_{i=1}^{n_A^\alpha}\left(\frac{\langle x_i^\alpha x_{i+1}\rangle}{x^2} + \frac{\langle x_i^\alpha x_{i+2}\rangle}{x^2} + \cdots\right)\right]$$

$$+ \frac{n_A^\beta}{n_A^\alpha + n_A^\beta}\left[1 + \frac{2}{n_A^\beta}\sum_{j=1}^{n_A^\beta}\left(\frac{\langle x_j^\beta x_{j+1}\rangle}{x^2} + \frac{\langle x_j^\beta x_{j+2}\rangle}{x^2} + \cdots\right)\right] \quad (55)$$

The superscript α denotes that the ith jump is from α to β. Obviously, in a long sequence of jumps

$$n_A^\alpha = n_A^\beta \quad (56)$$

For the evaluation of, for example, the first term in braces, we define Q_- as the probability that an α to β jump of the atom is reversed and Q_+ as the probability that the next tracer jump has the same direction as the previous one. The values P_+ and P_- represent similar quantities, but for a β to α

jump. Then, from Fig. 10,

$$\langle x_1 x_2 \rangle / x^2 = Q_+(+1) + Q_-(-1) = Q_+ - Q_-$$
$$\langle x_1 x_3 \rangle / x^2 = Q_+ P_+(+1) + Q_- P_-(+1) + Q_+ P_-(-1) + Q_- P_+(-1)$$
$$= (Q_- - Q_+)(P_- - P_+) \tag{57}$$

In the same way further terms are obtained. By definition,

$$q_A = Q_+ - Q_- \tag{58a}$$

and

$$p_A = P_+ - P_- \tag{58b}$$

where $-q_A$ can be interpreted as the net probability of reversing an α to β jump, and $-p_A$ of reversing a β to α jump. Substitution into Eq. (55) yields, after some algebra,

$$1/f_A = \tfrac{1}{2}[(1/f_A^\alpha) + (1/f_A^\beta)] \tag{59}$$

with

$$f_A^\alpha = (1 + q_A)/(1 - q_A) \tag{60a}$$
$$f_A^\beta = (1 + p_A)/(1 - p_A) \tag{60b}$$

Eq. (59) is similar to a result of Kikuchi and Sato (1972).

The next task is to find expressions for q_A and p_A. Let us therefore examine Manning's (1968, 1971) expressions for the random alloy (Section IV) more closely. From the Eqs. (25) and (29) it is easily shown that for the random alloy the probability of reversing a jump of an A atom is

$$\frac{w_A}{w_A + H_a} = \frac{w_A}{w_A + 2(1 - f_0)^{-1}(c_A f_A w_A + c_B f_B w_B)} \tag{61}$$

If we approximate $2(1 - f_0)^{-1} = 7.3$ by 7, we obtain

$$\frac{w_A}{w_A + H_a} = \frac{w_A}{w_A + 7(c_A f_A w_A + c_B f_B w_B)} \tag{62}$$

We interpret the latter results as follows. In the first instance the probability for an A atom to reverse its previous jump will be $w_A/[w_A + 7(c_A w_A + c_B w_B)]$, where the term in parentheses is the mean jump frequency of an (average)

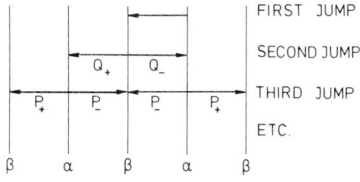

Fig. 10. Scheme of the jump probability for the case of diffusion in an ordered alloy. [From Bakker (1979).]

atom. However, when such an "average" atom jumps into the vacancy and then reverses this jump, the vacancy is again available for an exchange with the A atom under consideration, which has the same effect as if the other atom had not jumped at all. What really matters is the *effective* jump frequency of such an average atom ($f_A w_A$ for an A atom, $f_B w_B$ for a B atom): Then the vacancy really dissociates after the jump and the site, originally available for the reverse jump of the A atom under consideration, is definitely blocked. Bearing this idea in mind, let us consider the situation that arises after a jump of an A atom from the α to the β sublattice in the B2 structure. The net probability of reversing this jump will depend on the occupation of the other 7 nn positions of the vacancy. Thus by weighting these net probabilities with the probability of a certain local configuration one obtains for the average value

$$-q_A = \sum_{k=0}^{7} \binom{7}{k} (p_{VA}^{\alpha\beta})^{7-k} (p_{VB}^{\alpha\beta})^k \frac{w_A^\beta}{w_A^\beta + (7-k) f_A^\beta w_A^\beta + k f_B^\beta w_B^\beta} \quad (63)$$

and similar equations result for p_A, q_B, and p_B. Substituting these results into Eq. (60a) and Eq. (60b) and into similar equations for the B component, one obtains four simultaneous equations, which can be solved by simple iteration (Bakker, 1979). For the quantities $p_{VA}^{\alpha\beta}$, etc., and w_A^β, etc., results obtained by the pair approximation were used. Estimates of the correlation factors were calculated for all combinations of energy parameters, compositions, and temperatures for which computer simulation results were also available.

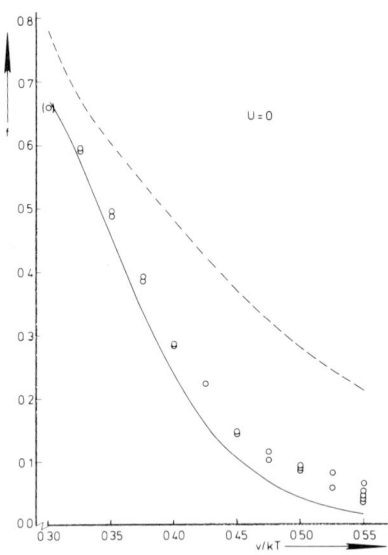

Fig. 11. The correlation factor versus v/kT for the equiatomic composition and $V_{AA} = V_{BB}$: (○) computer simulation results; (---) Kikuchi and Sato values; (—) random-walk approach. [From Bakker (1979).]

4. TRACER DIFFUSION IN CONCENTRATED ALLOYS   231

These estimates always showed a better agreement than the Kikuchi and Sato calculations and are, with the exception of situations in which w_A^α/w_A^β is extremely small, rather close to the "simulated" values. An example is given in Fig. 11.

Thus values for the correlation factors in better agreement with the computer simulation results than those obtained by the path probability method can be found by the use of a random-walk approach.[2] Nevertheless, within a simple interaction model a successful description is now available for all relevant quantities composing the diffusion coefficient in the ordered B2 structure. The averaged atom–vacancy exchange rate and the vacancy availability factor are predicted well by Kikuchi and Sato (1972), a random-walk calculation gives rather reliable results for the correlation factor. A drawback of the theoretical expressions in their present form is that direct application to experimental results is somewhat difficult. Yet some examples of such applications to real experiments will be discussed in Section VII. Furthermore, it must be emphasized that neither in the path probability method, nor in the random walk approach a presupposed unit diffusion process like the six-jump cycle is considered explicitly. On the other hand, it has been shown by computer simulation studies that the six-jump cycle contributes to the diffusion process in some measure. However, the six-jump-cycle mechanism is nothing but a monovacancy mechanism in which the vacancy has the opportunity to make jumps that result eventually in a six-jump cycle. The fact that the path probability method and the random-walk approach are able to reproduce the computer simulation results for exchange frequencies, vacancy availability factors, and correlation factors, respectively, shows that there is no need to consider the six-jump cycle as a separate mechanism. Thus the diffusion process in ordered structures can be described in terms of averaged quantities without considering the six-jump cycle explicitly. Apparently, the discussion of the occurrence of a six-jump cycle is somewhat irrelevant in the light of these more modern approaches.

H. ALLOYS WITH SHORT-RANGE ORDER ONLY

Similar approaches as were presented in Sections VI.E–VI.G have also been applied to diffusion in disordered alloys with short-range order. Kikuchi and Sato (1970) employed the path probability method for calculations on bcc disordered alloys. Simulated results were obtained by Bakker *et al.* (1976) on bcc alloys and Murch (1982a) on sc alloys. Stolwijk (1981) extended Manning's theory (1971) for correlation factors in bcc alloys. All this work concentrates mainly on correlation factors, because the other quantities in

---

[2] See footnote 1.

which the diffusion coefficient is partitioned are predicted well by results analogous to those described in Section VI.E.

Kikuchi and Sato (1970) obtained a correlation factor f_B for B atoms of the same form as postulated by Manning (1971) for the random alloy,

$$f_B = H_B/(H_B + 2\hat{w}_B) \tag{64}$$

Here $\hat{w}_B$ is given by an expression similar to Eq. (52) for short-range order only and H_B is the effective escape frequency of the vacancy after a vacancy–B atom exchange, given by

$$\frac{H_B}{\hat{w}_B} = 7\frac{q_A\hat{w}_A/(\hat{w}_A + \hat{w}_B) + q_B/2}{q_A\hat{w}_B/(\hat{w}_A + \hat{w}_B) + q_B/2} \tag{65}$$

Here q_A and q_B are short-range order parameters obtainable from a calculation using the cluster variation method. The above result is easily interpretable. Consider in the first instance the migration in a linear equivalent of a binary alloy

●　　　□　　　○
1　　　2　　　3
B tracer　Vacancy　A or B atom

and write the correlation factor in its usual form as a function of the net probability $-t_B$ to reverse a jump (see, e.g., Adda and Philibert, 1966)

$$f_B = (1 + t_B)/(1 - t_B) \tag{66}$$

By assuming permanent vacancy loss after a dissociative jump of the vacancy, $-t_B$ can be identified with the probability that the vacancy will exchange again with the tracer (1) on the next vacancy jump. Then

$$-t_B = q_A\frac{\hat{w}_B}{\hat{w}_A + \hat{w}_B} + \frac{q_B}{2} \tag{67}$$

where q_A is the probability that the other atom (3) is an A atom, q_B that it is a B atom. The other possibility is an exchange of the vacancy with the atom in position 3. So the probability of dissociation is

$$1 - (-t_B) = 1 + t_B = q_A\frac{\hat{w}_A}{\hat{w}_A + \hat{w}_B} + \frac{q_B}{2} \tag{68}$$

Comparison of Eq. (64) and Eq. (66) just yields

$$H_B/\hat{w}_B = (1 + t_B)/-t_B \tag{69}$$

Substitution of Eq. (67) and (68) into Eq. (69) gives Kikuchi and Sato's result

4. TRACER DIFFUSION IN CONCENTRATED ALLOYS 233

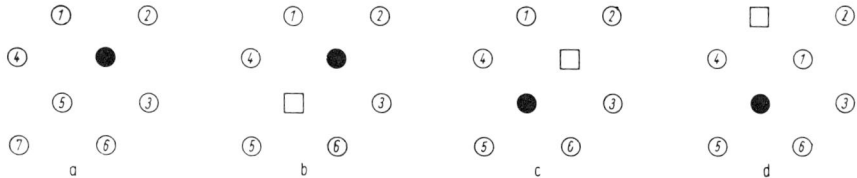

Fig. 12. Representation of the two-dimensional analog of a piece of bcc lattice around a particular atom A (tracer); (●) tracer; (□) vacancy; (a) before a vacancy has arrived at a nn position of the tracer; (b) a vacancy has arrived at a nn position of the tracer; (c) the vacancy has exchanged position with the tracer; (d) the vacancy has dissociated from the tracer. [From Stolwijk (1981).]

apart from a factor of 7. This factor accounts for the seven sites to which the vacancy can escape in the bcc structure.

Results obtained by computer simulation for the various quantities in 50%–50% bcc alloys are given in Fig. 8 (Bakker *et al.*, 1976). For other compositions, the reader is referred to the original paper. In simple cubic alloys diffusion was simulated by Murch (1982a) for different energy parameters and for the whole range of compositions. A remarkable result of this study is the extended linear regions of Arrhenius plots of the diffusion coefficient per unit vacancy concentration for all cases.

A random-walk approach to the calculation of correlation factors in disordered bcc alloys was used by Stolwijk (1981). The starting point was again Manning's Eq. (24) in the form

$$f_A = H_A/(2w'_{A*} + H_A) \quad (70)$$

and a similar expression for f_B. For the evaluation of w'_{A*} and the escape frequency H_A of a vacancy after an A atom–vacancy exchange, "historical" effects were taken into account. These effects can be understood by inspection of Fig. 12, which presents a two-dimensional analog of the bcc structure. After an exchange of a tracer A* with the vacancy, the tracer arrives at a site originally occupied by the atom numbered 5. This is either an A atom (probability p_{AA}) or a B atom (probability p_{AB}). Let us assume that this atom is a B atom, then the new neighborhood of A* (Fig. 12c) will, apart from this atom 5, consist of an (average) nn configuration corresponding to a B atom. Such effects are included in the calculation of w'_{A*}. Similarly, the surroundings of the vacancy in Fig. 12c will have, apart from the tracer atom, an average configuration corresponding to the A* atom, originally situated at this site. Therefore,

$$H_A = 7J_0(p_{AA}f'_{AA}w_A + p_{AB}f'_{AB}w_B) \quad (71)$$

with a similar expression for H_B. The quantity J_0 is included to guarantee

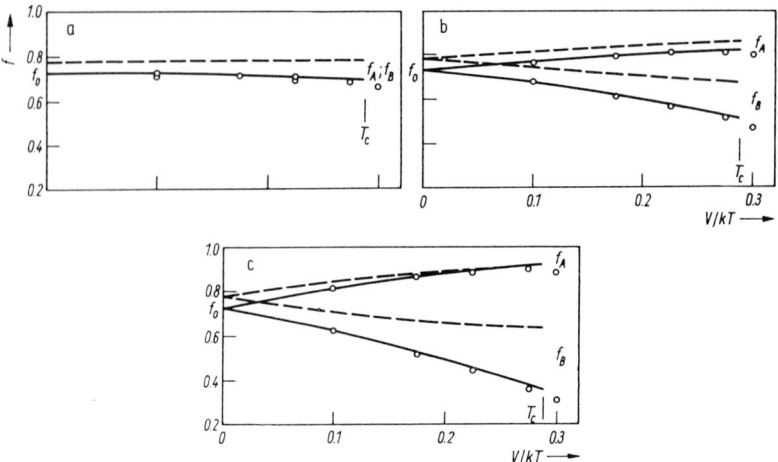

Fig. 13. Correlation factors for A and B atoms in the equiatomic composition of a binary bcc alloy with short-range order as a function of v/kT for three sets of interaction energies $U = (V_{BB} - V_{AA})/2v$. (○) computer simulation results (Bakker *et al.*, 1976); (---) following Kikuchi and Sato (1972); (—) random-walk approach. [From Stolwijk (1981).]

the correct numerical value for the correlation factor in the case of self-diffusion in pure metals. An interpretation of the expression for H_A has been discussed in Section VI.G. There it was suggested that, for example, the correlation factor f'_{AA} corresponds to the correlation factor for the A atoms (other than A*), surrounding the vacancy. This correlation factor has the form of Eq. (70) again. However, it should be noted that after an exchange of such an A atom with the vacancy (Fig. 12d) one of its nearest neighbors is the A* tracer, which affects the A-vacancy reexchange rate. This effect was also included in the calculations of Stolwijk (1981). Results of the calculations turned out to be almost "exact" within the "experimental" errors of computer simulation results for all compositions and energy parameters studied. For the 50%–50% alloy, the f values are given in Fig. 13 (solid lines) together with results from Kikuchi and Sato (1970), represented by dashed lines. Open circles are the results from diffusion simulation (Bakker *et al.*, 1976). Thus the situation for diffusion in disordered bcc alloys with short-range order is similar to that of diffusion in the ordered B2 structure: The path probability method predicts correct values for the vacancy–atom exchange rate and the vacancy availability factor; the correlation factors can be evaluated using a random-walk approach.[3]

---

[3] See footnote 1.

VII. Tracer Diffusion Experiments in Intermediate Phases

A. Intermediate Phases with the B2 (CsCl) Structure

In this section we will distinguish between two types of ordered phases crystallizing in the B2 structure. First, we will discuss tracer diffusion experiments in ordered alloys exhibiting substitutional disorder. In these solids both species of atoms are able to substitute on the "wrong" sublattices. These alloys may also undergo an order–disorder transition, where the ordered B2 structure transforms into the disordered bcc structure. Secondly, intermetallic compounds will be considered, where one of the components is able to substitute on the "wrong" sublattice, whereas structural vacancies are formed when the other component is added in excess. These solids may exhibit a special type of disorder at higher temperatures consisting of the formation of large amounts of vacancies on the one sublattice combined with antistructure atoms on the other one (Fig. 3).

1. Systems with Substitutional Disorder

The theoretical treatments, developed in Section VI, mainly apply to systems exhibiting substitutional disorder. Diffusion in some of these systems has been investigated, namely, β-CuZn and ordered FeCo. Both systems undergo an order–disorder transition at higher temperature. Also, we will consider the system β-AgMg, which exhibits long-range order up to the melting point, but in which the excess component is accomodated on the "wrong" sublattice. For β-CuZn and FeCo experiments were performed on the (nearly) equiatomic composition, whereas β-AgMg was studied as a function of concentration. The experimental results will be compared with computer simulations, bearing in mind that these simulations can be reproduced now by analytical theories (see Section VI).

Kuper *et al.* (1956) investigated the diffusion of both components in β-CuZn (~ 48 at. % Zn). This work has been reviewed previously (Girifalco, 1973; Adda and Philibert, 1966). The results are given in Fig. 14. The most remarkable feature is a change of slope of the Arrhenius plots somewhat above the ordering temperature. Such a change of slope could be reproduced qualitatively by computer simulation. This result, obtained by using rather arbitrary values for the interaction energies, is shown in Fig. 15a. Here the logarithm of diffusion coefficient per unit vacancy concentration is plotted (Bakker *et al.*, 1976). Indeed, the change of slope occurs somewhat above the ordering temperature. A remarkable fact is that in this example the change of slope is mainly due to the temperature dependence of the correlation factor, which becomes manifest from Fig. 15b. Peterson and Rothman (1970b, 1967) measured the isotope effects of copper and zinc in the disordered alloys (49 at. % Zn, 560.3°C) and in the ordered alloy (47 at. % Zn, 410°C, where the long-range order parameter $S = 0.7$). For the disordered alloy the ratio of both isotope effects turned out to be $(f \Delta K)_{Zn}/(f \Delta K)_{Cu} = 0.738 \pm 0.035$. By assuming $(\Delta K)_{Zn} = (\Delta K)_{Cu}$, the influence of the kinetic energy factor could be eliminated. The value obtained is in excellent agreement with Manning's prediction on the basis of the ratio of the diffusion coefficients $(D_{Zn}/D_{Cu} = 2.37)$, namely, $f_{Zn}/f_{Cu} = 0.732 \pm 0.023$ (see Section IV). Under the assumption above one obtains for the kinetic energy factor $\Delta K = 0.39 \pm 0.01$. At first sight it seems that passing the ordering temperature has no dramatic influence on the isotope effects: Within experimental error no dif-

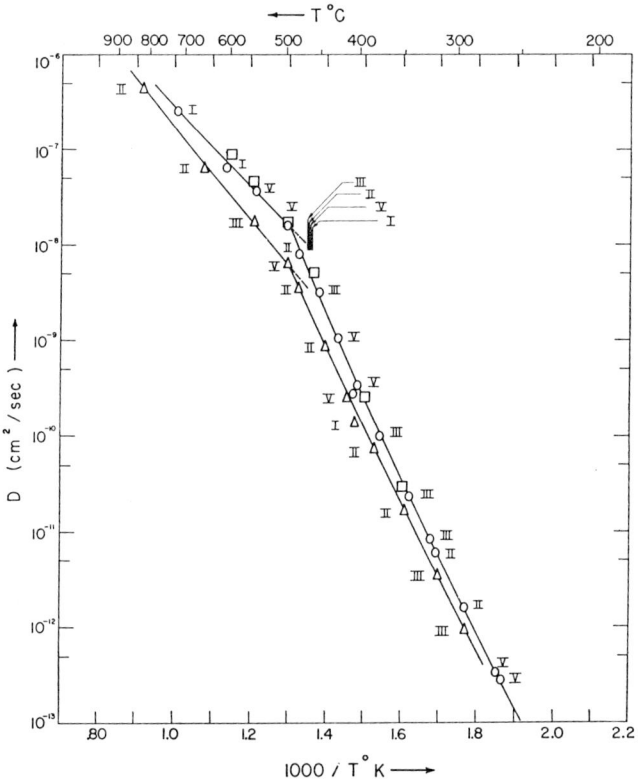

Fig. 14. Logarithm of the diffusion coefficients of Cu ($\triangle$), Zn ($\bigcirc$), and Sb ($\square$) in β-CuZn as a function of reciprocal temperature: (I) 45.65 at. % Zn; (II) 47.15 at. % Zn; (III) 48.00 at. % Zn; (V) 46.51 at. % Zn. Also, the transition temperatures are indicated. [From Kuper et al. (1956).]

4. TRACER DIFFUSION IN CONCENTRATED ALLOYS

ference was detectable for copper ($E_{Cu} = 0.325$ at 410°C in the 47 at. % alloy), while E_{Zn} decreased from 0.24 to 0.20. The authors argued that the reason for this is the relatively high degree of disorder in the ordered phase ($S = 0.7$). On the other hand, Manning's theory fails to account for the values obtained in the ordered alloy. It is, therefore, interesting to consider the results in the light of what is known theoretically on diffusion in ordered alloys. First, for ordered alloys the relation from Manning's model

$$c_A f_A + c_B f_B = f_0 \tag{32}$$

is no longer obeyed. For ordered compounds the left-hand side of the equation always yields values smaller than f_0. If we assume the kinetic energy factor ΔK to be not much different from that in the disordered alloy, indeed the left-hand side of Eq. (32) yields 0.68, obviously smaller than $f_0 = 0.727$. Let us now make an attempt to interpret the experimental data more quantitatively with the aid of the computer simulation results, represented by the symbols in Fig. 8. The temperature at which Peterson and Rothman (1970b, 1967) performed their isotope effect measurements in the ordered state corresponds to a reciprocal reduced temperature $1/T_r = 1.08$ used in these figures. Furthermore, at this temperature a ratio of diffusion coefficients $D_{Zn}/D_{Cu} \simeq 2.4$ was measured. A ratio of diffusion coefficients of about this magnitude is obtained from inspection of Fig. 8d, if one chooses a value for

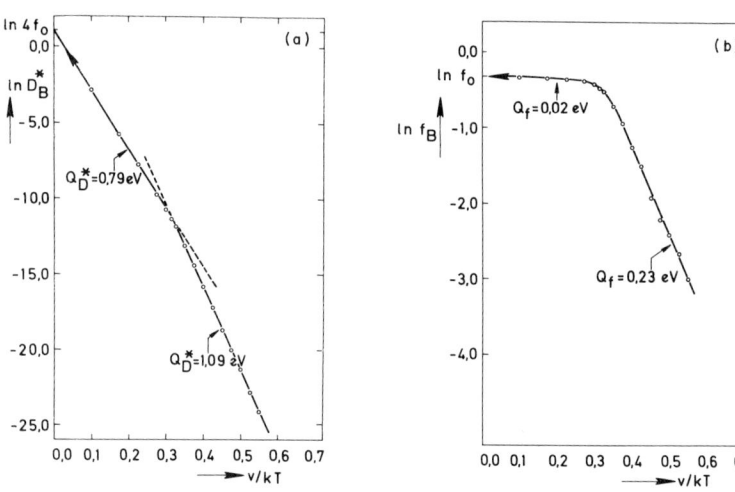

Fig. 15. (a) Logarithm of the diffusion coefficient per unit vacancy concentration in the B2 ordered and bcc disordered structure and (b) logarithm of the correlation factor in the B2 ordered and bcc disordered structure, as functions of v/kT as obtained by computer simulation; $c_B = 0.5$, $U = 0.0$. [From Bakker et al. (1976).]

the parameter U for zinc diffusion of

$$U(Zn) \equiv (V_{ZnZn} - V_{CuCu})/2v \simeq -0.6 \qquad (72a)$$

and, correspondingly, for copper diffusion

$$U(Cu) \equiv (V_{CuCu} - V_{ZnZn})/2v = +0.6 \qquad (72b)$$

Because of their definitions, $U(Zn)$ and $U(Cu)$ have necessarily equal absolute values, but opposite signs. The ratio of the diffusion coefficients dictates the U values. If one further assumes that the measured isotope effects in the ordered state are equal to $f_{Zn}\Delta K$ and $f_{Cu}\Delta K$, respectively, with a ΔK value of 0.39, as given above, the correlation factors can be calculated both for the ordered and disordered state. Then $f_{Cu} \simeq 0.83$ both for the disordered and ordered state at the temperatures at which the measurements were performed, whereas f_{Zn} decreases from 0.61 in the disordered state to 0.51 in the ordered state (here $1/T_r = 1.08$). By using a $U(Cu)$ value of $+0.6$, as given by Eq. (72b), it is observable from Fig. 8a that in passing the transition temperature scarcely any effect has to be expected on f_{Cu} in this temperature range, and that the simulated value for f_{Cu} has indeed a magnitude of about 0.8. For f_{Zn} we have to follow the symbols corresponding to a $U(Zn)$ value of -0.6. In the first place, it is observed from Fig. 8a that f_{Zn} decreases more steeply in the temperature interval. In the second place, it appears that also f_{Zn} has the right magnitude. Thus, remarkably, it turns out that the measurements presented here can be reproduced even quantitatively by the theory. Finally, one could ask whether the signs of the U values found in this way are reasonable. Then one has to realize that the quantities V_{CuCu} and V_{ZnZn} can tentatively be associated with the Cu–Cu and Zn–Zn interaction energies in the pure metals, respectively. In a nn interaction model, V_{CuCu} and V_{ZnZn} are a direct measure for the cohesive energies of the pure metals. From Eqs. (72a) and (72b) it turns out that V_{CuCu} is greater than V_{ZnZn}. The cohesive energy of pure copper is indeed greater than that of pure zinc! Therefore, for β-CuZn an interpretation of the diffusion coefficients, the isotope effects and, the cohesive energies of the pure metals gives, within the framework of the theory, a consistent picture. However, we recall emphatically the limitations of the model used in the simulation work and the difficulties in translating isotope effects into correlation factors. Therefore, an agreement with experiment may be somewhat fortuitous.

Fishman et al. (1970) obtained values for the isotope effect of iron and ratios of diffusion coefficients of both species in ordered equiatomic FeCo in the temperature interval 655–722°C (see Table IV). These measurements were complicated by the small penetration depths at these low temperatures. The experimental errors were, therefore, correspondingly large. The lowest

TABLE IV

ISOTOPE EFFECTS MEASURED BY MEANS
OF Fe^{55} AND Fe^{59} IN FeCo

T (°C)	E_{Fe}	D_{Fe}/D_{Co}
655	0.06 ± 0.20	1.0
683	0.30 ± 0.20	0.9
702	0.16 ± 0.20	1.2
721.5	0.30 ± 0.15	1.1

temperature corresponds to $1/T_r = T_c/T \simeq 1.08$. By comparing the values for the ratios of the diffusion coefficients of both species (Table IV) with computer simulation results (Fig. 8d) one would conclude that $U \simeq 0$. Then a correlation factor of ~0.6 would have to be expected at the ordering temperature and a somewhat lower value at the lowest temperature. Within experimental error and with a ΔK value similar to those found for self-diffusion in bcc metals, at least the results at the three higher temperatures could be compatible with computer simulation data. The value of $U \simeq 0$ would be qualitatively consistent with the fact that the cohesive energies of iron and cobalt are nearly equal. Of course, the same restrictions concerning the validity of such an interpretation hold as mentioned before for CuZn.

Diffusion of silver (Hagel and Westbrook, 1961, 1965; Domian and Aaronson, 1964) was studied over a wide composition range of β-AgMg. Magnesium diffusion was measured in the stoichiometric compound (Domian and Aaronson, 1965). Since this work has been reviewed before (Adda and Philibert, 1966), we will only mention some salient features. From x-ray and density measurements it turns out that on both sides of stoichiometry the excess component is substituted on the "wrong" sublattice. The compound is ordered up to the melting point. Isothermal values of the silver diffusion coefficient showed cusps as a function of composition at the equiatomic composition. At this composition the activation energy also exhibits a maximum. For the 50–50 at. % compound the ratio $D_{Mg}/D_{Ag} \simeq 1$. The authors explained their results by a six-jump cycle mechanism. Such a mechanism is, for the completely ordered stoichiometric compound, supported by calculations of Wynblatt (1967). By use of a Morse potential, Wynblatt showed that this mechanism requires a lower activation energy than a nnn exchange, whereas a divacancy has a tendency to split up into two silver vacancies and a silver atom on a "wrong" site. A remarkable fact is that Kikuchi and Sato (1969) demonstrated that the experimental results on β-AgMg agree well with the results of their calculations, in which no presupposed mechanism like the six-jump cycle mechanism enters. This may indicate that a decisive conclusion on the diffusion mechanism is difficult on the basis of the experimental material.

Thus for ordered phases, crystallizing in the B2 structure and exhibiting substitutional disorder, it appears that the theory developed in the Sections VI.E–VI.G gives a satisfactory description, even quantitatively, of the experimental data. There seems to be no necessity to invoke the six-jump cycle mechanism, explicitly, to explain the results.

2. *Systems with a Tendency to Form (Structural) Vacancies*

A number of intermediate phases with the B2 structure show a tendency to form structural vacancies: On one side of stoichiometry the excess component can be accommodated on the "wrong" sublattice, whereas on the other side structural vacancies (and antistructure atoms?) are formed to maintain the phase. For some compounds it is already known that at stoichiometry thermal vacancies are also easily formed. For example, stoichiometric CoGa, with a melting temperature of 1200°C, contains 5–6% of vacancies at 900°C (Berner *et al.*, 1975; Van Ommen *et al.*, 1981). Those thermal vacancies occur in combination with antistructure Co atoms in a ratio 2:1. Obviously, the presence of large amounts of vacancies will have a profound influence on the diffusion behavior of these phases. We will discuss a number of examples of such compounds.

The intermetallic compound β'-AuZn has been studied most extensively: Diffusion coefficients and isotope effects of both components have been

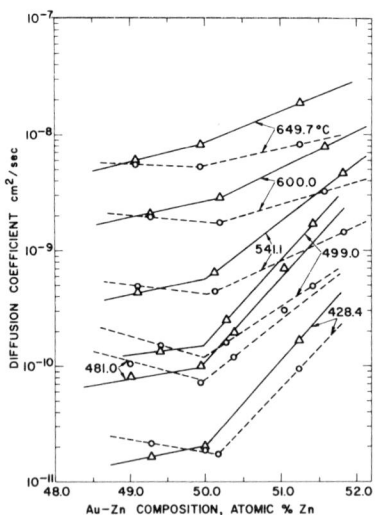

Fig. 16. Temperature and compositional dependence of diffusion coefficients of Au and Zn in β'AuZn: ($\triangle$) Zn^{65}; ($\bigcirc$) Au^{195}. [From Gupta *et al.* (1971).]

measured as a function of composition and temperature and also the influence of pressure on the diffusion coefficients has been investigated. The long-range order parameter in stoichiometric AuZn was found to equal 0.96 at 650°C, increasing to 0.98 at 500°C (Iwasaki and Uesugi, 1968). Gupta and Lieberman (1971) measured the simultaneous diffusion of gold and zinc in single-crystalline specimens of AuZn, having 49, 50, and 51 at. % Zn nominal composition. The temperature range was 428–650°C. The variation of the diffusion coefficients with composition at fixed temperatures is displayed in Fig. 16. The D_{Au} isotherms show distinct cusps at the equiatomic composition, whereas the isotherms representing the zinc diffusivity exhibit only a change in positive slope. This behavior could qualitatively be reproduced by computer simulation (Arnhold, 1981). Values of the activation energies and frequency factors are given in Table V. All values show a maximum at the stoichiometric composition and decrease rather sharply on the zinc-rich side. This may be explained by the fact that AuZn is a Hume–Rothery electron compound with a electron to atom ratio of about 1.5. Then the addition of an excess of zinc can be compensated for by the formation of structural vacancies on the gold sublattice, so that a part of the vacancy formation energy drops out of the diffusion activation energy (cf. Section VII.B). The authors explained their results by assuming a six-jump cycle for diffusion. On the zinc-rich side the vacancies on the Au sublattice outnumber those on the other sublattice. So nearly all six-jump cycles start with a vacancy on the gold sublattice. This means that gold and zinc atoms are displaced mainly by this cycle and therefore have the same activation energy. Furthermore, in such a cycle two zinc atoms and only one gold atom are displaced and the ratio $D_{0Zn}/D_{0Au} \simeq 2$. The agreement between this interpretation and the experimental values from Table V is rather satisfactory. When both types of cycles contribute to the diffusion process the measured activation energies are not necessarily equal. This was argued to be the case for the stoichiometric composition. Here the measured ratio of diffusion coefficients is consistent with a six-jump cycle mechanism; D_{Zn}/D_{Au} varies from 1 at lower temperature to 1.7 at higher temperature. On the gold-rich

TABLE V

ACTIVATION ENERGIES AND PREEXPONENTIAL FACTORS FOR ZINC AND GOLD DIFFUSION IN AuZn

c (at % Zn)	Q_{Zn} (eV)	D_{0Zn} (cm^2 s^{-1})	Q_{Au} (eV)	D_{0Au} (cm^2 s^{-1})
49	1.500	0.84	1.383	0.19
50	1.535	1.93	1.435	0.33
51	1.193	0.047	1.171	0.016

side, gold is substituted on the zinc sublattice and the importance of a six-jump cycle for gold diffusion diminishes in agreement with computer simulation results (Arnhold, 1981). This could explain why at gold-rich compositions the gold diffusion coefficient becomes larger than the zinc diffusion coefficient, as is observable from Fig. 16. Thus it seems that all results can be explained by a six-jump cycle mechanism. However, in the light of computer simulation results by Arnhold and Heumann (1982; Arnhold, 1981) discussed in Section VI.F, it is doubtful if a six-jump cycle can be the only diffusion mechanism occurring in an intermetallic compound.

Jeffery and Gupta (1972) investigated the influence of pressure on the diffusion coefficients of gold and zinc. From this type of measurement the activation volumes for diffusion are derived. Such an activation volume has to be conceived as the sum of the activation volumes for vacancy formation and migration,

$$\Delta V = \Delta V_f + \Delta V_m \tag{73}$$

where ΔV_f is the expansion of the crystal when a vacancy is formed, and ΔV_m is the expansion when an atom is in the saddle point position. From the scarce number of experiments on pure metals in which both ΔV_f and ΔV_m have been measured separately, it turns out that ΔV_m is substantially smaller than ΔV_f (Adda and Philibert, 1966). The results of Jeffrey and Gupta's study are tabulated in Table VI. First, it is noted that the zinc and gold activation volumes are equal within experimental error for each composition. For pure bcc metals ΔV is usually equal to 0.4–0.6 of the molar volume. For the stoichiometric composition of AuZn, ΔV is large (0.9 of the molar volume of 9.45 cm^3 mol^{-1}). The authors argued that this finding is consistent with the six-jump cycle mechanism, because the local disorder that occurs after the third jump in the six-jump cycle may give rise to a relatively large change in volume. In the off-stoichiometric gold-rich alloy, the six-jump cycle will contribute to the diffusion process to a lesser extent, resulting in a smaller value of ΔV. In the 51.2 at. % Zn intermetallic compound, structural vacancies would cause a decrease of ΔV to the value of the migration activation volume only, because no thermal vacancies have to be formed. Remark-

TABLE VI

ZINC AND GOLD ACTIVATION VOLUMES IN AuZn

c (at. % Zn)	ΔV_{Zn} (cm^3 mole^{-1})	ΔV_{Au} (cm^3 mole^{-1})
49.0	3.5 ± 0.6	2.8 ± 1.4
50.0	8.1 ± 1.0	9.4 ± 1.5
51.2	3.7 ± 1.6	3.4 ± 1.6

ably, the decrease of ΔV is nearly symmetrical on both sides of stoichiometry. The authors concluded that this result does not seem compatible with a vacancy defect structure in Zn-rich alloys, in contrast with previous interpretations.

Hilgedieck and Herzig (1982; Hilgedieck, 1981) measured the isotope effect both for gold and zinc diffusion in various compositions of the compound. The results can be summarized as follows. For $T = 482$–$602°C$, the isotope effect for gold is in the range 0.2–0.4 for compositions from 49–52 at. % Zn without any clear temperature or concentration dependence. The isotope effect for Zn is lower: 0.2 for 49 at. % Zn, 0.1 at stoichiometry, and 0.05 for the 52 at. % Zn alloy. On the basis of these measurements the zinc diffusion was assumed to occur by exchange of zinc atoms and adjacent vacancies. Because of the high degree of order such a mechanism is highly correlated, which explains the low values for the zinc isotope effect. From the independence of the gold isotope effect from composition and temperature a possible conclusion could be that the migration of gold occurs over its own sublattice. If a ΔK value of about 0.5 is assumed, the mean value of the gold isotope effect of 0.3 would be compatible with such a mechanism. However, the authors prefered an interpretation based on migration of zinc and gold atoms via nn jumps. They argued that there is no necessity to invoke the six-jump cycle in contrast with previous interpretations of Gupta and Lieberman (1971) and Jeffrey and Gupta (1972). In conclusion, in spite of all information available for diffusion in the β' phase of AuZn, a number of questions still remain unanswered.

Diffusion coefficients of both components in stoichiometric β-AuCd (Huntington et al., 1961), and in the compositions with 47.5, 49, and 50.5 at. % Cd (Gupta et al., 1967), showed a behavior similar to that in AuZn and were interpreted by the latter authors analogously.

The defect structure of the group VIII–group IIIA intermetallic compound β-NiAl has been discussed by Neumann et al. (1976), who evaluated relevant experimental information. It appears that in nickel-rich compounds the amount of vacancies is small. The excess nickel atoms substitute on the aluminum sublattice. At stoichiometry the vacancy concentration is some tenths of a percent, which is small compared with some other compounds of the same type like NiGa and CoGa. In the aluminum-rich β-NiAl alloys the number of structural vacancies corresponds to the excess of aluminum, e.g., 6% vacancies are found in the 53 at. % Al composition (see also Yang et al., 1978). The explanation for this is that the compound is a Hume–Rothery electron compound with an electron to atom ratio of 1.5. Hancock and McDonnell (1971) measured nickel diffusion in NiAl in the composition range 41.3–51.7 at. % Al at temperatures from 1000–1350°C. Isothermal diffusion coefficients, plotted as a function of concentration, showed cusps near stoichiometry, slightly shifted towards the aluminum-rich side. Arrhenius plots tended to be curved. When these plots were analyzed as straight lines the activation energies shown in Fig. 17 were obtained. In this figure results from an older study on Co diffusivity are also displayed (Berkowitz et al., 1954). The agreement between both investigations is satisfactory. It is seen that the presence of structural vacancies has a dramatic effect on the activation energy, which decreases sharply near the 50 at. % composition as aluminum is added. Thus the presence of structural vacancies strongly influences the activation energy, thereby making it clear that diffusion occurs by a vacancy mechanism. It is tempting to interpret the results by stating that in the nickel-rich alloys the activation energy is equal to the sum of the vacancy formation and migration energies, whereas in the region where structural vacancies outnumber the thermal ones only the migration energy has to be supplied for diffusion to take place. However, this does not indicate exactly which process is responsible for migration. The authors suggested a six-jump cycle as a possible mechanism, though, because of the high ordering energy in this system and the presence of large amounts of vacancies in the aluminum-rich alloys, second-nearest neighbor jumps could not be excluded. A change of the diffusion mechanism across the phase field was also considered as a possibility. A further investigation on NiAl was performed by Lutze-Birk and Jacobi (1975). Because of the unavailability of a suitable aluminum radioisotope, an indium tracer was diffused. Four compositions (42–54.9 at. % Al) were investigated at temperatures ranging from 1000–1400°C. Again the variation of the diffusion coefficient with concentration at fixed temperature is characterized by a minimum near stoichiometry. At this composition the activation energy also has its maximum. The results, together with corresponding values for nickel diffusion (Hancock and McDonnell, 1971) are given in Table VII. As a possible diffusion mechanism a dynamic equilibrium of the type (in self-explanatory notation)

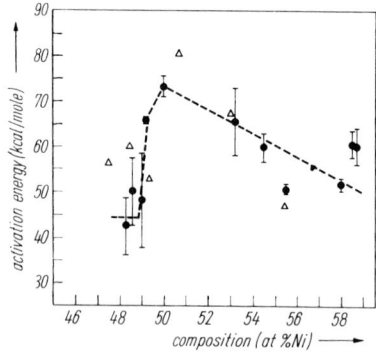

Fig. 17. Variation of activation energy with composition in NiAl: (△) Co diffusion (Berkowitz et al., 1954); (●) Ni diffusion. [From Hancock and McDonnell (1971).]

$V^{Ni} + Ni^{Al} \rightleftarrows V^{Al} + Ni^{Ni}$ was proposed, combined with migration of vacancies over their own sublattices.

From inspection of the phase diagrams of the system CoGa and NiGa, which are very similar, it becomes clear that these systems are closely related. Diffusion was measured in the electron compounds β-NiGa (Donaldson and Rawlings, 1976a) and CoGa (Stolwijk et al., 1977; Bose et al., 1979; Stolwijk et al., 1980). The difference between β-NiAl and these two intermediate phases is that thermal vacancies are formed more easily in the latter two compounds. The production of any two thermal vacancies is combined necessarily with the creation of one antistructure transition metal atom with the result that these compounds exhibit a special type of disorder. This combination of two vacancies with one antistructure atom is sometimes termed a "triple defect." Since below a certain temperature the mobility of vacancies is too low to attain thermodynamic equilibrium in a reasonable time, the alloys always contain these defects at room temperature. Apart from thermal vacancies structural vacancies also occur in the Ga-rich composition, which compensate the excess of gallium and preserve the electron to atom ratio of 1.5. Extensive studies have been performed on the content of structural and thermal vacancies (Berner et al., 1975; Donaldson and Rawlings, 1976b; Van Ommen et al., 1981; Van Ommen, 1982; also see for references Neumann et al., 1976). In spite of the presence of large amounts of vacancies (thermal and/or structural) at nearly all compositions and the low formation energy of thermal vacancies, no drastic effects of composition—as in NiAl—were found on the activation energy for diffusion. Isothermal diffusion coefficients of the transition metal depended only weakly on composition. The gallium diffusion coefficients revealed a cusp near stoichiometry. While Donaldson and Rawlings (1976a) measured diffusion coefficients of both components in β-NiGa over about three orders of magnitude and Bose et al. (1979) in β-CoGa over one or two orders of magnitude, the measurements of Stolwijk et al. (1980) extended over four to

TABLE VII

ACTIVATION ENERGIES AND PREEXPONENTIAL FACTORS FOR INDIUM AND NICKEL DIFFUSION IN NiAl

c (at. % Al)	D_{0In} (cm^2 s^{-1})	D_{0Ni} (cm^2 s^{-1})	Q_{In} (eV)	Q_{Ni} (eV)	D_{Ni}/D_{In}
42	1.83×10^{-4}	3.5×10^{-2}	1.761	2.242	4.4
50	3.98×10^{-3}	4.5	2.515	3.183	6.0
51.4	1.29×10^{-3}	1.0×10^{-3}	2.454	2.173	7.3
54.9	1.02×10^{-7}	—	0.963	—	—

six orders of magnitude. The latter investigation revealed that in reality the Arrhenius plots are clearly curved. Of course, this has important implication in the interpretation. Besides, the results on gallium diffusivity in the other experiments have been criticized on experimental grounds (Stolwijk et al., 1980). Stolwijk et al. found that, if only one flat surface of a cylindrical sample is electroplated with radioactive gallium, the measured diffusion coefficients are somewhat too large. This effect is due to the very fast surface diffusion of gallium over the sample surface and could be avoided by electroplating the whole surface of the sample with gallium tracer. Since the Arrhenius plots for diffusion in pure bcc metals also are usually curved, the curvatures found in CoGa are not surprising. Stolwijk et al. (1980) analyzed their results in terms of two exponentials. Remarkably, one of the activation energies for cobalt diffusion appeared within experimental error to equal one of the activation energies for gallium diffusion. For further analysis these two energies were put equal (Fig. 18). The other ("low" temperature) activation energies and frequency factors for cobalt diffusion were quite normal, and resemble diffusion by monovacancies in bcc metals like Ta, Nb, V, Mo, and Cr. The higher activation energy and D_0 values for cobalt are comparable to the high-temperature parameters for the bcc metals mentioned above. These are usually interpreted as being due to migration via divacancies. This fact inspired Stolwijk et al. (1980) to propose a modified divacancy mechanism (the so-called triple defect mechanism) for this diffusion process. Diffusion via these triple defects implies that the cobalt atoms migrate between the two sublattices, whereas gallium atoms only jump over their own sublattice. The latter is facilitated by the presence of the other vacancy of the divacancy. The process is presented in Fig. 19, parts a–c, in which the various exchange frequencies are also defined. Since diffusion of both cobalt and gallium occurs by the same defect and since migration of cobalt atoms is not possible without gallium jumps, the activation energies for both species tend to be the same for this diffusion mechanism. Mathematically, it is the correlation factor that accounts for this effect. Detailed calculations on correlation factors were carried out for the stoichiometric composition (Bakker et al., 1981). It was also demonstrated that the isotope effect parameter for this mechanism equals $f \Delta K$. When the triple-defect mechanism predominates and the ratio w_B^{11}/w_A^{12} (Fig. 19) is not too extreme, an isotope effect in the range 0.2–0.4 is predicted. Finally, the other ("high" temperature) activation energy for gallium is seen from Fig. 18 to be high for compositions containing 48–60% Co. A possible natural mechanism for this process could be the migration of gallium over its own sublattice by monovacancies.

Measurements on the FeAl system, which exhibits a number of ordered structures, were reported by Larikov et al. (1975). It is interesting to compare

4. TRACER DIFFUSION IN CONCENTRATED ALLOYS

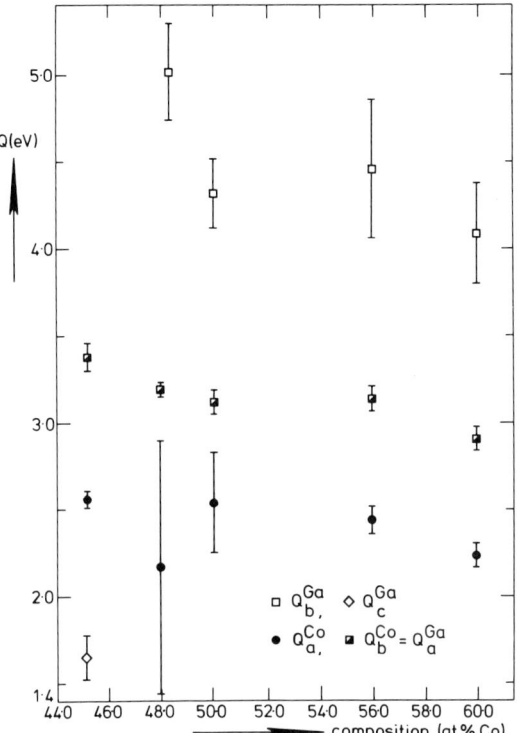

Fig. 18. The variation of the activation energies for Co and Ga diffusion as a function of composition in CoGa, analyzed following a two-exponential fit. The higher Co activation energy (Q_b^{Co}) is within experimental error equal to the lower Ga activation energy (Q_a^{Ga}). [From Stolwijk et al. (1980).]

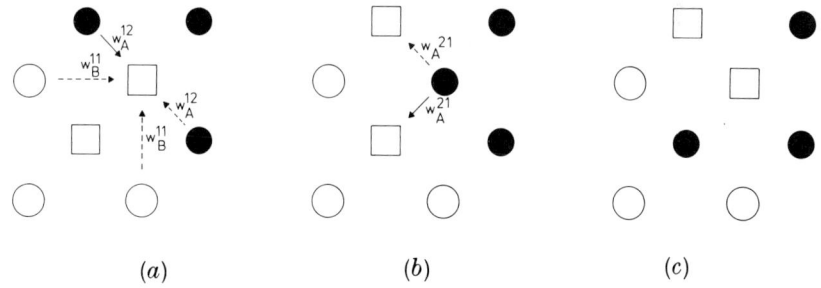

Fig. 19. A jump sequence of a divacancy in a two dimensional analog of the B2 structure: (●) A(Co) atoms; (○) B(Ga) atoms; (□) vacancies. All possible jumps in (a) and (b) are indicated by arrows labeled with the corresponding exchange frequencies. (a, c) The divacancy in the nn configuration. (b) The divacancy in the triple-defect configuration. [From Stolwijk et al. (1980).]

the results on the Fe 48.5 at. % compound with the measurements on CoGa. Straight line fits yielded

$$D(Fe) = 1.82 \times 10^4 \exp(-3.44/kT) \, cm^2 \, s^{-1}$$

$$D(Al) = 8.7 \times 10^3 \exp(-3.52/kT) \, cm^2 \, s^{-1}$$

So values for D_0 and Q are high, in complete agreement with the results on CoGa if analyzed in terms of a single exponential.

In conclusion, all measurements discussed in this section are interpreted in terms of migration of atoms by exchanges with vacancies. It is clear that a six-jump cycle mechanism can not be the only diffusion mechanism. Apart from atomic migration by exchanges with nearest-neighboring vacancies, sometimes jumping of atoms into next-nearest-neighboring vacancies is also suggested. For NiAl the activation energies of nickel and cobalt diffusion show a sharp decrease in aluminum-rich compounds, probably due to the formation of structural vacancies. In NiGa and CoGa large amounts of vacancies are present in nearly all compositions, because thermal vacancies are formed easily. This explains why in these compounds such a sharp decrease is absent. Accurate measurements over many orders of magnitude in D reveal that in CoGa the Arrhenius plots are curved. One could ask if such a curvature would be also detected in other compounds if the measurements would be extended over larger temperature regions. From a detailed analysis of the curved Arrhenius plots in CoGa it turns out that in this compound a divacancy diffusion mechanism is likely to play an important role. In spite of the occurrence of large numbers of vacancies in NiGa and CoGa the activation energies for diffusion are not particularly low. On the contrary, the activation energies and preexponentials are quite normal, or even high. High values for these quantities are also found in FeAl. It appears that diffusion in the compounds reviewed in this section is far from showing uniform features.

B. Intermediate Phases with the DO_3 (BiF_3 or Fe_3Si) Structure

In this section we shall describe tracer diffusion measurements in a number of intermediate phases with the DO_3 structure. The phases to be considered (β-Ni_3Sb, β-Cu_3Sb, and β-Cu_3Sn) are β Hume–Rothery phases (Wever and Frohberg, 1974). This explains the occurrence of structural vacancies in these compounds and their relationship to some of the equiatomic intermetallic compounds, which was reviewed in the previous section. However, by inspection of the crystallographic structure (Fig. 7), it is noticed that there is a marked difference to those compounds. The majority component is able to migrate between the α and γ sublattice without dis-

TABLE VIII

ACTIVATION ENERGIES AND PREEXPONENTIAL FACTORS
FOR NICKEL DIFFUSION IN β-Ni$_3$Sb

c (at. % Ni)	Q (eV)	$D_0 \times 10^4$ (cm^2 s^{-1})
75.0[a]	0.716	6.9
73.7	0.685	6.3
72.9	0.642	4.9
71.7	0.637	5.4

[a] This composition was shown to be outside the single phase region.

turbing the state of order and so a more directly interpretable effect of structural vacancies on the activation energy of diffusion is to be expected. The investigations comprise β-Ni$_3$Sb, existing in the Sb rich compositions, Cu-rich β-Cu$_3$Sn and β-Cu$_3$Sb existing on both sides of stoichiometry.

Heumann and Stüer (1966) diffused nickel tracers into four compositions of β-Ni$_3$Sb in the composition range 71.7–75.0 at. % Ni at temperatures from 600 to 1000°C. The mobility of the nickel atoms appeared to be extremely high. The activation energies turned out to be less than half the values to be expected on the basis of the melting point (Table VIII). By means of density and lattice parameter measurements the authors demonstrated that corresponding to the excess of antimony structural vacancies are formed. It is plausible that by adding antimony in excess of the stoichiometric composition the antimony extends its own sublattice, creating unoccupied sites in the nickel sublattices. Thus the formation of vacancies is not a thermally activated process and the activation energy, measured in a tracer diffusion experiment, equals the migration energy of vacancies only. If the nickel positions are conceived as octahedral and tetrahedral interstitial sites in an fcc Sb lattice

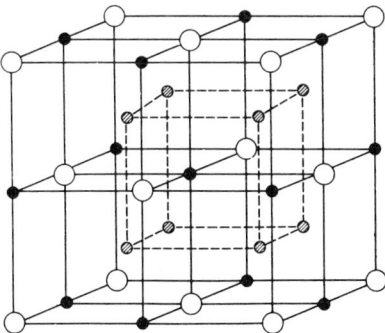

Fig. 20. The DO$_3$ (Fe$_3$Si) structure for binary compound A$_3$B: (●) A atoms on the γ sublattice; (dashed circles) A atoms on the α sublattice; (○) B atoms on the β sublattice. [From Heumann and Stüer (1966).]

(Fig. 20), it is clear that the nickel diffusion process is, in a sense, intermediate between substitutional and interstitial migration.

A study of copper diffusion in β-Cu$_3$Sb with compositions from 71 to 79 at. % Cu was made by Heumann et al. (1970) in the temperature interval 500–650°C. In contrast to β-Ni$_3$Sb, the single-phase region extends also to compositions on the copper-rich side. The mobility of copper is high in all compositions and the diffusion coefficient at fixed temperature increases with a factor of about two from the 79 at. % Cu to the 71 at. % Cu composition. The activation energies are, relative to the melting point, even lower than for Ni$_3$Sb (Table IX). The preexponential factors, given by Heumann et al. (1970) are probably a factor of 10 in error, because they do not reproduce the measured values of diffusion coefficients. In Table VIII we tentatively give values for D_0 a factor of 10 smaller than those from the original paper. Then for the Sb-rich composition the D_0 is smaller than for Ni$_3$Sb as is the activation energy. From density and lattice parameter measurements it turned out that structural vacancies exist over the whole composition range. The number of vacancies increases with antimony content, which is reflected by the decrease of the activation energy. Vacancies may also be present on the Sb sublattice or there is some substitution of Cu atoms on the Sb lattice in the whole composition range. It is argued that the diffusion process is comparable with that in Ni$_3$Sb.

Diffusion of both copper tracers and tin tracers in 83.4 at. % Cu and 79.8 at. % Cu β-Cu$_3$Sn was studied by Prinz and Wever (1980) over the temperature range 540–735°C. The authors found that even in these copper rich alloys structural vacancies and copper antistructure atoms occur. This is a consequence of the Hume–Rothery rule, requiring a number of 1.5 electrons per lattice site (Wever and Frohberg, 1974). The concentration of vacancies increases from about 0.2% below the 16.5 at. % Sn composition to 6% for the Cu$_3$Sn compound (Frohberg, 1981). Again this concentration is only weakly dependent on temperature, so the activation energies are

TABLE IX

ACTIVATION ENERGIES AND PREEXPONENTIAL FACTORS FOR COPPER DIFFUSION IN β-Cu$_3$Sb

c (at. % Cu)	Q (eV)	$D_0 \times 10^4$ (cm^2 s^{-1})[a]
79	0.454 ± 0.058	8.57
75	0.315 ± 0.044	1.99
71	0.252 ± 0.038	1.36

[a] See text.

4. TRACER DIFFUSION IN CONCENTRATED ALLOYS

TABLE X
ACTIVATION ENERGIES AND PREEXPONENTIAL FACTORS FOR COPPER AND TIN DIFFUSION IN β-Cu$_3$Sn

c (at. % Cu)	$D_{0Cu} \times 10^3$ (cm^2 s^{-1})	Q_{Cu} (eV)	$D_{0Sn} \times 10^2$ (cm^2 s^{-1})	Q_{Sn} (eV)
83.4	8.3 ± 1.7	0.86 ± 0.02	22 ± 19	1.22 ± 0.07
79.8	18 ± 3	0.85 ± 0.01	3.5 ± 1.6	1.11 ± 0.03

determined by the activation energies for migration only and are correspondingly low (Table X). The authors explained their results by diffusion of copper between the α and γ sublattice (Fig. 7). The diffusion of tin with a higher activation energy occurs via tin antistructure atoms between the α and γ sublattice.

It is clear that the diffusion behavior of the three compounds, discussed in this section, is similar. In all the compounds the diffusion is fast and the activation energies are low, due to the presence of structural vacancies.

C. INTERMEDIATE PHASES WITH THE L1$_2$ (Cu$_3$Au) STRUCTURE

The unit cell of the L1$_2$ (Cu$_3$Au) structure is presented in Fig. 21. Some older work on diffusion in this structure has been reviewed by Girifalco (1973).

Hancock (1971) performed diffusion experiments of nickel in Ni$_3$Al on both sides of stoichiometry. Annealing temperatures ranged from 920 to 1280°C. A striking feature is the lack of variation of the diffusion coefficient with composition. A fit of the data to the Arrhenius equation gives the values of Table XI. The values for the activation energies and frequency factors are quite normal and not much different from those of self-diffusion in pure nickel. In this structure nearest-neighbor jumps of a nickel atom to eight positions on the nickel sublattice are possible without any difficulty. On the basis of the results such a mechanism seems the most simple explanation of

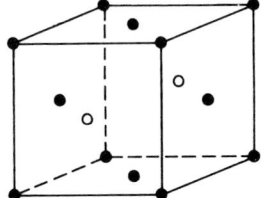

Fig. 21. The L1$_2$ superlattice (Cu$_3$Au): (●) Cu atoms; (○) Au atoms. [From Schapink (1969).]

TABLE XI

ACTIVATION ENERGIES AND PREEXPONENTIAL FACTORS
FOR NICKEL DIFFUSION IN NI_3AL

c (at. % Ni)	Q (eV)	D_0 (cm² s⁻¹)
76.2	3.173	4.41
74.7	3.140	1.00
73.2	3.111	3.11

the data. The fact that an excess of aluminum does not change the activation energy dramatically indicates that excess aluminum atoms are accommodated on the nickel sublattice. This is in contrast to the behavior of Ni_3Sb, etc., reviewed in the previous section. Diffusion of Ni in the stoichiometric compound was also investigated by Bronfin et al. (1975). At a number of temperatures nickel self-diffusion coefficients were also measured. These turned out to be always a factor of about 2 higher than those in the compound. The Arrhenius analysis for the compound yielded $D = 1.0 \exp(-2.971 \pm 0.043)/kT$ cm² s⁻¹, in fair agreement with Hancock's (1971) value. The authors explained their results by diffusion of nickel via nickel sites. The fact that a nickel atom has only eight nickel nns in this structure, in contrast to twelve in pure nickel, affects, in particular, the preexponential factor, which is about a factor of two smaller than for self-diffusion in pure nickel.

Ansel et al. (1979) measured manganese diffusion in three ordered compounds of the $MnPt_3$ intermediate phase. The parameters obtained for diffusion of this minority component are given in Table XII. The authors claimed that these results represent bulk diffusion. The extremely low values for the off-stoichiometric compositions were, on the basis of density measurements, attributed to the presence of structural vacancies: 4.6% vacancies plus antistructure Pt atoms in the Pt-rich composition and 1% vacancies and Mn antistructure atoms on the manganese-rich side.

TABLE XII

ACTIVATION ENERGIES AND PREEXPONENTIAL FACTORS
FOR DIFFUSION OF MANGANESE IN $MNPT_3$ OBTAINED
IN THE TEMPERATURE INTERVAL 750–1010°C

c (at. % Pt)	Q (eV)	D_0 (cm² s⁻¹)
82	0.55	2.1×10^{-10}
75	2.3	3×10^{-2}
65	0.59	2.3×10^{-10}

VIII. Conclusions

Considerable progress has been made in the field of tracer diffusion in concentrated alloys in the past fifteen years. This becomes clear when one compares the contents of this chapter with the state of affairs around the mid-1960s as reviewed by Adda and Philibert (1966). Not only has the number of experiments performed increased substantially, but also theories have been developed that describe the rather complicated cooperative character of the diffusion in alloys. Monte Carlo computer simulation studies have served not only as a test of the reliability of the theoretical results, but sometimes have inspired new analytical approaches. With only a few exceptions the lattice vacancy, exchanging with nearest-neighboring atoms, has been assumed to be responsible for atomic migration. An Arrhenius behavior of the diffusion coefficient as a function of temperature has usually been found. Empirical rules or Manning's random alloy model have been applied successfully to a number of primary phases. For a number of intermediate phases with substitutional disorder the diffusion behavior has also been shown to be consistent with theoretical predictions. The diffusion in intermediate phases with a special defect structure, where large amounts of (structural) vacancies occur, has been demonstrated to be strongly related to this defect structure. However, so far only alloys crystallizing in relatively simple crystallographic structures have been investigated. It is to be expected that accurate diffusion and isotope effect measurements over large temperature ranges will reveal more details of the diffusion processes in these alloys and thereby will give more information about the behavior of lattice defects. On the other hand, diffusion in intermediate phases with more complex crystallographic structures is still an unexplored field.

Acknowledgments

I thank Dr. V. Arnhold, Dr. G. E. Murch, Prof. A. S. Nowick, Prof. S. Radelaar, Dr. N. A. Stolwijk, Prof. G. de Vries, Dr. A. A. H. J. Waegemaekers, and Dr. A. van Winkel for helpful comments and improvements. I am obliged to my wife and Mrs. M. G. J. O. Mölders-Bos for typing the first and final draft, respectively.

References

Adda, Y., and Philibert, J. (1966). "La Diffusion Dans les Solides." Presses Universitaires de France, Paris.
Ansel, D., Barre, J., Meziere, C., and Debuigne, J. (1979). *J. Less-Common Met.* **65,** 1.
Arnhold, V. (1981). Thesis. Westfälische Wilhelms-Universität, Münster, West Germany.
Arnhold, V., and Heumann, T. (1982). To be published.
Askill, J. (1971). *Phys. Status Solidi A* **8,** 587.

Bakker, H. (1971). *Phys. Status Solidi B* **44**, 369.
Bakker, H. (1979). *Philos. Mag.* [*Part*] *A* **40**, 525.
Bakker, H., Stolwijk, N. A., and Hoetjes-Eijkel, M. A. (1981). *Philos. Mag.* [*Part*] *A* **43**, 251.
Bakker, H., and Stolwijk, N. A., Van der Meij, L., and Zuurendonk, T. J. (1976). *Nucl. Metall.* **20**, 96.
Beke, D. L., Gödény, I., Kedves, F. J., and Erdélyi, G. (1979). *J. Phys. Chem. Solids* **40**, 543.
Berkowitz, A. E., Jaumot, F. E., and Nix, F. C. (1954). *Phys. Rev.* **95**, 1185.
Berner, D., Geibel, G., Gerold, V., and Wachtel, E. (1975). *J. Phys. Chem. Solids* **36**, 221.
Borg, R. J., and Lai, D. Y. F. (1970). *J. Appl. Phys.* **41**, 5193.
Borovskiy, I. B., Marchukova, I. D., and Ugaste, Y. E. (1970). *Fiz. Met. Metalloved.* **29** (2), 308.
Bose, A., Frohberg, G., and Wever, H. (1979). *Phys. Status. Solidi A* **52**, 509.
Bowen, A. W., and Leak, G. M. (1970). *Metall. Trans.* **1**, 2767.
Bronfin, M. B., Balatov, G. S., and Drugova, I. A. (1975). *Fiz. Met. Metalloved.* **40** (2), 363.
Brown, A. M., and Ashby, M. F. (1980). *Acta Metall.* **28**, 1085.
Butrymowicz, D. B., Manning, J. R., and Read, M. E. (1977). INCRA Monograph V. Int. Copper Research Assn., New York.
Cermák, J., Cíha, K., and Kučera, J. (1980). *Phys. Status Solidi A* **62**, 467.
Chatterjee, A., and Fabian, D. J. (1970). *Scr. Metall.* **4**, 285.
de Bruin, H. J., Murch, G. E., Bakker, H., and Van der Meij, L. P. (1975). *Thin Solid Films* **25**, 47.
de Bruin, H. J., Bakker, H., and Van der Meij, L. P. (1977). *Phys. Status Solidi B* **82**, 581.
deGonzales, C. O., and deReca, N. E. W. (1971). *J. Phys. Chem. Solids* **32**, 1067.
Domian, H. A., and Aaronson, H. I. (1964). *Trans. AIME* **230**, 44.
Domian, H. A., and Aaronson, H. I. (1965). "Diffusion in Body Centered Cubic Metals," p. 209. Amer. Soc. Metals, Metals Park, Ohio.
Donaldson, A. T., and Rawlings, R. D. (1976a). *Acta Metall.* **24**, 285.
Donaldson, A. T., and Rawlings, R. D. (1976b). *Acta Metall.* **24**, 811.
Elcock, E. W. (1959). *Proc. Phys. Soc., London* **73**, 250.
Elcock, E. W., and McCombie, C. W. (1958). *Phys. Rev.* **109**, 605.
Elliot, R. P. (1965). "Constitution of Binary Alloys," 1st Suppl. McGraw-Hill, New York.
Fillon, J., and Calais, D. (1977). *J. Phys. Chem. Solids* **38**, 81.
Fishman, S. G., Gupta, D., and Lieberman, D. S. (1970). *Phys. Rev. B* **2**, 1451.
Frantsevich, I. N., Kalinovich, D. F., Kovenskii, I. I., and Smolin, M. D. (1969). *J. Phys. Chem. Solids* **30**, 947.
Frohberg, G. (1971). Habilitationsschrift. Technische Universität Berlin, West Germany. [Referred to in Arnhold, V. (1981).]
Frohberg, G. (1981). *Z. Metallkd.* **72**, 596.
Gardner, A. B., Sanders, R. L., and Slifkin, L. M. (1968). *Phys. Status Solidi* **30**, 93.
Gibbs, G. B., Graham, D., Tomlin, D. H. (1963). *Philos. Mag.* **8**, 1269.
Girifalco, L. A. (1964a). "Atomic Migration in Crystals." Ginn (Blaisdell), Boston, Massachusetts.
Girifalco, L. A. (1964b). *J. Phys. Chem. Solids* **25**, 323.
Girifalco, L. A. (1973). "Diffusion" (H. I. Aaronson, ed.), p. 185. Amer. Soc. Metals, Metals Park, Ohio.
Gödény, I., Beke, D., Kedves, F. J., and Groma, G. (1975). *Phys. Status Solidi A* **32**, 195.
Gruzin, P. L., and Fedorov, G. B. (1955). *Dokl. Akad. Nauk SSSR* **105**, 264.
Gupta, D., Lazarus, D., and Lieberman, D. S. (1967). *Phys. Rev.* **153**, 863.
Gupta, D., and Lieberman, D. S. (1971). *Phys. Rev. B* **4**, 1070.
Guttman, L. (1961). *J. Chem. Phys.* **34**, 1024.

Hagel, W. C., and Westbrook, J. H. (1961). *Trans. AIME* **221**, 951.
Hagel, W. C., and Westbrook, J. H. (1965). "Diffusion in Body Centered Cubic Metals," p. 197. Amer. Soc. Metals, Metals Park, Ohio.
Hancock, G. F. (1971). *Phys. Status Solidi A* **7**, 535.
Hancock, G. F., and McDonnell, B. R. (1971). *Phys. Status Solidi A* **4**, 143.
Hansen, M. (1958). "Constitution of Binary Alloys." McGraw-Hill, New York.
Hässner, A., and Lange, W. (1965). *Phys. Status Solidi* **8**, 77.
Herzig, C., Eckseler, H., Bussman, W., and Cardis, D. (1978). *J. Nucl. Mater.* **69/70**, 61.
Heumann, Th., and Stüer, H. (1966). *Phys. Status Solidi* **15**, 95.
Heumann, Th., Meiners, H., and Stüer, H. (1970). *Z. Naturforsch.* **25a**, 1883.
Hilgedieck, R. (1981). Thesis. Westfälische Wilhelms-Universität, Münster, West Germany.
Hilgedieck, R., and Herzig, C. (1982). *Z. Metallkde.* **74**, 38.
Hilliard, J. E., Averbach, B. L., and Cohen, M. (1959). *Acta Metall.* **7**, 86.
Hirano, K., and Cohen, M. (1972). *Trans. Jpn. Inst. Met.* **13**, 96.
Hoffmann, R. E., Turnbull, D., and Hart, E. W. (1955). *Acta Metall.* **3**, 417.
Huntington, H. B., Miller, N. C., and Nerses, V. (1961). *Acta Metall.* **9**, 749.
Irmer, V., and Feller-Kniepmeier, M. (1972). *J. Appl. Phys.* **43**, 953.
Iwasaki, H., and Uesugi, T. (1968). *J. Phys. Soc. Japan* **25**, 1640.
Jeffery, R. N., and Gupta, D. (1972). *Phys. Rev. B* **6**, 4432.
Kalinovich, D. F., Kovenskii, I. I., and Smolin, M. D. (1972). *Fiz. Tverd. Tela (Leningrad)* **14**, 3699.
Kalinovich, D. F., Kovenskii, I. I., and Smolin, M. D. (1973). *Fiz. Met. Metalloved.* **35** (6), 1315.
Kikuchi, R. (1977). *J. Phys. Colloq. (Orsay, Fr.)* **38**, C7-307.
Kikuchi, R., and Sato, H. (1969). *J. Chem. Phys.* **51**, 161.
Kikuchi, R., and Sato, H. (1970). *J. Chem. Phys.* **53**, 2702.
Kikuchi, R., and Sato, H. (1972). *J. Chem. Phys.* **57**, 4962.
Kinoshita, C., Tomokiyo, Y., and Eguchi, T. (1978). *Philos. Mag. B* **38**, 221.
Kučera, J., and Million, B. (1970a). *Metall. Trans.* **1**, 2603.
Kučera, J., and Million, B. (1970b). *Metall. Trans.* **1**, 2599.
Kučera, J., Million, B., and Plšková, J. (1972). *Phys. Status Solidi A* **11**, 361.
Kučera, J., Million, B. (1975). *Phys. Status Solidi A* **31**, 275.
Kučera, J., and Stránský, K. (1982). *Mater. Sci. Eng.* **52**, 1.
Kuper, A. B., Lazarus, D., Manning, J. R., and Tomizuka, C. T. (1956). *Phys. Rev.* **104**, 1536.
Larikov, L. N., Geichenko, V. V., and Fal'chenko, V. M. (1975). "Diffusion Processes in Ordered Alloys" (in Russian). Naukova Dumka Publ., Kieve, USSR. [Engl. transl.: by Amerind Publ., New Delhi, India (1981).]
Lazarus, D. (1960). *Solid State Phys.* **10**, 71.
Lazarus, D. (1965). "Diffusion in Body Centered Cubic Metals," p. 155. Amer. Soc. Metals, Metals Park, Ohio.
Le Claire, A. D. (1962). *Philos. Mag.* **7**, 141.
Lutze-Birk, A., and Jacobi, H. (1975). *Scr. Metall.* **9**, 761.
Mallard, W. C., Gardner, A. B., Bass, R. F., and Slifkin, L. M. (1963). *Phys. Rev.* **129**, 617.
Manning, J. R. (1968). "Diffusion Kinetics for Atoms in Crystals." Van Nostrand-Reinhold, Princeton, New Jersey.
Manning, J. R. (1971a). *Phys. Rev. B* **4**, 1111.
Manning, J. R. (1971b). *Z. Naturforsch.* **26a**, 69.
Manning, J. R. (1971c). *Proc. Conf. Atom. Trans. Solids Liquids* (A. Lodding and T. Lagerwall, eds.), p. 213. Verlag Z. Naturforsch., Tübingen, West Germany.
Mead, N. W., and Birchenall, C. E. (1957). *Trans. AIME* **209**, 874.
Mehrer, H. (1978). *J. Nucl. Mater.* **69/70**, 38.

Million, B. (1972). *Z. Metallkde.* **63**, 484.
Million, B. (1977). *Czech. J. Phys.* B **27**, 928.
Million, B., and Kučera, J. (1969). *Acta Metall.* **17**, 339.
Million, B., and Kučera, J. (1971). *Czech. J. Phys.* B **21**, 161.
Monma, K., Suto, H., and Oikawa, H. (1964a). *J. Jpn. Inst. Met.* **28**, 192.
Monma, K., Suto, H., and Oikawa, H. (1964b). *J. Jpn. Inst. Met.* **28**, 188.
Murch, G. E. (1982a). *Philos. Mag.* [*Part*] A **45**, 941.
Murch, G. E. (1982b). *Philos. Mag.* [*Part*] A **46**, 565.
Murch, G. E., and Rothman, S. J. (1981). *Philos. Mag.* [*Part*] A **43**, 229.
Murdock, J. F., and McHargue, C. J. (1968). *Acta Metall.* **16**, 493.
Neumann, J. P., Chang, Y. A., and Lee, C. M. (1976). *Acta Metall.* **24**, 593.
Obrtlík, K., and Kučera, J. (1979). *Phys. Status Solidi A* **53**, 589.
Okabe, T., Hines, A. L., and Hochmann, R. F. (1976). *J. Appl. Phys.* **47**, 49.
Peterson, N. L. (1975). "Diffusion in Solids. Recent Developments" (A. S. Nowick and J. J. Burton, eds.), p. 115. Academic Press, New York.
Peterson, N. L., and Rothman, S. J. (1967). *Phys. Rev.* **154**, 558.
Peterson, N. L., and Rothman, S. J. (1970a). *Phys. Rev.* B **1**, 3264.
Peterson, N. L., and Rothman, S. J. (1970b). *Phys. Rev.* B **2**, 1540.
Pontau, A. E., and Lazarus, D. (1979). *Phys. Rev.* B **19**, 4027.
Prinz, N., and Wever, H. (1980). *Phys. Status Solidi A* **61**, 505.
Radelaar, S. (1970). *J. Phys. Chem. Solids* **31**, 219.
Radelaar, S. (1982). Private communications.
Resing, H. A., and Nachtrieb, N. H. (1961). *J. Phys. Chem. Solids* **21**, 40.
Santos, E., and Dyment, F. (1975). *Philos. Mag.* **31**, 809.
Sato, H. (1970). *Phys. Chem. (NY)* **10**, 579.
Sato, H., and Kikuchi, R. (1983). *Phys. Rev.* B **28**, 648.
Schapink, F. W. (1969). Thesis. Technical Univ., Delft, The Netherlands.
Schoen, A. H. (1958). Ph.D. Thesis. Univ. of Illinois, Champaign. [Referred to in Gardner *et al.* (1968).]
Schoijet, M., and Girifalco, L. A. (1968). *J. Phys. Chem. Solids* **29**, 481.
Schoijet, M., and Girifalco, L. A. (1968), *J. Phys. Chem. Solids* **29**, 497.
Schoijet, M., and Girifalco, L. A. (1968). *J. Phys. Chem. Solids* **29**, 911.
Shewmon, P. G. (1963). "Diffusion in Solids" McGraw-Hill, New York.
Shires, P. J., Hines, A. L., and Okabe, T. (1977). *J. Appl. Phys.* **48**, 1734.
Stanley, J., and Wert, C. (1961). *J. Appl. Phys.* **32**, 267.
Stolwijk, N. A. (1981). *Phys. Status Solidi B* **105**, 223.
Stolwijk, N. A., Bakker, H., and Van Gend, M. (1980). *J. Phys. C* **13**, 5207.
Stolwijk, N. A., Spruijt, T., Hoetjes-Eijkel, M. A., and Bakker, H. (1977). *Phys. Status Solidi A* **42**, 537.
Stolwijk, N. A., Van Gend, M., and Bakker, H. (1980). *Philos. Mag.* [*Part*] A **42**, 783.
Ugaste, Yu. E. (1971). *Fiz. Met. Metalloved.* **31** (1), 59.
Van Ommen, A. H., Waegemaekers, A. A. H. J., Moleman, A. C., Schlatter, H., and Bakker, H. (1981). *Acta Metall.* **29**, 123.
Van Ommen, A. H. (1982). *Phys. Status Solidi A* **72**, 273.
Walter, C., and Peterson, N. L. (1969). *Phys. Rev.* **178**, 922.
Wanin, M., and Kohn, A. (1968). *Compt. Rend.* **267**, 1558.
Wever, H., and Frohberg, G. (1974). *Z. Metallkde.* **65**, 747.
Wynblatt, P. (1967). *Acta Metall.* **15**, 1453.
Yang, W. J., Lin, F., and Dodd, R. A. (1978). *Scr. Metall.* **12**, 237.

5

The Mathematical Analysis of Diffusion in Dislocations

A. D. LE CLAIRE AND A. RABINOVITCH*

MATERIALS DEVELOPMENT DIVISION
ATOMIC ENERGY RESEARCH ESTABLISHMENT
HARWELL, OXON, UNITED KINGDOM

	List of Symbols	257
I.	Introduction	259
II.	The Dislocation Model	261
III.	Solutions of the Diffusion Equations	266
	A. A Solution for c'	266
	B. An Approximate Solution for c	267
	C. The Exact Solutions for c	268
IV.	Properties of the Solutions	274
	A. Comparison of $Q(\eta, \varepsilon/\alpha)$ with $Q(\eta)$	274
	B. Analytic Representation of $Q^I(\eta)$ and $Q^{II}(\eta)$	277
	C. General Features of Diffusion Penetration Plots	280
	D. Total Amounts Absorbed or Desorbed during Diffusion under Case I Conditions	304
Appendix A.	Derivation of Eq. (39)	313
Appendix B.	The Poles of Eq. (39) for the Case of the Dislocation Array	314
Appendix C.	Numerical Considerations in the Calculation of $Q(\eta, \varepsilon/\alpha)$	314
Appendix D.	Numerical Considerations in the Calculation of $Q(\eta)$	315
Appendix E.	An Order of Magnitude Estimate of Δ	316
	References	316

List of Symbols (Those that appear only once in the text are not included in this list.)

$A(\alpha)$	Alpha-dependent parameter, describing tail slopes
a	Dislocation pipe radius
$c, c(r, y, t)$	Diffusant concentration outside dislocations

* Present address: Department of Physics, Ben Gurion University of the Negev, Beer-Sheva, 84105 Israel.

c', $c'(r, y, t)$	Diffusant concentration inside dislocation pipes
c_0	Constant surface concentration
$\langle c \rangle$, $\langle c(y, t) \rangle$	Total average concentration
$c_1(y, t)$	Dislocation free concentration
$c_2(r, y, t)$	Additional concentration outside dislocations
$\langle c'(y, t) \rangle$	Average concentration inside dislocation pipes
$\bar{c}$	Fourier transform of c
$\bar{\bar{c}}$	Fourier–Laplace transform of c
D	Crystal diffusion coefficient
D'	Dislocation diffusion coefficient
D_{eff}	Effective diffusion coefficient
D_{Hart}	Effective diffusion coefficient for large ε/α
d	Dislocation density (cm^{-2})
E	Lattice diffusion activation energy
E'	Dislocation diffusion activation energy
FLD	First linear domain (constant D_{eff} for small ε/α)
$I_0(x)$	Modified Bessel function of the first kind, zeroth order
$I_1(x)$	Modified Bessel function of the first kind, first order
$J_0(x)$	Bessel functions of the first kind, zeroth order
$J_1(x)$	Bessel functions of the first kind, first order
$K_0(x)$	Modified Bessel functions of the second kind, zeroth order
$K_1(x)$	Modified Bessel functions of the second kind, first order
k, k'	Defined by Eqs. (24a,b)
L	Half distance between nearest neighbor dislocations
$M^{\text{I,II}}$	Diffusion enhancement factor in the FLD; Cases I, II
$Q^{\text{I,II}}$	Average concentration, dislocation contribution factor; Cases I, II
$Q^{\text{I,II}}(\eta)$	Average concentration, dislocation contribution factor. Isolated dislocation approximation
$Q^{\text{I,II}}(\eta, \varepsilon/\alpha)$	Average concentration, dislocation contribution factor. General dislocation array
$q(y, t, d)$	Contribution per dislocation to $\langle c(y, t) \rangle$
R	Radius of the Wigner–Seitz cylinder, equal to effective half-distance between dislocations
r	Radial distance from a dislocation core
SLD	Second linear domain (constant D_{eff} for large ε/α)
t	Time
T_n	Function defined by Eq. (43)
U	Dislocation contribution factor for absorption or desorption
$U(\alpha, \Delta)$	Dislocation contribution factor for absorption or desorption; isolated dislocation approximation
$U(\alpha, \Delta, \varepsilon/\alpha)$	Dislocation contribution factor for absorption or desorption; dislocation array
V_1, V_2, V_3, V_4, V_5	Parameters for analytic representation of $Q^{\text{I}}(\eta)$ $Q^{\text{II}}(\eta)$
W	Total amount diffused into or out of sample with constant surface concentration c_0
x	Integration variable
y	Distance from surface
$Y_0(x)$	Bessel function of second kind, zeroth order
$Y_1(x)$	Bessel function of second kind, first order
z	Integration variable

5. MATHEMATICAL ANALYSIS OF DIFFUSION IN DISLOCATIONS

z_n	Roots of Eq. (44)
α	$a/(Dt)^{1/2}$
β	$(\Delta - 1)\alpha$
γ	Quantity of deposited diffusant per unit area; Case II
Δ	D'/D
ε^2	Volume fraction of material in dislocations, equal to $\pi a^2 d$
ε/α	$=(Dt)^{1/2}/R$
η	$y/(Dt)^{1/2}$
η_0	Value of η above which $Q(\eta, \varepsilon/\alpha)/Q(\eta) > 1$
θ	Function defined by Eq. (45a)
μ	Fourier conjugate of y
λ	Laplace conjugate of t
$v'(k, r, R)$	Function defined by Eq. (32) or (35)
ρ	r/R
σ_n	Function defined by Eq. (42)
σ_n'	Function defined by Eq. (49)
ϕ	Function defined by Eq. (45b)
$\psi(k, R)$	Function defined by Eq. (33) or (36)
Superscript I	Case I: diffusion with constant surface concentration
Superscript II	Case II: diffusion from a finite thin surface deposit

I. Introduction

There is now much evidence to support the belief that atomic migration in solids is very much more rapid along, or close to, dislocations than through the regular crystal lattice itself; The subject has been reviewed by Gibbs and Harris (1969), Gjostein (1970, 1973), and Balluffi (1970). Since all crystals contain dislocations, it follows that any measured bulk diffusion rate will in principle always contain a contribution from dislocation diffusion. This may be quite negligible for low dislocation densities, and expecially so at high temperatures, but it can become important at lower temperatures because of the lower activation energy for dislocation diffusion relative to that for volume diffusion. It may even be the dominant mode of transport in some diffusion processes observable at relatively low temperatures, like radiation damage recovery, precipitation, metal oxidation, etc. The study and understanding of dislocation diffusion is therefore a matter of some importance.

Perhaps one of the simpler and more generally applicable means of demonstrating fast dislocation diffusion, and of making some quantitative study of it, is from the results of tracer diffusion coefficient measurements by the conventional sectioning method. For diffusion into a sample from a thin source deposited on the surface, for example, results are usually represented as a plot of the logarithm of the average concentration $\langle c \rangle$ of diffused tracer in a cut section versus the square of the distance y of that section from the surface. Such plots usually show $\log\langle c \rangle$ initially decreasing effectively linearly with y^2, and from the slope is deduced an effective diffusion coefficient D_{eff}

$$D_{\text{eff}} = (4t\, \partial \ln\langle c\rangle/\partial y^2)^{-1} \tag{1}$$

D_{eff}, even in single crystals, can be observed to increase following a sufficient increase in the dislocation content of the samples, indicative of a fast dislocation contribution to the overall diffusion [e.g., Whitton and Kidson (1968), Morrison and Yuen (1971)]. But more striking are results obtained when measurements are extended to sufficient depths to reveal, beyond the linear region of the $\log\langle c\rangle$ versus y^2 plot, a usually nonlinear "tail" of much smaller average slope and often extending to considerable distances (See, for example, Figs. 22 and 23.) Usually, in this tail region $\log\langle c\rangle$ is linear in y. Because the concentrations in the tails are found to increase with increasing dislocation density, it is natural to attribute them to diffusion down and out of dislocations at rates sufficiently enhanced to carry material to significantly much greater depths than reached by volume diffusion alone.

There are several other experimental techniques that have been used to provide indication of fast dislocation diffusion. One of the more direct of these employs measurement of the rate of permeation of radioactive diffusant across thin wafers of deformed single crystals from one side to another [e.g., Hendrickson and Machlin (1954); Wuttig and Birnbaum (1966a); Baker et al. (1968)]. Such permeation has been observed to occur in times much too short for the transport to have been by volume or lattice diffusion alone, so that it can only be attributed to the dislocations introduced by the deformation providing paths for very rapid diffusion.

Equally direct are the results of electron microscope observations of the shrinkage of pores in Al foils while they are being annealed. Volin et al. (1971) noted that those pores which were connected by dislocation lines to the sample surface, and such pores were readily identifiable, shrank at a more rapid rate than those not so attached, indicating again that a dislocation provides a path for extra-rapid atom transport.

To the extent that low-angle grain boundaries can be adequately described geometrically as planar arrays of individual dislocations (Balluffi et al. (1982) and Chapter 6), the common observation of enhanced diffusion in such boundaries, and particularly the nature of its anisotropy, is also clear evidence of rapid dislocation transport (Balluffi, 1970).

Most of the early estimates of dislocation diffusion rates were in fact derived from measurements of low-angle grain boundary diffusion rates. The analyses that were, and still are, made assume that transport down and out of the constituent dislocations can be identified with that down and out of an equivalent homogeneous grain boundary slab with cross-sectional area the same as the sum of the component dislocation area (Turnbull and Hoffmann, 1954; Balluffi, 1970). This allows existing mathematical relations that describe diffusion in such boundaries (Fisher, 1951; Whipple, 1954; Suzuoka, 1964; Le Claire, 1963) to be employed to derive an effective diffusion coefficient for the component dislocations.

To such a procedure there are, however, some fairly obvious objections. In the first place, to represent diffusion in a planar array of dislocations by diffusion in an equivalent continuous slab is a poor approximation anyway, but especially so if the dislocations are as well separated as they need to be for the boundary to be representable in terms of isolated dislocations. Secondly, if the dislocations *are* close enough together for the equivalent boundary slab model to be a reasonable approximation, then interaction between them may endow the dislocations with diffusive and other properties signficantly different from those of isolated dislocations. In any event, results from grain boundary experiments cannot always be expected to accurately reflect the true diffusion properties of isolated dislocations.

For any more careful and detailed study of diffusion in the presence of dislocations, and especially of those not closely arrayed in boundaries, we need to consider the process of diffusion into, along, and out of individual dislocations. Considerable progress has recently been made in developing mathematical relations to describe such diffusion in several commonly occurring experimental situations (Le Claire and Rabinovitch, 1981, 1982, 1983, 1984) and a main object of this chapter is to describe in useful detail the near-rigorous methods used and the results obtained. They provide a more satisfactory basis than that previously available for making estimates of the contribution dislocations can make to measured diffusion rates, and for the analysis of dislocation tails to yield the parameters of dislocation diffusion.

II. The Dislocation Model

In the simplest model for discussing their diffusion properties, first introduced by Smoluchowski (1952), dislocations are considered as cylindrical pipes of radius a within which homogeneous isotropic diffusion occurs, according to Fick's laws, with a coefficient D' very much greater than the coefficient D for diffusion in the regular crystal beyond a. We denote this ratio D'/D as Δ.

More realistic models would recognize that the diffusion coefficient will in all probability vary with radial distance r from the dislocation core. For example, in terms of a vacancy mechanism, usually favored for dislocation diffusion in metals, diffusion is enhanced through an increased vacancy concentration near to a dislocation, arising from an attractive vacancy–dislocation interaction energy E_B, or through a reduced mobility enthalpy H_m, or most probably through both; both E_B and H_m are expected to vary with r and, therefore, also the diffusion coefficient $D'(r)$ (Gjostein, 1973). However, since there has been no clear indication of a suitably simple and universal form for $D'(r)$, all authors, apart from Luther (1965), have preferred to regard dislocation diffusion as adequately represented by the Smoluchowski step-function type model, and it is to this that we confine our attention. Luther (1965) investigated the consequences of taking $D' \propto r^{-n}$, with $n = 2$, 4, but the advantages of such an approach were not too apparent (see Wood et al., 1966).

As a model for discussing the influence of dislocations on diffusion, we consider diffusion from a surface $y = 0$ into a semi-infinite solid containing, as an idealization of the real condition (see p. 266), an array of parallel dislocations all normal to and ending in the surface. There are d dislocations per unit area, and we assume for mathematical convenience that they are regularly arranged, for example, such as in the hexagonal array shown in Fig. 1a. We may then divide the solid into identical regular hexagonal prisms, the axis of each being the axis of a dislocation pipe, and with each prism face parallel with and midway between a nearest neighbor pair of dislocations (Le Claire and Rabinovitch, 1983). The distance between these, $2L$, is related to d by $2\sqrt{3} \times L^2 d = 1$.

Outside the pipes, diffusion occurs from the surface with coefficient D. Down each pipe there is rapid diffusion of material at a rate determined by D', with loss from the sides at a rate governed by D again. The resulting distribution of concentration, illustrated by the section of a contour of constant concentration shown in Fig. 1b, will be the same within each prism and will have, within any y plane, the symmetry of the array. It follows that the gradient of concentration at and normal to any prism face must be zero: *there is no diffusion flux across any prism face*. This is a boundary condition to which solutions of the diffusion equations must be subject. However, in analogy with the Wigner–Seitz approximation in the electron theory of metals (see, for example, Sachs, 1963), we shall assume it sufficient to replace each prism by a coaxial cylinder of the same cross-sectional area, being of radius R, and require that the solution for the concentration c satisfy the condition

$$(\partial c/\partial r)_{r=R} = 0 \qquad (2)$$

In the hexagonal array, $R = (2\sqrt{3}/\pi)^{1/2} L$. Also, and for any array, $\pi R^2 d = 1$.

For short times, or low dislocation densities, such that the mean lattice diffusion length $(Dt)^{1/2}$ is substantially less than the dislocation spacing, the

5. MATHEMATICAL ANALYSIS OF DIFFUSION IN DISLOCATIONS

diffusion zones around the dislocations do not overlap; material diffusing down and out of one dislocation does not reach any other so that the dislocations are acting independently of one another. In particular, the boundary at $r = R$ is now irrelevant because $\partial c/\partial r$ will anyway become effectively zero at some $r < R$. Under these conditions $[(Dt)^{1/2} \ll R]$ we may seek solutions of the diffusion equations subject to the condition

$$\partial c/\partial r \to 0 \quad \text{as} \quad r \to \infty \tag{3}$$

instead of Eq. (2). Such solutions, for an effectively isolated dislocation (Le Claire and Rabinovitch, 1981, 1982), must of course be numerically identical with those derived, at the same $(Dt)^{1/2}$, from the general condition, Eq. (2). However, as we shall see, there are profound differences in their

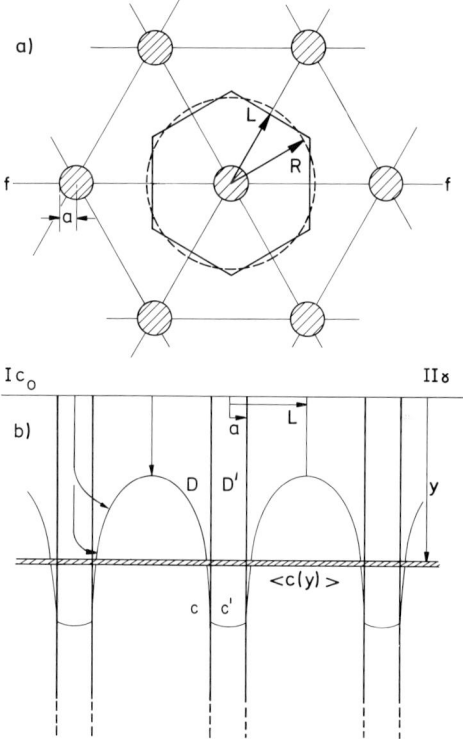

Fig. 1. (a) Plan view of a hexagonal array of dislocation pipes; pipe radius a, spacing $2L$. Bold lines: Wigner–Seitz-type cell, a hexagonal prism; broken lines: equivalent circular cylinder of same cross-sectional area, radius R. (b) Vertical section through f–f of Fig. 1(a), showing the trace of a contour of constant concentration; $\langle c(y) \rangle$ is the average concentration in the section, shown hatched, at depth y.

mathematical form. The solutions derived from Eq. (3) are of particular importance because from them numerical results for $(Dt)^{1/2} \ll R$ turn out to be very much more conveniently computed than from the general array solutions based on Eq. (2).

If c and c' represent respectively the concentrations of diffusant outside and inside a dislocation pipe, the equations to be solved are

$$D\left\{\frac{1}{r}\frac{\partial}{\partial r}\left(r\frac{\partial c}{\partial r}\right) + \frac{\partial^2 c}{\partial y^2}\right\} = \frac{\partial c}{\partial t} \qquad (4)$$

outside a pipe, i.e., in $a \le r \le R$, and

$$D'\left\{\frac{1}{r}\frac{\partial}{\partial r}\left(r\frac{\partial c'}{\partial r}\right) + \frac{\partial^2 c'}{\partial y^2}\right\} = \frac{\partial c'}{\partial t} \qquad (5)$$

inside a pipe, i.e., in $0 \le r \le a$. The solutions are subject to the boundary conditions at $r = a$ that there must be continuity of flux and of concentration, i.e.,

$$\left(D\frac{\partial c}{\partial r}\right)_{r=a^+} = \left(D'\frac{\partial c'}{\partial r}\right)_{r=a^-} \qquad (6)$$

and

$$c(r = a^+) = c'(r = a^-) \qquad (7)$$

together with the boundary condition [Eq. (2) or (3)], according to whether we seek the general array or the isolated dislocation solution, respectively.

For the diffusion of a solute species that segregates to dislocations with a segregation coefficient s, Eq. (7) may be replaced by $sc = c'$, but we consider only the case $s = 1$.

D and D' are being assumed isotropic and independent of concentration.

We shall consider two different experimental conditions at the surface $y = 0$. In Case I the concentration at the surface is maintained constant at a value c_0 for all $t \ge 0$:

Case I $\qquad\qquad c(y = 0, r, t) = c_0 \qquad (8)$

In Case II a very thin layer of diffusant, of quantity γ per unit area, is deposited on the surface at $t = 0$. The concentration at $t = 0$ can be written in terms of the Dirac δ function as

Case II $\qquad\qquad c(t = 0) = 2\gamma\delta(y) \qquad (9)$

The 2 is the normalizing constant that makes $\int_0^\infty c\,dy = \gamma$.

Case I is appropriate where, for example, diffusion occurs into the sample from the vapor phase: c_0 is then the concentration in the surface in equilibrium with the vapor. Case II is intended to simulate the conditions of a conventional "thin film" tracer diffusion measurement. However, it is only fully appropriate if there is in practice no rapid diffusion along the surface towards dislocations to compensate for the loss near $r = 0$ due to rapid diffusion down them. By tending to maintain the concentration near $r = 0$ such diffusion introduces an element of the Case I condition. Also, Case I may in practice be more appropriate if only very little of the deposited layer diffuses into the sample, as for example when there is only a very small solubility of diffusant, or when the time is short or the temperature low. In practice, neither Case I nor II may exactly represent the situation prevailing at the surface, although they probably do represent the limits between which any likely experimental condition will lie.

While the concentrations calculated for Case I and Case II of course differ it turns out, as we shall see, that the *gradient* of the $\log\langle c \rangle$ versus y dislocation tail is the same for both cases; at least this feature is uninfluenced by any uncertainty there may be about the surface conditions.

As we shall see, the solution for $c(r, y, t)$ will be of the form

$$c(r, y, t) = c_1(y, t) + c_2(r, y, t), \qquad a \leq r \leq R \tag{10}$$

where $c_1(y, t)$ is the standard expression for the concentration following diffusion in the absence of dislocations, under Case I or Case II conditions, and c_2 is the additional concentration outside dislocations due to rapid diffusion down and out of them. Because of the diffusion zone overlap, $c_2(r, y, t)$ and also the concentration $c'(r, y, t)$ inside dislocation pipes will in general be functions of the dislocation density.

We shall usually be concerned only with the *mean* concentration $\langle c(y, t) \rangle$ at a depth y after a time t, for this is what is measured in a serial sectioning experiment. It is given by

$$\langle c(y, t) \rangle = c_1(y, t) + dq(y, t, d) \tag{11}$$

where

$$q(y, t, d) = \int_0^a 2\pi r c'(r, y, t) dr + \int_a^R 2\pi r c_2(r, y, t) \, dr \tag{12}$$

When $(Dt)^{1/2} \ll R$ the upper limit on the second integral may be taken as ∞.

Another quantity of experimental interest is the total amount of material that has diffused into the sample under the Case I condition of a constant surface concentration c_0. This is simply

$$W = \int_0^\infty \langle c(y, t) \rangle^1 \, dy \tag{13}$$

and is easily evaluated once $\langle c(y,t)\rangle^I$ is known. Superscript I denotes a solution for Case I conditions.

The dislocations being considered are all straight and normal to the surface. In real crystals, of course, there is usually a much less simple geometry with dislocations in all directions of screw, edge, and mixed types that intersect to provide dislocation networks, forests, etc. There has been no mathematical discussion of such more complex configurations, but at least when $(Dt)^{1/2} \gg R$ the nature of the configurations can become largely irrelevant for many purposes; a statistical treatment then suffices, as we shall see (Section IV.C.2). Also, for shorter times, diffusion in from the surface $y = 0$ will be most enhanced by those dislocations with line components that are normal to the surface, so a model with dislocations only in this direction is not inappropriate for discussing such diffusion.

III. Solutions of the Diffusion Equations

A. A Solution for c'

Since we are assuming that $D' \gg D$ (see Appendix E), and since usually the diffusion length $(Dt)^{1/2} \gg a$, the concentration c' inside a dislocation will be nearly uniform radially. We need therefore be concerned only with its average value $\langle c'(y,t)\rangle$ at any depth y, and an approximate solution for $c'(r, y, t)$ may then be written as

$$c'(r, y, t) \simeq \langle c'(y, t)\rangle \simeq c'(a, y, t) = c(a, y, t) \qquad (14)$$

where the last equality is Eq. (7) again.

An exact relation entailing $\langle c'(y,t)\rangle$ is obtained by multiplying Eq. (5) for c' by $2\pi r$ and integrating with respect to r along the dislocation radius a; this gives

$$2\frac{D'}{a}\left(\frac{\partial c'}{\partial r}\right)_{r=a^-} + D'\frac{\partial^2 \langle c'(y,t)\rangle}{\partial y^2} = \frac{\partial \langle c'(y,t)\rangle}{\partial t} \qquad (15)$$

Using the boundary condition Eq. (6) with the relations in Eq. (14) gives the result

$$\frac{2}{a}D\left(\frac{\partial c}{\partial r}\right)_{r=a} + D'\frac{\partial^2 c(a, y, t)}{\partial y^2} = \frac{\partial c(a, y, t)}{\partial t} \qquad (16)$$

as a boundary condition at $r = a$ to be set upon the solution of Eq. (4) for c when Eq. (14) is taken as the solution for c'.

All authors have employed this, or a very similar, device to express the solution inside a (c') in terms of the solution outside (c).

The approximation of Eq. (14) can be more quantitatively justified as follows. From symmetry, $(\partial c'/\partial r) = 0$ at $r = 0$, so the leading term in a power series expansion of c' is of degree r^2. We may write

$$c'(r, y, t) = b_1(y, t) - \tfrac{1}{2} r^2 b_2(y, t) + \cdots \quad (17)$$

from which we obtain, using again the conditions of Eqs. (6) and (7)

$$c'(r, y, t) = c(a, y, t) \left\{ 1 + \frac{1}{2\Delta} \left[1 - \left(\frac{r}{a}\right)^2 \right] a \left(\frac{\partial \ln c}{\partial r}\right)_{r=a} + \cdots \right\} \quad (18)$$

Since Δ is expected to be of order 10^3 or more (see Appendix E), and if, as expected, $a(\partial \ln c/\partial r)_{r=a}$ is never greater than order unity, $c'(r, y, t)$ will vary only very little across $0 < r < a$, and so differ insignificantly from $c(a, y, t)$.

B. An Approximate Solution for c

Smoluchowski (1952) gave an approximate solution for diffusion down and out of an isolated dislocation under the Case I surface condition. This solution is of interest for its simplicity and we briefly outline the procedure and the assumptions made. These are the same as those employed by Fisher (1951) in treating the analogous grain boundary problem. (i) The concentration inside the pipe is assumed to be quasi-stationary so that $\partial \langle c' \rangle / \partial t = 0 = \partial c(a, y, t)/\partial t$ in Eqs. (15) and (16). This assumption was introduced following preliminary numerical solutions for Case I diffusion that indicated c' increasing at a rapidly diminishing rate so that for most of the time c' is near its current value. (ii) All diffusion out from a dislocation is assumed to be in the radial direction only so that $\partial^2 c/\partial y^2$ in Eq. (4) is omitted. Equation (4) then refers to radial diffusion from a cylindrical source of radius a and constant concentration $\langle c'(y, t) \rangle$, for which the solution is well known (Carslaw and Jaeger, 1959, p. 335). From this, $(\partial c'/\partial r)_{r=a} = \Delta^{-1}(\partial c/\partial r)_{r=a}$ is calculated for insertion into Eq. (15). With $\partial \langle c' \rangle / \partial t = 0$, Eq. (15) is then easily solved for $\langle c'(y, t) \rangle$ for the condition $\langle c'(0, t) \rangle = c_0$, and this completes the solution.

Because of assumption (i), the results only apply after long time and then, because of (ii), only in the dislocation tail region where $y \gg (Dt)^{1/2}$. In this region the calculations give a result that may be written

$$\frac{\partial \ln \langle c \rangle}{\partial y} = -\frac{A(\alpha)}{[(\Delta - 1)a^2]^{1/2}} \quad (19)$$

where $\alpha = a/(Dt)^{1/2}$ and

$$A^2(\alpha) = \frac{8}{\pi^2} \int_0^\infty \frac{e^{-z^2} \, dz}{z[J_0^2(z\alpha) + Y_0^2(z\alpha)]} \quad (20)$$

J_0 and Y_0 being Bessel functions of the first and second kind respectively, of order zero. The exact calculations (Le Claire and Rabinovitch, 1981) confirm this linear relation between $\ln \langle c \rangle$ and y, but with a different $A(\alpha)$ (see Section IV.C.3). Pavlov et al. (1964) and also Brebec (1965) have published tables of computed values of the integral in Eq. (20). Values of $A(\alpha)$ from Eq. (20) are included in Fig. 20.

C. The Exact Solutions for c

Equations (4) and (5) are solved via a Fourier–Laplace transformation with respect to the variables y and t. The transform of $c(r, y, t)$, for example, is

$$\bar{c}(r, \mu, \lambda) = \int_0^\infty \int_0^\infty c(r, y, t)_{\cos \mu y}^{\sin \mu y} \exp - \lambda t \, d\mu \, d\lambda \tag{21}$$

The $\sin \mu y$ is appropriate for surface condition I and the $\cos \mu y$ for condition II.

Since c and $\partial c/\partial y \to 0$ as $y \to \infty$, application of Eq. (21) to Eqs. (4) and (5) transforms them, for surface condition II [for example, Eq. (9)], into

$$\frac{1}{r}\frac{\partial}{\partial r}\left(r\frac{\partial \bar{c}}{\partial r}\right) - k^2 \bar{c} = -\frac{\gamma}{D} \tag{22}$$

and

$$\frac{1}{r}\frac{\partial}{\partial r}\left(r\frac{\partial \bar{c}'}{\partial r}\right) - k'^2 \bar{c}' = -\frac{\gamma}{D'} \tag{23}$$

where

$$k^2 = \mu^2 + \lambda/D, \tag{24a}$$

and

$$k'^2 = \mu^2 + \lambda/D' = (\mu^2(\Delta - 1) + k^2)/\Delta \tag{24b}$$

For type I surface conditions [Eq. (8)], the two equations for $\bar{c}$ and $\bar{c}'$ are the same but with $-\mu c_0/\lambda$ in place of $-\gamma/D$. We obtain results for both type I and II conditions, but outline procedures only for type II.

Equations (22) and (23) are Bessel equations; the general solution of Eq. (22) in $a \leq r \leq R$ is

$$\bar{c}^{II}(r, \mu, \lambda) = B_1 K_0(kr) + B_2 I_0(kr) + \gamma/Dk^2 \tag{25}$$

For Eq. (23), the general solution in $0 \leq r < a$ that remains finite as $r \to 0$ is

$$\bar{c}'^{II}(r, \mu, \lambda) = B_3 I_0(k'r) + \gamma/D'k'^2 \tag{26}$$

I_0 and K_0 are modified Bessel functions, of order zero, of the first and second kind respectively. Superscript II denotes a solution for surface condition II.

5. MATHEMATICAL ANALYSIS OF DIFFUSION IN DISLOCATIONS

As remarked in Section III.A, it is a sufficient approximation to the solution for c' to assume $c'(r, y, t) \simeq c(a, y, t)$ or, transformed,

$$\bar{c}'(r, \mu, \lambda) = \bar{c}(a, \mu, \lambda) \tag{27}$$

so the formal solution of Eq. (26) need not be further developed. However, it can be used to show that the approximation being employed is equivalent to assuming valid the relation

$$k'aI_0(k'a) = 2I_1(k'a) \tag{28}$$

where I_1 is the modified Bessel function of the first kind and first order (see Le Claire and Rabinovitch, 1981). This equality is indeed quite accurately obeyed for small $k'a$ ($\lesssim 0.5$) and, as will be apparent later, it is small values of $k'a$, and of ka, that dominate the integrations needed to invert the transformation.

The transform of the boundary condition, Eq. (16), is

$$\frac{2}{a}\left(\frac{\partial \bar{c}}{\partial r}\right)_{r=a} - \Delta k'^2 \bar{c}(a) + \frac{\gamma}{D} = 0 \tag{29}$$

With this, and with the transform of the general boundary condition [Eq. (2)] at $r = R$, viz., $(\partial \bar{c}/\partial r)_{r=R} = 0$, the constants B_1 and B_2 in Eq. (25) are evaluated to lead to the formal general solution for $\bar{c}^{II}$ for a dislocation array (Le Claire and Rabinovitch, 1983), as

$$\bar{c}^{II} = \frac{\gamma}{Dk^2} - \frac{\mu^2 \gamma (\Delta - 1) v'(k, r, R)}{\psi(k, R) \cdot Dk^2} \tag{30}$$

$$= \bar{c}_1^{II}(\mu, \lambda) + \bar{c}_2^{II}(r, \mu, \lambda) \tag{31}$$

In these equations

$$v'(k, r, R) = I_1(kR)K_0(kr) + K_1(kR)I_0(kr) \tag{32}$$

$$\psi(k, R) = I_1(kR)\left\{\frac{2k}{a}K_1(ka) + \Delta k'^2 K_0(ka)\right\}$$

$$- K_1(kR)\left\{\frac{2k}{a}I_1(ka) - \Delta k'^2 I_0(ka)\right\} \tag{33}$$

I_1 and K_1 are modified Bessel functions of order one, of the first and second kind, respectively. The values $\bar{c}_1^{II}$ and $\bar{c}_2^{II}$ are the transforms of c_1 and c_2 in Eq. (10).

For the particular case of an isolated dislocation, where the boundary condition of Eq. (3) can be employed, viz.,

$$(\partial \bar{c}/\partial r) \to 0, \quad \text{as} \quad r \to \infty, \tag{34}$$

B_2 in Eq. (25) must be zero, for $I_0(z)$ and its derivatives tend to ∞ with z. Evaluating B_1 from condition Eq. (29) then gives the same Eq. (30) for $\bar{c}^{II}$, but with much simpler v' and ψ, neither of which now, of course, depends on R. Thus *for an isolated dislocation* (Le Claire and Rabinovitch, 1981)

$$v'(k, r, R) \to v'(k, r, \infty) = K_0(kr) \tag{35}$$

and

$$\psi(k, R) \to \psi(k, \infty) = (2k/a)K_1(ka) + \Delta k'^2 K_0(ka) \tag{36}$$

We have now to derive $c = c_1 + c_2$ [Eq. (10)] by the standard integration for inverting a Fourier–Laplace transform. The inverse of $\bar{c}_1^{II} = \gamma/Dk^2$ [Eq. (31)] is just the well known result

$$c_1^{II}(\eta, t) = \frac{\gamma}{(\pi Dt)^{1/2}} \exp -\tfrac{1}{4}\eta^2 \tag{37}$$

where we introduce the reduced distance $\eta = y/(Dt)^{1/2}$. For c_2^{II}, we have to evaluate

$$c_2^{II}(r, \eta, t) = \frac{2}{\pi} \int_0^\infty \cos \mu y \, d\mu \left\{ \frac{1}{2\pi i} \int_{-i\infty + \tau}^{+i\infty + \tau} \bar{\bar{c}}_2^{II}(r, \mu, \lambda) e^{\lambda t} \, d\lambda \right\} \tag{38}$$

where τ is to the right of (i.e., greater than the real part of) all singularities of $\bar{\bar{c}}_2^{II}$.

The quantity in braces in Eq. (38) is just the Fourier transform $\bar{c}_2^{II}(r, \mu, t)$ of $c_2^{II}(r, y, t)$. Substituting for $\bar{\bar{c}}_2^{II}$ from Eqs. (30) and (31), writing $\lambda = Dk^2 - D\mu^2$ [Eq. (24a)] and applying a theorem on the Laplace transform of an integral with respect to t (see Appendix A), $\bar{c}_2^{II}(r, \mu, t)$ can be written in the convenient form

$$\bar{c}_2^{II} = -\gamma(\Delta - 1)\mu^2 e^{-\mu^2 Dt} \frac{1}{2\pi i} \int_{-i\infty + \tau}^{i\infty + \tau} \frac{(e^{-Dk^2 t} - 1)v'(k, r, R)}{Dk^2 \cdot \psi(k, R)} d(Dk^2) \tag{39}$$

To evaluate this integral requires careful consideration of the singularities of the integrand. The results for the general array case, with Eqs. (32) and (33) for v' and ψ, and for the isolated dislocation case with Eqs. (35) and (36), are quite different.

1. *Solutions for the Dislocation Array*[1]

As described in Appendix B, the integrand of Eq. (39) with Eqs. (32) and (33) written for v' and ψ is single valued and has a discrete series of poles

---

[1] Le Claire and Rabinovitch, 1983.

coming only from the zeros of ψ, all of which are simple and pure imaginary. We can therefore make standard application of the Cauchy theorem and write the line integral simply as $2\pi i$ times the sum of the residues at the poles. After evaluating c_2^{II} in this way, and after then carrying out the integrations over r defined by Eq. (12) to obtain $q(y, t, d)$, we arrive at the final result for $\langle c(\eta, t) \rangle$, Eq. (11). [See the appendix of Le Claire and Rabinovitch (1983) for details.] We write it in the form

$$\langle c(\eta, t) \rangle^{II} = \frac{\gamma}{(\pi Dt)^{1/2}} \{\exp -\tfrac{1}{4}\eta^2 + \varepsilon^2 Q^{II}\} \quad (40)$$

where $\varepsilon^2 = \pi a^2 d = a^2/R^2$ is the volume fraction of material in dislocation pipes and $Q^{II}(\eta, \varepsilon/\alpha)$ is given by

$$Q^{II}\left(\eta, \frac{\varepsilon}{\alpha}\right) = 2\pi^{1/2}\beta \int_0^\infty x^2 e^{-x^2} \cos \eta x \left\{\sum_n \frac{\sigma_n[\exp(-z_n^2) - 1]}{T_n}\right\} dx \quad (41)$$

In this equation

$$\sigma_n = J_1\left(\frac{z_n\alpha}{\varepsilon}\right)\left[Y_0(z_n\alpha) - \frac{2}{z_n\alpha}Y_1(z_n\alpha)\right]$$

$$- Y_1\left(\frac{z_n\alpha}{\varepsilon}\right)\left[J_0(z_n\alpha) - \frac{2}{z_n\alpha}J_1(z_n\alpha)\right] \quad (42)$$

$$T_n = \left[\frac{\phi}{J_1}\left(\frac{z_n\alpha}{\varepsilon}\right)\right] - \frac{\pi}{2}z_n\alpha(x^2\beta - z_n^2\alpha)$$

$$\times \left[J_1\left(\frac{z_n\alpha}{\varepsilon}\right)Y_1(z_n\alpha) - J_1(z_n\alpha)Y_1\left(\frac{z_n\alpha}{\varepsilon}\right)\right] \quad (43)$$

and the z_n are the roots of

$$Y_1\left(\frac{z\alpha}{\varepsilon}\right)\phi - J_1\left(\frac{z\alpha}{\varepsilon}\right)\theta = 0 \quad (44)$$

with

$$\theta = 2zY_1(z\alpha) + (x^2\beta - z^2\alpha)Y_0(z\alpha) \quad (45a)$$

$$\phi = 2zJ_1(z\alpha) + (x^2\beta - z^2\alpha)J_0(z\alpha) \quad (45b)$$

In these equations

$$\alpha = a/(Dt)^{1/2} \quad \text{and} \quad \beta = (\Delta - 1)\alpha \quad (46)$$

Q^{II} is a function not only of η, and of course of α and Δ, but also of the ratio ε/α. Since $\varepsilon = a/R$, this is just the ratio of the diffusion length $(Dt)^{1/2}$ to the half distance R between dislocations, i.e.,

$$\varepsilon/\alpha = (Dt)^{1/2}/R \tag{47}$$

and this is clearly a convenient parameter in terms of which to describe and discuss the effects of dislocation diffusion zone overlap on the progress of dislocation assisted diffusion.

Although we refer to it only once again, we include for completeness the expression for the concentration distribution around a dislocation in $a < r < R$. With $\rho = r/R$, this is

$$c_2^{II}(\rho, \eta, \varepsilon/\alpha) = -\frac{2\gamma}{(\pi Dt)^{1/2}} \pi^{1/2}\beta \int_0^\infty x^2 e^{-x^2} \cos \eta x \left\{ \sum_n \frac{\sigma_n'[\exp(-z_n^2) - 1]}{T_n} \right\} dx \tag{48}$$

with

$$\sigma_n' = Y_1\left(\frac{z_n\alpha}{\varepsilon}\right) J_0\left(\frac{z_n\alpha}{\varepsilon}\rho\right) - J_1\left(\frac{z_n\alpha}{\varepsilon}\right) Y_0\left(\frac{z_n\alpha}{\varepsilon}\rho\right) \tag{49}$$

For type I surface condition of diffusion from a constant concentration c_0 at $y = 0$, the corresponding solution for $\langle c(\eta, t) \rangle^{\mathrm{I}}$ is

$$\langle c(\eta, t) \rangle^{\mathrm{I}} = c_0 \{\mathrm{erfc}\, \tfrac{1}{2}\eta + \varepsilon^2 Q^{\mathrm{I}}\} \tag{50}$$

with

$$Q^{\mathrm{I}}(\eta, \varepsilon/\alpha) = -2\beta \int_0^\infty xe^{-x^2} \sin \eta x \left\{ \sum_n \frac{\sigma_n[\exp(-z_n^2) - 1]}{T_n} \right\} dx \tag{51}$$

We note that the solutions for the two surface conditions I and II are related by

$$\langle c(\eta, t) \rangle^{\mathrm{II}} = -\frac{\gamma}{c_0(Dt)^{1/2}} \frac{\partial}{\partial \eta} \langle c(\eta, t) \rangle^{\mathrm{I}} \tag{52}$$

For the total amount of material W that has diffused into the sample we calculate

$$\int_0^\infty \langle c(y, t) \rangle^{\mathrm{I}} \, dy = W = (Dt)^{1/2} \int_0^\infty \langle c(\eta, t) \rangle^{\mathrm{I}} \, d\eta \tag{53}$$

using Eqs. (50) and (51) for $\langle c(\eta, t) \rangle^{\mathrm{I}}$. η occurs in Q^{I} only as $\sin \eta x$ and it is easily proved that

$$\int_0^\infty \sin \eta x \, d\eta = x^{-1} + \pi \delta(x) \tag{54}$$

5. MATHEMATICAL ANALYSIS OF DIFFUSION IN DISLOCATIONS

$\delta(x)$ is Dirac's δ function; since $\int_0^\infty f(x)\delta(x)\,dx = \frac{1}{2}f(0), = 0$ in the present case, the second term in Eq. (54) does not contribute to the subsequent integration over x in Eq. (51). With the elementary integration of the erfc term, Eq. (53) gives

$$W = 2c_0[Dt/\pi]^{1/2}\{1 + \varepsilon^2 U\} \tag{55}$$

where

$$U(\alpha, \Delta, \varepsilon/\alpha) = -\pi^{1/2}(\Delta - 1)\alpha \int_0^\infty \exp(-x^2)\left\{\sum_n \frac{\sigma_n[\exp(-z_n^2) - 1]}{T_n}\right\} dx \tag{56}$$

2. Solution for an Isolated Dislocation[2]

With Eqs. (35) and (36) written for v' and ψ in Eq. (39), the method of Erdélyi and Kermack (1945) can then be used to show that the integrand of the equation now contains no singularities other than a branch point at the origin. The conventional treatment for such cases reduces the calculation to an integration along just the top and bottom sides of a "cut" along the negative real axis of the complex plane and over a small circle around the origin (Carslaw and Jaeger, 1959, pp. 304 and 335). Performing such an integration for Eq. (39), followed by the integrations over r [Eq. (12)] to obtain $q(y, t, d)$, leads to a result for $Q^{II}(\eta)$,

$$Q^{II}(\eta) = \frac{16}{\pi^{5/2}}(\Delta - 1)^2 \int_0^\infty x^4 e^{-x^2} \cos \eta x \, dx \int_0^\infty \frac{[\exp(-z^2) - 1]}{(\theta^2 + \phi^2)z^3} dz \tag{57}$$

for use in Eq. (40). We shall systematically use $Q(\eta, \varepsilon/\alpha)$ to denote solutions for the symmetric array and $Q(\eta)$ for the solutions for an isolated dislocation.

For the type I surface condition the corresponding quantity $Q^I(\eta)$ is

$$Q^I(\eta) = -\frac{16}{\pi^3}(\Delta - 1)^2 \int_0^\infty x^3 e^{-x^2} \sin \eta x \, dx \int_0^\infty \frac{[\exp(-z^2) - 1]}{(\theta^2 + \phi^2)z^3} dz \tag{58}$$

for use in Eq. (50).

We note that Eq. (52) also applies to the results for $c(\eta, t)$ written with Eqs. (57) and (58).

Using Eq. (58) with Eq. (50) for $\langle c(\eta, t)\rangle$ in Eq. (53), we obtain for isolated dislocations,

$$U(\alpha, \Delta) = -\frac{8(\Delta - 1)^2}{\pi^{5/2}} \int_0^\infty x^2 \exp(-x^2)\, dx \int_0^\infty \frac{[\exp(-z^2) - 1]}{(\theta^2 + \phi^2)}\frac{dz}{z^3} \tag{59}$$

for use in Eq. (55).

---

[2] Le Claire and Rabinovitch, 1981.

There have been a few previous attempts at the solution for this limiting case of diffusion down and out of an isolated dislocation, avoiding the approximations used by Smoluchowski (1952) and following a similar procedure to the above. Stark (1965) discussed approximate solutions for c_2^I and for $\langle c \rangle^I$, obtained through an evaluation of the inversion integral by the method of steepest descent, but his results were too incomplete to be of much practical use. Brebec (1965) reported the correct double integral expression for the concentration c_2^I as a function of y, t, and ρ, but he did not evaluate this or develop it into any more useful form. Mimkes and Wuttig (1970) and Mimkes (1973) have provided what was claimed to be an exact solution of the problem. However, in attempting to simplify the integrand of their inversion integral before evaluating it they made some unjustified mathematical assumptions that unfortunately vitiate their results. This is discussed in rather more detail in Le Claire and Rabinovitch (1981). Finally, Murch (1983) has very recently performed some Monte Carlo calculations of $\langle c \rangle^{II}$ that agree well with numerical values computed from Eqs. (40) and (57).

IV. Properties of the Solutions

A. Comparison of $Q(\eta, \varepsilon/\alpha)$ with $Q(\eta)$

$Q^I(\eta, \varepsilon/\alpha)$ and $Q^{II}(\eta, \varepsilon/\alpha)$ for diffusion into a dislocation array and $Q^I(\eta)$ and $Q^{II}(\eta)$ for diffusion into an isolated dislocation have been computed for several sets of values of α and $\Delta - 1$ and for a range of values of ε/α (Le Claire and Rabinovitch, 1981, 1982, 1983). Some mathematical points in connection with their computation are included in the Appendices C and D.

While Eqs. (41) and (51) for the $Q(\eta, \varepsilon/\alpha)$ are of course in principle appropriate for all values of ε/α, in practice their computation for very small ε/α presents difficulties. This is because the sum of residues becomes ever more slowly convergent as ε/α diminishes below about one; for $\varepsilon/\alpha \lesssim 0.1$ the computation time is excessive. However, as discussed in Section II, we expect $Q(\eta, \varepsilon/\alpha)$ to approach $Q(\eta)$ at low enough dislocation densities or short enough times, i.e., at low enough ε/α. It turns out that the two Qs become numerically identical, to four significant figures, for all $\varepsilon/\alpha \lesssim 0.4$, a value at which there are not yet too serious difficulties with the computation of $Q(\eta, \varepsilon/\alpha)$. Thus the isolated dislocation solutions, which provide no particular problems in their evaluation, are important and essential as providing the more convenient relations for computing $Q(\eta, \varepsilon/\alpha)$ at low ε/α, where it has become independent of ε/α.

Parts a, b, and c of Fig. 2 show for a few cases the ratio $Q(\eta, \varepsilon/\alpha)/Q(\eta)$ at various η, plotted as a function of ε/α. As this increases the ratio can be seen

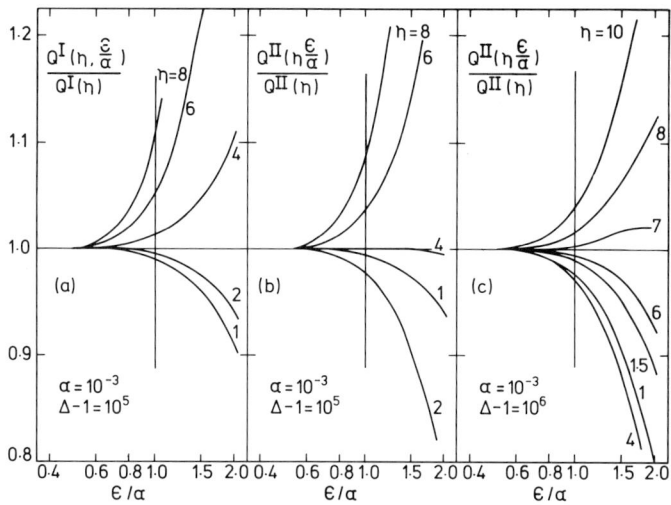

Fig. 2. Comparison of the $Q(\eta, \varepsilon/\alpha)$ and $Q(\eta)$ solutions. (a) Case I: $\alpha = 10^{-3}$, $\Delta - 1 = 10^5$; (b) Case II: $\alpha = 10^{-3}$, $\Delta - 1 = 10^5$; (c) Case II: $\alpha = 10^{-3}$, $\Delta - 1 = 10^6$.

in all cases to differ noticeably from unity only beyond about $\varepsilon/\alpha \approx 0.6$. At $\varepsilon/\alpha = 1$ the differences are still only a few percent, but as ε/α increases further $Q(\eta, \varepsilon/\alpha)$ starts differing substantially and increasingly from $Q(\eta)$.

The growing differences between the two reflect the expected effects of dislocation diffusion zone overlap. Beyond a certain η, say η_0, the ratio is greater than unity and the mean concentrations [Eqs. (40) or (50)] are therefore higher than would occur for the same density of dislocations without overlap, while nearer the surface, for $\eta < \eta_0$, the ratio is less than one and the concentrations therefore are smaller. Overlap is enhancing overall diffusion rates and reducing gradients more than would the same number of dislocations acting independently.

Figure 3 compares the general forms of the dependence of Q^I and Q^{II} on η. The curves are drawn for $\alpha = 10^{-2}$, $\Delta - 1 = 10^5$, and for small ε/α so that these Q are $Q(\eta)$. Both Q^I and Q^{II} first increase as η increases from zero, pass through a maximum and then decrease monotonically; Q^I is always positive and becomes zero at $\eta = 0$; Q^{II} has a zero and changes sign at the η for which Q^I is a maximum, in accordance with Eq. (52). The value of Q^{II} is negative at the smaller η because the finite amount of diffusant available with condition II is depleted from around the pipes by the rapid diffusion down them; the concentration is *less* here than in the absence of the dislocations so the second term of Eq. (40) has to be negative at small η. Where Q^{II} is positive, at the

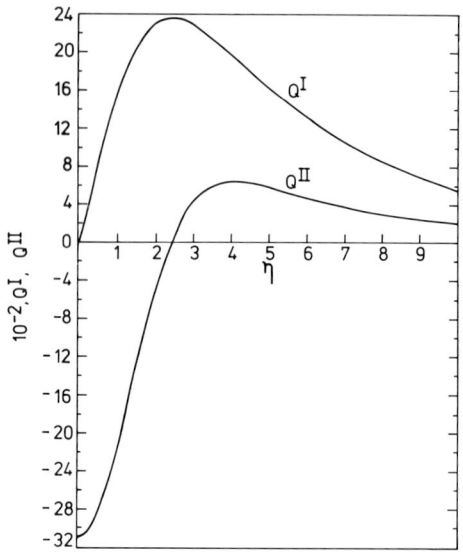

Fig. 3. The general behavior of Q^I and Q^{II} as functions of η, calculated for small ε/α; $\alpha = 10^{-2}$, $\Delta - 1 = 10^5$.

larger η, $\langle c \rangle$ is being increased by material diffusing out of the pipes into their surroundings.

Figure 4 shows how $Q^{II}(\eta, \varepsilon/\alpha)$ varies with ε/α. For $0 < \varepsilon/\alpha < 1$, Q^{II} changes very little, as Fig. 2 indicated, but for larger ε/α there is a progressive flattening of the curve of Q^{II} versus η; the negative values at smaller η and the maxima in the positive values both become numerically smaller with increase in ε/α.

Many of the features of Fig. 2c can be seen reproduced in Fig. 4. The bottom right-hand side of the figure shows the curves for $\eta \geq 5$ replotted with a much enlarged ordinate scale so as to indicate more clearly those intersections of $Q^{II}(\eta, \varepsilon/\alpha)$ with $Q^{II}(\eta)$ [$= Q^{II}(\eta, \varepsilon/\alpha = 0)$] that define η_0. These are crossed in the figure. Also evident is how the ratio $Q^{II}(\eta, \varepsilon/\alpha)/Q^{II}(\eta)$ for $\eta \gtrsim 3$ progressively increases with increasing η and how the same is true for the ratios of the *negative* values in the range $\eta \lesssim 2$. These two ranges of monotonic increase of the ratio with increase in η are, respectively, represented in Fig. 2c by the sequence $\eta = 4, 6, 8, 10$ and the pair $\eta = 1, 1.5$. The apparently irregular sequence of curves in Figs. 2b and 2c is thereby understood.

For plots of $Q^I(\eta, \varepsilon/\alpha)$ versus η there is a similar progressive flattening as ε/α increases beyond zero, with $Q^I(0, \varepsilon/\alpha)$ always zero. Q^I is everywhere positive and $Q^I(\eta, \varepsilon/\alpha)/Q^I(\eta)$ increases monotonically with η over the whole range, as exemplified in Fig. 2a.

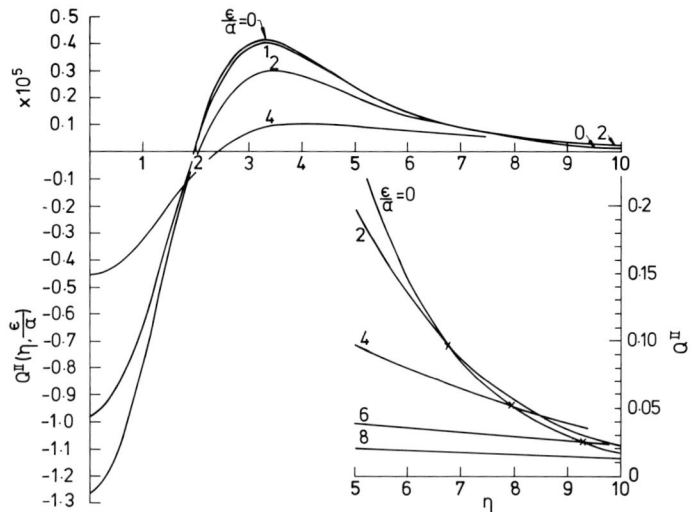

Fig. 4. $Q^{II}(\eta, \varepsilon/\alpha)$ plotted as a function of η for various ε/α; $\alpha = 10^{-3}$, $\Delta - 1 = 10^6$.

B. Analytic Representation of $Q^I(\eta)$ and $Q^{II}(\eta)$

For some purposes (see Section C, for example) it is convenient to have an analytic representation of Q^I and Q^{II}.

We have found that $Q^{II}(\eta)$ can be very well represented by the five parameter equation (Le Claire and Rabinovitch, 1982)

$$Q^{II}(\eta) = V_1 \exp(-\tfrac{1}{4}\eta^2) + V_2 \exp[-(V_3\eta)^2] + V_4 \operatorname{sech}(V_5\eta) \quad (60)$$

Except in a few unimportant cases at the larger η, this equation, with least-squares fitted values of V_1 to V_6, reproduces the computed values of Q^{II} to well within 1% for the range of values of α and $\alpha\beta$ studied (indicated in Fig. 5–8).

By integration, following Eq. (52), we obtain an equation that equally well represents $Q^I(\eta)$,

$$Q^I(\eta) = -\left\{ V_1 \operatorname{erf}\left(\tfrac{1}{2}\eta\right) + \frac{V_2}{2V_3} \operatorname{erf}(V_3\eta) + \frac{2V_4}{V_5\pi^{1/2}}\left[\tan^{-1}(e^{V_5\eta}) - \pi/4\right] \right\} \quad (61)$$

As $\eta \to \infty$, Q^I must become zero. Therefore, it follows from Eq. (61) that we must have

$$V_1 + \frac{V_2}{2V_3} + \frac{\pi^{1/2} V_4}{2V_5} = 0 \quad (62)$$

so that the Vs are not all independent.

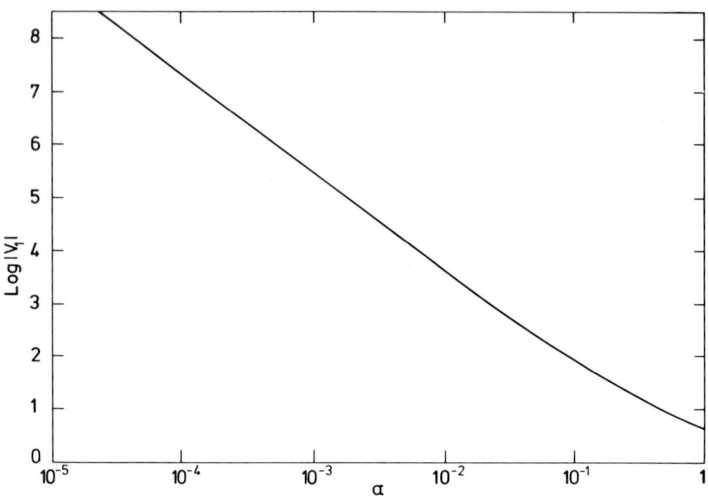

Fig. 5. V_1 of Eqs. (60) and (61) as a function of α; V_1 is always negative and effectively independent of $\alpha\beta$.

Equation (62) allows the expression in Eq. (61) for $Q^I(\eta)$ to be written in the slightly modified form, which may sometimes be more convenient,

$$Q^I(\eta) = V_1 \operatorname{erfc} \frac{1}{2}\eta + \frac{V_2}{2V_3} \operatorname{erfc} V_3\eta - \frac{2V_4}{\pi^{1/2}V_5}[\tan^{-1}(e^{V_5\eta}) - \pi/2]. \quad (61a)$$

The values of V_1, V_2, and V_4 are shown graphically in Figs. 5, 6, and 7.

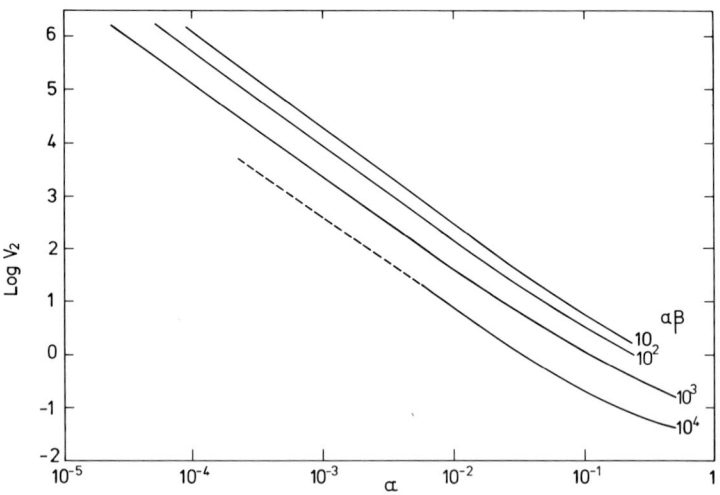

Fig. 6. V_2 of Eqs. (60) and (61) as a function of α for several values of $\alpha\beta$.

5. MATHEMATICAL ANALYSIS OF DIFFUSION IN DISLOCATIONS

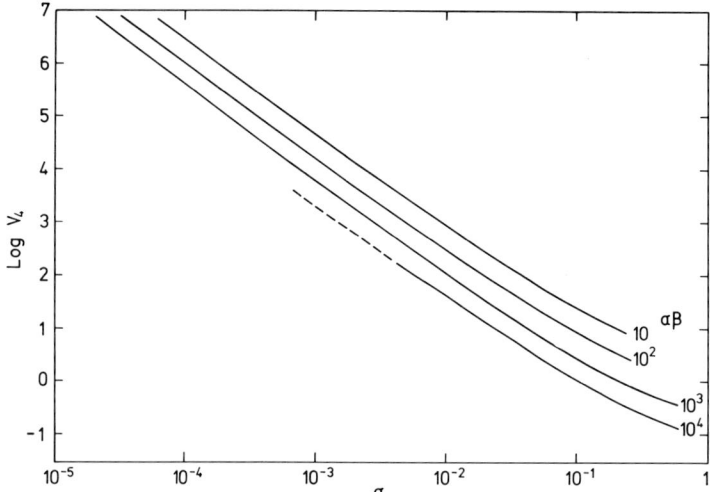

Fig. 7. V_4 of Eqs. (60) and (61) as a function of α for several values of $\alpha\beta$.

For the parameter range covered in the figures, V_1 is always negative, V_2 and V_4 positive; V_5, which is always positive, is reported in Fig. 8 as the ratio $2V_4/V_5\pi^{1/2}$; V_3 may be calculated from Eq. (62).

Usually $|V_1| \gg V_2$ and V_4. Thus the V_1 term is dominant at low η in Eq. (60) where it represents, V_1 being always negative, the depletion of

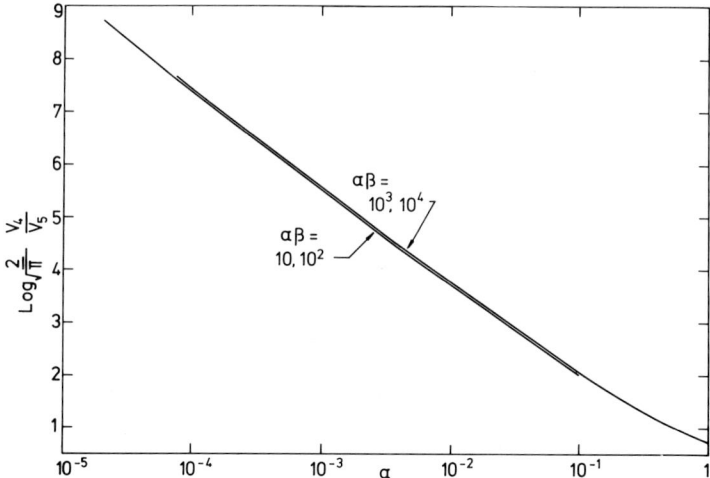

Fig. 8. The ratio $2V_4/\pi^{1/2}V_5$ as a function of α for several values of $\alpha\beta$. (There are slight differences between the values for 10^3 and for 10^4 and between those for 10 and 10^2, but they are too small to depict usefully.)

deposited material by rapid diffusion down the pipes. Since an area of order πDt around each dislocation is so depleted (when ε/α is small) we expect the order of magnitude relation for V_1,

$$\varepsilon^2 V_1 = d\pi\, Dt$$

Since $\varepsilon^2 = \pi a^2 d$, this is equivalent to $V_1 \approx \alpha^{-2}$. This is a fair approximation to the $V_1 \approx \alpha^{-1.84}$ deduced from Fig. 5.

The quantity U of Eq. (55) is also readily expressed in terms of the V_is. Using in Eq. (50) the Eq. (61a) value for $Q^{\rm I}$, the integration in Eq. (53) leads to

$$U(\alpha, \Delta) = V_1 + \frac{V_2}{4V_3^2} + 0.916\frac{V_4}{V_5^2}$$

where the 0.916 comes from numerical integration of $[\tan^{-1}(\exp V_5\eta) - \pi/2]$ over η from 0 to ∞.

For large η, Eq. (60) takes the asymptotic form

$$Q^{\rm II}(\eta) \to 2V_4 \exp - V_5\eta \tag{63}$$

and Eq. (61) the form

$$Q^{\rm I}(\eta) \to (2V_4/V_5\pi^{1/2})\exp - V_5\eta \tag{64}$$

That is, both $Q^{\rm I}(\eta)$ and $Q^{\rm II}(\eta)$ decrease exponentially with η at large η and with exactly the same constant V_5. In fact, this property was first identified in examination of the computed values of $Q^{\rm I}(\eta)$ and $Q^{\rm II}(\eta)$ (Le Claire and Rabinovitch, 1981) and helped to guide the choice of the mathematical forms to represent them, Eqs. (60) and (61).

No attempt has yet been made to find a similar mathematical representation of the $Q(\eta, \varepsilon/\alpha)$, with their additional ε/α parameters.

C. General Features of Diffusion Penetration Plots

Nearly all the more precise measurements of diffusion coefficients in single crystals have employed Case II source conditions and used serial sectioning to measure $\langle c \rangle$ as a function of y out to distances corresponding to at least one, and often two or more, powers of ten decrease in $\langle c \rangle$. If the experimental results show an acceptably linear relation between $\ln\langle c \rangle$ and y^2 the plot is said to be gaussian and there is then calculated from the slope an effective bulk or volume diffusion coefficient defined by Eq. (1).

When the dislocation density is low enough for the second term of Eq. (40) to be negligible over the range of y or η studied, plots of $\ln\langle c \rangle$ versus y^2 are exactly linear and $D_{\rm eff} = D$. When the second term is not negligible, so that dislocation diffusion contributes significantly to $\langle c \rangle$, $\ln\langle c \rangle$ will not in general vary linearly with y^2, and a single effective diffusion coefficient cannot therefore necessarily always be defined.

Figure 9 illustrates the nature of the penetration plots that can be observed, depending on the time of anneal, or ε/α, drawn for the case $\Delta - 1 = 10^7$ and for a fixed dislocation density corresponding to $\varepsilon^2 = 10^{-6}$. The plots are of $\log[\langle c \rangle (\pi Dt)^{1/2}/\gamma]$ versus η^2 for a decreasing sequence of α values, corresponding to a progressive increase in time. For illustrative purposes some actual times are shown, calculated for $D = 10^{-13}$ cm^2 s^{-1} and $a = 5 \times 10^{-8}$ cm.

The dashed line, $\varepsilon = 0$, is the plot for a sample free of dislocations and has $\partial \ln \langle c \rangle / \partial \eta^2 = -0.25$.

Curve 1 is for the shortest time considered, with $\alpha = 10^{-3}$ and $\varepsilon/\alpha = 1$. We see a near gaussian (near linear) initial portion, out to $\eta \approx 3$ or 4, followed by a well-defined dislocation tail of much smaller average slope starting at $\eta \approx 5$. Any plot for times such that $0 < \varepsilon/\alpha < 1$, taken to large enough η, would show the same two essential characteristics of clearly resolved gaussian and tail regions. The gaussian region defines, through Eq. (1), a $D_{\text{eff}} \geq D$ and is referred to as the "first linear domain" (FLD) of $\log \langle c \rangle$ versus η^2 plots.

Curve 2, for about twice the anneal time of curve 1 ($\varepsilon/\alpha = 1.43$), also shows a tail, but the initial gaussian portion of the plot is now already very much less in evidence.

A further threefold increase in anneal time takes us to curve 3, with $\varepsilon/\alpha = 2.5$, where it is no longer possible clearly to resolve two regions and for which $\log \langle c \rangle$ varies nowhere linearly either with η or η^2.

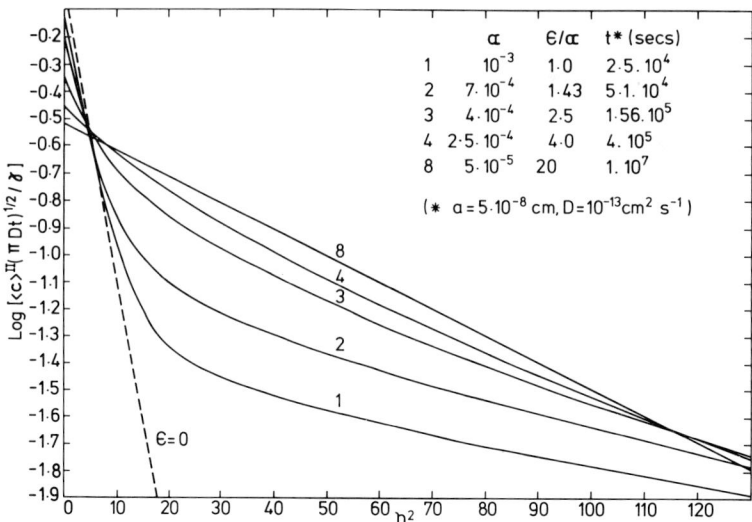

Fig. 9. Computed plots, $\log \langle c \rangle^{\text{II}}$ versus η^2, showing change with time for fixed $\varepsilon = 10^{-3}$, $\Delta - 1 = 10^7$.

Increasing further the anneal time produces a slow but progressive straightening of the $\log\langle c\rangle$ versus η^2 plots until, for $\varepsilon/\alpha \approx 10$ or greater (curve 8), the plots have become effectively linear again. The progression from curve 4 to curve 8 is shown in more detail in Fig. 10. These gaussian plots have much smaller slope than for $\varepsilon/\alpha \leq 1$, and so are associated with a much larger D_{eff}. As we shall see, this is very close to the dislocation enhanced diffusion coefficient discussed by Hart (1957),

$$D_{\text{Hart}} = D(1 + \varepsilon^2(\Delta - 1)) \tag{65}$$

The near-gaussian regions for $\varepsilon/\alpha \gtrsim 10$ define what is referred to as the "second linear domain" (SLD) of $\ln\langle c\rangle$ versus η^2 plots.

These same features may, of course, also be demonstrated in a series of plots for a fixed time of anneal of samples containing successively larger dislocation densities. Figure 11 shows such a series for $\alpha = 10^{-4}$, $\Delta - 1 = 10^7$ and for dislocation densities starting, where $\varepsilon/\alpha = 1$, at a value for which $\varepsilon^2 = 10^{-8}$. This is 10^2 times smaller than for the previous example so there are no dislocation tails shown in this figure; they would appear at values of $\langle c\rangle$ about 10^2 times lower than represented on the ordinate. We see again at $\varepsilon/\alpha = 1$ a plot that may be difficult to distinguish experimentally from linear, at least out to $\eta \approx 3$ or 4, and so which defines a unique D_{eff}. As ε/α increases from unity there is a progressive increase in curvature, continuing in this example to about $\varepsilon/\alpha \approx 4$ to 6 or so, whereafter the curvature diminishes

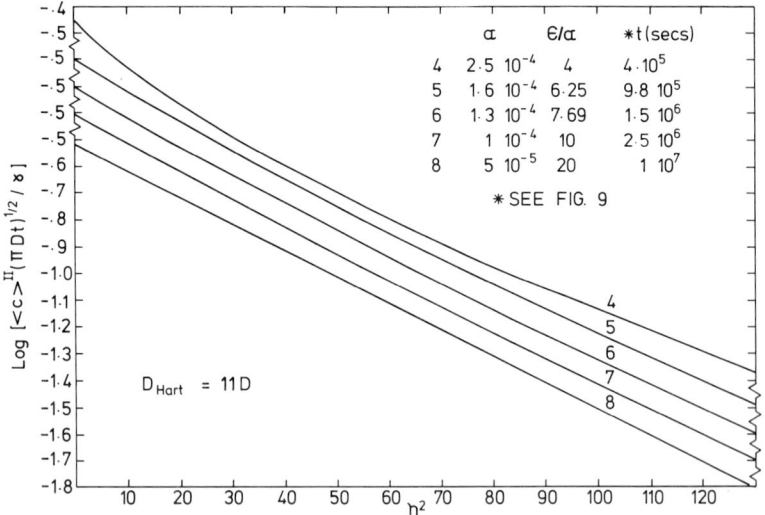

Fig. 10. Same as Fig. 9, showing progression from curve 4 to curve 8.

progressively until, at ε/α beyond 10 or so, it again becomes barely perceptible so that experimental plots can define once more a unique D_{eff}.

For diffusion under constant-concentration Case I conditions, D_{eff} is most simply calculated from the slope of a penetration plot of $\text{erfc}^{-1}(\langle c \rangle / c_0)$ versus y, i.e.,

$$D_{\text{eff}} = (4t)^{-1} \left(\frac{\partial \, \text{erfc}^{-1}(\langle c \rangle / c_0)}{\partial y} \right)^{-2} \tag{66}$$

A study of sets of such penetration plots would indicate the same general features over the first two or so powers of 10 decrease in $\langle c \rangle$ as for Case II, of a first linear domain at $\varepsilon/\alpha \lesssim 1$ associated with dislocation tails and a second linear domain at the larger $\varepsilon/\alpha \gtrsim 10$.

We now discuss in more detail the characteristics and properties of the first and second linear domains and of the dislocation tails. However, we shall confine discussion of D_{eff} in the FLD to diffusion under Case II conditions only, for Case I conditions have been so very much less commonly employed in precise measurements of diffusion coefficients.

1. *The First Linear Domain*

a. *Calculation of D_{eff}*. We are to calculate D_{eff}, as defined by Eq. (1) with $\langle c \rangle$ expressed as in Eq. (40), when the dislocation density is low enough, or the time short enough, for ε/α to lie in the range 0 to about 1. Figures 9 and 11

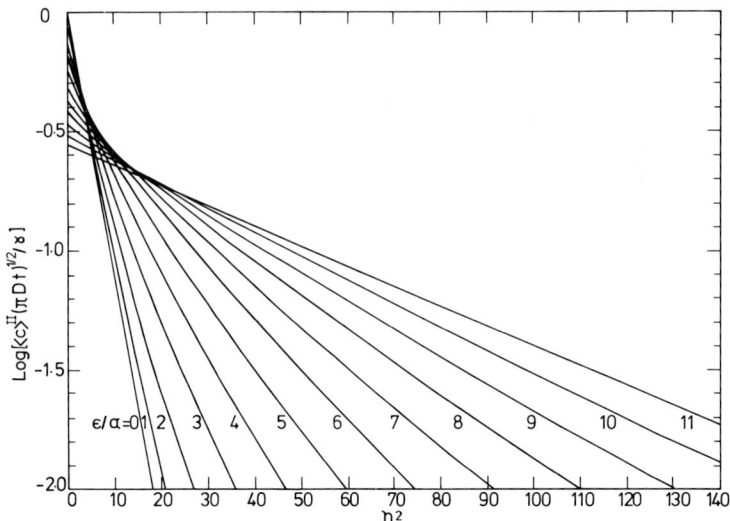

Fig. 11. Computed plots, $\log \langle c \rangle^{\text{II}}$ versus η^2, showing change with increasing dislocation density for fixed $\alpha = 10^{-4}$, $\Delta - 1 = 10^7$.

indicated that for this range experimental plots of $\log\langle c\rangle$ versus η^2 might be sufficiently linear, at least over the one or two orders of magnitude change in $\langle c\rangle$ that correspond with $0 < \eta < 3$ or 4, to define a unique D_{eff}. In this range, also, $Q^{\text{II}}(\eta,\varepsilon/\alpha)$ is effectively independent of ε/α (overlap negligible) and is well represented by the isolated dislocation solution $Q^{\text{II}}(\eta)$. We can therefore make use of the analytic representation of $Q^{\text{II}}(\eta)$ (Section IV.B) to obtain an expression for the penetration plot slopes. From Eqs. (1), (40), and (60) we find

$$\frac{\partial \ln \langle c \rangle}{\partial \eta^2} = -0.25[1 + \varepsilon^2 Q^{\text{II}}(\eta)\exp(\tfrac{1}{4}\eta^2)]^{-1}[1 + \varepsilon^2(V_1 + \exp(\tfrac{1}{4}\eta^2)$$
$$\times \{4V_2 V_3^2 \exp[-(V_3\eta)^2]$$
$$+ (2V_4 V_5/\eta)\tanh(V_5\eta)\operatorname{sech}(V_5\eta)\})] \qquad (67)$$

The slope is now, of course, not constant, but changes with η. Figure 12 illustrates Eq. (67) for η over the practically important range of $0 < \eta < 5$ and for various ε^2 in the range $0 < \varepsilon/\alpha < 1$, for the case $\alpha = 10^{-4}, \Delta - 1 = 10^8$.

For $\varepsilon^2 = 0$ the slope has the ideal constant value of -0.25, corresponding to the slope $-\tfrac{1}{4}Dt$ of the practical plot of $\log\langle c\rangle$ versus y^2. As the dislocation density begins to increase and contribute to $\langle c\rangle$, the slope, especially at the smaller η^2, remains quite close to -0.25 until ε^2 rises to reach and exceed, in this case, a magnitude of $\approx 10^{-9}$ ($\varepsilon/\alpha \approx 0.3$). Between these values of ε^2 the slope changes relatively little over the common experimental span of $0 <$

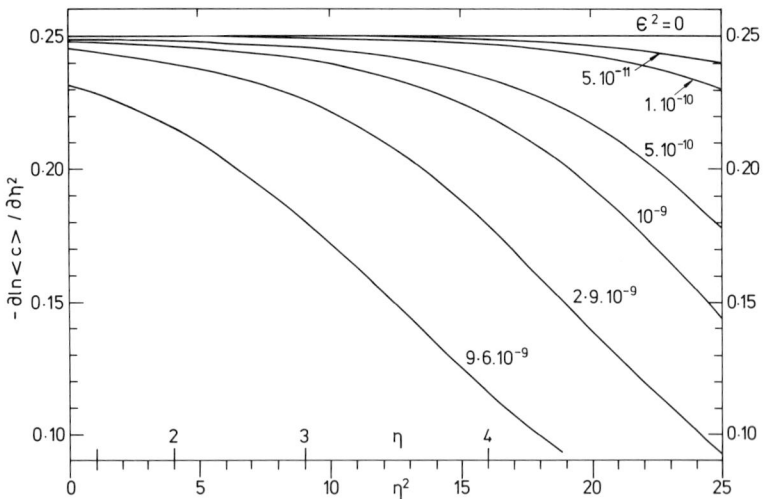

Fig. 12. The gradient $\partial \ln\langle c\rangle/\partial \eta^2$ as a function of η^2 for $\alpha = 10^{-4}, \Delta - 1 = 10^8$ and values of ε^2 indicated on the curves [Eq. (67)].

$\eta < \approx 3$ or 4; the experimental accuracy may often be insufficient to reveal the mild curvatures that in fact must exist in the penetration plots, and they will then pass for linear to yield a D_{eff} slightly greater than D.

As ε^2 continues to increase the change of slope with η^2 becomes more marked, and to an extent that sufficient curvature may be evident in carefully measured $\log\langle c \rangle$ versus y^2 plots as to make impracticable any reliable derivation of a unique D_{eff} from them.

As an example, Fig. 13 shows the plot for the lowest curve of Fig. 12, $\varepsilon^2 = 9.61 \times 10^{-9}$. Even for this case, where ε/α is near unity, measurements extending only to $\eta \approx 3$ (1 cycle of 10 decrease in $\langle c \rangle$) might still pass for linear, unless very precise, because of the only slight curvature over this range. But pursued to larger η, measurements may reveal the then larger curvature and so discourage attempts to calculate any D_{eff}.

We may redefine the "first linear domain" (FLD) as the range of ε, or of ε/α, starting from zero, for which plots of $\log\langle c \rangle$ versus y^2 are not experimentally distinguishable from linear, or gaussian, and so define a unique D_{eff}. Such conditions will usually prevail for $\varepsilon/\alpha \lesssim 1$, at least when the y or η range is not too extensive, but it will be clear that no precise general upper limit can be set to the range because of its experimental subjectivity.

Equation (67) is somewhat cumbersome. Since we are concerned in FLD with relatively small dislocation densities where the dislocation contribution

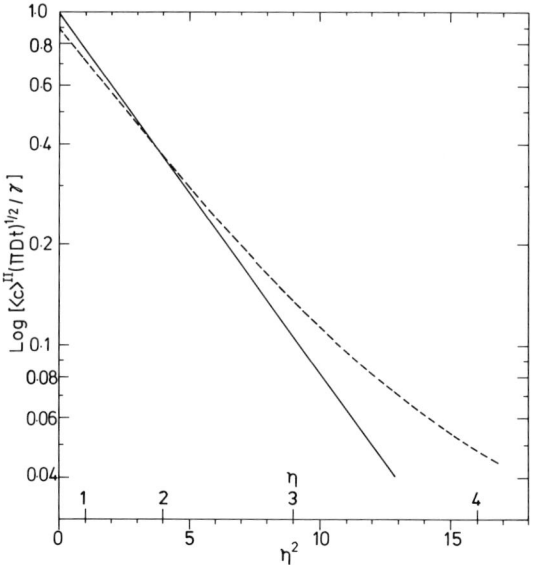

Fig. 13. Penetration plots, $\log\langle c \rangle$ versus η^2. Broken curve: $\alpha = 10^{-4}$, $\Delta - 1 = 10^8$, and $\varepsilon^2 = 9.61 \times 10^{-9}$; full curve: pure lattice diffusion, $\varepsilon^2 = 0$.

to $\langle c \rangle$, the second term in Eq. (40), is often small compared with the first, it may for many purposes be sufficient to use just an expansion of Eq. (67) to first order in ε^2. In particular, this will be quite sufficient for the important problem of estimating at what dislocation density D_{eff} may first begin significantly to exceed the true lattice D that it frequently purports to represent.

Expanding the denominator of Eq. (67) we obtain the first order result,

$$\frac{\partial \ln \langle c \rangle}{\partial \eta^2} = -\frac{1}{4}\left(1 - \varepsilon^2 \exp\frac{\eta^2}{4}\left\{V_2[1 - V_3^2]\exp(-\eta^2 V_3^2)\right.\right.$$
$$\left.\left. + V_4 \operatorname{sech}(V_5\eta)\left[1 - \frac{2V_5}{\eta}\tanh(V_5\eta)\right]\right\} + \cdots\right), \quad (68)$$

This is valid for $\varepsilon^2 Q^{\text{II}} \exp\frac{1}{4}\eta^2$ small compared with unity. This condition is equivalent to $(\varepsilon/\alpha)^2$ small compared with unity, for $Q^{\text{II}} \approx V_1 \exp -\frac{1}{4}\eta^2$ at small η and $V_1 \approx \alpha^{-2}$ (see Section IV.B). Thus Eq. (68) is a very good approximation for $\varepsilon/\alpha \lesssim 0.3$ or 0.4 say.

We may ignore the two terms in square brackets in Eq. (68); they usually differ little from unity because $(2V_5/\eta) \tanh V_5\eta \to 2V_5^2$ for small $V_5\eta$ and both V_3 and V_5 are very much less than unity for the α and $\alpha\beta$ values of present interest. Equation (68) then becomes

$$\frac{\partial \ln\langle c \rangle^{\text{II}}}{\partial \eta^2} = -\frac{1}{4}(1 - \varepsilon^2 M^{\text{II}} + \cdots)$$

where

$$M^{\text{II}} = Q^{\text{II}}(\eta)\exp(\tfrac{1}{4}\eta^2) - V_1 \quad (69)$$

so that

$$D_{\text{eff}} = D(1 + \varepsilon^2 M^{\text{II}} + \cdots) \quad (70)$$

We find that M^{II} is always positive, as expected, and that it increases monotonically with η. At $\eta = 0$, $M^{\text{II}} = V_2 + V_4$.

Although we are not discussing here any similar representation of D_{eff} for diffusion under Case I surface conditions, as defined by Eq. (66), we may note the result that at $\eta = 0$, (Le Claire and Rabinovitch, 1982)

$$M^{\text{I}}(\eta = 0) = -2(V_1 + V_2 + V_4) \quad (71)$$

Figure 14 shows M^{II} for $\eta = 3$ plotted as a function of α for various Δ. The value $\eta = 3$ was chosen for illustration because $\eta^2 = 9$ is near the center of the η^2 range when observations are over the typical span $0 < \eta < 4$; the slope at $\eta = 3$ is therefore probably close to the mean experimental slope of a set of measured points being taken as linear. The appendix to

5. MATHEMATICAL ANALYSIS OF DIFFUSION IN DISLOCATIONS

Le Claire and Rabinovitch (1982) contains a table of $Q^{II}(\eta)$ from which, with Fig. 5 for V_1, values of M^{II} may be calculated for other values of η.

We may note the following properties of these values of M^{II}:

(i) M^{II} increases with decreasing α so that enhancement of diffusion by dislocations at the relatively low densities of the FLD increases with increase in anneal time. This is in contrast with the larger enhancements associated with the SLD, or Hart domain, which are independent of anneal time (see Section IV.C.2).

(ii) For most of the α range shown the FLD values of M^{II} are less than the SLD or Hart values of $\Delta - 1$ [see Eq. (65)]. However, at low enough α, corresponding roughly to $\alpha^2(\Delta - 1)$ becoming less than unity, the M^{II} values in Figure 14 actually exceed $\Delta - 1$.

(iii) At the larger α, M^{II} increases as Δ decreases, but as α becomes smaller (time progressing) the behavior eventually changes towards M^{II}

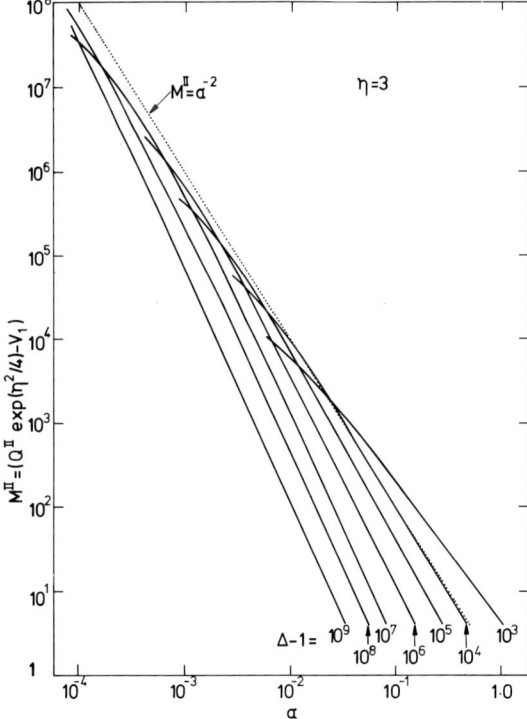

Fig. 14. The enhancement factor M^{II} for the first linear domain [Eq. (69)] as a function of α for several values of $\Delta - 1$, evaluated for $\eta = 3$.

decreasing with decreasing Δ. At small enough Δ, below 10^3 evidently, M^{II} *must* decrease as Δ decreases because M^{II} must be zero for all α at $\Delta = 1$.

(iv) For all Δ, except $\Delta = 10^3$ at large α, M^{II} satisfies the condition that

$$M^{II} \leq \alpha^{-2} \tag{72}$$

The dotted line in Fig. 14 represents the relation $M^{II} = \alpha^{-2}$.

b. *Condition for a Significant Difference between D_{eff} and D.* Diffusion coefficients can be estimated, under favorable conditions, to about $\pm 5\%$ say, so we may regard any D_{eff} more than $\approx 10\%$ greater than D as significantly enhanced by dislocation diffusion. From Eq. (70), this requires a dislocation density such that

$$\varepsilon^2 > 0.1/M^{II} \tag{73a}$$

or

$$d > 1.13 \times 10^{13}/M^{II} \tag{73b}$$

if we take $a = 5 \times 10^{-8}$ cm. For any given d, D_{eff} is more likely to exceed the true D the longer the anneal time, for M^{II} increases with time. Table I shows some values of d above which dislocation enhancement may become significant, calculated from Eq. (73b) for a range of values of α and Δ. The Δ were chosen, following Balluffi (1970) (see Appendix E) as appropriate for the temperature range 0.3–0.5 times the melting point, covering the lower end of the range over which self-diffusion measurements have been made and where dislocation enhancement is more likely; the range of $(Dt)^{1/2}$, or α, embraces values usually met with in lower temperature measurements. Since a value of $d \sim 10^6$ cm^{-2} is usually quoted as typical for well-annealed metal single crystals we see that, for this range of Δ, dislocation enhancement in

TABLE I

The Dislocation Density d (cm^{-2}) above which Dislocation Diffusion Begins, According to Eq. (73b), to Enhance Bulk Diffusion Coefficients

$(Dt)^{1/2}$ (cm)	α	Δ		
		10^6	10^8	10^{10}
5×10^{-6}	10^{-2}	3×10^9	3×10^{10}	4×10^{11}
5×10^{-5}	10^{-3}	2×10^7	4×10^7	4×10^8
5×10^{-4}	10^{-4}	6×10^5	3×10^5	7×10^5

such crystals will be unimportant provided anneal times are restricted so that $(Dt)^{1/2} \lesssim 10^{-4}$ cm.

A similar result follows directly from Eqs. (72) and (73). From these, there is no enhancement provided $\varepsilon^2 < 0.1 \alpha^2$, that is, provided $Dt \lesssim 0.1/\pi d$. With $d = 10^6$ cm^{-2} this condition becomes $(Dt)^{1/2} \lesssim \sqrt{3} \cdot 10^{-4}$ cm.

Table II lists some data from careful measurements of metal self-diffusion coefficients at lower temperatures: All the $(Dt)^{1/2}$ are below, or well below, 10^{-4} cm so, provided d was in fact no greater than $\sim 10^6$ cm^{-2}, there is no reason to expect there to have been any significant dislocation enhancement of the coefficients measured in any of these experiments.

These considerations are relevant in discussing the slight but significant positive curvature observed in the self-diffusion Arrhenius plots of many fcc metals. Wuttig and Birnbaum (1966b) and others have suggested that this may be due to a dislocation enhancement of D that is increasing with decreasing temperature; this possibility can only be rejected when the dislocation density is known to be *less* than the right-hand side of Eq. (73) at all temperatures in the range studied.

Bakker and Backus (1973) have also considered the influence of dislocations on measured self-diffusion coefficients. However, they employ a criterion for dislocation influence that does not directly refer to the measured gradient that defines D_{eff} and, furthermore, base their calculations on an erroneous solution (Mimkes and Wuttig, 1970; see Le Claire and Rabinovitch, 1981). Detailed comparison of the two approaches is thus not profitable but they nevertheless reach a conclusion similar to the above, that only

TABLE II

Lower Temperature Self-Diffusion Measurements: T Range, $(Dt)^{1/2}$, and $\alpha(a = 5 \times 10^{-8}$ cm$)$

Metal	T range (°C)	$(Dt)^{1/2}$ (cm)	α	Reference
Cu	300–630	2×10^{-7}–2×10^{-5}	2×10^{-1}–3×10^{-3}	Maier (1977)
Ni	542–922	8×10^{-7}–1×10^{-5}	6×10^{-2}–5×10^{-3}	Maier et al. (1976)
Au	286–411	1×10^{-6}–9×10^{-5}	4×10^{-2}–6×10^{-3}	Rupp et al. (1969)
Au	350–460	2–8×10^{-5}	3×10^{-3}–6×10^{-4}	Gainotti and Zecchina (1965)
Ag	270–500	8×10^{-7}–4×10^{-5}	6–1×10^{-3}	Lam et al. (1973)
Ag	358–580	3×10^{-5}–1×10^{-4}	2×10^{-3}–4×10^{-4}	Backus et al. (1974)
Ag	307–560	2×10^{-6}–6×10^{-5}	2×10^{-2}–8×10^{-4}	Bihr et al. (1978)
Ag	255–380	3×10^{-8}–5×10^{-7}	2–1×10^{-1}	Savitski (1963)

dislocation densities in excess of $10^6 \sim 10^7$ cm^{-2} will have any significant influence on D_{eff} for values of Δ in the range of Table I.

2. The Second Linear Domain

We are concerned here with the calculation of D_{eff}, as defined by Eq. (1), for those nearly linear penetration plots determined after times sufficiently long, or with samples of sufficiently high dislocation density, so that $\varepsilon/\alpha \geq 10$ or so.

It was pointed out by Hart (1957) that when ε/α was sufficiently large, so that the diffusion length sufficiently exceeds the dislocation spacing, there could be a sampling by diffusing atoms of many dislocation and crystal regions sufficient for the diffusion overall to appear fickian and describable by a unique D_{eff}. Hart estimated that D_{eff} should tend towards a mean of D' and D, weighted respectively by the volume fractions of dislocation pipe and of regular crystal lattice, i.e.,

$$D_{\text{eff}} \to D_{\text{Hart}} = \varepsilon^2 D' + (1 - \varepsilon^2)D$$
$$= D[1 + \varepsilon^2(\Delta - 1)] \quad (74)$$

D_{eff} has been calculated as a function of η^2 from the slopes of the plots shown in Fig. 11. Figure 15 shows the ratio $D_{\text{eff}}/D_{\text{Hart}}$ plotted against η^2 for

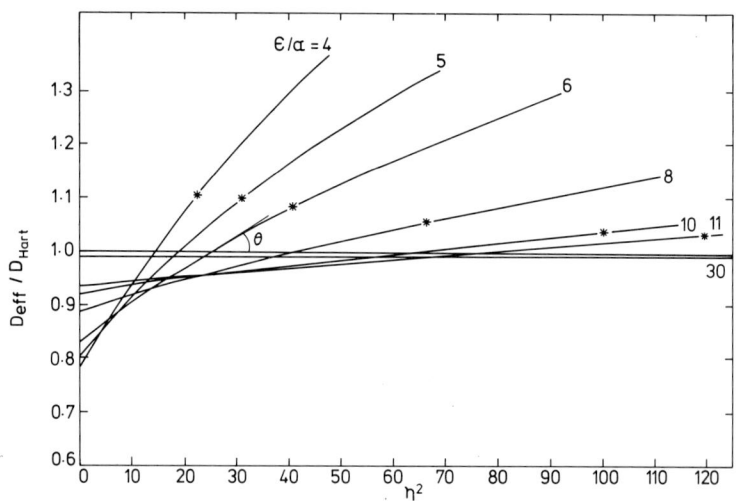

Fig. 15. The ratio $D_{\text{eff}}/D_{\text{Hart}}$ as a function of η^2 for various ε/α, calculated from the plots of Fig. 11; $\alpha = 10^{-4}$, $\Delta - 1 = 10^7$. The asterisks indicate values of η^2 at which $\delta \log\langle c \rangle = -1$.

5. MATHEMATICAL ANALYSIS OF DIFFUSION IN DISLOCATIONS 291

various ε/α. The stars are placed at those values of η^2 at which $\langle c \rangle$ has fallen by a factor 10 from its value at $\eta = 0$.

For $\varepsilon/\alpha = 30$ (not included in Fig. 11), D_{eff} and D_{Hart} are for all practical purposes identical, D_{eff} being nearly constant and at most only 1% smaller than D_{Hart} over the whole range shown.

For $\varepsilon/\alpha = 10$ or 11 there is a small but perceptible increase in $D_{\text{eff}}/D_{\text{Hart}}$ with η^2, reflecting the very slight, but experimentally usually imperceptible, positive curvature in $\log\langle c \rangle$ versus η^2 still persisting at these ε/α values. However, being a few percent below one at low η^2 and a similar few percent above one at the larger η^2, the *average* value of $D_{\text{eff}}/D_{\text{Hart}}$ may be much closer to one when D_{eff} is obtained from a mean slope determined over a suitable range of η^2. Figure 15 suggests that such a range should correspond roughly to about $1\frac{1}{2}$ factors of 10 decrease of $\langle c \rangle$ which, fortunately, is quite representative of much experimental practice.

For smaller values of ε/α the larger curvatures, that Fig. 11 indicates are to be expected in $\log\langle c \rangle$ versus η^2 plots, may well deter the experimenter from determining a D_{eff}. However, experimental scatter may conceal curvature to the extent that a slope is determined of a plot that is taken to be linear. Figure 15 shows that the resulting mean value of D_{eff} will, even for ε/α as small as 4, be quite close to D_{Hart} provided the mean slope is determined for, again, about a $1\frac{1}{2}$ factors of 10 decrease in $\langle c \rangle$. But even without this proviso, it is to be noted that over the range shown, derived values of D_{eff} never differ by more than about 30% from D_{Hart}.

The same conclusion follows from a study of Fig. 10, showing in more detail the changes that occur in the penetration plots during the 25-fold increase in anneal time between curves (4) and (8) of Fig. 9, corresponding to a change in ε/α from 4 to 20. While there is the progressive straightening already commented upon, there is clearly little change with time of the *mean* slope, showing again that mean values of D_{eff} need not differ very much from D_{Hart}, equal to $11D$ in this case, even though determined at times too small for $(Dt)^{1/2}$ to be *very* much greater than $2R$.

We may now refine the definition of the second linear domain (SLD) as the range of ε/α from infinity down to the smallest for which plots of $\log\langle c \rangle$ versus y^2 are not distinguishable experimentally from linear and so can yield a unique D_{eff}. We have seen that the SLD starts at $\varepsilon/\alpha \approx 10$ so that for $(Dt)^{1/2}$ no greater than about 5 times the dislocation spacing, dislocation and regular lattice diffusion are sufficiently mixed for D_{eff} to be within a percent or two of D_{Hart}.

The lower limit for the SLD cannot be more precisely set because of its obvious dependence on the experimental precision being practiced.

Diffusion under conditions where $D_{\text{eff}} \approx D_{\text{Hart}}$ was called "type A" diffusion by Harrison (1961); by an approximate calculation he deduced

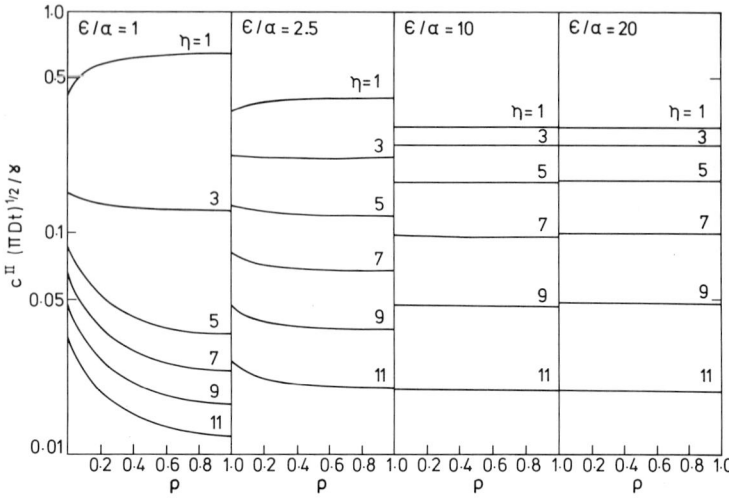

Fig. 16. Concentration $c^{II}(\rho, \eta, t)$ around a dislocation pipe plotted versus $\rho = r/R$ for fixed $\varepsilon = 10^{-3}$, $\Delta - 1 = 10^7$, i.e., same conditions as for Figs. 9 and 10 [see Eqs. (10) and (48)].

the relations

$$Dt \gtrsim 10^5 R^2$$

or (75)

$$\varepsilon/\alpha \gtrsim 10^{2.5}$$

as the necessary condition for observing it. The present exact calculations replace Eq. (75) by $\varepsilon/\alpha \gtrsim 10$ and so confirm previous suggestions (Morrison, 1965; Ruoff and Balluffi, 1963; Gupta and Asai 1973) that the condition in Eq. (75) is a very much too restrictive limitation on the valid and useful application of Eq. (74). This equation has often been employed, and with apparent success, in the analysis of measured diffusion coefficients that were believed to have been enhanced by dislocation diffusion even though the condition in Eq. (75) was believed to be far from satisfied. The present calculations justify such procedure.

The usual application is to plot $\log[(D_{\text{eff}} - D)/D]$ versus $1/T$; from Eq. (74) this should be linear with a slope from which can be deduced the difference $E - E'$ between the activation energies E for lattice and E' for dislocation diffusion. Such linear plots have been demonstrated for cases where it is by no means clear that ε/α is near to or greater than 10, yet from them quite plausible values of E' have been deduced. For example, from dislocation enhanced self-diffusion Morrison and Yuen (1971) obtained $E' = 0.68E$ for Ag and Lai and Morrison (1970) report $E' = 0.67E$ for Au.

It was noted by Harrison (1961) that Type A, or SLD, diffusion, being

5. MATHEMATICAL ANALYSIS OF DIFFUSION IN DISLOCATIONS

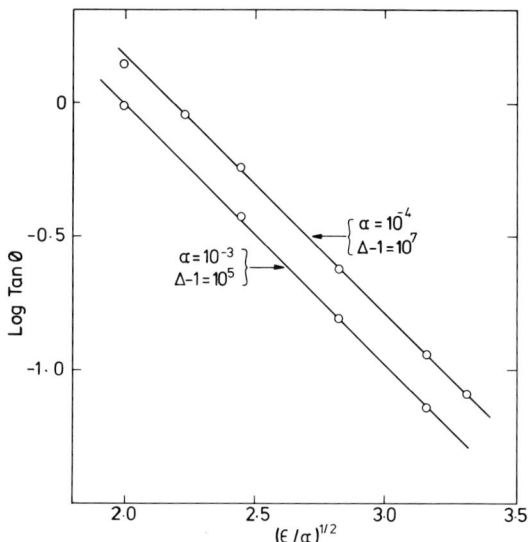

Fig. 17. The linear dependence of log tan θ on $(\varepsilon/\alpha)^{1/2}$ for two sets of fixed α and Δ.

representable as gaussian with a unique D_{eff}, should entail a near uniform concentration of diffusant within each section, i.e., $c(r, y, t)$ should be effectively independent of r. Figure 16, showing c plotted as a function of $\rho = r/R$ for fixed ε and Δ, computed from Eqs. (48) and (49), indicates that this is indeed the case. For $\varepsilon/\alpha = 1$ or less, the concentration changes markedly with ρ and in the manner expected. The changes are already much reduced at the later time represented by $\varepsilon/\alpha = 2.5$ and for ε/α greater than about 10, c has indeed become effectively uniform over $0 < \rho < 1$.

We have attempted to identify some parameter to characterize the approach to the SLD, or to Hart-like behavior, as ε/α increases. One possibility is the angle θ at which a line of $D_{\text{eff}}/D_{\text{Hart}}$ versus η^2 (Fig. 15) intersects the line for this ratio equal to one; as θ becomes smaller the penetration plots will be more nearly gaussian, D_{eff} will be less variable and its value will approach D_{Hart} more closely. Figure 17 shows, as an empirical result, how log tan θ varies quite linearly with $(\varepsilon/\alpha)^{1/2}$ for fixed α and Δ. A more detailed understanding of the functional dependence of θ on α, Δ, ε/α, etc., would be of some interest.

3. Dislocation Tails

a. *Mathematical Properties.* As indicated by the plots of Fig. 9, dislocation tails can be observed when ε/α is less than about unity and when measurements have been extended to values of η such that the first term of Eq. (40), or of Eq. (50), has become negligible compared with the second. In other words,

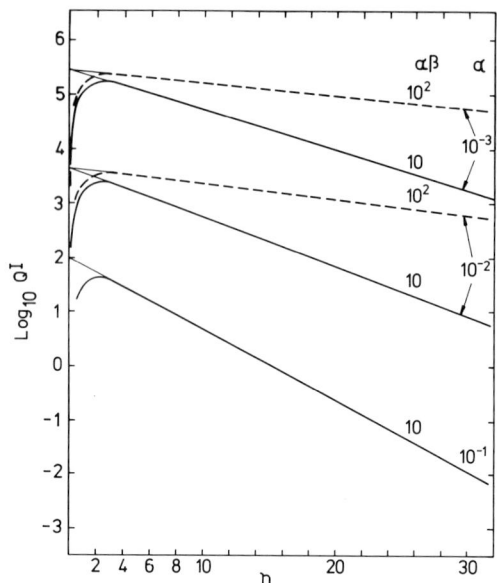

Fig. 18. Computed plots of log $Q^I(\eta)$ versus η for $\alpha = 10^{-1}$, 10^{-2}, 10^{-3} with $\alpha\beta = 10$, (full lines) and $\alpha\beta = 10^2$, (broken lines). Small ε/α.

the concentration in the tails is due to material that has diffused down and out of the dislocations to depths well below that reached by volume diffusion alone and under conditions of no diffusion zone overlap. Their properties are wholly determined by $Q^I(\eta)$ or by $Q^{II}(\eta)$.

Figures 18 and 19 show log $Q^I(\eta)$ and log $Q^{II}(\eta)$ plotted versus η for several values of the parameters α and $\alpha\beta$; these illustrate many of the properties to be discussed.

(i) Beyond $\eta \approx 4$ or 5 the plots are effectively linear for $\alpha \lesssim 1$, the change in slope over the range $\eta = 6$ to 30 being barely perceptible for $\alpha \lesssim 10^{-1}$ and amounting only to 1–2% for $\alpha \simeq 1$ with the larger $\alpha\beta$ studied. As α increases much above one the plots become noticeably nonlinear with a negative curvature. This is as expected. For very large α, which implies very little diffusion other than within the pipes, $\ln Q^{II}$ must become proportional to $-\eta^2/4\Delta$ and Q^I proportional to $\mathrm{erfc}(\eta/2\Delta^{1/2})$.

(ii) For a given α and $\alpha\beta$ the gradients of these linear regions are exactly the same for Q^I and for Q^{II}. The gradient depends strongly on $\alpha\beta$ but only very weakly on α, and can be conveniently represented as

$$\frac{\partial \ln Q^I}{\partial \eta} = \frac{\partial \ln Q^{II}}{\partial \eta} = -\frac{A(\alpha)}{(\alpha\beta)^{1/2}} \qquad (76)$$

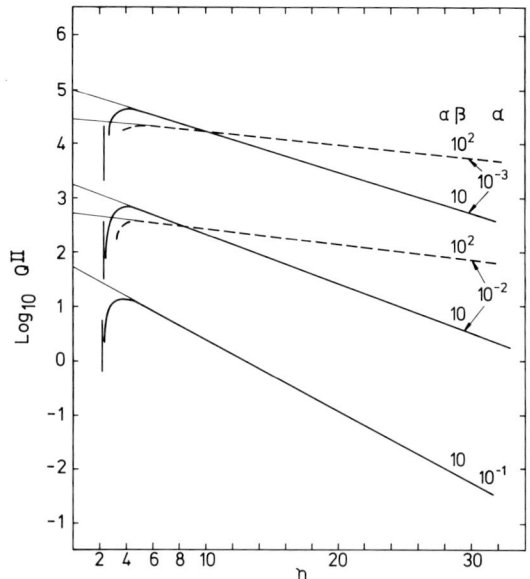

Fig. 19. Computed plots of log $Q^{II}(\eta)$ versus η for $\alpha = 10^{-1}$, 10^{-2}, and 10^{-3} with $\alpha\beta = 10$ (full lines) and $\alpha\beta = 10^2$ (broken lines), at small ε/α. Vertical lines at left indicate values of η at which Q^{II} changes sign (see Fig. 3).

A is of order unity and a slowly varying function of α with very weak dependence on $\alpha\beta$. These properties are illustrated in Fig. 20, which shows A plotted versus α. For definiteness, A was calculated from the mean slope over the range $6 \leq \eta \leq 30$.

From Eqs. (63) and (64) we obtain

$$-A(\alpha)/(\alpha\beta)^{1/2} \doteq V_5$$

Exact equality is not expected in this relation because V_5 defines an asymptotic slope as $\eta \to \infty$ while $A(\alpha)$ is calculated from a mean slope over a finite range of η.

Experimentally, dislocation tails are always observed to be accurately linear when plotted as $\ln\langle c \rangle$ versus y, and it will be evident that their slope is that given by Eq. (76). Experimental values of $\alpha\beta$ calculated from it turn out to lie in the range 1 to 10^4, to which the calculations of $A(\alpha)$ for Fig. 20 were therefore restricted. The two ends of this range may be considered to represent limiting types of observable tails. With $A \approx 0.7$–0.8 say, $\alpha\beta = 1$ relates to a very steep tail with, for example, a concentration change of as large as 10^2 (near the maximum usually measurable in a tail) occurring over as short a range of η as ~ 5 to 6. At the other limit, $\alpha\beta = 10^4$ represents a near-horizontal

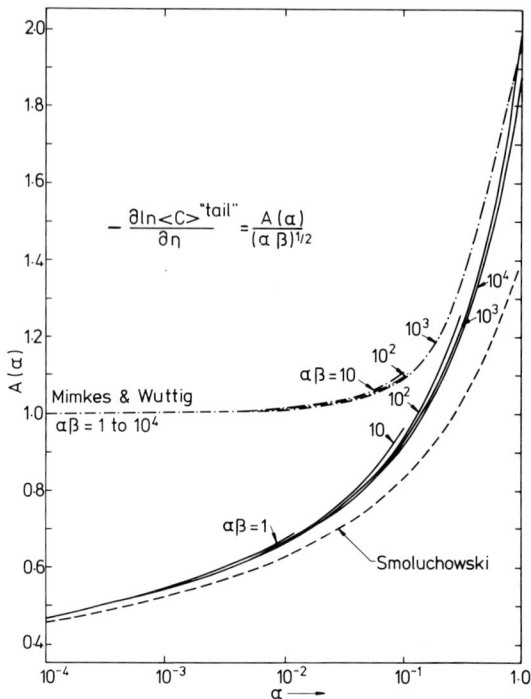

Fig. 20. The quantity $A(\alpha)$ of Eq. (76) plotted as a function of α for various $\alpha\beta$.

tail with, for example, $\langle c \rangle$ only halving over a range ~ 100 of η. The calculations for each $\alpha\beta$ were then further restricted, because of the underlying assumption that $\Delta \gg 1$, to those values of α small enough to ensure $\Delta \geq 10^3$ to 10^4. Thus $\alpha = 1$ is the largest value considered. In practice α always turns out to be less than one (assuming $a \approx 10^{-7}$ cm) and Δ greater than 10^3.

Earlier approximate calculations by Smoluchowski (1951) and by Mimkes and Wuttig (1970) also yield the same result as Eq. (76) but with different values for $A(\alpha)$. These are included in Fig. 20 for comparison. The Smoluchowski values are fairly close to the true values, and especially so at low α as might be expected from the nature of the Smoluchowski approximations. The Mimkes and Wuttig values differ considerably from the true, except for $\alpha \gtrsim 1$.

(iii) Extrapolated back to $\eta = 0$, the linear portions of the plots in Fig. 18 intersect the ordinate at values, $\log Q_{\text{ex}}^l$, which decrease rapidly with increasing α but which are nearly independent of $\alpha\beta$ for a given α. These properties are illustrated in Fig. 21, showing computed values of $\log Q_{\text{ex}}^l$ plotted versus α. For $\alpha \lesssim 0.1$, Q_{ex}^l is quite well represented by the remarkably

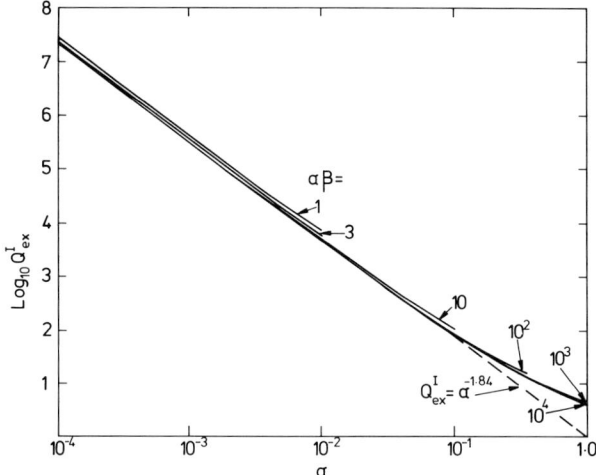

Fig. 21. Log Q_{ex}^I plotted against α. Log Q_{ex}^I is the intercept with the ordinate made when the linear tails of log Q^I versus η plots (Fig. 18) are extrapolated back to $\eta = 0$.

simple expression

$$Q_{ex}^I \simeq \alpha^{-1.84} \tag{77}$$

As can be seen from Fig. 19, the same extrapolation of the Q^{II} plots yields intercepts, log Q_{ex}^{II}, that depend both on α and $\alpha\beta$. However, Q_{ex}^{II} can always be expressed in terms of Q_{ex}^I, for it is simple to show from Eqs. (52) and (76) that

$$Q_{ex}^{II} = \frac{\pi^{1/2} A(\alpha)}{(\alpha\beta)^{1/2}} Q_{ex}^I \tag{78}$$

Q_{ex}^I and Q_{ex}^{II} are also simply related to the V_4 and V_5 of Section IV.B. Thus from Eqs. (63) and (64)

$$Q_{ex}^I \doteq \frac{2V_4}{\pi^{1/2} V_5} \doteq \alpha^{-1.84} \tag{77a}$$

and

$$Q_{ex}^{II} \doteq 2V_4 \tag{78a}$$

The relationships in Eq. (77a) are well illustrated by the very close similarity of Figs. (8) and (21).

We now use these properties to discuss the treatment of experimental results.

b. *Dislocation Tail Analysis.* From Eq. (76) the experimentally measured tail slope is, for either surface condition,

$$\frac{\partial \ln \langle c \rangle}{\partial y} = -\frac{A(\alpha)}{[(\Delta - 1)a^2]^{1/2}} \qquad (79)$$

So, from the slope can be determined the diffusion coefficient ratio $\Delta = D'/D$, provided a is known, but obviously the results depend strongly on what value is used for a. There are no definitive or undisputed measurements of a but a common assumption, at least for metals, is $a = 5 \times 10^{-8}$ cm. Because of electrostatic effects, larger values probably prevail in ionic crystals (e.g., Yan *et al.*, 1977) and there are experimental indications that this may in fact be true (see below).

The quantity $(\Delta - 1)a^2$ is of course determinable with much less uncertainty than Δ alone, because $A(\alpha)$ depends only weakly on α. For example, a change in α from 10^{-4} to 10^{-2} alters the value of $(\Delta - 1)a^2$ derived from a measured slope by only a factor ≈ 2.

Because $A(\alpha)$ is only a very slowly varying function of α, and so an even more slowly varying function of time, the slopes of dislocation tails are almost independent of the anneal time. This is in marked contrast with similar tails due to fast diffusion in grain boundaries where slopes are proportional to $t^{-1/4}$ (Le Claire, 1963). Thus, measurements of the slope as a function of time provide a means of distinguishing experimentally between tails due to diffusion down isolated and independently acting dislocations and tails due to diffusion down dislocations closely arrayed into boundaries or walls. The $t^{-1/4}$ dependence for boundary tails was observed some time ago by, for example, Stark and Upthegrove (1966) for self-diffusion in Pb bicrystals; that the slopes of dislocation tails are nearly independent of time has only very recently been demonstrated by Ho (1982) and Ho and Pratt (1983) for ^{22}Na diffusion in single crystals of NaCl (see Fig. 22).

The dislocation tail can also provide an estimate of the dislocation density d if experimental accuracy is sufficient to allow an extrapolation of the tail back to $y = 0$, or $\eta = 0$, to intersect the ordinate at $\langle c \rangle_{\text{Int}}$ say (see Figs. 22 and 23).

For type I conditions, from Eq. (50),

$$\langle c \rangle_{\text{Int}}^{\text{I}} = c_0 \varepsilon^2 Q_{\text{ex}}^{\text{I}} \qquad (80)$$

Using for Q_{ex}^{I} the result of Eq. (77) and recalling that $\varepsilon^2 = \pi a^2 d$, we obtain

$$d^{\text{I}} = \frac{(\langle c \rangle_{\text{Int}}^{\text{I}}/c_0)}{\pi a^{0.16}(Dt)^{0.92}} \qquad (81)$$

For type II conditions, from Eq. (40)

$$\langle c \rangle_{\text{Int}}^{\text{II}} = c(0)\varepsilon^2 Q_{\text{ex}}^{\text{II}} \qquad (82)$$

5. MATHEMATICAL ANALYSIS OF DIFFUSION IN DISLOCATIONS

where

$$c(0) = \frac{\langle c(\eta = 0) \rangle^{\text{II}}}{[1 + \varepsilon^2 Q^{\text{II}}(\eta = 0)]} \quad (82a)$$

and $\langle c(\eta = 0) \rangle^{\text{II}}$ is the measured concentration at $y = 0$. Since $\varepsilon^2 Q^{\text{II}}(\eta = 0)$ is always negative but fractional, for $\varepsilon/\alpha < 1$, [see Le Claire and Rabinovitch (1982)] $c(0)$ is in general greater than $\langle c(\eta = 0) \rangle$. However, for low dislocation densities the difference between the two may not be important. From Eqs. (76)–(78) and (82) we then obtain

$$d^{\text{II}} = \frac{[\langle c \rangle^{\text{II}}_{\text{Int}}/c(0)]}{\pi^{3/2} a^{0.16} (Dt)^{1.42} (\partial \ln \langle c \rangle / \partial y)_{\text{Tail}}} \quad (83)$$

The concentration ratios, the tail gradients and $(Dt)^{1/2}$ in these equations can all, in favorable circumstances, be determined to within a few percent, so errors in these quantities are not too important; only a is imprecisely known, but estimates of d from Eqs. (81) or (83), according to the experimental surface conditions prevailing, will not be too sensitive to the value used for a because of its being raised to the power 0.16. Thus, for example, even an uncertainty of 10^2 in a produces an uncertainty in d of only a factor 2.

In principle, Eqs. (81) and (83), rewritten, could be used to derive estimates of a, given reliable values of the dislocation density. For example,

$$a^{\text{I}} = \left[\frac{\langle c \rangle_{\text{Int}}/c_0}{\pi d (Dt)^{0.92}} \right]^{6.25} \quad (84)$$

However, a^{I} is obviously now very sensitive to d because of the 6.25 power to which it is raised; exceedingly precise values of d would be necessary for reliable values of a to be derived. Dislocation densities are frequently determined by etch pit counts but the technique cannot always be relied upon to reveal every dislocation. Dislocation densities tend therefore to be underestimated so that a would tend to be overestimated. The effects can be profound. For example, if d is underestimated by a factor 3 because of etching inefficiency, which is not an untypical value (Ahlborn, 1979, Springer, 1971) the deduced value of a would be nearly 10^3 times larger than the true value. Thus, quite misleading indications of the effective radii of dislocations could ensue from any uncritical estimate of them from dislocation densities and tail diffusion data.

Equation (83) can be used to provide another relation for testing for dislocation enhancement of diffusion under type II surface conditions. Combining Eqs. (72) and (73a) with Eq. (83) we find there can have been no significant enhancement of D when

$$\langle c \rangle^{\text{II}}_{\text{Int}} / \langle c(y = 0) \rangle \lesssim 0.18 a^{0.16} (Dt)^{0.42} (\partial \ln \langle c \rangle / \partial y)_{\text{Tail}} \quad (85)$$

However, larger values of $\langle c \rangle^{\text{II}}_{\text{Int}} / \langle c(y = 0) \rangle$ do not necessarily indicate that there *is* enhancement; only more detailed examination of the data can establish this.

Finally, we may now consider in a little more detail the experimental conditions necessary for observing dislocation tails.

We have already seen that tails are only observable for $\varepsilon/\alpha \lesssim 1$ or so; for larger ε/α the several dislocation diffusion zones overlap to an extent sufficient for tail characteristics to be lost. Thus there is a fairly sharp *upper* limit on the range of anneal times for tail observation, given by $t \lesssim 1/\pi Dd$. The *lower* limit is set by the sensitivity of the measurements, for concentrations in the tail increase with increase in t, as evident from Eqs. (81) and (83), and to observe a tail requires that at least $\langle c \rangle_{\text{Int}}$ be detectable.

For case I conditions, Eqs. (77) and (80) give

$$\langle c \rangle^{\text{I}}_{\text{Int}}/c_0 = (\varepsilon/\alpha)^2 \alpha^{0.16} \tag{86}$$

If l is the smallest fraction of the surface concentration $[c_0$ or $c(0)]$ that can be measured, then to observe tails requires that

$$1 \gtrsim \varepsilon/\alpha \gtrsim l^{1/2}\alpha^{-0.08} \approx 2l^{1/2} \tag{87}$$

where the 2 is a rough mean of $\alpha^{-0.08}$ for typical values of α. In terms of anneal time, Eq. (87) may be written

$$4l \lesssim \pi dDt \lesssim 1 \quad \text{(Case I)} \tag{88}$$

There is also an additional condition to be met that $\alpha\beta$ be somewhat larger than unity so as to provide a tail of moderate slope such that $\langle c \rangle$ does not decrease too rapidly below the limit of measurability. In a plot of $\ln\langle c \rangle$ versus y this requires that $(\Delta - 1)a^2$ is sufficiently large, but this is a condition outside experimental control.

For Case II conditions, Eqs. (77), (78), and (82) give

$$\frac{\langle c \rangle^{\text{II}}_{\text{Int}}}{c(0)} = \left(\frac{\varepsilon}{\alpha}\right)^2 \frac{\pi^{1/2} A(\alpha)}{(\alpha\beta)^{1/2}} \alpha^{0.16} \geq l \quad \text{(Case II)} \tag{89}$$

Since $(\pi^{1/2}A(\alpha))^{1/2} \approx 1$, the condition for observing tails in this case becomes

$$1 \gtrsim \varepsilon/\alpha \gtrsim l^{1/2}(\Delta - 1)^{1/4}\alpha^{0.42} \quad \text{(Case II)} \tag{90}$$

4. Application to Experimental Data

In this section we discuss some experimental results that illustrate the use of many of the relations already derived.

a. *Self-Diffusion of Na in NaCl.*[3] Ho (1982) measured by serial sectioning, following diffusion under type II conditions, the self-diffusion coefficient of

---

[3] Ho (1982); Ho and Pratt (1983).

^{22}Na in very pure (<1 ppm total impurity) single crystals of NaCl. Figure 22 shows two of his penetration plots at effectively the same temperature but for times that differ by more than a factor of 4.

Excellent dislocation tails are to be seen, with $\ln\langle c\rangle$ quite accurately linear in y and with slopes, in keeping with remarks following Eq. (79), that are very nearly the same. In fact, the ratio of the slope of "Run 4" to that of "Run 5" is 0.93 and is wholly accounted for by the change in $A(\alpha)$ [Eq. (79) and Fig. 20] over the four-fold change in t; the ratio of the $A(\alpha)$ is 0.92.

Equation (79) and Fig. 20 were used to calculate from the slopes the mean result $(\Delta - 1)a^2 = 7.33 \times 10^{-5}$ cm^2 at 505–506°C. At this temperature, $D = 1.45 \times 10^{-11}$ cm^2 s^{-1}. Since it seems unlikely that D' can much exceed a value of $\approx 10^{-5}$ cm^2 s^{-1}, typical of liquid state diffusion, there is an upper limit of $\approx 10^6$ on Δ at this temperature; the measured value of $(\Delta - 1)a^2$ then indicates a value no less than $a \approx 10^{-5}$ cm^2 s^{-1} for the dislocation radius. We assume these values for a and Δ in what follows.

From the slopes and the intercept values, $\langle c \rangle_{\text{Int}}^{\text{II}}$, the dislocation densities were calculated from Eq. (83) to give $d = 2.3 \times 10^3$ cm^{-2} for both the samples of "Run 4" and "Run 5." Etch pit counts gave d of the order of 10^4 cm^{-2} for similar samples. The agreement is not unsatisfactory, especially as etch pit counts might be expected to give a larger value than the calculated d, which gives only the effective number of dislocations normal to the surface.

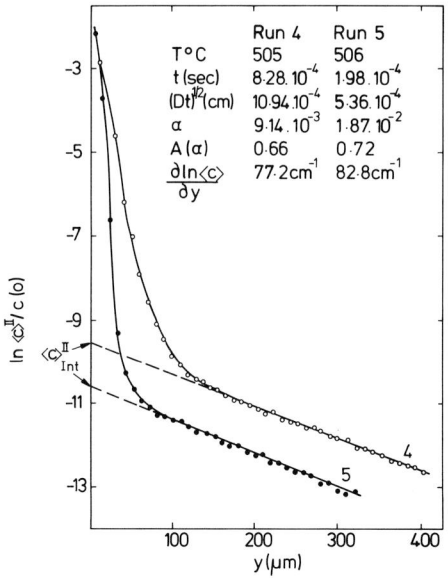

Fig. 22. Self-diffusion of ^{22}Na in NaCl single crystals. Penetration plots, $\ln\langle c\rangle$ versus y, for two different anneal times at the "same" temperature (Ho, 1982).

With $d = 2.3 \times 10^3$ cm^{-2} and the $(Dt)^{1/2}$ shown in Fig. 22, $\varepsilon/\alpha = 9.3 \times 10^{-2}$ for "Run 4," 4.5×10^{-2} for "Run 5," values that are consistent with Eq. (90).

With $\alpha = 9.14 \times 10^{-3}$ and $\Delta = 10^6$, Fig. 14 gives an estimate of $M^{II} \approx 10^4$. Thus $\varepsilon^2 M^{II} \approx 7 \times 10^{-3}$, so that from Eq. (70) there was no significant enhancement of D by the low dislocation density prevailing. An application of Eq. (85) leads to the same conclusion.

The above and other values of $(\Delta - 1)a^2$ measured by Ho and Pratt are the first reliable estimates of this quantity for cation dislocation diffusion in any ionic crystal. They were used by Ho and by Ho and Pratt to estimate, via the Nernst–Einstein equation, the contribution σ_{dis} of dislocation transport to the total ionic conductivity, which in NaCl is almost totally cationic. The relation for σ_{dis} is

$$\sigma_{dis} = \frac{Ne^2 D}{kTf} \pi[a^2(\Delta - 1)]d \tag{91}$$

where N is the number density of cations, f the correlation factor and e, k, and T have their usual meanings. For the very pure crystals used, σ_{dis} was found to be a major component of the directly measured total conductivity in the extrinsic region. For less pure crystals, the extrinsic conductivity would be determined proportionately more by the impurity content, less by dislocation transport, but that there would seem to be two contributions to extrinsic transport, with the dislocation contribution totally ignored previously, might be at least in part responsible for the notorious variability of the defect energies quoted in the literature (e.g., Barr and Lidiard, 1970; Fredericks, 1975; Ho, 1982) for the extrinsic region. There is a need for a thorough reappraisal of the situation.

b. *Diffusion of Ga in Ge.*[4] The circles in Fig. 23 represent measurements of $\log\langle c \rangle^I$ against y obtained by Ahlborn for diffusion of Ga from a constant concentration source (case I) into single-crystal Ge at 713°C. From more extensive measurements the lattice $D = 1.66 \cdot 10^{-15}$ cm^2 sec^{-1}, giving $\alpha = 1.62 \times 10^{-3}$, if we take $a = 5 \times 10^{-8}$ cm.

There is a well-defined linear tail for $\eta \gtrsim 6$, from the slope of which is calculated, via Eq. (79), the value $\alpha\beta = 3.78$. From this we derive $\Delta = 1.44 \times 10^6$.

The dislocation density is estimated, from Eq. (81) and the observed values of $\langle c \rangle_{Int} = 8.87 \times 10^{17}$ and $c_0 = 1.1 \times 10^{20}$ atoms cc^{-1}, as $d = 7.5 \times 10^6$ cm^{-2}. Ahlborn reports an etch pit density of 6.7×10^6, but considers the total dislocation density to be about three times larger because of etch inefficiency. Since the d calculated from the tail is the effective number normal

---

[4] Ahlborn (1979).

to the surface, which is about a third of the total, the present estimate and the author's are in very good agreement.

With the above parameters $\varepsilon^2 = 5.9 \times 10^{-8}$ and $\varepsilon/\alpha = 0.15$, in the range expected, Eq. (87), for well resolved dislocation tails.

Also, calculating for the parameters derived values for V_1, V_2 and V_4 we find from Eq. (71) that $\varepsilon^2 M^1(\eta = 0) \approx 9 \times 10^{-3}$. Even allowing for a rather larger $\varepsilon^2 M^1$ at a more appropriate η, = 3 say, the dislocation enhancement of D, Eq. (70), in this example is not expected to be more than a few percent. This is confirmed by the directly determined value of $D_{\text{eff}} = 1.77 \times 10^{-15}$ cm^2 s^{-1} [Eq. (66)], only a little above D, and by there being among other samples only few percent differences in D_{eff} for two to five fold changes in etch pit density.

The full curve in Fig. 23 represents $\langle c \rangle^I$, as expressed by Eq. (50), calculated from values of $Q^I(\eta)$ computed from Eq. (58) for the parameters α and Δ quoted above. There is a very good fit to the experimental points except,

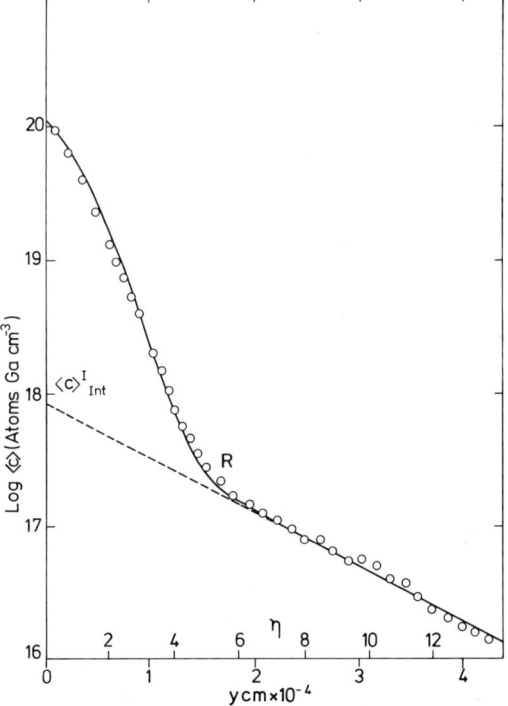

Fig. 23. Diffusion of Ga in Ge single crystal. $T = 713°C$, $t = 5.784 \times 10^5$ s. Circles: experimental data (Ahlborn, 1979). Full curve: $\langle c \rangle^I$ from Eq. (50). See text (Section IV.C.4.b) for parameter values.

perhaps, near the root of the tail, R in the figure, where the experimental slope may be changing less abruptly than that of the theoretical curve. This rounding-off feature has been observed in other comparisons of computed with experimental results (see Le Claire and Rabinovitch, 1982) and may be due to the dislocations being not the ideal regularly arranged pipes, all normal to the surface, assumed in the calculations.

D. Total Amounts Absorbed or Desorbed during Diffusion under Case I Surface Conditions

The total amount W of diffusant that in time t diffuses into unit area of a perfect semi-infinite solid is well known to be (Crank, 1975)

$$W = 2c_0(Dt/\pi)^{1/2} \tag{92}$$

More generally, if the solid contains initially a uniform concentration c_1, the relation

$$W = 2|c_0 - c_1|(Dt/\pi)^{1/2} \tag{93}$$

gives the amount diffusing into ($c_0 > c_1$) or out of ($c_0 < c_1$) unit area of the surface.

Measurements of W provide a convenient method for determining chemical diffusion coefficients for diffusants that can readily be supplied to or removed from a sample in some vapor form, such as the common gaseous elements and the metalloids. With these, the constant surface concentration is usually easily established. In certain cases the method can be used too for self-diffusion coefficient measurements when c_0 and c_1 will then be isotope compositions in otherwise chemically homogeneous samples; it is then known as the isotope exchange method and has been used, in particular, to measure anion diffusion coefficients in several ionic materials (Barr et al., 1965; Dawson and Barr, 1967; Dawson 1968; Williams and Barr, 1973. Because of its high sensitivity, measurements can be extended down to relatively low temperatures where appreciable diffusion coefficient enhancements, thought to be due to dislocation diffusion, have been observed.

As shown in Section III.C.1, the above equations become modified when dislocations are present; Eq. (93), for example, is then

$$W = 2|c_0 - c_1|(Dt/\pi)^{1/2}(1 + \varepsilon^2 U) \tag{94}$$

For the general dislocation array, $U = U(\alpha, \Delta, \varepsilon/\alpha)$ and is given by Eq. (56); Eq. (59) gives the relation for $U(\alpha, \Delta, \varepsilon/\alpha < 1) = U(\alpha, \Delta)$ valid for small dislocation densities or short times ($\varepsilon/\alpha < 1$) and that is best calculated from the isolated dislocation solution.

We now consider the numerical values of U and their time dependence and how Eq. (94) can be applied to analyze measurements of dislocation-enhanced absorption and desorption. The same relations may formally describe absorption and desorption in any solid containing linear regions of very high diffusivity and so may find wider application.

1. *Limiting Behavior at Very Short and Very Long Times*

Elementary considerations lead to simple forms of U in the limits of very short ($\alpha \to \infty$) and very long ($\alpha \to 0$) times. For definiteness we consider just diffusion *into* a sample.

a. *Very Short Times (Large α).* In times so short that the diffusion length $(Dt)^{1/2}$ is but a small fraction of the dislocation radius, i.e., $\alpha > 10^2$ or so, there is relatively very little diffusion laterally out from dislocations; W is then nearly just the sum of two terms like Eq. (93), one for the regular crystal regions and one for the dislocation pipe regions of the surface, and may be written

$$W = 2(c_0 - c_1)\left[(1 - \varepsilon^2)\left(\frac{Dt}{\pi}\right)^{1/2} + \varepsilon^2\left(\frac{D't}{\pi}\right)^{1/2}\right]$$

$$= 2(c_0 - c_1)\left(\frac{Dt}{\pi}\right)^{1/2}[1 + \varepsilon^2(\Delta^{1/2} - 1)] \qquad (95)$$

It is implied of course in writing Eq. (95) that $(Dt)^{1/2}$ is also very much less than the dislocation spacing R, i.e., $\alpha \gg \varepsilon$. This inequality is always satisfied for the large α being considered and the usual dislocation densities. (More rigorously, this condition should be written $(Dt)^{1/2} \ll R - a$, or $\alpha > \varepsilon/(1 - \varepsilon)$, but this refinement will rarely be necessary.)

Comparing Eqs. (94) and (95) we have the result

$$U \to \Delta^{1/2} - 1 \quad \text{as} \quad \alpha \to \infty \qquad (96)$$

b. *Very Long Times (Small α).* After sufficiently long times, when $(Dt)^{1/2}$ has become many times the spacing between dislocations, the concentration variations due to the dislocations become smoothed out and diffusion proceeds, at least for not too large η, in a manner describable by a unique effective diffusion coefficient D_{eff}. In Section C.2 we found that for $\varepsilon/\alpha \gtrsim 10$, D_{eff} was for practical purposes idential with D_{Hart}, as expressed by Eq. (74). Writing this D_{eff} for D in Eq. (93) and equating the result to the W expressed in Eq. (94), we obtain

$$U \to \{[1 + \varepsilon^2(\Delta - 1)]^{1/2} - 1\}/\varepsilon^2 \quad \text{as} \quad \alpha \to 0 \qquad (97)$$

as the long-time limit expression for U. We expect this to be valid for all $\varepsilon/\alpha \gtrsim 10$. In this limit, W again increases in proportion to $t^{1/2}$.

For the smaller values of Δ and ε, such that also $\varepsilon^2(\Delta - 1) \ll 1$, Eq. (97) simplifies to the constant value, independent of ε^2,

$$U \to \tfrac{1}{2}(\Delta - 1) \qquad (98)$$

although this limit is unlikely to be of wide interest because the dislocation enhancement is small when it applies.

2. Computed Values of U[5]

Parts a, b, c, and d of Fig. 24 show how $\varepsilon^2 U$ in Eq. (94) varies with time, through α, for various values of $\Delta - 1$ and over ranges of ε^2 of practical interest; no values of $\varepsilon^2 U < 10^{-2}$ are shown, for these are unlikely to be of experimental interest, and this determines the lowest ε^2 considered.

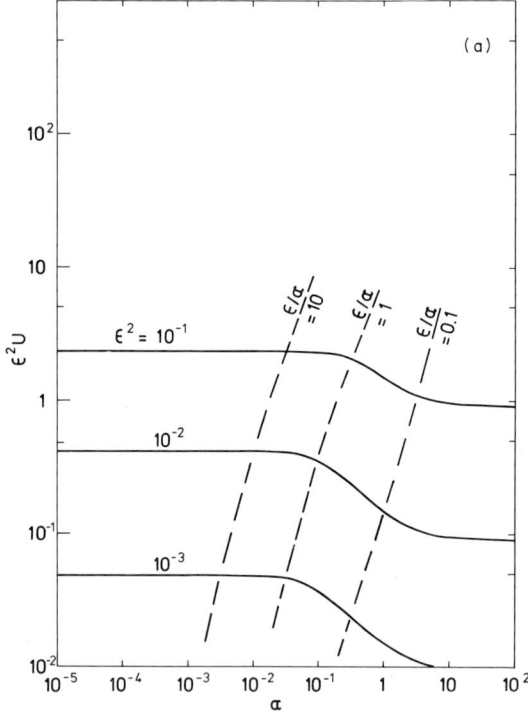

Fig. 24. Computed values of $\varepsilon^2 U$ [Eqs. (56) or (59)] plotted versus α for various ε^2: (a) $\Delta - 1 = 10^2$, (b) $\Delta - 1 = 10^4$, (c) $\Delta - 1 = 10^6$, (d) $\Delta - 1 = 10^8$.

[5] Le Claire and Rabinovitch, 1984.

For short times (α large) $\varepsilon^2 U$ has effectively the asymptotic value, calculated from Eq. (96), for all $\alpha \gtrsim 10^2$.

In the range $\sim 10 < \alpha < \sim 1$, $\varepsilon^2 U$ begins to increase quite rapidly with increasing time as material begins to diffuse appreciably out of the dislocations into surrounding crystal. The rise continues to about $\varepsilon/\alpha \approx 1$, when the rate of rise begins to diminish as dislocation diffusion zones begin to overlap. By the time that $\varepsilon/\alpha \gtrsim 10$, where $D_{\text{eff}} \doteq D_{\text{Hart}}$, $\varepsilon^2 U$ has become nearly constant again with time and accurately follows the values calculated from Eq. (97).

Equation (98) is illustrated by the approach of the lowest curve in each of the four parts of Fig. 24 to the value 5×10^{-2} at the lowest value of α.

While the total amount diffused [W, Eq. (94)] always increases linearly with $t^{1/2}$ at very short and at very long times, where U is constant, there is an intermediate time range where U is increasing with time and here W will increase more rapidly than as $t^{1/2}$, and especially so when $\varepsilon^2 U$ is appreciably greater than unity. We may roughly estimate from the figures that, at its

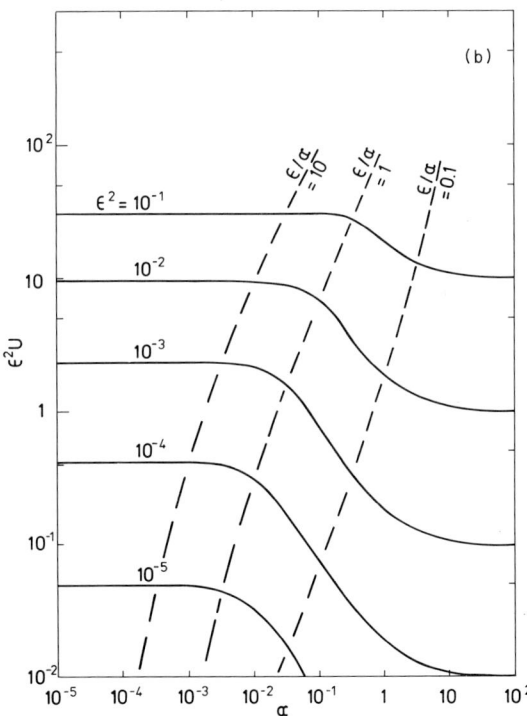

Fig. 24. (*Continued*)

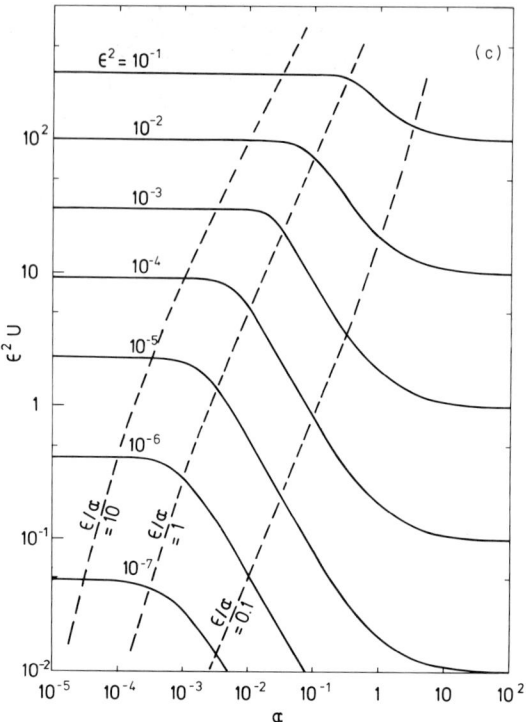

Fig. 24. (*Continued*)

maximum, $d \log U / d \log t^{1/2} \approx 0.85$; it follows from Eq. (94) that, with $\varepsilon^2 U \gg 1$, the maximum rate of change of W with time is as $t^{0.925}$.

No departures due to dislocations from a $W \sim t^{1/2}$ type relationship in an absorption, desorption or isotope exchange experiment seem ever to have been reported, probably because such experiments have never entailed high enough dislocation densities, or have not been conducted for times long enough to reach values of α small enough ($\lesssim 0.5$), for such departures to be easily observable. There has never, in fact, been much incentive for long anneals in this type of experiment because plots of W^2 versus t can usually be established to give slopes with sufficient precision for determining D_{eff}, the usual objective of the measurement, in times that are short relative to those necessary for other types of diffusion measurement. Whereas the latter entail αs of 10^{-2}–10^{-4} or so (see, e.g., Table II), in absorption or desorption measurements α may be as large as 1 or more and the steeply rising portions of the $\varepsilon^2 U$ versus α plots are not reached; parameters remain in the short-time domain.

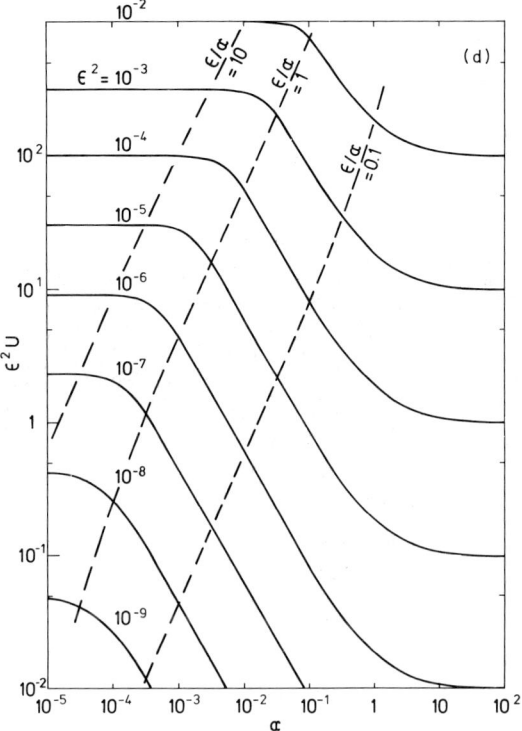

Fig. 24. (*Continued*)

Data therefore do not yet exist to provide illustration of the results obtained in this section over the full range of α, from $W \sim t^{1/2}$, through W varying more rapidly than $t^{1/2}$, to, at the longest times, where $W \sim t^{1/2}$ again.

In the next section we shall indicate how the results of the calculations can be employed to discuss and interpret experimental data, taking as an illustrative example the careful measurements by Dawson (1968) and Dawson and Barr (1967) of anion diffusion in KBr single crystals.

3. *An Example: Anion Diffusion in KBr*

Dawson (1968) measured by the isotope exchange method over a wide range of temperature, 400–700°C, the self-diffusion coefficient of Br in single crystals of KBr. These were first irradiated to create ^{82}Br and then isothermally annealed in initially inactive Br_2 vapor. Measurements were made of the growth of activity in the vapor, which is equivalent to W. There was no evident departure from linearity noted in any of the plots of W^2 versus t

from the slopes of which D_{eff} were calculated. However, the Arrhenius plot of D_{eff} showed a distinct positive curvature consistent with there being an increasing relative enhancement of diffusion by dislocations with decreasing temperature. Dawson and Barr (1967) summarized the results as

$$D_{\text{eff}}(\text{cm}^2\,\text{s}^{-1}) = 3\cdot 10^4 \exp(-2.61eV/kT)$$
$$+ 3\cdot 10^{-3} \exp(-1.49eV/kT) \quad (99)$$

where the first term is the crystal D and the second represents the enhancement due to dislocations. At 460°C, we obtain $D_{\text{eff}} = 2.04\cdot 10^{-13}$ with $D = 3.39\cdot 10^{-14}$ cm² s⁻¹, so that from the relation $D_{\text{eff}} = D(1 + \varepsilon^2 U)^2$, which is evident from Eq. (94), we obtain $\varepsilon^2 U = 1.45$.

At this same temperature of 460°C one inactive crystal was annealed for 19.5 hr in active ^{82}Br vapor and then sectioned to determine the penetration plot of $\ln\langle c \rangle$ versus y. This showed a well-defined linear dislocation tail from which was deduced the value of $(\Delta - 1)a^2 = 3.03\cdot 10^{-6}$ cm².

Knowing $(\Delta - 1)a^2$ and D, and also the dislocation density d, a graphical analysis of experimental results may be effected as follows. We calculate for time t during an isotope exhange experiment the fixed quantities

$$(\Delta - 1)\alpha^2 = (\Delta - 1)a^2/Dt \quad (100)$$

and

$$\varepsilon/\alpha = (\pi d \cdot Dt)^{1/2} \quad (101)$$

Now for each $\Delta - 1$ there is a corresponding α which, from Eq. (101), defines in turn a value of ε. From these can be calculated, or read from curves like parts a, b, c and d of Fig. 24, the corresponding $\varepsilon^2 U$. From the graph of $\varepsilon^2 U$ versus $\Delta - 1$ one can determine the values of $\Delta - 1$, and then of a, that correspond with the experimental $\varepsilon^2 U$.

Unfortunately, the dislocation densities of the crystals used by Dawson in the isotope exchange experiments are not recorded. (They were probably quite variable, judging from the variations in D_{eff} from crystal to crystal at the same temperature so that the second term of Eq. (99) represents enhancement for some mean d.) Accordingly, we have calculated in the above way values of $\varepsilon^2 U$ at 460°C as a function of $\Delta - 1$ for several d and for $t = 50$ and 240 min, these being the times at which the first and last measurements of W were recorded in the isotope exchange experiment reported at this temperature. The results are shown in Fig. 25, on which is also drawn the experimental value of $\varepsilon^2 U = 1.45$ at 460°C.

It would seem that the experiments could be well accounted for if we assume $d = 10^7$ cm⁻²; this gives a mean $\Delta - 1 \approx 2\cdot 10^4$ with a corresponding $a \approx 1.23\cdot 10^{-5}$ cm, and these are all reasonable values. That for a

is much larger than the $5 \cdot 10^{-8}$ cm commonly assumed for metals, but larger values are expected anyway for ionic crystals because of charge effects (Yan et al., 1977). As mentioned in Section IV.C.4.a, Ho's measurements in NaCl also indicated similar values for a.

However, note how sensitive the deduced values of a, and of $\Delta - 1$, are to the value used for d. For example, increasing d by only a factor of 2 to $2 \cdot 10^{-7}$ cm^{-2} increases $\Delta - 1$ to $\approx 2 \cdot 10^6$ and reduces a to $\approx 1.2 \cdot 10^{-6}$ cm. A similar but rather greater sensitivity to d was encountered in Section IV.C.3.b in the values of a derived from dislocation tail slopes and intercepts.

With $a = 1.23 \cdot 10^{-5}$ cm, α ranges from 1.22 at 50 min to 0.56 at 240 min; for such α it is evident from Fig. 24 that $\varepsilon^2 U$ is just beginning to increase with time. It is for this reason too that the plots in Fig. 25 for different times at fixed d are separated. Thus, for the conditions of these isotope exchange experiments at 460°C, and if $d = 10^7$ cm^{-2} as assumed, W^2 should not vary exactly linearly with t. It is of interest to enquire how discernible such nonlinearity might be, for Dawson reported only linear behavior.

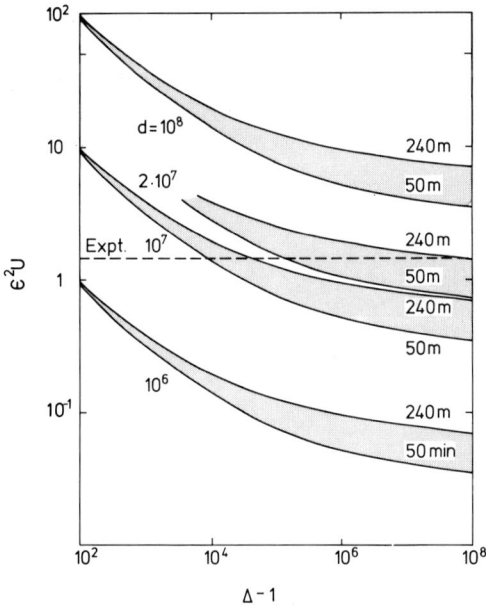

Fig. 25. $\varepsilon^2 U$ as a function of $\Delta - 1$ for various dislocation densities d, calculated for $(\Delta - 1)a^2 = 3.03 \times 10^{-6}$ cm^2, determined for Br diffusion in KBr at 460°C (Dawson, 1968). "Expt" is the mean experimental $\varepsilon^2 U$ from isotope-exchange measurements between 50 and 240 min at 460°C.

The time dependence of W^2 is wholly contained in the factor $t(1 + \varepsilon^2 U)^2$. Figure 26 shows this quantity, calculated at four times for $a = 1.23 \cdot 10^{-5}$ cm, $\Delta - 1 = 2 \cdot 10^4$ and $d = 10^7$ cm^{-2} and plotted against t. There is quite evidently a positive curvature over the range 50–240 min, as expected, but the extent to which the four calculated points depart from the best straight line through them is comparable with the deviation due to scatter of data in many experimental plots of W^2 versus t. Thus, plots for conditions such as these may often pass for linear, when they are not precisely so, and have a D_{eff} calculated from their then mean slope. Rather precise measurements might be necessary to distinguish a plot from linear, at least over a limited range of times.

There is some scatter in Dawson's 460°C measurements that could be just about enough to have concealed an intrinsic curvature of about the extent seen in Fig. 26. A calculated plot of less curvature would have resulted if d had been assumed to be less than 10^7 cm^{-2}, as can be seen from Fig. 25: the smaller d the less the separation of the 50 and 240 min curves at the experimental value of $\varepsilon^2 U$, as required for smaller curvature. Thus, less curvature and a surer compatibility with Dawson's measurements might be achieved with a d just a little smaller than that which has been assumed.

It will be clear that in experimental studies of dislocation effects on diffusion by methods such as isotope exchange, a full analysis of results requires that measurements of W should be supplemented always by sectioning experiments to reveal the dislocation tail for obtaining $(\Delta - 1)a^2$, and that these should be carried out on the same sample as used for the W

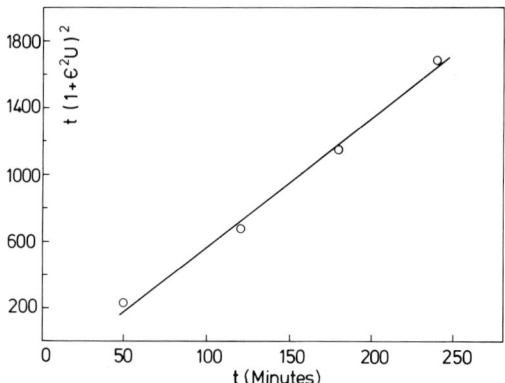

Fig. 26. A small departure from linearity in W^2 versus t plots. The circles represent four values at 460°C of the time dependent part of W^2, namely, $t(1 + \varepsilon^2 U)^2$, calculated for $d = 10^7$ cm^{-2}, $\Delta - 1 = 2 \times 10^4$, and $a = 1.23 \times 10^{-5}$ cm.

5. MATHEMATICAL ANALYSIS OF DIFFUSION IN DISLOCATIONS

measurements or an equivalent sample, so that a realistic d can at the same time be determined from the tail intercept (Eq. 81).

Appendix A. Derivation of Eq. (39)

The Laplace transform ($\mathscr{L}$) of $\bar{c}_2^{II}(r, \mu, t)$ is

$$\mathscr{L}_{(\lambda)}\{\bar{c}_2^{II}(r, \mu, t)\} = \bar{\bar{c}}_2^{II}(r, \mu, \lambda) = \int_0^\infty \bar{c}_2^{II}(r, \mu, t) e^{-\lambda t}\, dt$$

Using Eq. (24a), this can be written as

$$\bar{\bar{c}}_2^{II}(r, \mu, \lambda) = \int_0^\infty [\bar{c}_2^{II}(r, \mu, t) e^{\mu^2 Dt}] e^{-k^2 Dt}\, dt = \mathscr{L}_{(D,k^2)}\{\bar{c}_2^{II} e^{\mu^2 Dt}\} \quad \text{(A.1)}$$

i.e., $\bar{\bar{c}}_2^{II}(r, \mu, \lambda)$ is also the Laplace transform of $\bar{c}_2^{II} e^{\mu^2 Dt}$ with respect to Dk^2. Now, by Eq. (30),

$$\bar{\bar{c}}_2^{II} = \frac{1}{Dk^2} G$$

where

$$G = -\frac{\mu^2 \gamma (\Delta - 1) v'}{\psi}$$

which is to be regarded as the Laplace transform, with respect to Dk^2, of some function $g(t, \mu)$, namely

$$g(t, \mu) = \frac{1}{2\pi i} \int_{\tau - i\infty}^{\tau + i\infty} G e^{Dk^2 t}\, d(Dk^2) \quad \text{(A.2)}$$

Hence

$$\bar{\bar{c}}_2^{II} = \frac{1}{Dk^2} \mathscr{L}_{(Dk^2)}\{g\} \quad \text{(A.3)}$$

Using a theorem given, for example, in Carslaw and Jaeger (1959, p. 300), viz.,

$$\frac{1}{Dk^2} \mathscr{L}_{(Dk^2)}\{g(t, \mu)\} = \mathscr{L}_{(Dk^2)}\left\{\int_0^t g(t', \mu)\, dt'\right\} \quad \text{(A.4)}$$

we obtain from Eqs. A.1–A.4 that

$$\bar{c}_2^{II}(r, \mu, t) e^{\mu^2 Dt} = \int_0^t \left[\frac{1}{2\pi i} \int_{\tau - i\infty}^{\tau + i\infty} G e^{Dk^2 t}\, d(Dk^2)\right] dt$$

Changing the order of integration, and integrating over t leads to Eq. (39).

Appendix B. The Poles of Eq. (39) for the Case of the Dislocation Array

There are two possible sources for poles in the integrand of Eq. (39), namely, the terms Dk^2 and ψ [Eq. (33)] in its denominator. There is in fact no singularity at $k = 0$, because of the term $[\exp(-Dk^2t) - 1]$ in the numerator. As for ψ, its zeros had fortunately already been identified by Jaeger (1942); our ψ is a special case of one of several function considered by him. A little development of Eqs. (7)–(10) of his paper yields the result that the zeros of ψ are all simple and pure imaginary and so occur at points k_n in k-space,

$$k_n = \pm i u_n$$

the u_n being the solutions of

$$\psi(\pm i u_n, R) = 0$$

With z_n as the roots of Eq. (44), $u_n = z_n/(Dt)^{1/2}$.

Appendix C. Numerical Considerations in the Calculation of $Q(\eta, \varepsilon/\alpha)$

Numerical computation of $Q(\eta, \varepsilon/\alpha)$ from Eqs. (41) or (51) entails no particular problems and the only comments that need to be made are (i) to recall that the sum of residues becomes very slowly converging for small $\varepsilon/\alpha \lesssim 0.1$, with the consequences discussed in Section IV.A, and (ii) to indicate the means employed to identify approximate values of the zeros z_n for initiating the computer program that calculates them accurately. Sufficient indication of these can be achieved by considering the zeros of the limiting forms of Eq. (44).

Thus for $z\alpha \to 0$

$$\phi \to x^2\beta$$

$$\theta \to -\frac{2x^2\beta}{\pi}\left(h + \frac{2}{x^2\alpha\beta}\right) \tag{C.1}$$

where $h = -[\Gamma + \ln(z/2)]$, Γ being Euler's constant.

If also $z\alpha/\varepsilon \to 0$,

$$Y_1\left(\frac{z\alpha}{\varepsilon}\right) \to -\frac{2\varepsilon}{\pi z\alpha}$$

$$J_1\left(\frac{z\alpha}{\varepsilon}\right) \to \frac{z\alpha}{2\varepsilon} \tag{C.2}$$

(see, e. g., McLachlan, 1955).

The zeros of Eq. (44) then become

$$z_n^2 \approx \frac{2(\varepsilon/\alpha)^2}{(2/x^2\alpha\beta) + h} \tag{C.3}$$

For small enough values of $x^2\alpha\beta$ there is then a zero located at

$$z \approx (\varepsilon/\alpha)x(\alpha\beta)^{1/2} \tag{C.4}$$

At vanishingly small values of x, $\phi \to 0$ and the zeros of Eq. (44) become the zeros just of $J_1(z\alpha/\varepsilon)$, i.e.,

$$z\alpha/\varepsilon = 0, \quad 3.8317, \quad 7.0156, \quad \ldots$$

etc.

These results suggest that the zeros are separated by $\approx \pi\varepsilon/\alpha$ with the first zero being small, unless ε/α is very large.

Appendix D. Numerical Considerations in the Calculation of $Q(\eta)$

Numerical calculation of $Q^{I,II}(\eta)$ from Eqs. (57) and (58) has the problem that the integral over z is only logarithmically convergent. To overcome this, the z integral was calculated as the sum of three integrals, namely,

$$\int_0^\omega \frac{z^2\,dz}{(\theta^2 + \phi^2)z^3} + \int_0^\omega \frac{(1 - e^{-z^2} - z^2)}{(\theta^2 + \phi^2)z^3}\,dz + \int_\omega^\infty \frac{(1 - e^{-z^2})\,dz}{(\theta^2 + \phi^2)z^3}$$

where ω was chosen to be sufficiently small for asymptotic values ($z \to 0$) of θ and ϕ to be good approximations in the first integral while sufficiently large for there to be no problems from the lower limit in evaluating the third integral. For $\alpha\omega \ll 1$, the first integral can be evaluated analytically to give

$$\int_0^\omega \frac{dz}{(\theta^2 + \phi^2)z} = \frac{\pi}{2x^4\beta^2} \tan^{-1}\frac{(\pi/2)x^2\alpha\beta}{(h_1 x^2\alpha\beta + 2)}$$

where $h_1 = -(\Gamma + \ln\tfrac{1}{2}\alpha\omega)$, Γ being Euler's constant. The value of the second integral, following expansion of the exponential, is easily seen to be very small in comparison and in practice was usually negligible. Calculating the third integral now posed no difficulty. Actually, in many cases, with $\omega \approx 10^{-3}$, the first integral turned out to be the major contributor to the total z integral.

Appendix E. An Order of Magnitude Estimate of Δ

By definition, and assuming Arrhenius-type behavior,

$$\Delta = D'/D = (D'_0/D_0)\exp - (E' - E)/k_B T \qquad \text{(E.1)}$$

where D'_0 and D_0 are the preexponential factors for the dislocation and bulk diffusion constants respectively, E' and E are the appropriate activation energies, and k_B is Boltzmann's constant. From Balluffi's (1970) review of experimental measurements of D' for metals, E'/E can be roughly estimated at ≈ 0.6, and $D'_0/D_0 \approx 1$, if we assume $a = 5 \times 10^{-8}$ cm. Using the empirical relation $E = 34 T_F$ cal mol^{-1}, where T_F is the melting temperature in degrees K, we obtain

$$\Delta \approx \exp(6.8\, T_F/T) \qquad \text{(E.2)}$$

Table III gives a few of the order of magnitude estimates of Δ that Eq. (E.2) provides.

TABLE III

Order-of-Magnitude Estimates of Δ

T/T_F	1/4	1/3	1/2	2/3	1
Δ	6.5×10^{11}	7.2×10^8	8×10^5	2.7×10^4	9×10^2

References

Ahlborn, K. (1979). *J. Phys. (Paris) Colloq. C, Suppl.* 6, **40**, 185.
Backus, J. G. E. M., Bakker, H., and Mehrer, H. (1974). *Phys. Status Solidi B* **64**, 151.
Baker, C., Wuttig, M., and Birnbaum, H. K. (1968). *Conf. Suppl. Jpn. Inst. Metals* **9**, 268.
Bakker, H., and Backus, J. G. E. M. (1973). *Phys. Status Solidi A* **19**, 537.
Balluffi, R. W. (1970). *Phys. Status Solidi* **42**, 11.
Balluffi, R. W., Brokman, A., and King, A. H. (1982). *Acta Metal.* **30**, 1453.
Barr, L. W., and Lidiard, A. B. (1970). "Physical Chemistry–An Advanced Treatise," Vol. 10, Chap. X. Academic Press, New York.
Barr, L. W., Morrison, J. A., and Schroeder, P. A. (1965). *J. Appl. Phys.* **36**, 624.
Bihr, J., Mehrer, H., and Maier, K. (1978). *Phys. Status Solidi A* **50**, 171.
Brebec, G. (1965). Rapport CEA-R.2831, Commisariat a l'Energie Atomique, Saclay, France.
Carslaw, H. S., and Jaeger, J. C. (1959). "Conduction of Heat in Solids," 2d Ed. Oxford Univ. Press (Clarendon), London and New York.
Crank, J. (1975). "The Mathematics of Diffusion." Oxford Univ. Press (Clarendon), London and New York.
Dawson, D. K. (1968). Thesis, Univ. of Reading, United Kingdom.
Dawson, D. K., and Barr, L. W. (1967). *Proc. Br. Ceram. Soc.* No. 9, p. 171.

5. MATHEMATICAL ANALYSIS OF DIFFUSION IN DISLOCATIONS 317

Erdélyi, A., and Kermack, W. O. (1945). *Proc. Cambridge Philos. Soc.* **41**, 74.
Fredericks, W. J. (1975). "Diffusion in Solids–Recent Developments" (A. S. Nowick and J. J. Burton, eds.), Chap. 8. Academic Press, New York.
Fisher, J. C. (1951). *J. Appl. Phys.* **22**, 74.
Gainotti, A., and Zecchina, L. (1965). *Nuovo Cimento B* **40**, 295.
Gibbs, G. B., and Harris, J. E. (1969). Interfaces Conference, Australian Inst. Metals. Butterworths, Melbourne.
Gjostein, N. A. (1970). "Physicochemical Measurements in Metals Research" (R. A. Rapp and R. F. Bunshah, eds.), Vol. 4, Pt. 2, p. 405. Wiley, New York.
Gjostein, N. A. (1973). "Diffusion," Chap. 9, p. 241. Amer. Soc. Metals, Metals Park, Ohio.
Gupta, D., and Asai, K. W. (1973). *Phys. Rev. B* **7**, 586.
Harrison, L. G. (1961). *Trans. Faraday Soc.* **57**, 1191.
Hart, W. E. (1957). *Acta Metall.* **5**, 597.
Hendrickson, A. A., and Machlin, E. S. (1954). *J. Met.* **6**, 1035.
Ho, Y. K. (1982). Thesis. Imperial College, London.
Ho, Y. K., and Pratt, P. L. (1983). *Radiat. Eff.* **75**, 183.
Jaeger, J. C. (1942). *Proc. R. Soc. NSW* **75**, 130.
Lai, C. T., and Morrison, H. M. (1970). *Can. J. Phys.* **48**, 1548.
Lam Nghi, Q., Rothman, S. J., Mehrer, H., and Nowicki, L. J. (1973). *Phys. Status Solidi B* **57**, 225.
Le Claire, A. D. (1963). *Br. J. Appl. Phys.* **14**, 351.
Le Claire, A. D., and Rabinovitch, A. (1981). *J. Phys. C* **14**, 3863.
Le Claire, A. D., and Rabinovitch, A. (1982). *J. Phys. C* **15**, 3455 & 5727 (corrigendum to the Appendix).
Le Claire, A. D., and Rabinovitch, A. (1983). *J. Phys. C* **16**, 2087.
Le Claire, A. D., and Rabinovitch, A. (1984). *J. Phys. C* **17**, 991.
Luther, L. C. (1965). *J. Chem. Phys.* **43**, 2213.
Maier, K. (1977). *Phys. Status Solidi A* **44**, 567.
Maier, K., Mehrer, H., Lessman, E., and Schule, W. (1976). *Phys. Status Solidi B* **78**, 689.
McLachlan, N. W. (1955). "Bessel Functions for Engineers." Oxford Univ. Press (Clarendon), London and New York.
Mimkes, J. (1973). *Phys. Status Solidi B* **58**, K31.
Mimkes, J., and Wuttig, M. (1970). *Phys. Rev. B* **2**, 1619.
Morrison, H. M. (1965). *Philos. Mag.* **12**, 985.
Morrison, H. M., and Yuen, L. S. (1971). *Can. J. Phys.* **49**, 2704.
Murch, G. E. (1983). *Diffusion Defect Data* **32**, 1.
Pavlov, P. V., Panteleev, V. A., and Maiorov, A. V. (1964). *Sov. Phys. Solid State* **6**, 305. (There are errors in this paper: Each of the two equations given for series expansions of the integral (p. 307) need to be divided on the right-hand side by 4.)
Ruoff, A. K., and Balluffi, R. W. (1963). *J. Appl. Phys.* **34**, 1848.
Rupp, W., Ermert, V., and Sizmann, R. (1969). *Phys. Status Solidi* **33**, 509.
Sachs, M. (1963). "Solid State Theory." McGraw-Hill, New York.
Savitski, A. V. (1963). *Fiz. Met. Metalloved.* **16**, 886.
Smoluchowski, R. (1952). *Phys. Rev.* **87**, 482.
Springer, E. (1971). *Z. Metallkd.* **62**, 298.
Stark, J. P. (1965). *J. Appl. Phys.* **36**, 3938.
Stark, J. P., and Upthegrove, W. R. (1966). *Trans. Am. Soc. Met.* **59**, 479.
Suzuoka, T. (1961). *Trans. Jpn. Inst. Met.* **2**, 25.
Turnbull, D., and Hoffman, R. E. (1954). *Acta Metall.* **2**, 419
Volin, T. E., Lie, K. H., and Balluffi, R. W. (1971). *Acta Metall.* **19**, 263.

Whipple, R. T. P. (1954). *Philos. Mag.* **45,** 1225.
Whitton, J. L., and Kidson, G. V. (1968). *Can. J. Phys.* **46,** 2589.
Williams, S. R., and Barr, L. W. (1973). *J. Phys. (Paris) Colloq. C,* Suppl. 9, **34,** 173.
Wood Van, E., Glasser, M. L., and Austin, A. E. (1966). *J. Chem. Phys.* **45,** 1079.
Wuttig, M., and Birnbaum, H. K. (1966a). *Phys. Rev.* **147,** 495.
Wuttig, M., and Birnbaum, H. K. (1966b). *Phys. Status Solidi* **13,** K15.
Yan, M. F., Cannon, R. M., Bowen, H. K., and Coble, R. L. (1977). *J. Am. Ceram. Soc.* **60,** 120.

6

Grain Boundary Diffusion Mechanisms in Metals[*]

R. W. BALLUFFI

DEPARTMENT OF MATERIALS SCIENCE AND ENGINEERING
MASSACHUSETTS INSTITUTE OF TECHNOLOGY
CAMBRIDGE, MASSACHUSETTS

I.	Introduction	320
II.	The Diffusion Spectrum	321
III.	Present Knowledge of the Structure of Grain Boundaries and Their Line and Point Defects	322
	A. Perfect Grain Boundaries	322
	B. Line Defects (Dislocations) in Grain Boundaries	337
	C. Point Defects in Grain Boundaries	338
	D. Supporting Experimental Results	343
IV.	Review of Experiments Relevant to the Atom Jumping Mechanism in Boundaries	348
	A. Diffusional Mass Transport along Boundaries during Diffusional Creep and Sintering	348
	B. Diffusional Mass Transport along Boundaries during Electromigration and Thermomigration	351
	C. Diffusion-Induced Grain Boundary Migration	352
	D. Effect of Hydrostatic Pressure and Activation Volume for Boundary Diffusion	353
	E. Correlation Factor and Kinetic Energy of the Jumping Atom during Boundary Self-Diffusion	354
V.	Model for Atom Jumping in Boundaries	355
VI.	Influence of Boundary Structure on Boundary Diffusion	360
VII.	Diffusion along Migrating Boundaries	365
VIII.	Model for Grain Boundaries as Point Defect Sources and Sinks and Comparison with Experimental Observations	367
IX.	Conclusions	372
	References	374

[*] This chapter is based on the 1982 Institute of Metals Lecture. American Institute of Mining, Metallurgical, and Petroleum Engineers.

I. Introduction

During the past few decades it has been clearly demonstrated (Gjostein, 1973; Martin and Perraillon, 1975, 1980; Peterson, 1980, 1982) that substitutional atoms generally diffuse more rapidly along grain boundaries than through the crystal lattice at temperatures appreciably below the melting point. Also, such "fast" grain boundary diffusion can produce a net diffusional transport of atoms along boundaries (Ashby, 1972; Burton, 1977) and, in addition, boundaries can act as sources and sinks for diffusing atoms (Balluffi, 1980), and hence generate or destroy a net current of atoms.

It has also been realized that the mixing of atoms and the transport of mass that may be produced by these phenomena often play key roles in a host of important technologies, particularly those involving fine-grained materials at relatively low temperatures. For example, diffusional "short-circuiting" of foreign solute atoms along transverse grain boundaries in layered electronic devices can destroy the performance of the device (Poate *et al.*, 1978), while the stress-motivated diffusional transport of atoms along grain boundaries in polycrystalline materials can produce substantial diffusional creep (Ashby, 1972; Burton, 1977).

As a result of these developments, there has been a strong motivation to understand more about the basic mechanisms responsible for these grain boundary diffusional processes. The purpose of this chapter is to review our present knowledge of this subject. A complication that arises in attempting such a task is the fact that a wide range of ostensibly different types of grain boundaries exists. Particular categories which have been used to describe boundaries include, for example: *small-angle boundaries* (i.e., boundaries with misorientations $\lesssim 20°$, which consist of arrays of discrete and clearly recognizable lattice dislocations); *large-angle boundaries* (i.e., boundaries with misorientations $\gtrsim 20°$, where any lattice dislocation structure has been less evident); *special boundaries* (i.e., boundaries such as twins with particularly good lattice matching); *general boundaries* (i.e., more average boundaries of the type typically found in polycrystalline material); etc. We shall show that recent advances allow us to discuss, for the first time, the structures of all of these interfaces within the context of a common framework that clarifies a number of aspects of the subject. A goal of the present work is, therefore, to discuss the diffusional processes and their relationships to the boundary structure within this framework. In doing this we shall restrict ourselves to the diffusion of atoms that are substitutional in the lattice, since very little is known at present about the grain boundary diffusion of small interstitial impurity atoms.

Progress in understanding the basic aspects of fast grain boundary diffusion has been limited in the past because the narrow widths of grain boundary cores and the extremely small number of atoms involved made the observation and measurement of grain boundary kinetic processes difficult. However, the recent development of high-resolution microscopic techniques and computer simulation methods has led to important advances in our understanding of the structure of grain boundaries and their point and line defects and of kinetic processes in the cores of grain boundaries. In addition, several new and significant kinetic measurements and observations have been made, and the grain boundary diffusion data base has been expanded considerably. It therefore seems a propitious time to attempt a review of the subject. Recent related reviews may be found in the papers by Gjostein (1973), Martin and Perraillon (1975, 1980) and Peterson (1980, 1982).

II. The Diffusion Spectrum

The experimentally determined diffusion spectrum for fcc metals, as deduced by Gjostein (1973) several years ago, is shown in Fig. 1 in the form

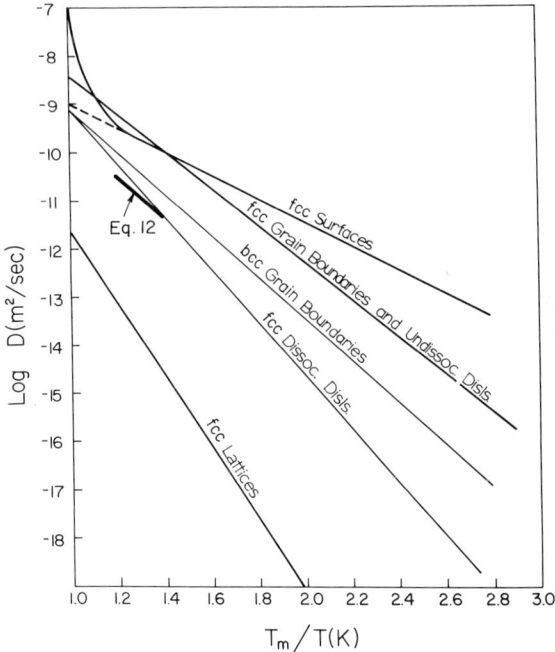

Fig. 1. Diffusivity spectrum for metals as deduced by Gjostein (1973). Curves represent averaged data, and a reduced reciprocal temperature scale is used in an effort to normalize the data.

of an Arrhenius-type plot employing a reduced reciprocal temperature scale T_M/T, where T_M is the melting temperature. The surface diffusion data represent average data for a variety of surfaces in a variety of metals. The grain boundary diffusion data represent average data for a wide variety of large-angle grain boundaries. The undissociated lattice dislocation diffusion data refer specifically to discrete edge dislocations in small angle tilt boundaries that are probably not significantly dissociated into partial dislocations and stacking fault ribbons, while the dissociated lattice dislocation data refer to dislocations that are dissociated in that way. It may be seen that at temperatures significantly lower than T_M

$$D_s > D_b \simeq D_d(\text{undiss}) > D_d(\text{diss}) > D_l \qquad (1)$$

where D_s, D_b, D_d, and D_l represent surface, grain boundary, lattice dislocation, and lattice diffusivities, respectively. These results are consistent with our intuitive feeling that the constraints placed upon atomic jumping should increase as we go from left to right in Eq. (1) and that the diffusivities should therefore decrease progressively. (For example, we might expect greater constraints in the cores of lattice dislocations that are relaxed by dissociation than in those which are not.) A particularly interesting result is the observation that $D_b \simeq D_d(\text{undiss})$, which indicates that the cores of nondissociated lattice dislocations are rather similar to the cores of large-angle boundaries with respect to diffusion processes. Also included in Fig. 1 is a curve that shows average diffusivity data for a number of grain boundaries in bcc metals.

III. Present Knowledge of the Structure of Grain Boundaries and Their Line and Point Defects

Any discussion of the detailed atomic mechanisms for the fast grain boundary diffusion documented in Section II must begin with a review of our present knowledge of the atomic structure of the grain boundaries themselves.

A. Perfect Grain Boundaries

A grain boundary is the region where two periodic structures (i.e., the two adjoining crystal lattices) meet, and there is now considerable reason to believe that many (and probably all) grain boundaries therefore possess ordered structures that may be profitably described in terms of a formal geometrical framework, or "grain boundary crystallography," developed by Bollmann (1970) and others. This crystallography can be made evident

6. GRAIN BOUNDARY DIFFUSION MECHANISMS IN METALS

by considering the steps involved in the construction of a boundary in a cubic material. The results may be easily generalized to include noncubic materials (Balluffi et al., 1982b). Begin by imagining crystals 1 and 2, which will adjoin the desired boundary, as rigid crystals made up of "black" and "white" atoms, respectively, that extend throughout all of space and coincide at an origin. Now rotate crystal 2 around the origin into the crystal misorientation desired for the grain boundary. At this point the two infinite, rigid, interpenetrating crystals form the so-called "dichromatic pattern" (Pond and Bollmann, 1979) and their lattice points coincide on a superlattice, i.e., the coincidence site lattice (Bollmann, 1970), or CSL, which is essentially the "beat pattern" produced in space by the two crystal lattices. An example is given in Fig. 2(a), corresponding to a 53.1° rotation around [001] of one fcc crystal with respect to another to form the $\Sigma = 5$ CSL. (Σ is the reciprocal of the fraction of atoms in coincidence.) Next, pass the desired plane of the boundary through the dichromatic pattern and discard all black atoms from one side of the plane and all white atoms from the other to produce a bicrystal. The result of this operation is illustrated in Fig. 2(b), where a bicrystal containing a symmetric [001] tilt boundary has been produced by discarding appropriate atoms from both sides of the plane PP'. Finally, allow the entire bicrystal system to relax to a final configuration of minimum energy. In this step atoms in the core will ajust their positions, and crystal 2 may translate with respect to crystal 1 by a rigid body translation t.

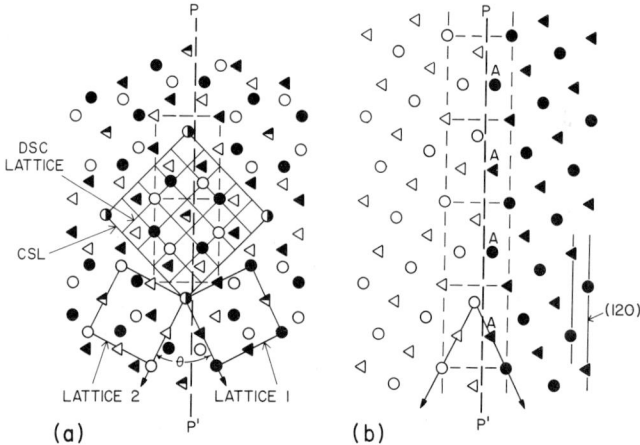

Fig. 2. (a) Dichromatic pattern, coincidence site lattice (CSL) and DSC lattice formed by rotating one fcc crystal with respect to another around [001] by $\theta = 53.1°$. Structures are viewed along [001]. Different symbols indicate atoms in the two crystal lattices that possess ABAB... stacking along [001]. (b) Rigid bicrystal containing grain boundary along PP' obtained from structure in (a) by discarding appropriate atoms from both sides of PP'.

By considering this method of boundary construction it is evident that there are generally nine variables (degrees of freedom) which must be specified in the description of a boundary in a material with a nonchiral structure (Kalonji, 1982). These include three variables to describe crystal misorientation, three variables to describe the grain boundary plane, and three variables to describe the rigid body translation t. Unfortunately, the existence of such a large number of variables greatly complicates the problem of dealing with grain boundary structure.

The relaxed atomic structure of the particular boundary in Fig. 2b is shown in Fig. 3a for the case of copper, as calculated by Sutton (1981) and Sutton and Vitek (1983) by the method of molecular statics. In this type of calculation it is assumed that the atoms interact via a pairwise central force potential, and the minimum energy configuration of the bicrystal ensemble is calculated by computer techniques (Gehlen et al., 1972; Arsenault et al., 1976; Lee, 1981). All of the energy is taken to be potential energy, i.e., no kinetic energy is present, and the calculation therefore refers to the situation at 0 K. A wide variety of interatomic potentials have been used in calculations of this type, including Morse, Born–Mayer, pseudo-, and empirical potentials. The method is clearly approximate, and its advantages and limitations with different forms of the pairwise potential have been discussed at length elsewhere (Gehlen et al., 1972; Arsenault et al., 1976; Lee, 1981). Despite these limitations it has been widely used to calculate the core structure of grain boundaries, and there is considerable experimental evidence, which is discussed below, that it is capable of producing reasonably realistic models of the atomistic core structures of grain boundaries in simple metals.

The core structure in Fig. 3a was calculated using an empirical pair potential fitted to the properties of copper. The atoms labeled by the As in Fig. 2b were removed in order to relieve crowding in the core, and it is seen that the periodicity of the relaxed structure is identical to the periodicity of the plane in the CSL (Fig. 2a) parallel to the boundary plane. The core structure also possesses a number of additional general features that are typical of all of the core structures calculated by molecular statics employing various pairwise potentials which have appeared in the literature. First, the core is relatively narrow, i.e., no more than a few atom distances in thickness. Second, the core structure is relatively dense, i.e., it is only slightly less dense than the perfect crystal lattice, and the average number of near neighbors surrounding each atom is only slightly lower than in the perfect crystal. This comes about because the atoms act almost as hard spheres at small separations due to the strong repulsive forces that set in when their ion cores impinge on one another. Some excess volume must then inevitably exist in the core, since it is impossible to pack such atoms into this region at the same high density as in the perfect crystal because of the constraints imposed

6. GRAIN BOUNDARY DIFFUSION MECHANISMS IN METALS

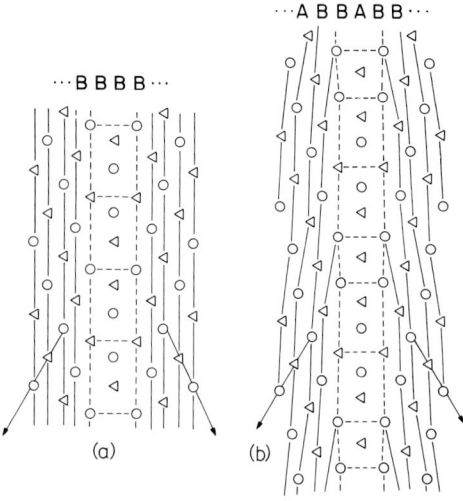

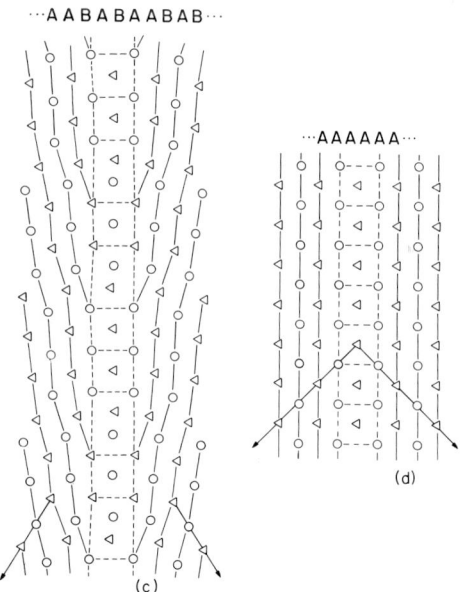

Fig. 3. Relaxed structures of series of [001] symmetric tilt boundaries in copper calculated by method of molecular statics by Sutton (1981) and Sutton and Vitek (1983): (a) $\Sigma = 5$ ($\theta = 53.1°$) boundary shown in Fig. 2b in rigid configuration; (b) $\Sigma = 17$ ($\theta = 61.9°$) boundary; (c) $\Sigma = 37$ ($\theta = 71.1°$) boundary; (d) $\Sigma = 1$ (90°) boundary.

by the adjoining crystal lattices. Also, a variety of atomic groupings and localized configurations exist in which some atoms are in loose environments and others are squeezed in one direction or another.

There is considerable evidence (Balluffi *et al.*, 1981a; Brokman and Balluffi, 1981) that boundaries of relatively short wavelength periodicity, which are parallel to relatively dense planes of the CSL (e.g., the boundary in Fig. 3a), are often of relatively low energy compared with more general boundaries with longer periodicities. When the crystal misorientation deviates from the special misorientation producing the dense CSL the boundary can then minimize its energy by preserving patches of the special low energy boundary and introducing an array of grain boundary dislocations (GBDs), which accommodates the difference in misorientation between the special misorientation and the actual boundary misorientation. The result is a structure consisting of an array of intrinsic GBDs embedded in a boundary structure of relatively low energy. In most cases the intrinsic GBDs are "perfect" dislocations in the sense that the boundary structure is the same on either side of them. Partial GBDs can appear in certain special cases (Pond, 1977), but these will not be discussed here. From a geometrical standpoint, the Burgers vectors of the perfect GBDs can be any vector translations that preserve the dichromatic pattern mentioned above. Bollmann (1970) first pointed out that all such vectors can be gathered together to form the so-called DSC lattice. A section of this lattice for the $\Sigma = 5$ boundary illustrated in Fig. 3a is shown in Fig. 2a. The primitive vectors of the DSC lattice are smaller than the primitive crystal lattice vectors (for $\Sigma > 1$) and tend to decrease in magnitude as the CSL becomes coarser. In fact, for the simple cubic structure, the DSC lattice is the reciprocal lattice of the CSL (Smith and Pond, 1976). In most cases the boundary selects relatively small vectors of the DSC lattice as Burgers vectors in order to decrease the elastic energy of the GBD array.

An example of such a boundary structure is shown in Fig. 3b. In this case the boundary is a [001] tilt boundary possessing a tilt angle that is larger by 8.8° than the misorientation of the $\Sigma = 5$ (i.e., low Σ) boundary illustrated in Fig. 3a. This structure was again calculated for copper by Sutton (1981) and Sutton and Vitek (1983) and is indeed seen to consist of an array of perfect edge GBDs separated by patches characteristic of the $\Sigma = 5$ boundary structure. Each GBD corresponds to the termination of two (120) planes and possesses a Burgers vector $\mathbf{b} = \frac{1}{5}[210]$ that is seen to be a small but nonprimitive vector of the $\Sigma = 5$ DSC lattice shown in Fig. 2a. It is noted that the core is again relatively narrow and dense. However, the lattice regions adjoining the core are now visibly strained due to the presence of the GBDs.

In additional work Sutton (1981) and Sutton and Vitek (1983) have used the molecular statics method to calculate the relaxed core structures of several extensive series of different types of tilt boundaries. These calculations have revealed further important aspects of the detailed core structure and provide further support for the applicability of the general boundary model described above. They find that the cores of certain relatively low Σ tilt boundaries in each of these series are actually comprised of a uniform array of a single type of "structural unit." (A structural unit is defined as a small group of atoms arranged in a characteristic configuration.) These boundaries, which are termed "favored" boundaries, are, therefore, extremely uniform in structure and consist of a contiguous sequence of primary (lattice) dislocations whose long-range elastic stress fields cancel almost exactly. The residual stress field is, therefore, highly localized at the core and relatively weak. All other boundaries, which possess misorientations that are intermediate between those of the favored boundaries, possess structures which may be described as mixtures of the units making up the two nearest favored boundaries. The structures of these intermediate (or "unfavored") boundaries are derived from a simple rule of mixing of the units of the nearby favored boundaries. Furthermore, the description in terms of mixtures of structural units corresponds directly to one in terms of GBDs. In the case of unfavored tilt boundaries, which are intermediate between two favored tilt boundaries consisting of all A- and B-type units, respectively, the structure of boundaries near the favored A-type boundary consists of B units embedded in an array of a larger number of A units. In such cases, each minority B unit is the core of a GBD having a Burgers vector belonging to the DSC lattice of the favored A-type boundary. For misorientations near that of the favored B-type boundary, the structure consists of A units embedded in an array of a larger number of B units, and a minority A unit acts as the core of a GBD having a Burgers vector belonging to the DSC lattice of the favored B-type boundary. At some intermediate misorientation the numbers of A and B units are equal. The cores of these GBDs are therefore quite narrow, i.e., localized, and, very importantly, they retain their identity at small spacings. These results therefore provide strong support for the idea that many (and probably all) boundary structures are ordered.

Selected examples of a few boundaries in such a series are shown in Fig. 3. The favored boundaries are the $\Sigma = 1$ ($\theta = 90°$) boundary (Fig. 3d) and the $\Sigma = 5$ ($\theta = 53.1°$) boundary (Fig. 3a), which has been discussed previously. The core of the $\Sigma = 1$ boundary consists of a continuous array of crystal lattice units (i.e., A units) and can be represented as ...AAAAA.... The $\Sigma = 5$ boundary consists of a continuous array of units of a different type (i.e., B units) and can be represented by ...BBBBB.... On the other hand, the $\Sigma = 17$ ($\theta = 61.9°$) and $\Sigma = 37$ ($\theta = 71.1°$) intermediate boundaries, illustrated in Figs. 3(b) and 3(c), consist of mixtures of these units. The $\Sigma = 17$

boundary is closer to the $\Sigma = 5$ boundary in misorientation and is seen to consist of A units embedded in an array of a larger number of B units, i.e., ...ABBABB..., while the $\Sigma = 37$ boundary is closer to the $\Sigma = 1$ boundary and consists of B units embedded in an array of a larger number of A units, i.e., ...AABABAABAB.... Furthermore, each minority A unit in the $\Sigma = 17$ boundary is the core of a $\Sigma = 5$ DSC lattice GBD as we have already noted, and each minority B unit in the $\Sigma = 37$ boundary is the core of a $\Sigma = 1$ DSC lattice GBD. In the latter case, two (110) planes terminate at each B unit and the GBD Burgers vectors is $\frac{1}{2}$[110], which is a lattice vector as expected, since the $\Sigma = 1$ DSC lattice is identical to the crystal lattice. The structures of various boundaries with misorientations between those of the favored $\Sigma = 5$ and $\Sigma = 1$ boundaries are shown as a function of θ in Fig. 4. Once the favored boundaries are chosen the structures of all intermediate boundaries are uniquely determined (Sutton, 1981; Sutton and Vitek, 1983; Sutton *et al.*, 1981). As may be seen, the structures of the $\Sigma = 13$ and $\Sigma = 41$ boundaries correspond to ...ABAB... and ...AAABAAAB..., respectively.

These results demonstrate for the first time the detailed variation of core structure as the boundary misorientation changes from the so-called small-misorientation angle to large-misorientation angle regimes. For example, the $\Sigma = 41$ boundary with $\theta = 77.3°$ possesses the structure ...AAABAAAB... and therefore consists of a stack of well-separated lattice edge dislocations characteristic of a small-misorientation angle boundary. Further decreases

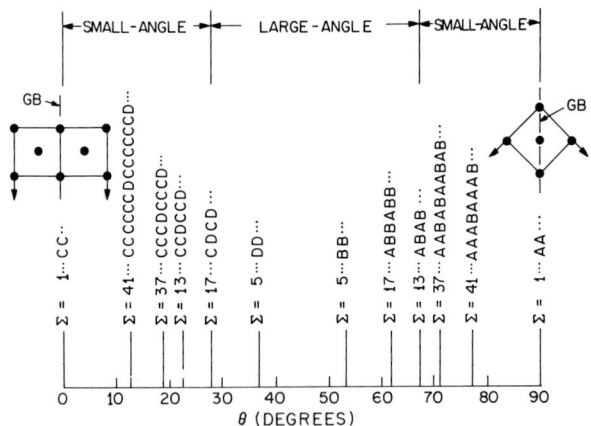

Fig. 4. Core structures of [001] symmetric tilt boundaries in copper in terms of structural units. Calculated by molecular statics by Sutton (1981), Sutton and Vitek (1982), and Wang *et al.* (1982).

in the misorientation angle would, of course, lead to greater separations of the B units, i.e., the lattice dislocations. On the other hand, as the misorientation is increased the spacing decreases until the $\Sigma = 13$ ($\theta = 67.4°$) boundary is reached, where the structure is ...ABABABA.... This misorientation may be somewhat arbitrarily regarded as the small-misorientation angle boundary limit, since it marks the point at which a lattice dislocation description of the boundary is no longer the preferred description.

The origin of the structural units in the relaxed boundaries may be visualized in all cases by examining the rigid dichromatic pattern for the corresponding boundary. For example, as seen in Fig. 2, the structural units of the relaxed $\Sigma = 5$ boundary shown in Fig. 3 can be delineated in incipient and distorted form (shown dashed) in the dichromatic pattern for that boundary. The final forms that the units assume in the relaxed boundary cores depend upon the choices of atoms removed from the dichromatic pattern, and the displacements that occur when the bicrystals are allowed to relax to their equilibrium configurations. Additional illustrations of this for the present [001] tilt boundaries have been given elsewhere by Balluffi et al. (1982).

Sutton (1981) has also pointed out that it is likely that the same general structural unit model applies to twist boundaries. For example, the relaxed structures of the $\Sigma = 5$ ($\theta = 36.9°$) and $\Sigma = 25$ ($\theta = 16.3°$) [001] twist boundaries in copper, as calculated by Bristowe and Crocker (1978), are shown in Fig. 5. It may be seen, as first pointed out by Sutton (1981), that the $\Sigma = 5$ boundary can be described in terms of all one type of structural

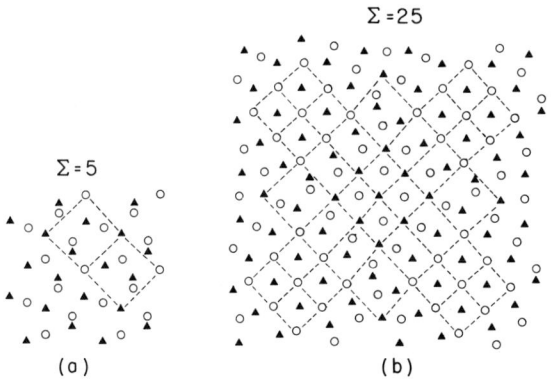

Fig. 5. Relaxed core structures of [001] twist boundaries in copper calculated by Bristowe and Crocker (1978) using molecular statics. Shown are the first 002 planes of crystals 1 and 2 adjoining the boundary viewed along twist axis. Structural units are shown as dashed lines. (a) $\Sigma = 5$ ($\theta = 36.9°$) boundary; (b) $\Sigma = 25$ ($\theta = 16.3°$) boundary.

unit. On the other hand, the $\Sigma = 25$ boundary can be described in terms of a mixture of (1) the $\Sigma = 5$ boundary units, (2) what are clearly $\Sigma = 1$ boundary units, and (3) a third type of unit similar to that appearing in the $\Sigma = 5$ tilt boundary (Fig. 3a). These same units are also clearly discernible in incipient and distorted form in the rigid dichromatic patterns for these same boundaries illustrated in Fig. 6. These results suggest that the $\Sigma = 5$ and $\Sigma = 1$ twist boundaries may be favored and that intermediate boundaries may contain their units. The $\Sigma = 25$ boundary, which may be classified as a small-angle boundary, is known to consist of a square grid of $\Sigma = 1$ screw dislocations in a pattern corresponding to the square grid of $\Sigma = 5$ units in Fig. 5b. As pointed out by Sutton (1981), this structure can therefore be interpreted on the following basis. The $\Sigma = 5$ units are the intersections of a crossed grid of $\Sigma = 1$ screw GBDs. The $\Sigma = 1$ units comprise patches of perfect lattice preserved by the $\Sigma = 1$ screw GBDs, and the units of the third

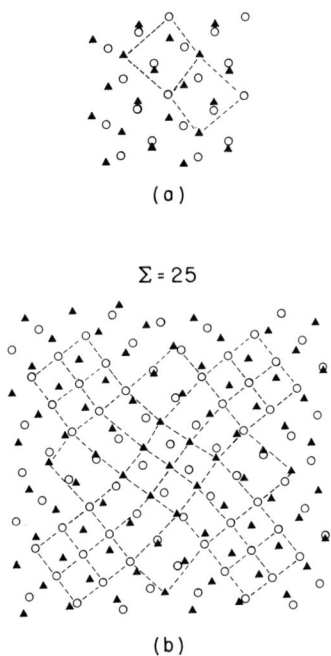

Fig. 6. Dichromatic patterns formed by juxtaposing two rigid (002) planes of the fcc crystal lattice at $\Sigma = 5$ and $\Sigma = 25$ misorientations. The plane of filled triangles is located $a/2$ below the plane of open circles. [From Balluffi *et al.* (1982).]

type are the cores of the $\Sigma = 1$ screw GBDs. Work is currently in progress to determine the stress fields near [001] twist boundaries in order to ascertain whether the structural units described above (Fig. 5) can be used to give a physically correct description of a wide range of [001] twist boundaries.

As already pointed out, the structure units can be delineated in incipient and distorted forms in the rigid dichromatic patterns. The subsequent relaxations then tend to improve the uniformity of the favored boundary units. However, the relaxations may not be sufficiently strong to make all structural units of a given type essentially identical. In such cases the structural units in different locations in the relaxed boundary may be significantly distorted, and the idealized structural unit model is only approximated. Nevertheless, the model remains as the most viable model for representing the systematics of the way in which the core structure changes with the crystal misorientation. A discussion of these points has been given by Balluffi and Bristowe (1984).

The above results suggest that all boundaries may possess ordered structures consisting of ordered arrays of structural units. Presumably, a considerable number of favored boundaries exists in the multidimensional space required to describe the macroscopic parameters of all grain boundaries, and all other boundaries can be made up of suitable mixtures of units of these boundaries. The structural unit model is seen explicitly to be equivalent to the CSL secondary GBD model. In fact, the model and its relation to the CSL secondary GBD model was clearly described on purely geometrical grounds as early as 1968 by Bishop and Chalmers (1968). However, a demonstration of the physical applicability of this model in regimes far from dense CSL misorientations in tilt boundaries has been provided for the first time by the computer calculations described above. Further calculations of the applicability of the model to twist boundaries and also more general boundaries possessing both tilt and twist components are required at this point.

Finally, an additional phenomenon not considered until now may be expected at very small misorientations in a number of cases. As the spacing of the lattice dislocations making up the boundary becomes relatively large the lattice dislocations may dissociate into extended configurations involving partial dislocations and patches of stacking faults. In the limit of zero angle the configuration of each dislocation will, of course, approach that of an isolated lattice dislocation. Li and Chalmers (1963) have shown that the interactions in a wall tend to inhibit dissociation and that the degree of dissociation tends to increase as the dislocation spacing in the wall increases. In the particular case of gold it has been shown experimentally (Darby and Balluffi, 1977) that the lattice dislocations in small-angle [001] tilt boundaries of the present type dissociate on {111} planes by adopting the configuration illustrated in Fig. 7. Here, each lattice dislocation adopts a "hill and valley" structure on (111) planes, which enables it to dissociate into partial dislocations and patches of stacking fault. Of course, the extent of such a dissociation will depend sensitively on the stacking fault energy. For gold, extensive dissociations of this type have been observed at small misorientation angles, i.e., $\sim 3°$, and other aspects of the results suggest that dissociation may persist to at least some extent for misorientations as high as $\sim 10°$. In summary, it therefore appears that the transition from large-angle boundaries to small-angle boundaries to very small-angle boundaries may be quite complex. Boundaries with $\Sigma = 1$ are always favored boundaries, and discrete lattice dislocations will begin to appear in the cores of boundaries in the form of structural units when the misorientation leaves the large-angle regime and approaches the small-angle regime. When the lattice dislocations become widely separated in the very small-angle regime their configurations may then change due to dissociations of various types that depend upon the energies of the faults that are produced.

Unfortunately, no systematic atomistic calculations of the core structures of lattice dislocations in grain boundaries have been made as they make this transition. However, a number of calculations of the core structures of isolated lattice dislocations have been made using the molecular statics method. A number of calculated structures in which point defect

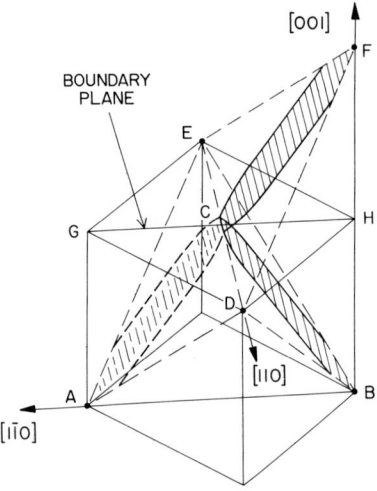

Fig. 7. Dissociation of lattice dislocation in small-angle [001] tilt boundary into partial dislocations and ribbons of stacking faults lying on {111} planes in fcc structure. Average direction of dislocation is along FB, i.e., the [001] tilt axis. [From Darby and Balluffi (1977).]

properties have also been studied are listed in Table I. The configurations vary widely and depend upon the Burgers vector, the dislocation line direction ξ, and the type of lattice. Again, as in the case of large-angle grain boundaries, the core structures consist of bad material in which the atomic density is slightly lower than in the lattice and the degree of coordination slightly lower. Also, some atoms are in loose environments and others are squeezed in one direction or another. Varying degrees of dissociation may occur depending upon the variables mentioned previously. An example (Miller et al., 1981) is shown in Fig. 8, which illustrates the core structure of an isolated edge dislocation in bcc iron with $\mathbf{b} = \frac{1}{2}[111]$ and $\xi = 1/\sqrt{6}\,[11\bar{2}]$. In this case three extra lattice planes are associated with the edge dislocation, as shown.

Only one rather unsatisfactory calculation has been made of the core structure of well-separated lattice dislocations arrayed in a small-angle grain boundary. Dahl et al. (1972) calculated the structure of a section of a 6° nonsymmetric [001] tilt boundary in fcc iron containing two lattice edge dislocations with $\mathbf{b} = \frac{1}{2}\langle 110 \rangle$ running along [001]. The core of only one dislocation was shown completely, and it appears to be nondissociated. However, since the computational model was small this result may have been determined by the border conditions imposed, and the result is therefore

TABLE I

CALCULATED MAXIMUM VACANCY AND/OR INTERSTITIAL BINDING ENERGIES TO CORES OF LATTICE DISLOCATIONS, AND FORMATION ENERGIES IN LATTICE

Metal	b	ξ	Binding energy (kJ mole^{-1})		Formation energy (kJ mole^{-1})		Reference
			$E_d^b(v)$ max	$E_d^b(i)$ max	$E_l^f(v)$	$E_l^f(i)$	
Mo	$\frac{1}{2}[111]$	$1/\sqrt{6}\,[11\bar{2}]$	−113	—	234	—	Miller et al. (1981)
Mo	$\frac{1}{2}[111]$	$1/\sqrt{2}\,[1\bar{1}0]$	−211	—	234	—	Miller et al. (1981)
Fe(bcc)	$\frac{1}{2}[111]$	$1/\sqrt{6}\,[11\bar{2}]$	−68	—	132	—	Miller et al. (1981)
Fe(bcc)	$\frac{1}{2}[111]$	$1/\sqrt{2}\,[1\bar{1}0]$	−118	—	132	—	Miller et al. (1981)
Fe(bcc)	[100]	[010]	−68	∼ −340	132	449	Bullough and Perrin (1968)
Cu	$\frac{1}{2}[1\bar{1}0]$	$1/\sqrt{6}\,[\bar{1}\bar{1}2]$	−24	−77	113	296	Schiffgens and Ashton (1974)
Cu	$\frac{1}{2}[1\bar{1}0]$	$1/\sqrt{2}\,[1\bar{1}0]$	−8^a −19^b	∼ −39	113	∼290	Doyoma and Cotterill (1968)
Cu	$\frac{1}{2}[1\bar{1}0]$	$1/\sqrt{6}\,[\bar{1}\bar{1}2]$	−9^a −19^b	∼ −39	113	∼290	Doyoma and Cotterill (1968)
Cu	$\frac{1}{2}[1\bar{1}0]$	$1/\sqrt{6}\,[\bar{1}\bar{1}2]$	−42	−112	∼97	—	Perrin et al. (1972)

^{a} Dislocation dissociated.
^{b} Dislocation nondissociated.

of questionable significance. We shall not attempt any further review of these discrete dislocation core structures in the present paper.

So far, we have only considered boundary structures at 0 K in the absence of thermal energy. Fortunately, a few calculations have been made of the core structures of large-angle boundaries at $T > 0$ by using the method of molecular dynamics (Gehlen et al., 1972; Arsenault et al., 1976; Lee, 1981). This method is similar to the molecular statics method, except that the atoms are allowed to vibrate and the bicrystal possesses kinetic as well as potential energy. The boundary structures calculated under these conditions have the same general features as those calculated by molecular statics at temperatures well below the melting point. However, phase changes may occur in the boundary in principle (Hart, 1972; Balluffi, 1979), and a very recent calculation (Cicotti et al., 1982) indicates that the boundary structure may approach that of the liquid at the melting temperature. However, little is known at present about boundary structure at temperatures near the melting point, and further work is required on that point. Interesting details about the thermal vibrations of the boundary atoms at moderate temperatures have emerged from the molecular dynamics calculations. For example, Hashimoto et al. (1980a, 1980b, 1981) have calculated the local phonon

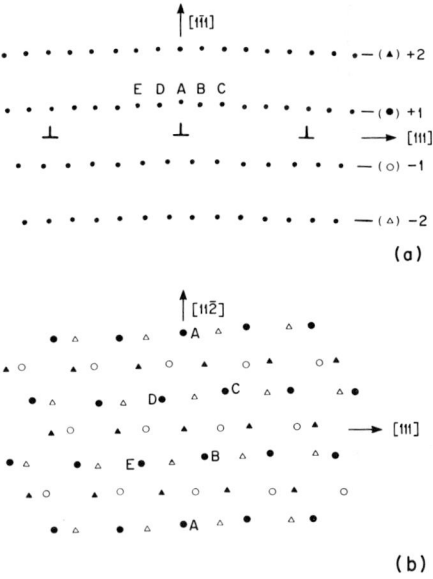

Fig. 8. Core structure of isolated edge dislocation in bcc iron with $\mathbf{b} = \frac{1}{2}[111]$ and $\xi = 1/\sqrt{6}[11\bar{2}]$, calculated by molecular statics by Miller et al. (1981). (a) Slip plane edge-on; (b) Slip plane in plane of paper.

6. GRAIN BOUNDARY DIFFUSION MECHANISMS IN METALS

density of states for different atoms in several boundaries, and the results for a $\Sigma = 11$ [110] symmetric tilt boundary in aluminum are shown in Figs. 9–11. The structure of the boundary is illustrated in Fig. 9. The core consists of a regular array of units (capped trigonal prisms), and a small excess volume is present. The phonon local density of states at each of the atoms indicated in Fig. 9 is shown in Fig. 10 along with the corresponding density of states for an atom in the perfect crystal lattice. It may be seen that the atoms in the core vibrate with quite different frequencies than those in the lattice. On average, there is a shift to lower frequencies consistent with the fact that the core atoms are generally less tightly bound. This is especially apparent for atom A, which exhibits a large low-frequency resonance peak at about half the maximum frequency. However, certain atoms are tightly squeezed in one or more directions, and this leads to relatively high-frequency vibrations. For example, for the B atom there is a high-frequency local mode component that is evidently due to the squeezing corresponding to the small distance between B and B'. In general, the spectra are seen to approach the lattice atom spectrum more closely as the distance from the core increases. In certain cases detailed information regarding the anisotropy of the vibrations has been obtained by analyzing the vibrational components in the x, y, and z directions. The results generally indicate considerable anisotropy, dependent upon the atomic environment. Also, the general shift towards lower frequencies is consistent with the result that the root mean square

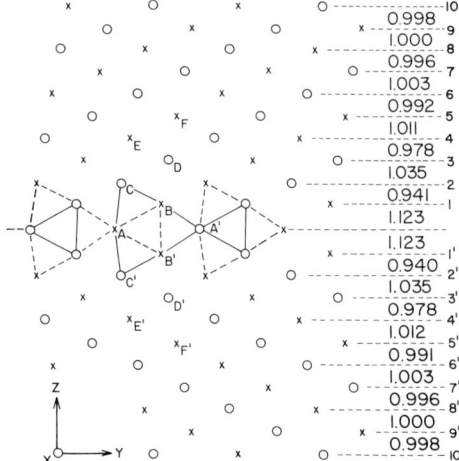

Fig. 9. Relaxed core structure of $\Sigma = 11$ [110] symmetric tilt boundary in aluminum. Numbers on right indicate interplanar spacings relative to the bulk value. Calculated by molecular dynamics by Hashimoto *et al.* (1981).

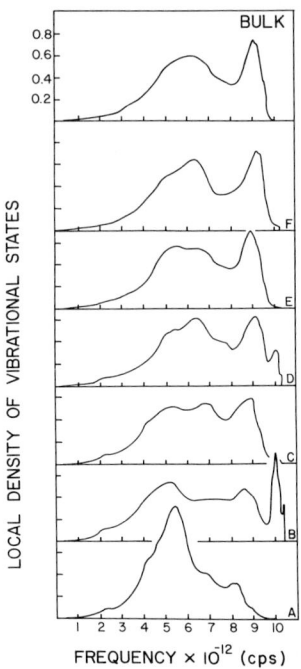

Fig. 10. Phonon local density of states at atoms A–F in core of boundary illustrated in Fig. 9. Corresponding values for atom in bulk at top. Calculated by molecular dynamics by Hashimoto et al. (1981).

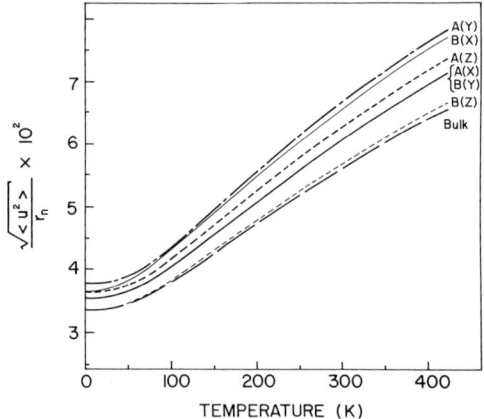

Fig. 11. Temperature dependence of the square root of the mean square displacements in x, y, and z directions of atoms A and B in the boundary illustrated in Fig. 9; r is the nearest-neighbor distance in bulk. Displacements of atom in bulk also shown. [From Hashimoto et al. (1981).]

6. GRAIN BOUNDARY DIFFUSION MECHANISMS IN METALS

vibrational displacements of the core atoms is larger in general than that of the lattice atoms. These results are shown in Fig. 11 for atoms A and B.

B. LINE DEFECTS (DISLOCATIONS) IN GRAIN BOUNDARIES

We have already seen that dislocations in the form of intrinsic GBDs are generally present as part of the equilibrium structure of grain boundaries. Additional GBDs, i.e., extrinsic GBDs, may also be present (Schober and Balluffi, 1971; Baluffi et al., 1972). These GBDs are extra dislocations that happen to be present in the boundary in more or less disorganized fashion as a result of the history of the specimen and that do not act in a systematic way to accommodate the crystal misorientation characteristic of the boundary as a whole. They may be produced, for example, by the impingement and dissociation of lattice dislocations (Darby et al., 1978) as illustrated in Fig. 12(a). The Burgers vectors of extrinsic perfect GBDs are again vectors of the DSC lattice. [However, in certain special cases, partial extrinsic GBDs may exist (Sun and Balluffi, 1982).] A schematic representation of the core structure of an extrinsic edge GBD in a $\Sigma = 5$ symmetric [001] tilt boundary is shown in Fig. 13 in the framework of the DSC lattice which serves as a particularly useful framework for revealing GBDs (Hirth and Balluffi, 1973). An important feature of many GBDs is revealed in Fig. 13, i.e., a step in the boundary plane at the GBD. Such a step appears whenever the translational shift in the dichromatic pattern produced by introducing the GBD has a nonzero component normal to the boundary plane (King, 1982; Balluffi et al., 1982). In general, this will be the case, and a step will be associated with a GBD. We note that no steps are associated with the GBDs in Figs. 2 and 3, and that these are therefore special cases.

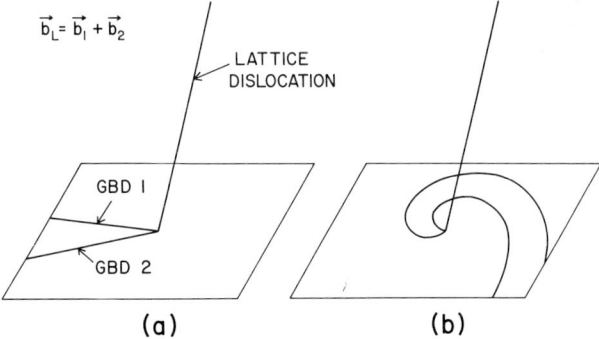

Fig. 12. (a) Dissociation of lattice dislocation into extrinsic GBDs in grain boundary. Burgers vector reaction shown near top. (b) Development of GBD spiral configuration by climb of GBDs in boundary plane.

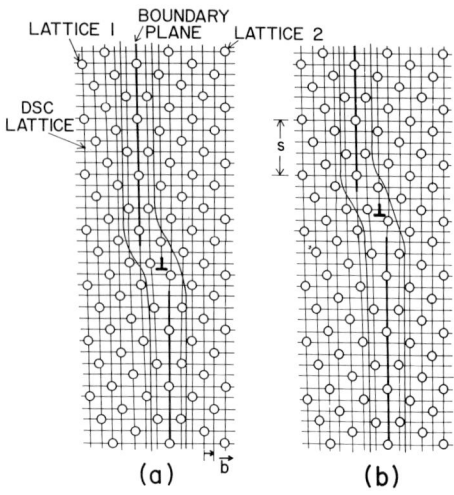

Fig. 13. (a) Schematic representation of core structure of extrinsic GBD in $\Sigma = 5$ symmetric [001] tilt boundary. GBD shown in framework of DSC lattice. Boundary step is present at the GBD due to a translation of the dichromatic pattern. (b) Same GBD after climb by distance s due to the annihilation of one vacancy (per lattice plane parallel to paper).

C. Point Defects in Grain Boundaries

Molecular statics calculations of the structures of vacancies and interstitials in the cores of a variety of large-angle grain boundaries have been carried out recently (Brokman et al., 1981; Balluffi et al., 1981b; Kwok et al., 1981). The perfect grain boundary structure was first calculated as described previously, and then an atom was either removed or added at a selected site in the core to form a vacancy or interstitial. The system was then relaxed in order to determine the equilibrium configuration of the defect. In general, the description of such point defects in grain boundaries is considerably more complicated than in the perfect lattice, since the less regular structure of the core makes the defect structure more difficult to specify. Also, the structure of the point defect varies with position in the core, and, in addition, an infinite number of different types of grain boundaries exists depending upon the type of metal, the crystal misorientation, the grain boundary plane orientation, etc. Despite these complications it has been found that vacancies and interstitials exist at various sites in boundaries as bona fide point defects in the form of missing or extra atoms respectively.

An example of a vacancy in a $\Sigma = 5$ [001] twist boundary in bcc iron is shown in Fig. 14. In this case the vacancy remains as a clearly recognizable vacant atomic site in the core structure. The relaxations around it are rela-

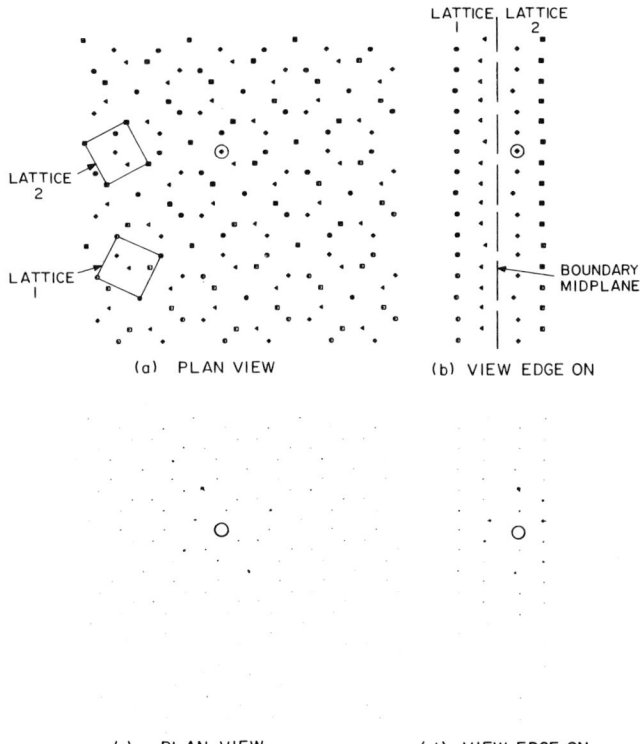

Fig. 14. Displacement field around a vacancy in $\Sigma = 5$ [001] twist boundary ($\theta = 36.9°$) in bcc iron calculated by molecular statics by Brokman et al. (1981). (a) and (b) Plan and edge-on views of relaxed boundary structure before insertion of vacancy at encircled atom. (c) and (d) Plan and edge-on views of atomic relaxation around the vacancy. Each atomic displacement is represented by a vector displacement projected on the plane of the paper.

tively small but are still considerably larger than those that would occur around a vacancy in the perfect lattice. A vacancy in a $\Sigma = 5$ [001] symmetric tilt boundary in copper is shown in Fig. 15. In this example the relaxation is considerably larger and the vacancy is split between atomic sites in the core. A wide variety of other configurations was found (Brokman et al., 1981a,b; Kwok et al., 1981) in other boundaries, including cases where the displacements were larger and spread over wider regions than those in Figs. 14 and 15. In general, the defect configuration produced by the relaxation depended on the variables mentioned above. In metals represented by "hard" interatomic potentials, e.g., tungsten, the vacancies tended to produce more widely spread displacements in the core, whereas for "softer" metals, e.g., copper, the displacements were more localized. However, in

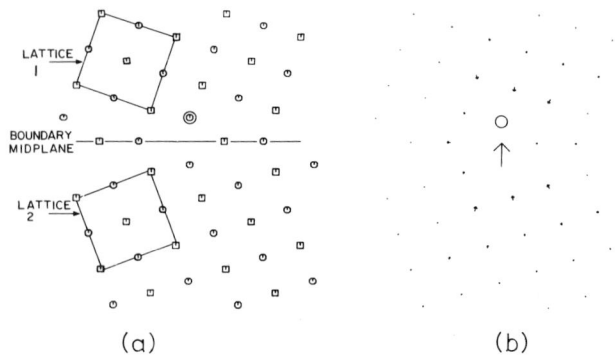

Fig. 15. Displacement field around a vacancy in $\Sigma = 5$ [001] tilt boundary ($\theta = 36.9°$) in fcc copper calculated by molecular statics by Brokman et al. (1981). (a) Edge-on view of relaxed boundary structure before insertion of vacancy at encircled atom. (b) Atomic relaxations around the vacancy. Each atomic relaxation displacement is represented by a vector displacement projected on the plane of the paper.

all cases the vacancy remained as a distinguishable point defect in the irregular core structure. The binding energy of the vacancy to the core was found to vary from site to site in a grain boundary and was generally attractive. An attractive binding energy equal to -58 kJ mole^{-1} was found (Brokman et al., 1981) for the vacancy in Fig. 14, and the binding energies for vacancies at the sites labeled A, B, C, and D in the $\Sigma = 5$ [001] tilt boundary in bcc iron shown in Fig. 16 are given in Table II. In other work, Ingle et al. (1976) have calculated the structures and energies of vacancies in various sites in

TABLE II

CALCULATED VACANCY AND INTERSTITIAL BINDING ENERGIES AT BOUNDARY SITES INDICATED IN FIG. 16 ALONG WITH CALCULATED FORMATION ENERGIES IN THE LATTICE[a]

Site	Binding energy (kJ mole^{-1})		Formation energy (kJ mole^{-1})	
	$E_b^b(v)$	$E_b^b(i)$	$E_l^f(v)$	$E_l^f(i)$
A (boundary)	-1.9	-211	—	—
B (boundary)	-40	-211	—	—
C (boundary)	-8.7	-139	—	—
D (boundary)	-17	-234	—	—
I (boundary)	—	-355	—	—
Lattice	—	—	130	458

[a] From Balluffi et al. (1981).

the $\Sigma = 3\langle 110\rangle\{112\}$ symmetric tilt boundary (coherent twin) in bcc iron and found attractive binding energies approaching 19 kJ mole^{-1}.

An example of an interstitial centered at the A site in the $\Sigma = 5$ [001] tilt boundary in bcc iron illustrated in Fig. 16 is shown in Fig. 17 (see caption for details). As might be expected, the introduction of an extra atom in this way produces large outward atomic displacements in its vicinity. The calculated binding energies of this interstitial and other interstitials located on the sites B, C, D, and I are listed in Table II. These binding energies are generally considerably larger than those for the vacancies, particularly at

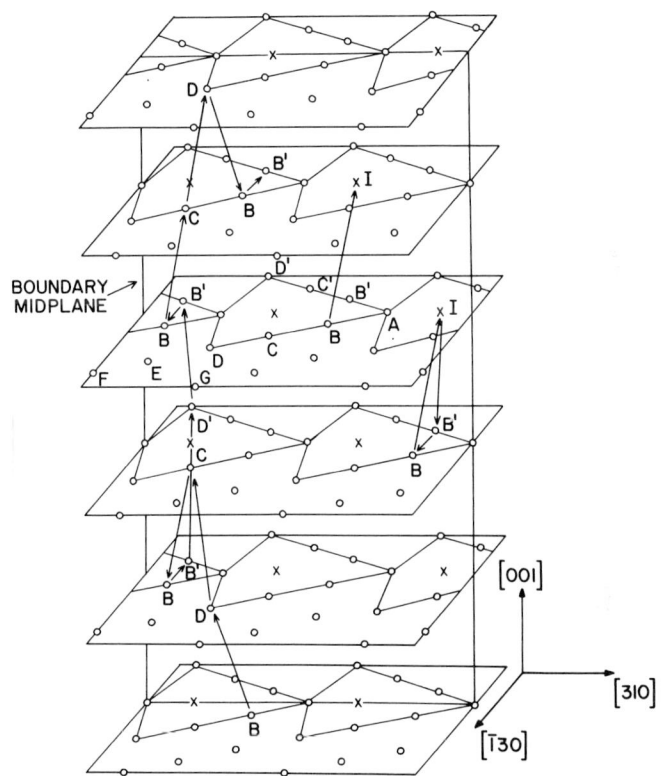

Fig. 16. Structure of $\Sigma = 5$ [001] tilt boundary ($\theta = 36.9°$) in bcc iron calculated by molecular statics and dynamics by Brokman et al. (1981) and Balluffi et al. (1981). In the typical vacancy jumping sequence shown at left, a vacancy inserted at site B preferentially jumps among the boundary sites A, B, C, and D rather than into the sites E, F, or G, which are further away from the boundary midplane. The arrow in the center shows an atom at B jumping into the interstitial site I, thus creating a boundary interstitial and a boundary vacancy. The sequence on the right shows the observed interchange of atoms at B and B′ via a ring mechanism involving the interstitial site at I. The ratio of the scale used in the drawing is $[\bar{1}30]:[310]:[001] = 1:1:5$.

site I, which is centered in the largest empty space (hole) in the core structure. Nevertheless, the interstitial formation energies are approximately twice as large (or larger) than those for vacancies with the exception of the very tightly bound interstitial on the I site.

The writer is not aware of any further relevant calculations for point defects in boundary cores. However, a number of similar calculations have been made for point defects in the cores of isolated lattice dislocations, and

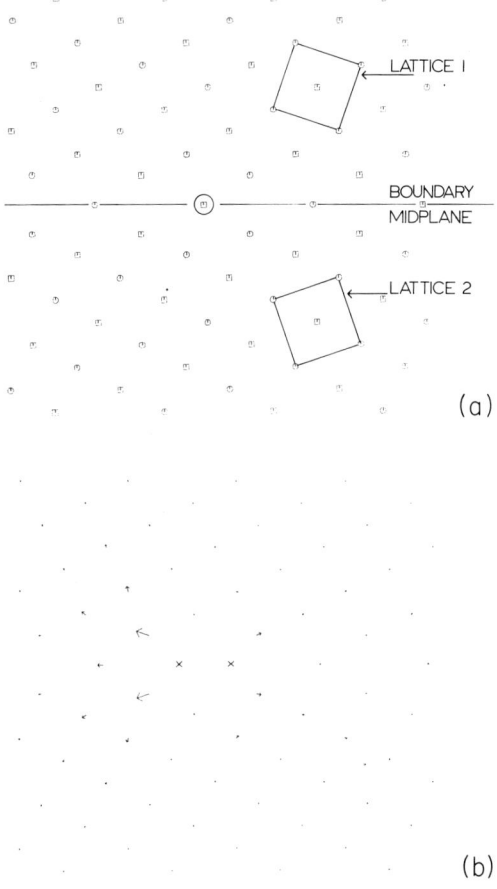

Fig. 17. Displacement field around an interstitial in $\Sigma = 5$ [001] tilt boundary ($\theta = 36.9°$) in bcc iron (see Fig. 16) calculated by molecular statics. (a) Edge-on view of relaxed boundary structure before insertion of interstitial at encircled site (which corresponds to the site labeled A in Fig. 16). (b) Atomic relaxations around the interstitial. Each atomic relaxation displacement is represented by a vector displacement projected on the plane of the paper. The interstitial was inserted by first slightly displacing the atom already occupying the A site.

some of the results are summarized in Table I. As expected, the point defect configurations and binding energies varied from site to site, and binding energies could be either attractive or repulsive. The maximum attractive binding energies are listed in Table I and are seen to vary widely. A finding of particular interest is the result that the maximum binding energies of vacancies to nondissociated dislocations is larger than to dissociated dislocations in copper. Again, the interstitial binding energies always exceed those of the vacancies. However, the minimum formation energies for the interstitials are approximately twice as large (or larger) than those for the vacancies.

D. Supporting Experimental Results

Essentially all of the detailed information about the structure of grain boundaries and their defects presented above was obtained from atomistic calculations using the method of molecular statics. We now point out that there is a considerable body of experimental work that is generally consistent with a number of the calculated results. However, in some cases detailed agreement is lacking, and it is clear that specific grain boundary structure cannot be calculated with assurance. Nevertheless, the experimental

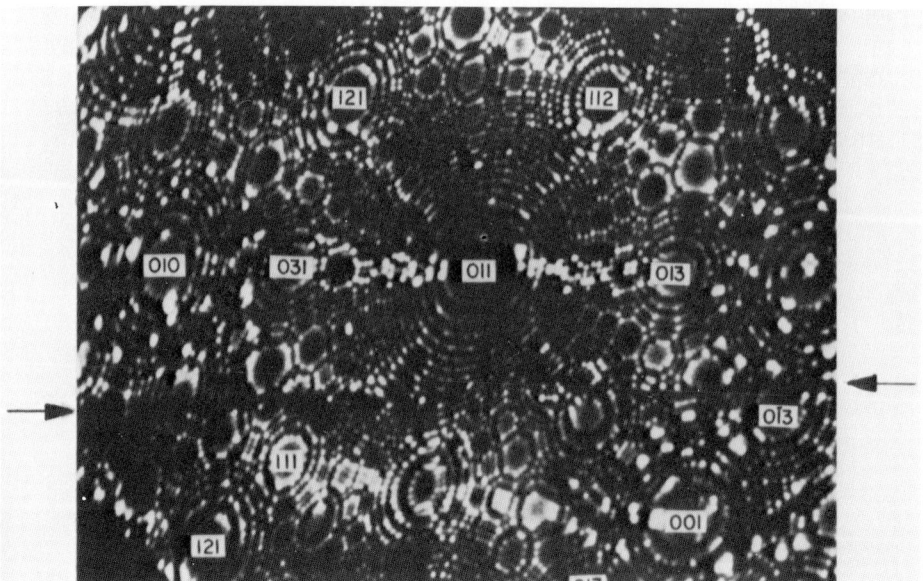

Fig. 18. Grain boundary (between arrows) observed in field ion microscope image of tungsten bicrystal. [From Müller and Tsong (1969).]

work lends credence to the general features of the calculated structures and the methods used to obtain them.

Direct observations of the general width of boundaries can be made by observing the intersection of boundaries with the free surfaces of specimens in the field ion microscope. The results (e.g., Fig. 18) indicate that the core is quite narrow, in agreement with the calculated results. Unfortunately, information about the detailed atomic arrangement in the core cannot be obtained by this technique because of the perturbing effects of the surface and electric field.

Recently, it has become possible to obtain direct images of the core structure of tilt boundaries in the electron microscope by the method of direct lattice imaging. An example (Hashimoto *et al.*, 1980a) is shown in Fig. 19 for a $\Sigma = 11$ [110] tilt boundary in gold along with the structure calculated by molecular statics. It may be seen that the main features of the calculated and observed structures correlate reasonably well.

The rigid body translation of lattice 1 with respect to lattice 2, i.e., the vector **t**, has been measured and also calculated for two boundary types in aluminum (Pond and Vitek, 1977). Satisfactory agreement between the results was obtained.

Further information about the detailed atomic arrangement in grain boundaries has been obtained using electron and x-ray diffraction techniques. It has been shown (Balluffi *et al.*, 1972; Sass and Bristowe, 1980) that the ordered structure of a grain boundary causes it to act essentially as a diffraction grating and that it therefore produces a unique diffraction pattern. In principle, it should therefore be possible to obtain information about the structure by analyzing the intensities of the scattered beams. In general, the

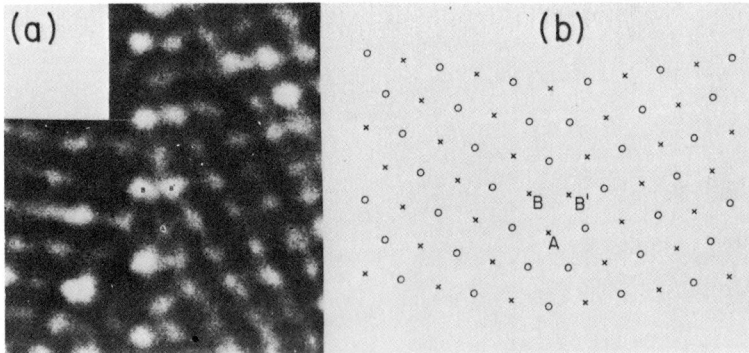

Fig. 19. (a) Electron microscope lattice image of $\Sigma = 11$ [110] symmetric tilt boundary in gold. (b) Structure of same boundary calculated by means of molecular statics. [From Hashimoto *et al.* (1980).]

diffracted intensity from the thin grain boundary region tends to appear in reciprocal space in regions near the lattice reflections and is concentrated on lattice points of the reciprocal CSL in the form of narrow relrods, which run perpendicular to the grain boundary plane and have lengths that are reciprocally related to the "thickness" of the grain boundary (Brokman and Balluffi, 1983). The appearance of this scattered intensity from a boundary in a plane in reciprocal space parallel to the boundary plane is shown in Fig. 20.

Experimental studies of the x-ray scattering from a number of boundaries have been compared (Sass and Bristowe, 1980) with the scattering predicted from the structures calculated by molecular statics. In almost all cases agreement has been found for the strongly scattered beams even though some discrepancies have been found for weaker reflections. The results indicate that the calculated structures in most cases reproduce the main features of the actual structures.

Extensive observations of the existence of intrinsic GBD networks have been made (Balluffi et al., 1972; Smith and Pond, 1976; Balluffi et al., 1981a) in boundaries possessing misorientations near low Σ misorientations. The networks have been revealed through electron microscopy by means of diffraction contrast, and the results in all cases appear to be consistent with the CSL model for grain boundaries and the types of structures generally predicted by the calculations of Sutton (1981) and Sutton and Vitek (1982). An example of a network of screw GBDs accommodating a twist misorientation from a $\Sigma = 13[001]$ twist misorientation in a boundary in gold is shown

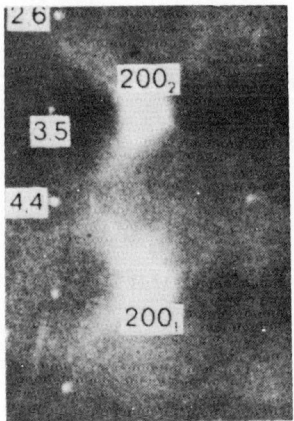

Fig. 20. X-ray diffraction pattern in 200 region of reciprocal space from core structure of $\Sigma = 13\,[001]$ twist boundary ($\theta = 22.6° \pm 0.1$) in gold. Film oriented parallel to grain boundary plane. Strong 200 lattice reflections near center. Relatively weak grain boundary reflections indexed 2,6; 3,5; etc. [From Sass and Bristowe (1980).]

in Fig. 21. So far, networks have only been detected in boundaries with misorientations in the near vicinity of low Σ misorientations. No intrinsic GBD structure has been detected in boundaries possessing large deviations in misorientation. This result is most likely due to the close spacings of the GBDs that are required in the latter cases. Extrinsic GBDs have also been observed (Schober and Balluffi, 1971; Balluffi et al., 1972; Darby et al., 1978) in many boundaries. GBDs of this type may also be seen in the boundary of Fig. 21 superimposed on the finely spaced intrinsic GBD network.

Correlations between calculated and measured values of the grain boundary energy should also be cited. The calculation of boundary energy has been more difficult than that of structure, since the main features of the calculated structures have proven to be less sensitive to the details of the method of calculation (e.g., choice of interatomic potential, cutoff in the potential, etc.) than the calculated energies. In many cases calculated energies have been unreliable to perhaps 50%. Nevertheless, in some cases apparently good agreement between calculated and experimental results has been obtained. Calculated and measured results (Hasson and Goux, 1971) for [100] and

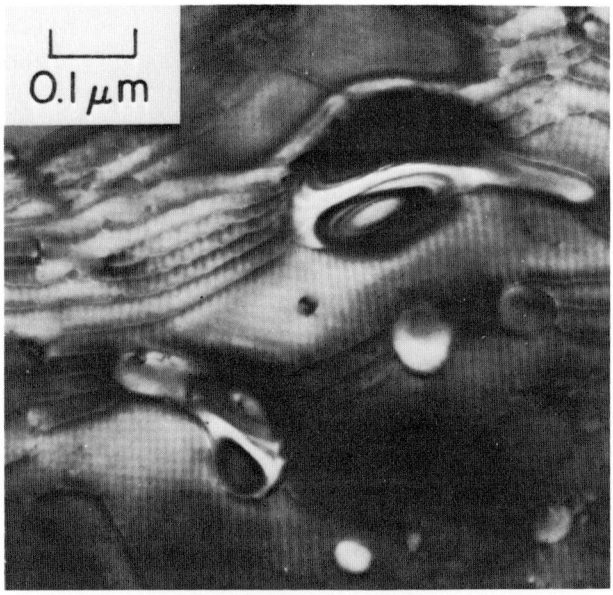

Fig. 21. Intrinsic and extrinsic GBDs in [001] twist boundary in gold possessing a small twist deviation from the exact $\Sigma = 13$ ($\theta = 22.6°$) misorientation. Intrinsic GBDs correspond to square network of finely spaced screw GBDs. Extrinsic GBDs correspond to irregular coarser lines superimposed upon the network.

6. GRAIN BOUNDARY DIFFUSION MECHANISMS IN METALS

[110] symmetric tilt boundary energies in aluminum are shown in Fig. 22. A direct comparison of these results cannot be made since the measured results are relative free energies measured near the melting point at 650°C while the calculated results are absolute enthalpies corresponding to 0 K. Estimated entropies (Hasson and Goux, 1971) are in the range 0.2–0.3 J m^{-2}K^{-1}, and, therefore, the calculated free energies of the high-angle boundaries would be near $\sim 600 - (923)0.25 = 369$ J m^{-2}K^{-1}. This value is reasonably consistent with the experimental value 340 J m^{-2}K^{-1} at 653 K reported by Bolling (1968). Furthermore, the general shapes of the calculated and measured energy versus misorientation curves for the [001] and [110] boundaries seem consistent. Unfortunately, nothing definitive

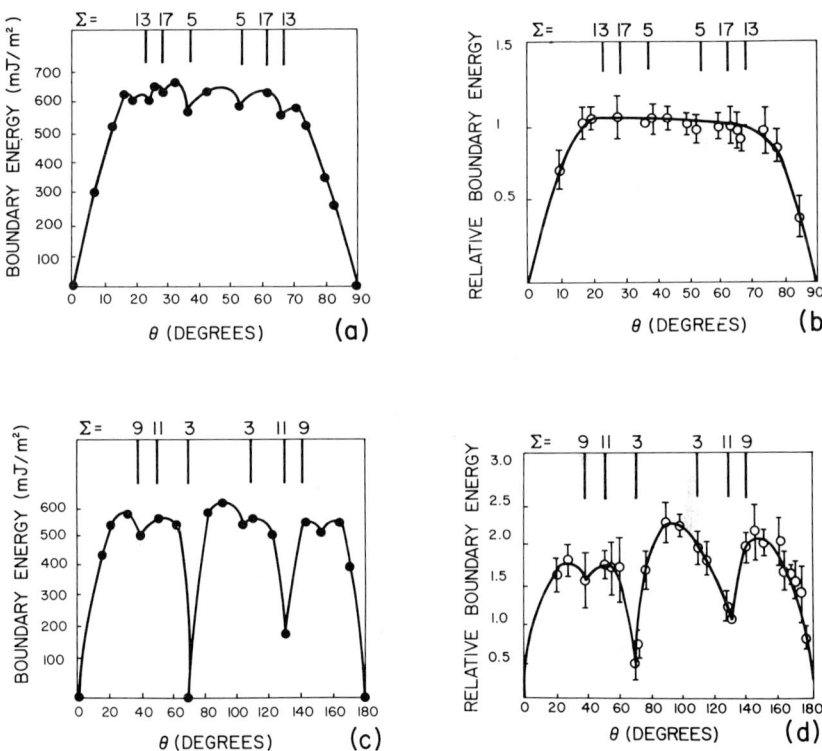

Fig. 22. Calculated and measured energies for symmetric tilt boundaries in aluminum. Calculated values (absolute enthalpies) obtained by method of molecular statics. Measured values (relative free energies) determined experimentally at 650°C. (a) and (b) Calculated and measured values, respectively, for [001] boundaries; (c) and (d) calculated and measured values, respectively, for [110] boundaries. [From Hasson and Goux (1971).]

can be concluded about the detailed structure of the cusps in view of the lack of reliable entropies for all the calculated boundaries and the appreciable uncertainty in the measured free energies.

Finally, we mention that it will be demonstrated below in Section V that calculated values of the grain boundary self-diffusion coefficient based on an atomistic model of the type just described seem to be in reasonable agreement with experiment.

IV. Review of Experiments Relevant to the Atom Jumping Mechanism in Boundaries

We now take up the problem of determining the mechanism of fast atom jumping in boundaries and begin with a review of a number of the relevant experimental results.

A. Diffusional Mass Transport along Boundaries during Diffusional Creep and Sintering

Numerous experiments (Burton and Greenwood, 1970; Ashby, 1972; Burton, 1977) have shown that stress-motivated diffusion currents are established along the grain boundaries of metal polycrystals subjected to applied stresses as illustrated in Fig. 23a. These diffusion currents are established between sources and sinks in the boundary network and cause the polycrystal to change shape in response to the applied stress. This mode of deformation can dominate all others under conditions of low temperature, small grain size, and low stress as shown in the deformation map for silver (Ashby, 1972) in Fig. 23b (see caption for details).

During the sintering of polycrystals containing pores, atoms tend to diffuse along the grain boundaries into the pores under the influence of capillarity (Alexander and Balluffi, 1957) as illustrated in Fig. 24a. The diffusion fluxes originate at sources in the boundaries, and the result is a net transport of atoms into the pores which shrinks the pores and causes the grains adjoining each boundary to move towards each other. This mechanism of sintering can dominate all others at low temperatures as shown in the sintering map for silver (Ashby, 1974) in Fig. 24b (see caption for details).

These experiments demonstrate clearly that grain boundary diffusion must occur by a defect mechanism which allows a net diffusional flux of atoms to occur. Furthermore, sources and sinks supporting the flux divergence must be present in the boundaries.

6. GRAIN BOUNDARY DIFFUSION MECHANISMS IN METALS

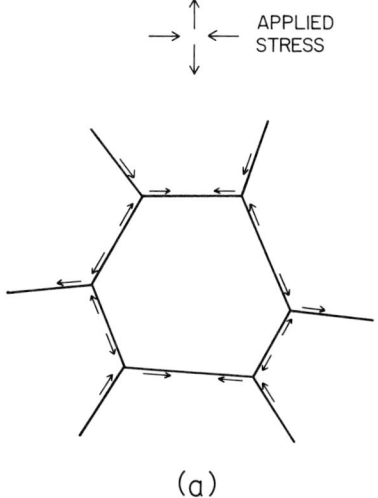

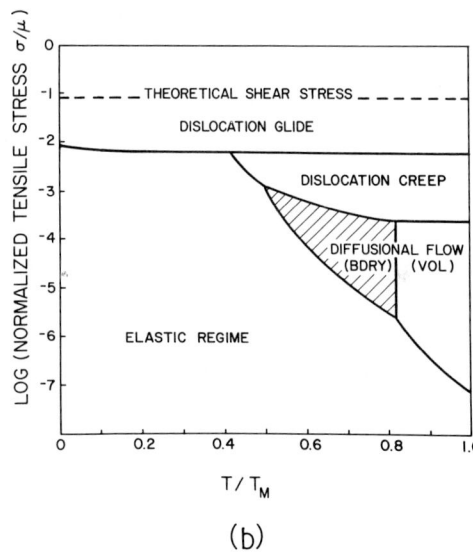

Fig. 23. (a) Schematic diagram showing stress-motivated grain boundary diffusional currents established during diffusional creep of metal polycrystal. (b) Deformation map for polycrystalline silver; grain size = 32 μm, strain rate = 10^{-8} s^{-1}. Mechanism corresponding to diffusional flow along boundaries dominates in hatched area of map. [From Ashby (1972).]

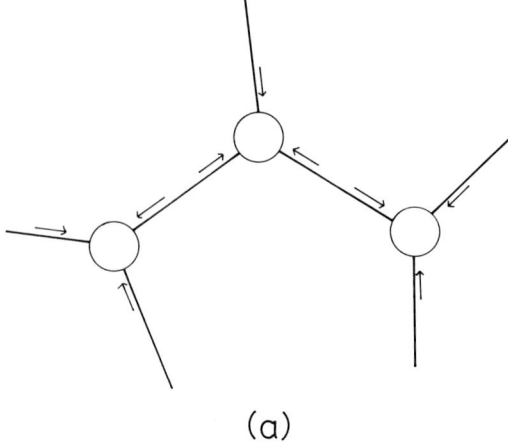

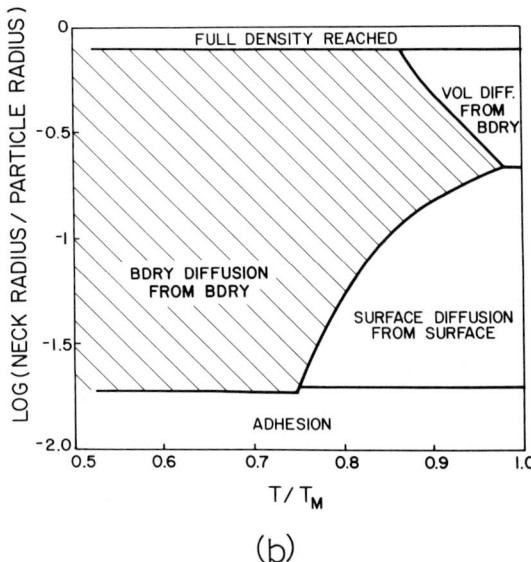

Fig. 24. (a) Schematic diagram showing capillarity-motivated grain boundary diffusional currents established during sintering of metal polycrystal containing voids. (b) Sintering map for aggregate of silver spheres of radius 10 μm. Mechanism corresponding to diffusional flow along grain boundaries dominates in hatched area of map. [From Ashby (1974).]

B. Diffusional Mass Transport along Boundaries during Electromigration and Thermomigration

Numerous electromigration experiments with single crystals and polycrystals (d'Heurle and Rosenberg, 1973) have shown that a net flux of atoms may be induced to diffuse along grain boundaries by the passage of an electrical current. In simple electron-conducting metals the atoms migrate towards the positive electrode, and clear evidence for the net transport of atoms is obtained from observations of voids in regions losing atoms and growths of varying types in regions gaining atoms (d'Heurle and Rosenberg, 1973). An example of the structural changes produced by electromigration in a polycrystalline thin film conducting "stripe" of gold (Blech and Kinsbron, 1975) is shown in Fig. 25. In this case the atoms migrated towards the positive electrode, and the main mechanism of transport was grain boundary diffusion.

A net flux of atoms may also be induced to diffuse along grain boundaries by the passage of a thermal current. For example, Johns and Blackburn (1975) have carried out thermomigration experiments with lead single crystals and polycrystals. Temperature gradients of $3-6 \times 10^4$ K m^{-1} were employed in the temperature range of 475–595 K. Mass fluxes from hot to cold were observed, and at the lower temperatures an increased flux was observed in the polycrystals, which may be attributed to a net flux of atoms occurring along the grain boundary network.

Both of these types of experiments offer further evidence for a defect mechanism for grain boundary diffusion.

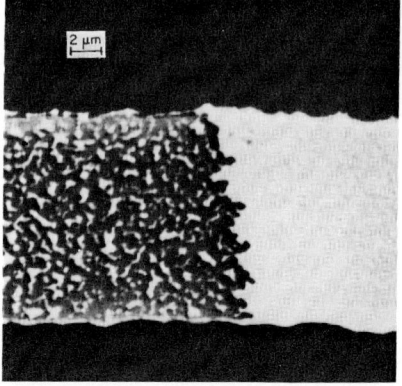 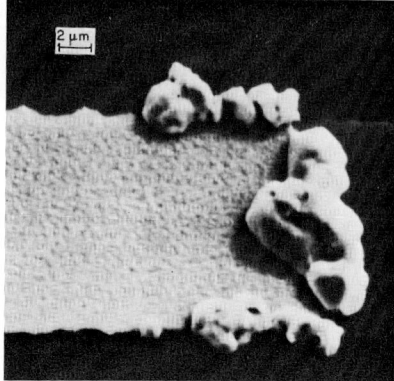

Fig. 25. Structural changes produced by electromigration in a polycrystalline, thin film, conducting "stripe" of gold. Cathode side at left, anode side at right. [From Blech and Kinsbron (1975).]

C. Diffusion-Induced Grain Boundary Migration

Evidence has been obtained recently (Balluffi and Cahn, 1981) that grain boundaries can be induced to migrate under certain conditions when solute atoms are diffused along them. This process, which has been termed diffusion-induced grain boundary migration (DIGM), is illustrated schematically in Fig. 26. If solute atoms are diffused into a region of a boundary from some source as in Fig. 26a under conditions where the lattice diffusion length is relatively short, the boundary is observed to migrate with a velocity v leaving behind an alloyed zone in its wake. Alternatively, if solute atoms are diffused out of a boundary to some sink as in Fig. 26b, the boundary is again observed to migrate leaving behind a zone with reduced alloy content (termed dealloyed) in its wake. The overall motivation for such an effect must be the change in chemical potential of the diffusing atoms when they engage in either solid solution mixing or unmixing as a result of the combined grain boundary diffusion and migration.

Experiments show that a net number of atoms is either added to or removed from the alloyed or dealloyed zone by grain boundary diffusion during DIGM. Associated with this effect is either an expansion or contraction of volume elements in this region in a direction perpendicular to the boundary plane. When zinc is diffused into massive copper specimens from the vapor phase along grain boundaries transverse to the surface to produce alloyed zones consisting of the α-phase by DIGM, it is observed (Cahn, 1980; Balluffi and Cahn, 1981) that whiskers and hillocks are formed on

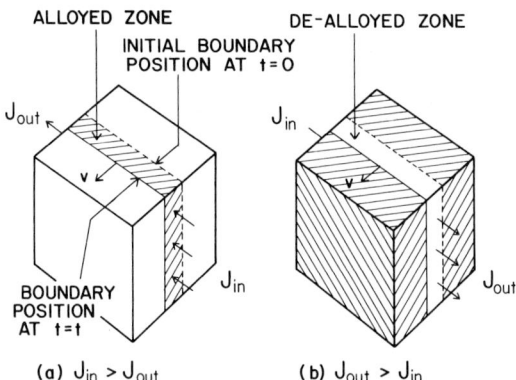

Fig. 26. Schematic diagram illustrating diffusion induced grain boundary migration (DIGM). (a) Solute atoms are deposited during inward grain boundary diffusion, and the boundary migrates while leaving behind an alloyed zone. (b) Solute atoms are removed during outward grain boundary diffusion, and the boundary migrates while leaving behind a dealloyed zone.

the surface. This result provides evidence for the existence of a compressive stress parallel to the surface due to the tendency of the specimen to expand in a direction perpendicular to the grain boundaries as a result of the addition of a net number of atoms to the specimen by boundary diffusion. (Here, only the surface region of the specimen is penetrated as a result of the grain boundary diffusion, and any transverse expansion of the alloyed regions is resisted by the nondiffused bulk of the specimen.) This result can be explained only if zinc atoms diffuse into the boundary more rapidly than copper atoms diffuse out. When the reverse experiment is carried out by diffusing zinc out of a massive α-brass specimen containing transverse boundaries into the vapor phase via DIGM, extensive porosity is observed (Cahn, 1980; Balluffi and Cahn, 1981) in the dealloyed zone. This result provides evidence for a net loss of atoms by unequal boundary diffusion. Again, we must conclude that the effect is due to the more rapid diffusion of zinc than copper in the grain boundary.

In other work, evidence has been found that the average apparent rate of DIGM is reduced when the restraints on expansions or contractions normal to the boundary plane are increased. In thin film solid–solid couples consisting of evaporated layers of copper and gold, and silver and gold, it is found (Pan and Balluffi, 1982) that the rate of DIGM of boundaries transverse to the plane of the thin film is reduced when any dimensional changes normal to the boundaries are restrained by leaving the thin film specimens attached to massive substrates.

These results indicate that the two substitutional components of a binary alloy diffuse at different rates in grain boundaries. This phenomenon leads to unequal fluxes of the two species and a net diffusional mass flow similar to that which occurs in the Kirkendall effect in bulk diffusion. This type of unequal diffusion requires a defect mechanism, and in the case of grain boundary diffusion the grain boundaries themselves provide the sources or sinks for the defect fluxes which support the process.

D. Effect of Hydrostatic Pressure and Activation Volume for Boundary Diffusion

Further information about grain boundary diffusion can be obtained by measuring the effects of hydrostatic pressure (Peterson, 1980, 1983). If grain boundary self-diffusion occurs by a defect mechanism, we can write, in general (Shewmon, 1963),

$$D_b = g_b f_b \tilde{\alpha}_b^2 \cdot \tilde{N}_b \cdot \tilde{\Gamma}_b \qquad (2)$$

where g_b is the geometrical factor, f_b the correlation factor, $\tilde{\alpha}_b^2$ the effective squared jump distance in the boundary, $\tilde{N}_b$ the effective equilibrium defect concentration, and $\tilde{\Gamma}_b$ the effective defect jump frequency. (We use "effective" quantities here, since the diffusion must occur by the jumping of atoms between different types of sites in the boundary core, and, therefore, suitably averaged effective quantities are required.)

Since

$$\tilde{\Gamma}_b = \tilde{\Gamma}_{bo} \exp(-\tilde{G}_b^m/kT) \tag{3}$$

and

$$\tilde{N}_b = \tilde{N}_{bo} \exp(-\tilde{G}_b^f/kT) \tag{4}$$

where $\tilde{G}_b^m$ and $\tilde{G}_b^f$ are the effective free energies of defect migration and formation respectively, and $\tilde{\Gamma}_{bo}$ and $\tilde{N}_{bo}$ are effective preexponential factors, we may write

$$D_b = g_b f_b \tilde{\alpha}^2 \tilde{N}_{bo} \tilde{\Gamma}_{bo} \exp[-(\tilde{G}_b^m + \tilde{G}_b^f)/kT] \tag{5}$$

Since, in general, $(\partial G/\partial P)_T = V$, we have

$$\tilde{V}_b = \tilde{V}_b^m + \tilde{V}_b^f = -kT\left(\frac{\partial \ln D_b}{\partial P}\right)_T + kT\left(\frac{\partial \ln[g_b f_b \tilde{\alpha}_b^2 \tilde{N}_{bo} \tilde{\Gamma}_{bo}]}{\partial P}\right)_T \tag{6}$$

where $\tilde{V}_b^m$ and $\tilde{V}_b^f$ are the effective activation volumes for defect migration and formation, respectively, and $\tilde{V}_b$ is the overall effective activation volume for the self-diffusion. Since the second term on the right side is only a few percent of the first (Peterson, 1980, 1983), the quantity $\tilde{V}_b$ can be determined from measurements of the pressure dependence of D_b. A measurement of this type has been performed for general large-angle grain boundaries in silver (Martin et al., 1967) with the result $\tilde{V}_b = 1.1 \pm 0.2\,\Omega$ (Ω is the atomic volume). For lattice self-diffusion, which occurs by a vacancy mechanism, $V_1 = 0.9\,\Omega$ (Bonanno and Tomizuka, 1965). Therefore, the activation volumes for lattice diffusion by a vacancy mechanism and grain boundary diffusion are found to be the same for this case within the estimated error. Peterson (1980, 1982) has suggested that this result may be taken as evidence that the boundary diffusion occurs by a vacancy mechanism involving vacancies with properties similar to those described in Section III.C.

E. Correlation Factor and Kinetic Energy of the Jumping Atom during Boundary Self-Diffusion

The product of the correlation factor f and the fraction of the total kinetic energy of the saddle point configuration associated with the jumping atom ΔK has been determined for lattice self-diffusion by use of the relation (Peterson, 1980, 1983).

$$f_1 \Delta K_1 = \frac{(D_1^\alpha/D_1^\beta) - 1}{(m_\beta/m_\alpha)^{1/2} - 1} \tag{7}$$

6. GRAIN BOUNDARY DIFFUSION MECHANISMS IN METALS

where D_l^α and D_l^β are the lattice diffusivities of two isotopes α and β of the same element with masses m_α and m_β. Robinson and Peterson (1972) have measured D_b^α/D_b^β for isotope self-diffusion along general grain boundaries in polycrystalline samples and also along [001] symmetric 16° tilt boundaries in silver. Assuming that Eq. (7) holds for grain boundary diffusion, they then calculate $f_b \Delta K_b = 0.46 \pm 0.03$ independent of temperature or boundary type. This latter procedure cannot be completely justified, however, since Eq. (7) may not hold exactly for grain boundary diffusion because the symmetry conditions required for its derivation for lattice diffusion are not fulfilled for boundary diffusion. Nevertheless, the results are suggestive and imply that most probably neither f_b nor ΔK_b can be smaller than ~ 0.46, since neither quantity can exceed unity. This has implications with respect to boundary diffusion mechanisms. The probable result that $f_b > \sim 0.46$ suggests that the grain boundary diffusion is relatively uncorrelated, i.e., it occurs in the core region by means of a network of alternate jump paths in two or three dimensions, which causes decorrelation of an atom and the point defect with which it has just exchanged by a successful jump. For example, the diffusion cannot occur along a single string of boundary sites, since this would lead to a much smaller value of f_b. The result that $\Delta K_b > \sim 0.46$ implies that a typical atomic jump does not involve the simultaneous motion of a large number of atoms amongst which the kinetic energy of the saddle point configuration is shared. Instead, it appears that the process is more localized and strongly involves only a few atoms.

V. Model for Atom Jumping in Boundaries

We concluded above that atom jumping in boundaries must occur by a defect mechanism. Also, we have seen that vacancies and interstitials can exist in boundaries as identifiable point defects with lower formation energies than in the lattice. However, the formation energies of interstitials in the boundary are generally considerably larger than those for vacancies except in a few special cases. It now remains to consider the kinetics of the exchanges between point defects and atoms in boundaries.

Only one fully dynamic set of calculations has appeared in the literature of diffusional atomic jump processes in a large-angle grain boundary. In this work (Balluffi et al., 1981b; Kwok et al., 1981) the molecular dynamics method was used to simulate vacancy migration in the [001] $\Sigma = 5$ tilt boundary in bcc iron discussed above and illustrated in Fig. 16. A number of isothermal "diffusion runs" were made in which a single vacancy was introduced in the boundary and observed to migrate as a function of time. The following results were obtained:

(i) The vacancy migrated in the core of the grain boundary by executing jumps between a variety of sites. Essentially, all jumps occurred between sites of the types labeled A, B, C, and D in Fig. 16. (Note that the A', B', C', and D' sites are equivalent sites because of the boundary symmetry.) The jumping was therefore confined almost entirely to the core of the grain boundary. A typical trajectory is shown on the left of Fig. 16. The number of times a vacancy jumped into a given type of site during a run at 1300 K involving 195 jumps is shown in Table III along with a calculated value of the vacancy binding energy at that site. The results indicate a clear correlation between the vacancy binding energy at different sites and the frequency with which a site received jumps; i.e., the larger the binding energy the more frequently it received jumps. The sites E, F, and G are at distances from the grain boundary midplane where the binding energy was relatively small, and they therefore received very few jumps.

(ii) The vacancy migration took place predominantly along the tilt axis rather than perpendicular to it (Fig. 16).

(iii) The effective vacancy jump frequency (due to all jumps) obeyed Eq. (3) in the form:

$$\tilde{\Gamma}_b(v) = \tilde{\gamma}_0(v) \exp(-\tilde{E}_b^m(v)/kT) \tag{8}$$

with $\tilde{\gamma}_0(v) = 4.85 \times 10^{13}$ s^{-1} and $\tilde{E}_b^m(v) = 49$ kJ mole^{-1}. Here, $\tilde{E}_b^m(v)$ is the effective vacancy migration energy and $\tilde{\gamma}_0(v)$ is the preexponential factor, which can be written as $\tilde{\gamma}_0(v) = \tilde{z}\tilde{v}_0 \exp(\tilde{S}_b^m(v)/k)$, where $\tilde{z}$ = effective coordination number, $\tilde{v}_0$ is an effective "attempt frequency," and $\tilde{S}_b^m(v)$ is the effective vacancy migration entropy. If $\tilde{z} = 8$, the quantity $\tilde{v}_0 \exp(\tilde{S}_b^m(v)/k) = 6.06 \times 10^{12}$ s^{-1} is about equal to the lattice Debye frequency (i.e., 7.1×10^{12} s^{-1}), as might be expected if effects due to the entropy of migration are not large as seems likely.

(iv) Atoms on B sites occasionally jumped into interstitial I sites by a process illustrated in the center of Fig. 16, which is essentially the thermally activated formation of a Frenkel pair. Inspection of Table II shows that this type of Frenkel pair must have a considerably lower formation energy than any other possible Frenkel pair, and this is evidently the reason for its occurrence. The vacancy formed in this way often diffused away leaving the interstitial behind at I. Furthermore, the interstitial at I remained completely immobile and could only be eliminated by mutual annihilation with a neighboring vacancy. The immobility of the interstitial at I is readily understood on the basis of the results in Table II where it is seen that the formation energies of the interstitial at the other interstitial sites are larger by at least 122 kJ mole^{-1}. The activation energy for the interstitial to migrate must therefore be larger than 122 kJ mole^{-1}, and this process was therefore

TABLE III

NUMBER OF VACANCY JUMPS INTO VARIOUS BOUNDARY SITES (FIG. 16) DURING COMPUTER SIMULATED VACANCY DIFFUSION RUN INVOLVING 195 JUMPS AT 1300 K.[a] BINDING ENERGY OF VACANCY AT EACH SITE IS ALSO GIVEN

	Site						
	A	B	C	D	E	F	G
Number of jumps	3	126	20	32	7	6	1
$E_b^b(v)$ (kJ mole^{-1})	-1.9	-40	-8.7	-17	—	—	—

[a] From Balluffi et al. (1981).

not observed. These results indicate therefore that any interstitial in this grain boundary will be strongly trapped at I sites and will therefore be rendered immobile and incapable of promoting self-diffusion.

(v) The interchange of atoms at B and B' by the process illustrated on the right side of Fig. 16 was observed occasionally. In this process an atom in site B jumped into site I in the adjacent boundary plane, followed by an atom in site B' jumping into the newly created vacancy in the site B. The process was then completed by the interstitial in the site I jumping into the vacancy in site B'. This sequence was found to occur more frequently as the temperature increased. The process does not contribute to diffusion because the two atoms involved remain trapped in the same pair configuration, thus producing no net matter transport relative to the other atoms.

We now derive an approximate expression for the boundary self-diffusion coefficient on the basis of a vacancy exchange model using the calculated vacancy parameters and show that it appears to be consistent with experimental data. Using Eqs. (2), (4), and (8) and the relation $\tilde{N}_{bo}(v) = \exp(\tilde{S}_b^f(v)/k)$ we obtain

$$D_b(v) = \tilde{D}_{bo}(v) \exp(-\tilde{Q}_b(v)/kT) \tag{9}$$

where

$$\tilde{Q}_b(v) = \tilde{E}_b^f(v) + \tilde{E}_b^m(v) \tag{10}$$

and

$$\tilde{D}_{bo}(v) = g_b f_b \tilde{\alpha}_b^2 \tilde{\nu}_0(v) \exp(\tilde{S}_b^f(v)/k) \tag{11}$$

Consider first the activation energy $\tilde{Q}_b(v)$. Because the equilibrium vacancy population in the various sites in the core should obey a Boltzmann distribution we can take the concentration at each site as proportional to $\exp(-E_b^f(v)/kT)$. By assuming that the constant of proportionality is the same for all sites, we can estimate the total effective vacancy concentration in the core (to within the constant of proportionality) by using the data in Table II. The effective vacancy formation energy can then be found from the temperature dependence of the total concentration. The result of this procedure is $\tilde{E}_b^f(v) \simeq 97$ kJ mole^{-1}. Because the effective migration energy is $\simeq 49$ kJ mole^{-1}, we obtain $\tilde{Q}_b(v) \simeq 146$ kJ mole^{-1}.

Consider next the magnitude of the effective preexponential factor, $D_{bo}(v)$. The geometrical factor g_b should be of order $\simeq 1/3$. Since there is a variety of vacancy jumps that occur in three dimensions in the boundary (i.e., it does not consist of a single row of jump sites), we expect the correlation factor f_b to be of order 0.5. Also, it is reasonable to take $\tilde{\alpha}_b^2 = [0.85 \, a_0]^2$ and $\tilde{S}_b^f(v)/k = 2$. Using these values and $\tilde{\gamma}_0(v) = 4.85 \times 10^{13}$ s^{-1}, we obtain:

$$D_b(v) = 3.6 \times 10^{-6} \exp(-146/kT) \text{ m}^2 \text{ s}^{-1} \qquad (12)$$

The value of $\tilde{Q}_b(v) \simeq 146$ kJ mole^{-1} is significantly lower than the measured value of 240 kJ mole^{-1} for lattice self-diffusion in bcc iron, and the preexponential factor is in the range generally expected for this quantity on simple physical grounds. The calculated magnitudes of the diffusivity are in remarkably good agreement with expected measured values for boundaries in bcc metals as seen in Fig. 1. The results given by Eq. (12) are seen to agree with the average values to sell within an order of magnitude. We also note that the vacancy migration was much more rapid along the tilt axis than perpendicular to it, in agreement with experimental self-diffusion data for tilt boundaries (see Section VI).

In contrast to the behavior of vacancies in the boundary core, interstitials appear to be relatively immobile and would therefore not be expected to contribute significantly to boundary self-diffusion rates. The molecular dynamics simulation demonstrated the immobility of the interstitial at the I site (Fig. 16). Also, the results in Table II indicate that $\tilde{E}_b^m(i) \gtrsim 122 \, kJ \, mole^{-1}$ and $\tilde{E}_b^f(i) \simeq 102$ kJ mole^{-1}, and, therefore, $\tilde{Q}_b(i) \gtrsim 224$ kJ mole^{-1}. This result suggests that boundary self-diffusion by an interstitial mechanism is probably slower than even lattice self-diffusion.

We mention at this point a further study by Pontikis (1982) of the self-diffusion mechanism in a symmetric [001] $\Sigma = 5$ tilt boundary in the fcc structure. The methods used were similar in many respects to those employed by Balluffi et al. (1981) and Kwok et al. (1981b). Molecular dynamics were again employed to study vacancy migration and interstitial phenomena in the boundary core. All particles interacted via a pairwise Lennard–Jones potential fitted to argon. The phenomena observed were qualitatively similar to those observed by Balluffi et al. (1981b) and Kwok et al. (1981), including vacancy jumping in the core, the spontaneous formation of Frankel pairs, and the immobility of the interstitial due to strong trapping. The results provide further support for a predominantly vacancy mechanism for boundary diffusion. They also suggest that this result is not strongly sensitive to the lattice crystal structure.

In other work with boundaries, Ingle and Crocker (1978) studied vacancy migration in a $\Sigma = 3\,[110]$ tilt boundary (coherent twin) in bcc iron. As already mentioned, vacancies are attracted to sites in such boundaries with binding energies approaching 19 kJ mole^{-1} with $E_l^f(v) = 132$ mJ mole^{-1} (Ingle et al., 1976). Migration energies between various sites were calculated using a molecular statics method in which an atom is forced to move quasistatically between two adjacent empty sites in a continuously relaxed atomic environment. The results showed that migration energies for a number of jumps were substantially lower than in the lattice for which $E_l^m(v) = 66$ kJ mole^{-1}. It was concluded that vacancy migration along the boundary would be easy compared with the lattice. Unfortunately, no corresponding calculations for interstitials were carried out. However, inspection of the boundary structure indicates that it is highly unlikely that interstitials could exist anywhere with reasonably low formation energies, thus ruling out an interstitial mechanism. Faridi and Crocker (1980) have also calculated vacancy migration energies in the $\Sigma = 3\,[110]$ tilt boundary (coherent twin) in fcc copper. In this case the migration energies were up to 10% lower than in the lattice indicating somewhat faster boundary self-diffusion than in the lattice. Again, an interstitial mechanism can be ruled out.

Finally, one set of molecular statics calculations has been reported for vacancy migration along the cores of isolated lattice dislocations. Miller et al. (1981) used the same quasi-static method to calculate energies for various vacancy jumps in the cores of the $\mathbf{b} = \frac{1}{2}[111]; \boldsymbol{\xi} = 1/\sqrt{6}\,[11\bar{2}]$ and $\mathbf{b} = \frac{1}{2}[111]; \boldsymbol{\xi} = 1/\sqrt{2}\,[1\bar{1}0]$ lattice dislocations in bcc iron and molybdenum already mentioned in Section III.A (see also Table I). They found jumps in each dislocation that were either more difficult or easier than in the lattice. By considering likely sequences of jumps for diffusion along the core they concluded that the overall migration process along the core would be more difficult than in the lattice. However, the vacancy binding energies to these cores are relatively large (see Table I) causing Miller et al. (1981) to predict overall activation energies for boundary self-diffusion to be appreciably lower than for the lattice. They conclude that the vacancy mechanism is more likely than the interstitial mechanism for fast core diffusion along these dislocations. Their conclusion that vacancy migration along the cores is actually slower than in the lattice is of particular interest since it is the only suggestion in the literature of this possibility. This result must be treated with some caution, however, since all jump paths were not evaluated, and since quasi-static calculation of the migration energy is an approximation which requires further testing by direct comparison with the more realistic dully dynamic method.

We may conclude from all of the results that the fast atomic jumping along grain boundaries (and lattice dislocations) in metals must occur in the vast majority of cases by a vacancy mechanism. This conclusion is in agreement with all of the atomistic calculations performed to date and also appears to be consistent with the experimental results cited in Section IV. The interstitial mechanism is generally ineffective because of the relatively high formation energy for interstitials in most boundary sites.

VI. Influence of Boundary Structure on Boundary Diffusion

In a series of classic experiments Turnbull and Hoffman (1954) studied self-diffusion in [001] symmetric tilt boundaries in silver parallel to the tilt axis as a function of crystal misorientation. Their data, expressed in terms of the parameter $p \equiv D_b \delta_b$, where δ_b is the effective "width" of the boundary, are shown in Fig. 27. The results show that p increases monotonically with θ up to $\theta = 28°$ and that the activation energy is not greatly different for the different boundaries. A plot of p (employing arbitrary units) versus θ at $T = 441°C$, which is near the center of the temperature range studied, is presented in Fig. 28 and shows that p increases approximately linearly with θ over the entire range of θ studied. This behavior can be simply understood in terms of the structural unit model for [001] tilt boundaries. Wang et al.

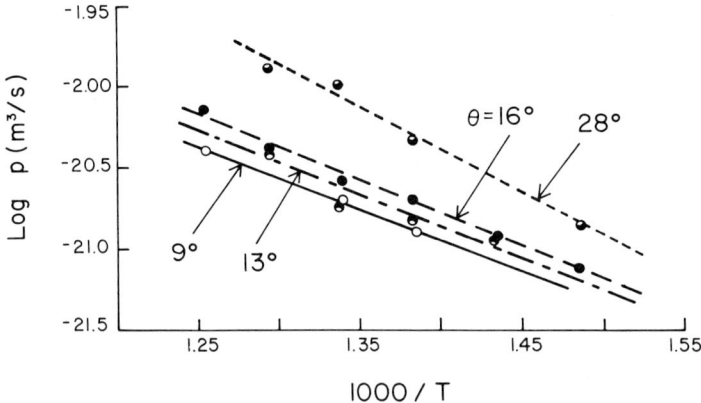

Fig. 27. Arrhenius plot for measured grain boundary diffusion parameter $p \equiv D_b \delta_b$ for [001] symmetric tilt boundaries in silver; p is measured along tilt axis, θ is the tilt angle. [From Turnbull and Hoffman (1954).]

6. GRAIN BOUNDARY DIFFUSION MECHANISMS IN METALS

(1982) have calculated the core structures of [001] tilt boundaries in copper possessing tilt angles in the range $0 \leq \theta \leq 36.9°$, with the results indicated schematically in Fig. 4. The favored boundaries are seen to be the $\Sigma = 1$ and $\Sigma = 5$ boundaries made up of all C and D units, respectively. Intermediate boundaries with $\theta \geq 10°$ consist of appropriate mixtures of the C and D units, and equal numbers of the two units are present in the $\Sigma = 17$ boundary. Assuming similarity between copper and silver we can use these results to construct an idealized, simple model for the Turnbull–Hoffman data. Because the units are arranged in columns parallel to the tilt axis, we may write p in the general form:

$$p = \sum_i A_i D_i \qquad (13)$$

where A_i is the area of suitably chosen patches of boundary of type i per unit length of boundary in a plane perpendicular to the tilt axis, and D_i is the corresponding diffusivity. The sum, of course, is over all patches making up the boundary. It can be shown (Balluffi and Brokman, 1983) that the value of p for all boundaries between the $\Sigma = 1$ and $\Sigma = 17$ boundaries can be expressed as a linear combination of the values of p for the $\Sigma = 1$ and $\Sigma = 17$ boundaries and that the value of p for all boundaries between the $\Sigma = 17$ and $\Sigma = 5$ boundaries can be expressed as a linear combination of the values of p for the $\Sigma = 17$ and $\Sigma = 5$ boundaries. A plot of p versus θ for boundaries between the $\Sigma = 1$ and $\Sigma = 5$ misorientations then appears as two almost perfectly straight line segments between the values of p for the $\Sigma = 1$, 17, and 5 boundaries as illustrated schematically in Fig. 28. This result is seen to be in reasonable agreement with the available measurements.

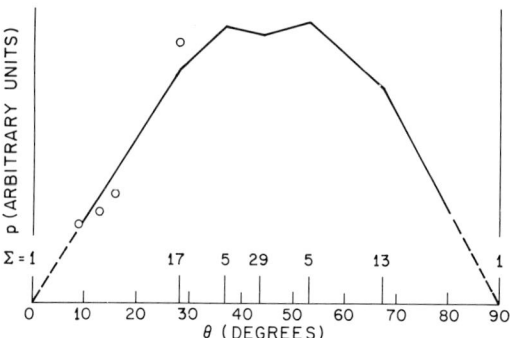

Fig. 28. Expected general form of plot of grain boundary diffusivity parameter p versus θ for symmetric [001] tilt boundaries according to structural unit model (see text); p is measured along tilt axis, θ is the tilt angle. First segment at left is fit to experimental data taken from Fig. 27. Remainder of curve is purely schematic.

More generally, it can be shown (Balluffi and Brokman, 1983) for the entire series of [001] tilt boundaries that p for this idealized model should vary linearly with θ between each favored boundary and each intermediate boundary made up of equal numbers of favored boundary units (such as the previous $\Sigma = 17$ boundary, which lies between the $\Sigma = 1$ and $\Sigma = 5$ favored boundaries). Using the structural information given in Fig. 4, we then expect the p versus θ curve over the full range of θ to have a form of the general type illustrated in Fig. 18. (It is emphasized that the curve drawn in Fig. 28 is purely schematic and that the actual p versus θ curve is presently unknown. In addition, the model must be regarded as an approximation since, as previously pointed out, the favored units exhibit some distortions.) The curve is drawn dashed below $\theta = 10°$, since, as previously explained, it is likely that the lattice dislocations in a low stacking fault energy metal such as silver dissociate in this regime and their cores then no longer correspond to D units. Unfortunately, no diffusivity measurements were made in this regime. However, diffusivities along dissociated dislocations are expected to be lower than along undissociated dislocations (Fig. 1), and we would therefore predict measured values of p for silver to fall below the dashed curve at small values of θ. (We also note here that the slower self-diffusion along dissociated dislocations is consistent with the lower calculated vacancy binding energies to dissociated dislocations listed in Table I.) It should also be evident that the structural unit model offers an explanation for the observation (Fig. 1) that averaged diffusion rates along the cores of undissociated dislocations are similar to averaged rates along grain boundaries as indicated by Fig. 1. This is due to the fact that the structural units making up the cores of the undissociated dislocations also appear in many large-angle boundaries, and, in addition, are generally similar to other structural units characteristic of many boundaries.

The structural unit model just described for silver is, of course, essentially equivalent to the "dislocation pipe" diffusion model originally invoked by Turnbull and Hoffman (1954) to explain their data. However, the structural unit model provides a framework for understanding the diffusional behavior over a wider range of misorientation and is, of course, considerably more specific at angles near 28°, where the dislocation cores, i.e., the D units become very closely spaced.

Herbeuval et al. (1973) have measured relative diffusion rates along the tilt axis of [110] symmetric tilt boundaries in aluminum with the results shown in Fig. 29. A number of aspects of these results may also be understood in terms of the structural unit model as first pointed out by Sutton (1981) and Sutton and Vitek (1982). Calculations for these boundaries indicate that the $\Sigma = 1$ ($\theta = 0°$), $\Sigma = 3$ ($\theta = 70.53°$), $\Sigma = 11$ ($\theta = 129.52°$), $\Sigma = 27$ ($\theta = 148.41°$), and $\Sigma = 1$ ($\theta = 180°$) boundaries are favored, and the $\Sigma = 9$ ($\theta = 141.06°$) and $\Sigma = 51$ ($\theta = 157.16°$) are the intermediate bound-

6. GRAIN BOUNDARY DIFFUSION MECHANISMS IN METALS

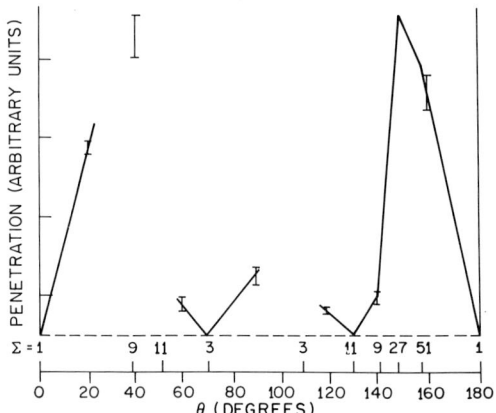

Fig. 29. Measured relative grain boundary diffusion rates in [110] symmetric tilt boundaries in aluminum. Penetration is measured along tilt axis, θ is the tilt angle. Data from Herbeuval et al. (1973). Straight line segments drawn in accordance with structural unit model (see text).

aries with equal numbers of favored units. The $\Sigma = 3$ ($\theta = 70.53°$) favored boundary is the well-known completely-coherent {111} planar twin boundary which has been shown (Barnes, 1960; Leymonie, 1963) in general to possess a very low diffusivity. As Faridi and Crocker (1980) proved, vacancy migration energies in the core of this boundary are at most only slightly lower than in the lattice. In addition, it seems quite certain that vacancy binding energies should be small, thereby explaining the observed relatively slow diffusion rate. As illustrated in Fig. 30, the structural units making up the $\Sigma = 27$ ($\theta = 148.41°$) boundary are found (Sutton, 1981) to be much

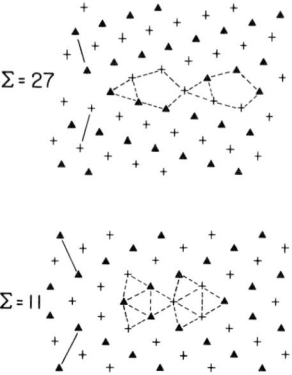

Fig. 30. Relaxed structures of [110] symmetric tilt boundaries in aluminum calculated by method of molecular statics by Sutton (1981) and Sutton and Vitek (1983). $\Sigma = 27$ ($\theta = 148.41°$) and $\Sigma = 11$ ($\theta = 129.52°$) boundaries shown.

more open (i.e., loosely packed) than the calculated units making up the $\Sigma = 11$ ($\theta = 129.52°$) boundary, which appear to be relatively closely packed. Therefore, we expect the diffusion rates along the $\Sigma = 3$ ($\theta = 70.53°$) and $\Sigma = 11$ ($\theta = 129.52°$) boundaries to be relatively slow, and the diffusion rate along the $\Sigma = 27$ ($\theta = 148.41°$) boundary to be relatively fast. Segments of the overall diffusivity versus θ curve consistent with this information, the structural unit model, and the original data are shown in Fig. 29. It is interesting to note that the asymmetry of the curve about each favored boundary misorientation is to be expected, and is simply due to differences in the types of structural units that are mixed into the units of each favored boundary on either side of it.

Several experiments have shown that boundary diffusion rates in symmetric tilt boundaries are usually faster along the tilt axis than perpendicular to it. In the case of silver self-diffusion along [001] symmetric tilt boundaries Hoffman (1956) observed p parallel to the tilt axis $p_{\parallel}$ to be considerably larger than p perpendicular $p_{\perp}$, as shown in Fig. 31. These results can again be interpreted on the basis of the structural unit model. Again assuming similarity between copper and silver, it is seen that the anisotropy becomes immeasurably large in the small-angle regime of θ, where the spacing between rows of D units (see Fig. 4) running parallel to [001] becomes larger than about two lattice parameters. (Note that two C units take up a length in the boundary equal to one lattice parameter.) This anisotropy is easily understood since the transverse diffusion is then essentially "open-circuited" by the regions of good crystal lattice, i.e., by the C units between the D units.

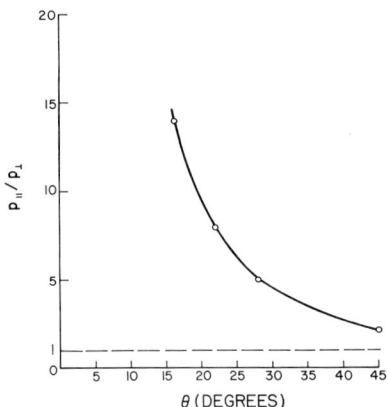

Fig. 31. Measured ratio of the grain boundary diffusion parameter, $p_{\parallel}$ (for diffusion parallel to the tilt axis), divided by the parameter $p_{\perp}$ (for diffusion perpendicular to the tilt axis), versus tilt angle θ for [001] tilt boundaries in silver. [Data from Hoffman (1956).]

However, transverse leakage through the C units apparently begins when the distance between rows of D units, i.e., lattice dislocations, becomes smaller than about this spacing. It is interesting that the anisotropy persists above the $\Sigma = 5$ ($\theta = 36.87°$) misorientation, where all of the lattice (C units) have disappeared. In this regime the structure consists of mixtures of B and D units that are arranged in rows parallel to the tilt axis, just as were the C and D units in the smaller-angle boundaries. Presumably, the observed anisotropy could then be a result of differences in the diffusivities of the two types of units. However, it could also be due to a diffusional anisotropy in the units themselves. This was found to be the case in the dynamic simulation of the $\Sigma = 5$ [001] 36.9° tilt boundary in bcc iron already described in Section V. This boundary is a favored boundary consisting of only one type of unit, and faster diffusion along the tilt axis was seen to be due to the preferred jumping of vacancies in the structural units along this axis. Relatively fast diffusion along the tilt axis of large-angle [110] tilt boundaries has also been observed (Herbeuval et al., 1973), and it therefore seems to be a common feature of tilt boundary diffusion.

These results indicate that a number of features of the dependence of boundary diffusion rates on boundary parameters can be understood on a semiquantitative basis in terms of boundary core structure and the structural unit model at this time. It appears that previous efforts, e.g., Gjostein (1973), to understand fast boundary diffusion on the basis of unusually rapid diffusion along the cores of intrinsic GBDs cannot be correct in general. Since the cores of the GBDs have been shown to be minority structural units, it is clear, as pointed out by Sutton (1981), and Sutton and Vitek (1983) that diffusion rates along GBD cores may be either faster or slower than elsewhere in the boundary depending upon the particular atomic structures of the units. Also, in a number of cases, suggested plots of diffusion rates versus misorientation have appeared, e.g., Fig. 4.19 of Gleiter and Chalmers (1972), which feature sharp cusps at certain special misorientations. We may conclude that these features must be incorrect. Further understanding will require additional calculations of boundary structures and the mobilities and concentrations of the point defects which may be present.

VII. Diffusion along Migrating Boundaries

Recently, it has been suggested (Hillert and Purdy, 1978; Smidoda et al., 1978) that diffusion rates along migrating boundaries may be several orders of magnitude faster than along stationary boundaries. Hillert and Purdy

(1978) reported diffusion rates of zinc along migrating boundaries in bcc iron which were possibly $\sim 10^4$ larger than those expected along normal stationary boundaries, whereas Smidoda *et al.* (1978) reported diffusion rates of aluminum along boundaries in a Ni–Cr–Al solid solution which were possibly $\sim 10^2$ to 10^4 larger than expected for stationary boundaries. This is an important result, if true, since it would suggest large differences between the structures of migrating and stationary boundaries and would require a whole new approach to the analysis of the many situations which involve diffusion along migrating boundaries.

In considering these results it should first be pointed out that it is difficult to understand such a phenomenon on the basis of our current ideas about boundary structure and boundary diffusion. We have seen that the grain boundary core consists of a slab of bad material in which the atoms are considerably less uniformly arranged than in the lattice. Furthermore, these atoms possess relatively high mobilities at temperatures at which grain boundary migration can occur at significant rates. In fact, as Gjostein (1973) has shown, the grain boundary self-diffusivity tends to approach that of the liquid at the melting point! Since grain boundary relaxation times must therefore be very short at boundary migration temperatures, it is difficult to understand how the grain boundary structure can be sufficiently perturbed by the migration to cause an increase in the diffusivity as large as several orders of magnitude. The currently most plausible models for boundary migration (Smith *et al.*, 1980) involve the motion of line defects (steps/GBDs) across the otherwise unperturbed boundary. Changes in the boundary diffusivity, according to this migration model, would only be of order unity. Rae (1981) has presented experimental evidence that there are no apparent differences in the core structures of GBDs in moving and stationary grain boundaries in aluminum. This result is consistent with the conclusion that no large differences in intrinsic boundary structure are to be expected.

Second, there is extensive counterevidence available that indicates that diffusion rates along plastically deformed or migrating boundaries are essentially the same as along stationary boundaries. As early as 1962, Blackburn and Brown (1962) found that the boundary diffusivity of silver in copper was essentially unaffected by simultaneous plastic shearing of the boundary region. This interesting result indicates that diffusion along a boundary perturbed by plastic deformation is not significantly affected at diffusion temperatures, as indeed might be expected on the basis of the above discussion. de Reca and Pampillo (1975) found self-diffusivities along migrating boundaries in nickel undergoing grain growth to be in good agreement with values for static boundaries reported in the literature. Mittemeijer and Beers (1980) and also Grovenor (1982) studied diffusion along migrating boundaries in copper–nickel thin film specimens and found no evidence that the boundary diffusion rates were any higher than those expected along static boundaries. Hintz (1981) has informed the author of a literature search in collaboration with Wolfgang Gust for cases in which the stationary boundary diffusivity and the diffusivity along a discontinuous precipitation reaction front could be compared for the same solute–solvent pair. In the roughly eight systems for which data exist there was no evidence of a 2–4 order of magnitude increase in diffusivity due to boundary migration.

Third, it should be noted that in no case did Hillert and Purdy (1978) or Smidoda *et al.* (1978) perform an experiment in which the diffusion rate in a migrating boundary was compared directly with the rate in the same type of stationary boundary in the same alloy using the same experimental technique. It is well known that large systematic errors may be present in measurements of boundary diffusion, and, in addition, the results may be sensitive to alloying effects and intrinsic structural effects. Gust *et al.* (1982) have found unusually large boundary diffusivities for indium along certain stationary boundaries in nickel. They suggest that diffusion along certain stationary boundaries can be unusually rapid in this system and that the rapid boundary diffusion observed by Hillert and Purdy, and Smidoda *et al.*, could have been due to the fact that the diffusivities along the migrating boundaries studied by them are simply intrinsically large irrespective of whether the boundaries are migrating or not. In this respect it is noted that rapidly migrating boundaries may also be rapidly diffusing boundaries and that such boundaries may receive the most attention in experiments.

We may conclude, therefore, that it is unlikely that diffusion rates along migrating boundaries are orders of magnitude larger than along stationary boundaries. However, there is a current need to confirm this conclusion by means of careful experiments involving both stationary and migrating boundaries of the same type in the same system.

VIII. Model for Grain Boundaries as Point Defect Sources and Sinks and Comparison with Experimental Observations

At this point we have concluded that grain boundary diffusion in the vast majority of metals occurs by a vacancy mechanism. This allows the net diffusional transport of mass along grain boundaries during phenomena such as diffusional creep and sintering (Section IV.A) and the unequal boundary diffusion of the two species in binary systems such as occurs, for example, during diffusion induced grain boundary migration (Section IV.C). In general, grain boundary sources and/or sinks for atoms (vacancies) are required to support these processes, and as a final topic, we now discuss models for high-angle grain boundaries as point defect sources/sinks.

All models that have been proposed are based, either explicitly or implicitly, on the existence of perfect GBDs that climb in the boundary plane by creating/annihilating the point defects. The most recent detailed description and discussion of this model has been given by Balluffi (1980). The model has the attractive feature that the atomic structure of the boundary is conserved as the GBDs climb, which is consistent with our expectation that every grain boundary possesses a unique atomic structure of minimum energy.[1] Furthermore, jogs on the GBDs provide unique locations where point defects can be created/annihilated by means of small fluctuations with a total energy change which is very close to the formation energy of the point defects in the boundary. This model, therefore, bears a close resemblance to models for single crystal growth by condensation/evaporation (Hirth and Pound, 1963). In the crystal growth case atoms are added/removed at kinks in surface steps that sweep across the surface. Therefore, there is a direct analogy between the GBDs and the surface steps, and the source and sink problem may be regarded essentially as a problem of crystal growth at an internal surface, i.e., the boundary.

The operation of a grain boundary as a sink, for example, is generally expected to be a highly complex process involving a number of steps which may involve diffusion of the point defects to the boundary, diffusion of the defects in the boundary, and ultimate annihilation at jogs on the GBDs. Factors which will be of importance in determining the sink efficiency of a boundary, therefore, include at least the following:

(a) the densities and distributions of the GBDs and associated jogs which are available;

(b) the ease (rate) with which point defects are created/destroyed at the jogs;

(c) the structure, energy, and diffusivity of the defects in the lattice and the grain boundary (For example, if the defects are strongly bound to the grain boundary core, and if they diffuse rapidly there, the core will act as an efficient collector which will deliver the defects to the GBDs, and, hence, increase the overall efficiency.);

(d) the nature of the point defect diffusion field adjoining the boundary (For example, as this field becomes larger in extent, the efficiency of the source/sink should become greater, i.e., less sensitive to local difficulties

---

[1] Models of the same general type could also be constructed using partial GBDs. Such GBDs might, for example, separate patches of boundary possessing slightly different energies. The characteristics of such models would not differ greatly from those based on perfect GBDs and hence will not be considered explicitly here.

in creating/destroying the defects at the boundary, since more time is then consumed in diffusing from/to the boundary.).

A schematic illustration of the climb of an edge dislocation in a grain boundary due to the annihilation of a vacancy is shown in Fig. 13. We note that the continued climb of this dislocation and its associated step by vacancy annihilation would cause the boundary to migrate relative to both crystal lattices and would also cause crystal 2 to translate relative to crystal 1 by the Burgers vector.

A wide range of structures involving climbing dislocations may develop in different source/sink situations depending upon such variables as the boundary type, the extrinsic GBD content of the boundary, the magnitude of the driving force for climb, the amount of climb which has occurred, etc. At the time of the review by Balluffi (1980) there was still some questions about the nature of the core structure of GBDs in high Σ boundaries. However, the recent work by Sutton (1981) and Sutton and Vitek (1983) with the structural unit model makes it likely that GBDs with localized cores exist in all grain boundaries, and, therefore, that the GBD climb model holds under all conditions.

In the special case of favored boundaries (which, of course, are taken to be free of intrinsic GBD structure) extrinsic GBDs may be present, and these may expand in length by the spiral growth mechanism illustrated in Fig. 12. This is the grain boundary analog of the well known Frank model (Hirth and Pound, 1963) for crystal growth, which involves the motion of surface ledge spirals associated with lattice dislocations that intersect the surface. Extrinsic GBDs may also be produced by the nucleation and growth of GBD loops in the boundary plane by point defect aggregation. An experimentally observed (King and Smith, 1980) example of this is shown in Fig. 32, where triangular GBD interstitial type loops are nucleating and growing on a favored [110] 70.5° tilt boundary (coherent twin) in copper. In this case, excess interstitials are being annihilated during the continuous nucleation, growth, and eventual impingement of the loops. In general, higher thermodynamic driving forces are required to nucleate loops in this way than to cause already existing extrinsic GBDs to climb as in Fig. 12.

In the case of nonfavored boundaries (which, of course, contain arrays of intrinsic GBDs) it is possible to have climb occur in the arrays in various ways that are described in detail by Balluffi (1980) or to have again the climb of extrinsic GBDs. An experimentally observed (Komem et al., 1972) example of this is seen in Fig. 33, which shows the climb of extrinsic GBDs in a nonfavored [001] twist boundary in gold near the $\Sigma = 5$ misorientation (see legend for details).

The experimentally determined abilities (efficiencies) of various grain boundaries to act as sources/sinks for vacancies under different conditions have also been discussed by Balluffi (1980). The experiments involved: (i) the observation of vacancy precipitate denudation adjacent to grain boundaries in quenched or cooled metals; (ii) measurements of diffusional creep kinetics where boundaries acted as sources/sinks; and (iii) measurements

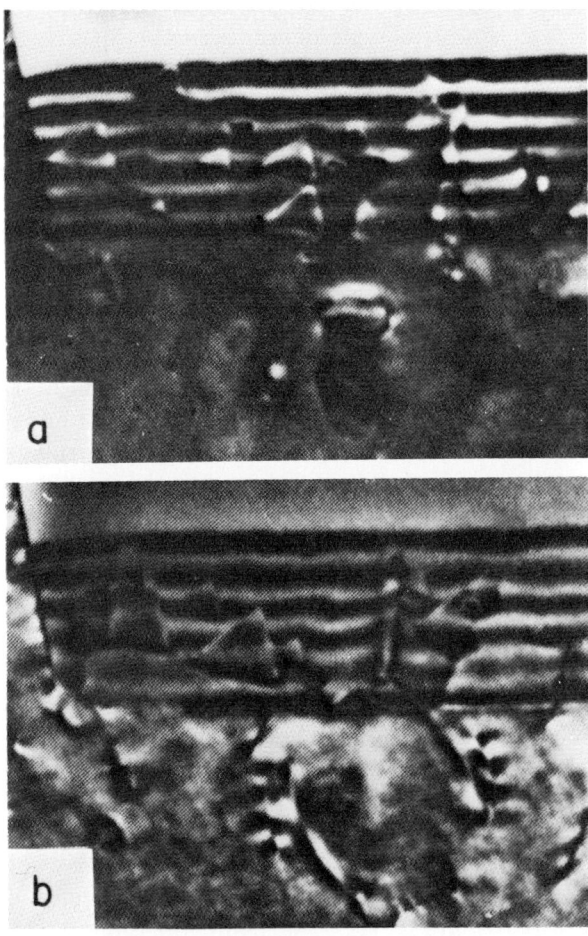

Fig. 32. [110] $\theta = 70.5°$ tilt boundary (coherent twin) in copper acting as sink for irradiation-produced interstitials by nucleation and growth of triangular GBD loops. Specimen electron irradiated in situ in high voltage electron microscope at 170°C. (a) and (b) show earlier and later views of same boundary area. [From King and Smith (1980).]

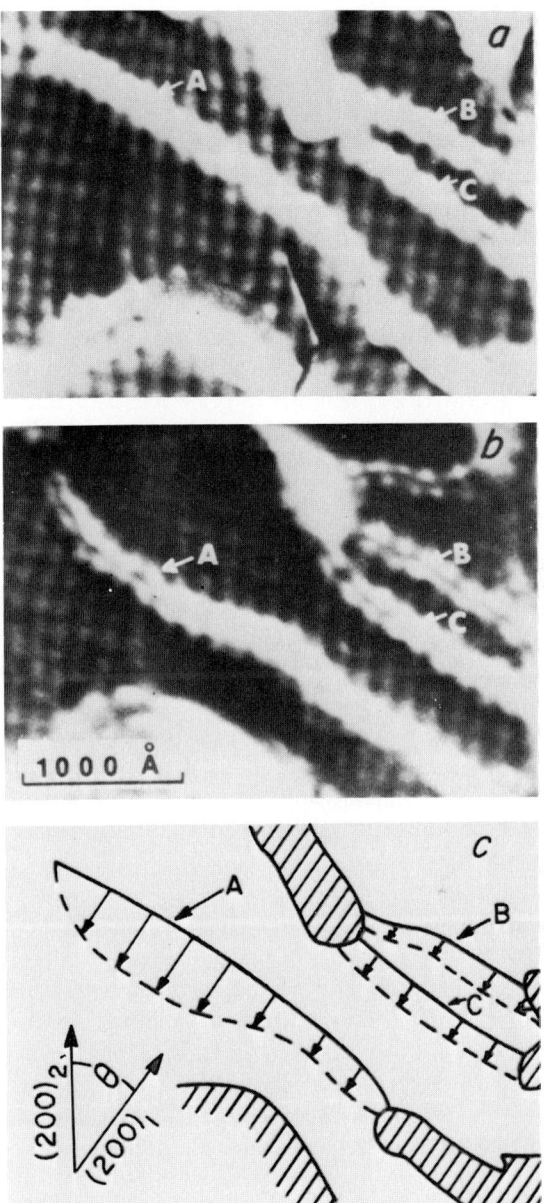

Fig. 33. [001] $\theta = 36.9°$ twist boundary in gold acting as a sink for irradiation-produced interstitials by climb of extrinsic GBDs. (a) GBD structure before irradiation; (b) same area as in (a) after irradiation with energetic gold ions; (c) schematic diagram showing climb of the extrinsic GBDs at A, B, and C. [From Komen et al. (1972).]

of sintering kinetics where boundaries acted as vacancy sinks. The results indicate the following:

(a) Nonfavored, high Σ boundaries operate as highly efficient vacancy sources/sinks in a number of situations in the presence of vacancy chemical potentials as low as $|\mu_v| = 10^{-3}$ kJ mole^{-1};

(b) However, threshold effects may appear in certain cases with generally occurring boundaries (which must include many nonfavored, high Σ boundaries) when $|\mu_v|$ is in the range 10^{-3}–10^{-2} kJ mole^{-1};

(c) A limited number of experiments indicates that the source/sink efficiency tends to be lower in certain special low Σ boundaries than in more general boundaries. Special boundaries with $3 \leq \Sigma \leq 15$ have been found to be practically inoperative as vacancy sources when $|\mu_v| = 2 \times 10^{-3}$ kJ mole^{-1} even though more general boundaries remained operative. All boundaries, including the special $\Sigma = 3$ coherent twin boundary, become operative, at least as vacancy sinks, at very high chemical potentials (i.e., $|\mu_v| = 70$). However, the efficiency of at least the special $\Sigma = 3$ coherent twin boundary remains lower than that of non-favored high Σ boundaries.

It appears that a number of these results can be understood semi-quantitatively on the basis of the GBD source/sink model. Threshold effects can result from difficulties in forcing the GBDs to climb in boundaries and from the necessity to nucleate GBD loops in certain cases. Since the Burgers vectors tend to increase as Σ decreases, it is more difficult to nucleate GBD jog pairs and also loops as Σ decreases. Also, the rate of point defect diffusion in the boundary may be relatively low for special low Σ boundaries, e.g., Fig. 29. Further discussion of this topic is given by Balluffi (1980).

IX. Conclusions

(i) Fast grain boundary diffusion in metals occurs by the fast jumping of atoms predominantly in the bad material making up the cores of grain boundaries.

(ii) The cores of large-angle boundaries are relatively narrow, i.e., no more than a few atom distances thick. The core structure is relatively dense, i.e., only slightly less dense at most than the perfect crystal, and the average number of near-neighbors around each atom is only slightly lower than in the perfect crystal. The core structures of many (and probably all!) boundaries can be described as mixtures of structural units (i.e., small characteristic groups of atoms that may be somewhat distorted in various ways) in ordered arrays, and in this sense may be regarded as ordered. A variety of atomic environments, therefore, generally exists in which some are loose and others exhibit squeezing in one direction or another. Local phonon frequencies are generally lower in the core than in the lattice, and resonant and local modes appear. The cores of a very few special boundaries, e.g., coherent twins, are again describable in terms of the structural unit model but are more ordered and lattice-like.

(iii) The cores of small-angle boundaries consist of regular arrays of discrete lattice dislocation cores. When undissociated, the cores of these dislocations are composed of structural units of the same type that occur in larger-angle boundaries and, in fact, a continuum of structures, describable in terms of the structural unit model, occurs as the transition from small- to large-angle boundaries is made.

(iv) The cores of grain boundaries can be described alternately in terms of an intrinsic secondary grain boundary dislocation (GBD) model, in which the GBDs possess Burgers vectors that are vectors of Bollmann's DSC lattice. However, the cores of the secondary GBDs turn out to be structural units which are present in minority concentrations in the overall mixture of structural units present in the boundary core. Hence, the structural unit model and the GBD model are actually two ways of describing the same structure. Both models apply over all crystal misorientations and involve a continuum of structures.

(v) Grain boundary cores are capable of supporting line defects in the form of intrinsic GBDs [see (iv) above] or extrinsic GBDs. Extrinsic GBDs are dislocations that are present as a result of specimen history and do not act in a systematic way to accommodate the crystal misorientation characteristic of the boundary as a whole. The dislocation cores consist of structural units and hence are localized.

(iv) Grain boundary cores are capable of supporting point defects, i.e., vacancies or interstitials, at various sites as bona fide point defects in the form of distinguishable missing or extra atoms in the relatively irregular structure.

(vii) It is concluded that fast grain boundary diffusion occurs by a vacancy mechanism in the vast majority of boundaries including small- and large-angle boundaries. The atom jumping occurs by vacancy exchange between a variety of sites in the core. An interstitial mechanism seems to be excluded in the majority of cases, because of the relatively high energy of formation of interstitials in most sites in the core. This result is consistent with: (a) numerous experiments indicating that boundary diffusion can transport net mass and hence must occur by a defect mechanism; (b) experiments which indicate that the two species in a binary system diffuse at different rates in boundaries; (c) measurements of hydrostatic pressure and isotope effects on boundary diffusion; and (d) results of computer simulation studies employing molecular statics and dynamics.

(viii) A number of aspects of the influence of boundary structure on boundary diffusion can be understood semiquantitatively on the basis of the structural unit core model. These include, for example, effects due to the changes of boundary structure brought about by changes in tilt angle in tilt boundaries and the observed anisotropy of boundary diffusion in tilt boundaries.

(ix) It is concluded that recent claims that diffusion along migrating boundaries may be several orders of magnitude faster than along corresponding stationary boundaries are most probably incorrect.

(x) Models for grain boundaries as point defect sources/sinks involve the climb of GBDs in the boundary plane. The overall operation of a boundary as a source/sink is generally expected to be a complex process involving the diffusion of the point defects from/to the boundary, the diffusion of the defects in the boundary, and the creation/annihilation of the defects at jogs on the GBDs. The source/sink efficiency of the boundary may vary considerably depending upon the details of these processes. The GBDs necessary for climb may be GBDs already present in the boundary or GBDs produced by the precipitation of point defects, and a wide range of behavior is therefore possible. There is experimental evidence that source/sink efficiencies tend to be lower for very special boundaries (e.g., twins) than for more general boundaries and also that they fall off when the driving chemical potential of the nonequilibrium point defects decreases.

Acknowledgments

The writer would like to thank A. P. Sutton, G-J. Wang, and V. Vitek for permission to discuss some of the results of their calculations of the structures of [001] tilt boundaries before publication and for helpful discussions regarding the structural unit model. He thanks M. B. Hintz for information regarding diffusion rates along migrating boundaries. He is indebted to N. L. Peterson for valuable discussions centered on the question of the mechanism of grain boundary diffusion. Finally, he thanks his coworkers, P. D. Bristowe, A. Brokman, and A. H. King, for many helpful suggestions and comments. This work was supported by the U.S. Department of Energy under Contract No. DE-AC02-78ER05002.

References

Alexander, B. H., and Balluffi, R. W. (1957). *Acta Metall.* **5,** 666.
Arsenault, R. J., Beeler, J. R., Jr., and Simmons, J. A., eds. (1976). *Nucl. Metall.* **20,** Parts 1 and 2.
Ashby, M. F. (1972). *Acta Metall.* **20,** 887.
Ashby, M. R. (1974). *Acta Metall.* **22,** 275.

Balluffi, R. W. (1979). *In* "Interfacial Segregation" (Johnson, W. C., and Blakely, J. M., eds.), p. 193. Amer. Soc. Metals, Metals Park, Ohio.
Balluffi, R. W. (1980). *In* "Grain Boundary Structure and Kinetics," p. 297. Amer. Soc. Metals, Metals Park, Ohio.
Balluffi, R. W., and Bristowe, P. D. (1984). *Surf. Sci.* In press.
Balluffi, R. W., and Brokman, A. (1983). *Scr. Metall.* **17,** 1027.
Balluffi, R. W., and Cahn, J. W. (1981). *Acta Metall.* **29,** 493.
Balluffi, R. W., Bristowe, P. D., and Brokman, A. (1982a). *In* "Advances in Ceramics," Amer. Ceram. Soc., Columbus, Ohio. In press.
Balluffi, R. W., Bristowe, P. D., and Sun, C. P. (1981a). *J. Am. Ceram. Soc.* **64,** 23.
Balluffi, R. W., Brokman, A., and King, A. H. (1982b). *Acta Metall.* **30,** 1453.
Balluffi, R. W., Brokman, A., Sutton, A. P., and Vitek, V. (1982c). Unpublished research.
Balluffi, R. W., Komem, Y., and Schober, T. (1972). *Surf. Sci.* **31,** 68.
Balluffi, R. W., Kwok, T., Bristowe, P. D., Brokman, A., Ho, P. S., and Yip, S. (1981b). *Scr. Metall.* **15,** 951.
Balluffi, R. W., Sass, S. L., and Schober, T. (1972). *Philos. Mag.* **26,** 585.
Barnes, R. S. (1960). *Philos. Mag.* **5,** 635.
Bishop, G. H., and Chalmers, B. (1968). *Scr. Metall.* **2,** 133.
Blackburn, D. A., and Brown, A. F. (1962). *In* "Radioisotopes in the Physical Sciences and Industry," Vol. I, p. 145. International Atomic Energy Agency, Vienna.
Blech, I. A., and Kinsbron, E. (1975). *Thin Solid Films* **25,** 327.
Bolling, G. F. (1968). *Acta Metall.* **16,** 1147.
Bollmann, W. (1970). "Crystal Defects and Crystalline Interfaces." Springer-Verlag, New York.
Bonanno, F. R., and Tomizuka, C. T. (1965). *Phys. Rev. A* **137,** 1264.
Bristowe, P. D., and Crocker, A. G. (1978). *Philos. Mag. A* **38,** 487.
Brokman, A., and Balluffi, R. W. (1981). *Acta Metall.* **29,** 1703.
Brokman, A., and Balluffi, R. W. (1983). *Acta Metall.* **31,** 1639.
Brokman, A., Bristowe, P. D., and Balluffi, R. W. (1981). *J. Appl. Phys.* **52,** 6116.
Bullough, R., and Perrin, R. C. (1968). *In* "Dislocation Dynamics" (Rosenfield, A. R., Hahn, G. T., and Bement, A. L., Jr., eds.), p. 175. McGraw-Hill, New York.
Burton, B. (1977). "Diffusional Creep of Polycrystalline Materials." Trans Tech, Bay Village, Ohio.
Burton, B., and Greenwood, G. W. (1970). *Met. Sci. J.* **4,** 215.
Cahn, J. W. (1980). National Bureau of Standards, Gaithersburg, Maryland, private communication.
Cicotti, G., Guillope, M., and Pontikis, V. (1982). Centre d'Etudes Nucleaires de Saclay, France. Unpublished research.
Dahl, R. W., Jr., Beeler, J. R., Jr., and Bourquin, R. D. (1972). *In* "Interatomic Potentials and Simulation of Lattice Defects" (Gehlen, P. C., Beeler, J. R., Jr., and Jaffee, R. I., eds.), p. 673. Plenum, New York.
Darby, T. P., and Balluffi, R. W. (1977). *Philos. Mag.* **36,** 53.
Darby, T. P., Schindler, R., and Balluffi, R. W. (1978). *Philos. Mag. [Part] A* **37,** 245.
de Reca, N. W., and Pampillo, C. A. (1975). *Scr. Metall.* **9,** 1355.
d'Heurle, F. M., and Rosenberg, R. (1973). *In* "Physics of Thin Films" (Hass, G., Francombe, M. H., and Hoffman, R. W., eds.), Vol. 7, p. 257. Academic Press, New York.
Doyoma, M., and Cotterill, R. M. J. (1968). *Trans. Jpn. Inst. Met. Suppl.* **9,** 55.
Faridi, B. A., and Crocker, A. G. (1980). *Philos. Mag. [Part] A* **41,** 137.
Gehlen, P. C., Beeler, J. R., Jr., and Jaffee, R. I., eds. (1972). "Interatomic Potentials and Simulation of Lattice Defects." Plenum, New York.

Gjostein, N. A. (1973). "Diffusion." Amer. Soc. Metals, Metals Park, Ohio.
Gleiter, H., and Chalmers, B. (1972). "High Angle Grain Boundaries," p. 88. Pergamon, New York.
Grovenor, C. R. M. (1982). *Scr. Metall.* **16,** 317.
Gust, W., Hintz, M. B., Lodding, A., Odelius, H., and Predel, B. (1982). *Acta Metall.* **30,** 75.
Hart, E. W. (1972). *In* "The Nature and Behavior of Grain Boundaries" (Hu, H., ed.), p. 155. Plenum, New York.
Hashimoto, M., Ichinose, H., Ishida, Y., Yamamoto, R., and Doyoma, M. (1980a). *Jpn. J. Appl. Phys.* **19,** 1045.
Hashimoto, M., Ishida, Y., Yamamoto, R., and Doyoma, M. (1980b). *J. Phys. F* **10,** 1109.
Hashimoto, M., Ishida, Y., Yamamoto, R., and Doyoma, M. (1981). *Acta Metall.* **29,** 617.
Hasson, G. C., and Goux, C. (1971). *Scr. Metall.* **5,** 889.
Herbeuval, I., Biscondi, M., and Goux, C. (1973). *Mem. Sci. Rev. Metall.* **70,** 39.
Hillert, M., and Purdy, G. R. (1978). *Acta Metall.* **26,** 333.
Hintz, M. B. (1981). Department of Metallurgical Engineering, Michigan Technological Univ., private communication.
Hirth, J. P., and Balluffi, R. W. (1973). *Acta Metall.* **21,** 929.
Hirth, J. P., and Pound, G. M. (1963). "Condensation and Evaporation." Macmillan, New York.
Hoffman, R. E. (1956). *Acta Metall.* **4,** 97.
Ingle, K. W., Bristowe, P. D., and Crocker, A. G. (1976). *Philos. Mag.* **33,** 663.
Ingle, K. W., and Crocker, A. G. (1978). *Philos. Mag.* **37,** 297.
Johns, R. A., and Blackburn, D. A. (1975). *Thin Solid Films* **25,** 291.
Kalonji, G. (1982). Ph.D. Thesis. Massachusetts Institute of Technology, Cambridge, Massachusetts.
King, A. H. (1982). *Acta Metall.* **30,** 419.
King, A. H., and Smith, D. A. (1980). *In* "Grain Boundary Structure and Kinetics," p. 331. Amer. Soc. Metals, Metals Park, Ohio.
Komem, Y., Petroff, P., and Balluffi, R. W. (1972). *Philos. Mag.* **26,** 239.
Kwok, T., Ho, P. S., Yip, S., Balluffi, R. W., Bristowe, P. D., and Brokman, A. (1981). *Phys. Rev. Lett.* **47,** 1148.
Lee, J. K., ed. (1981). "Interatomic Potentials and Crystalline Defects." Amer. Inst. Metall. Eng., Warrendale, Pennsylvania.
Leymonie, C. (1963). "Radioactive Tracers in Physical Metallurgy." Chapman and Hall, London.
Li, J. C. M., and Chalmers, B. (1963). *Acta Metall.* **11,** 243.
Martin, G., Blackburn, D. A., and Adda, Y. (1967). *Phys. Status Solidi* **23,** 223.
Martin, G., and Perraillon, B. (1975). *J. Phys. Colloq. (Orsay, France)* (C4) **36,** 165.
Martin, G., and Perraillon, B. (1980). *In* "Grain Boundary Structure and Kinetics," p. 239. Amer. Soc. Metals, Metals Park, Ohio.
Miller, K. M., Ingle, K. W., and Crocker, A. G. (1981). *Acta Metall.* **29,** 1599.
Mittemeijer, E. J., and Beers, A. M. (1980). *Thin Solid Films* **65,** 125.
Müller, E. W., and Tsong, T. Z. (1969). "Field Ion Microscopy," p. 254. Elsevier, New York.
Pan, J. D., and Balluffi, R. W. (1982). *Acta Metall.* **30,** 861.
Perrin, R. C., Englert, A., and Bullough, R. (1972). *In* "Interatomic Potentials and Simulation of Lattice Defects" (Gehlen, P. C., Beeler, J. R., Jr., and Jaffee, R. I., eds.), p. 509. Plenum, New York.
Peterson, N. L. (1980). *In* "Grain Boundary Structure and Kinetics," p. 209. Amer. Soc. Metals, Metals Park, Ohio.
Peterson, N. L. (1983). *Int. Metall. Rev.* **28,** 65.

Poate, J. M., Tu, K. N., and Mayer, J. W., eds. (1978). "Thin Films–Interdiffusion and Reactions." Wiley, New York.
Pond, R. C. (1977). *Proc. R. Soc. London, Ser. A* **357,** 471.
Pond, R. C., and Bollmann, W. (1979). *Philos. Trans. R. Soc. London* **292,** 449.
Pond, R. C., and Vitek, V. (1977). *Proc. R. Soc. London, Ser. A* **357,** 453.
Pontikis, V. (1982). *J. Phys. Colloq. (Orsay, France)* (C6) **43,** 65.
Rae, C. M. F. (1981). *Philos. Mag. [Part] A* **44,** 1395.
Robinson, J. T., and Peterson, N. L. (1972). *Surf. Sci.* **31,** 586.
Sass, S. L., and Bristowe, P. D. (1980). *In* "Grain Boundary Structure and Kinetics," p. 71. Amer. Soc. Metals, Metals Park, Ohio.
Schiffgens, J. O., and Ashton, D. H. (1974). *J. Appl. Phys.* **45,** 1023.
Schober, T., and Balluffi, R. W. (1971). *Philos. Mag.* **24,** 165.
Shewmon, P. G. (1963). "Diffusion in Solids." McGraw-Hill, New York.
Smidoda, K., Gottschalk, W., and Gleiter, H. (1978). *Acta Metall.* **26,** 1833.
Smith, D. A., and Pond, R. C. (1976). *Int. Metall. Rev.*, No. 205, 61.
Smith, D. A., Rae, C. M. F., and Grovenor, C. R. M. (1980). *In* "Grain Boundary Structure and Kinetics," p. 337. Amer. Soc. Metals, Metals Park, Ohio.
Sun, C. P., and Balluffi, R. W. (1982). *Philos. Mag. [Part] A* **46,** 63.
Sutton, A. P. (1981). Ph.D. thesis. Univ. of Pennsylvania, Philadelphia.
Sutton, A. P., Balluffi, R. W., and Vitek, V. (1981). *Scr. Metall.* **15,** 989.
Sutton, A. P., and Vitek, V. (1983). *Philos. Trans. R. Soc. London, Ser. A* **309,** 1; **309,** 37; **309,** 55.
Turnbull, D., and Hoffman, R. E. (1954). *Acta Metall.* **2,** 419.
Wang, G-J., Sutton, A. P., and Vitek, V. (1982). Department of Materials Science and Engineering, Univ. of Pennsylvania. Private communication.

7

Simulation of Diffusion Kinetics with the Monte Carlo Method

GRAEME E. MURCH

MATERIALS SCIENCE AND TECHNOLOGY DIVISION
ARGONNE NATIONAL LABORATORY
ARGONNE, ILLINOIS

I.	Introduction	379
II.	Tracer Diffusion	381
	A. The Jump Frequency	382
	B. The Tracer Correlation Factor	389
III.	Ionic Conductivity	407
	The Drift Mobility	408
IV.	Chemical Diffusion	412
	The Chemical Diffusion Coefficient	412
V.	Conclusions	423
	References	424

I. Introduction

Shortly after the appearance of the first commercial electronic computers, Mayer and Ulam[1] independently suggested the use of machine-generated random numbers to direct the simulation of a physical process. This type of procedure came to be known under the colorful and all-embracing name of Monte Carlo, because of the analogy of such a simulation process with a game of chance, in particular, roulette.

Many years ago King (1951) suggested the use of the Monte Carlo method for simulating random walks of particles. Unfortunately, the limited Monte Carlo calculations performed at that time put such large demands on the modest computational facilities then available that progress was essentially stifled. However, by 1960, technological progress in computer design, engineering, and software had advanced to the point at which relatively small Monte Carlo calculations became a reality. In what has become a classic paper, Flinn and McManus (1961) reported the first comprehensive Monte Carlo simulations based entirely on a postulated solid-state diffusion process. That paper was soon followed by a number of others dealing with diffusion-related topics such as alloy ordering kinetics and radiation damage annealing by point defect migration (Beeler and Delaney 1963; Beeler, 1964; Scholz, 1966; Young and Elcock, 1966; Mehrer, 1969a,b). Calculations in a similar vein have continued up to the present day (see, for example, the conference proceedings edited by Arsenault *et al.* (1976). But as late as 1970, almost twenty years after the original suggestion made by King (1951), no Monte Carlo calculations had been made of quantities familiar in the "random walk" theory of tracer diffusion, ionic conductivity, and chemical diffusion in solids. By this time, too, exact analytical solutions to many topical diffusion problems were proving virtually impossible to obtain with the available analytical methods. One can cite, for example, problems such as diffusion in concentrated and ordered alloys, nonstoichiometric compounds, fast ionic conductors, and glasses. The time was ripe for completely new approaches: One of these has turned out to be Monte Carlo simulation of diffusion kinetics.

In 1971, Bennett and Alder reported on a Monte Carlo study of vacancy "persistency" effects in hard-sphere gases. In the course of that study they calculated tracer correlation factors for both mono- and divacancy diffusion in the fcc lattice. The values obtained were in very good agreement with analytically determined results. The paper probably marks the starting point from which the field of Monte Carlo computer simulation of tracer diffusion, ionic conductivity, and chemical diffusion in solids has evolved. This review is an attempt to provide an exposition of the development of this field.

It is important to draw a careful distinction between the Monte Carlo random-walk calculations described here, which are based implicitly on *discrete* hopping models, and the Monte Carlo calculations of defect formation energies, etc., in *continuous* space systems. These calculations share no more in common than the fact that both use machine-generated random

[1] See Metropolis *et al.* (1953).

numbers. The latter calculations have been reviewed by Bennett (1975); more recent studies are included in Chapter 8 by Jacucci.

II. Tracer Diffusion

In the conventional random-walk theory of isotropic solid-state diffusion, it is customary to express the tracer diffusion coefficient D^* as

$$D^* = a^2 \Gamma f / \gamma \tag{1}$$

where a is the jump distance, Γ the mean atomic jump frequency, f the well-known Bardeen and Herring (1952) correlation factor, and γ a geometric constant depending on the dimension. Let us focus on Γ and f.

A detailed calculation of Γ requires consideration of the dynamics of the jumping process and the equilibrium defect concentration. Progress along analytical lines for both classical and quantum systems has been reviewed by Franklin (1975). For classical systems, the computer simulation methods of molecular dynamics and molecular statics have been especially popular, and we refer to the review by Bennett (1975) and Chapter 8 by Jacucci. The Monte Carlo method described in this chapter is formulated toward the kinetics of the diffusion process, i.e., the random-walk aspects of the diffusion process. It would appear at first to be an inappropriate technique for the study of Γ. Nonetheless, by assuming a discrete hopping model we can introduce the jumping atom's environment, by way of bonds or interaction energies (both at a lattice site and, for that matter, at the saddle point), into a factor that forms part of Γ. In this way, a simple model for diffusion in alloys or nonstoichiometric compounds can be studied rigorously with the present Monte Carlo method. This forms the subject of Section II.A.

The importance of f comes about because of the possibility of identifying the diffusion mechanism through either the measurement of the isotope effect (Peterson, 1975) or the Haven ratio (Le Claire, 1970; Murch, 1982g), both of which depend on f. A detailed calculation of f requires a study of the various possible tracer-atom trajectories. A very large body of literature has grown up around the analytical calculation of f, and we refer to the excellent treatise on the earlier calculations by Le Claire (1970). In many modern instances one is interested in the calculation of f for complex diffusion mechanisms or where the jump frequency can vary from site to site, both of which occur, for example, in nonstoichiometric compounds. The analytical calculations then become so tedious and inevitably require some numerical analysis for their solution that we might just as well address such problems from the point of view of a direct computer simulation. The Monte Carlo method provides a particularly convenient and general means of performing this task. In Section II.B the various ways that have been devised to extract f from Monte Carlo simulations are reviewed.

A. The Jump Frequency

For metals and simple stoichiometric binary compounds one usually writes the jump frequency Γ as

$$\Gamma = Zv \exp(-G^m/kT)\theta_d \tag{2}$$

where Z is the coordination number, v the fundamental lattice vibration frequency, G^m the free energy of migration, and θ_d the concentration of defects. With respect to computer simulation in continuous space, one can, in principle, calculate G^m with the molecular dynamics or molecular statics methods, and one can also calculate θ_d with the molecular dynamics or Monte Carlo methods (Bennett, 1975). In solids with order (short or long range), such as found in substitutional alloys, fast ionic conductors, and nonstoichiometric compounds, one is required to generalize Eq. (2) because Γ varies from site to site. One can invoke a discrete hopping approximation so that one rewrites Eq. (2) as (for, say, the unary system, i.e., the lattice gas)

$$\Gamma = Zv \exp(-E_0^m/kT)WV \tag{3}$$

where E_0^m is the energy of migration of an isolated atom, V the mean vacancy availability factor, which is a more general quantity than θ_d, and W is referred to as the mean bond-breaking factor or transition probability in the presence of a neighboring vacancy. The factor W depends directly on the environment of the jumping species. The choice for W at the microscopic level is restricted to those transition probabilities that satisfy detailed balance (Binder, 1979). A common choice is to write, in the case of say nearest neighbor (nn) interactions of strength ϕ_{nn},

$$W_i = \exp(Z_i\phi_{nn}/kT), \qquad Z_i\phi_{nn} \leq E_0^m \tag{4}$$

where Z_i is the number of occupied nearest neighbor sites about the atom at site i. In this case, only the environment at the initial state is included. This will be an appropriate assumption for potentials that decrease rapidly with distance. Another choice for W_i, which is somewhat more realistic in many cases, is to factor the potential into the saddle point as well (Singer

7. SIMULATION OF DIFFUSION KINETICS

and Peschel, 1980), but this has not been used in Monte Carlo simulations so far. The principal problem with transition probabilities like Eq. (4) is that, for attractive interactions especially, a large amount of computer time is expended in waiting for jumps. Some authors, such as Kutner *et al.* (1982), have tried to avoid this problem to a large extent by writing W_i as

$$W_i = 1 - \tanh(\Delta E/2kT) \tag{5}$$

where ΔE is the energy change of the system that would result if the jump took place. This choice is obviously much more "efficient" than Eq. (4), but it is not, unfortunately, a realistic transition probability for most modeling work using a lattice gas.

Once equilibrium has been established both V and W can be calculated in the usual way for a mechanical property (Valleau and Whittington, 1976) as exemplified here by the quantity A,

$$\langle A \rangle = \frac{\sum_\Omega A(\Omega) \exp(-E(\Omega)/kT)}{\sum_\Omega \exp(-E(\Omega)/kT)} \tag{6}$$

where $E(\Omega)$ is the energy of configuration Ω. Thus any transition probability satisfying detailed balance could be used, e.g., the well-known nondiffusive one of Metropolis *et al.* (1953), where jumps of any distance are allowed. In this case, V and W would be calculated in a hypothetical sense. Alternatively, V and W could be calculated directly in an actual simulation of nearest neighbor jumps by using say Eq. (4). No difference has been found between the two methods.

There has been a very large number of calculations of V and W in lattice gas systems. (In such systems discrete atom hopping is assumed between neighboring sites.) For the purpose of modeling ionic conductivity in fluorite related oxides, Barker and Knop (1971) calculated the equivalent of $V \cdot W$ for the vacancy mechanism in a simple cubic lattice gas with various site exclusion rules. Their results were collected for an unequilibrated array and cannot be judged wholly reliable. Murch (1975a) calculated the equivalent of $V \cdot W$ for the interstitialcy mechanism in the anion sublattice of CaF_2 with nearest neighbor exclusion between interstitials; a similar calculation was also performed for a defect complex in the same lattice. These calculations were intended to simulate O diffusion in UO_{2+x}. Murch *et al.* (1974) calculated the equivalent of $V \cdot W$ for a vacancy mechanism for iron diffusion in the $Fe_{1-\delta}S$ for various site exclusion models. Murch and Thorn (1977a,b) calculated V and W for the honeycomb lattice gas for nearest neighbor interactions (attractive and repulsive) for the cases in which all sites were equivalent a priori and in which there was site inequivalence. These calculations were intended to model Na diffusion in β''- and β-alumina. Results from these calculations showed the essential validity of the analytical calculations of Sato and Kikuchi (1971) who used the path probability method. Some results of these calculations are illustrated in Figs. 1 and 2, where we have presented results from Murch and Thorn (1977a) on the behavior of V and W in the honeycomb lattice gas with nearest-neighbor interactions. Given Eq. 4 and the sign of the interaction energy, the general behavior of these quantities follows what is intuitively obvious. Other calculations include the determination of V and W for the fcc lattice gas with attractive interactions (Murch and Thorn, 1977c), with repulsive interactions (Murch and Thorn, 1979a), and the simple cubic lattice gas with attractive and repulsive interactions (Murch and Thorn, 1977d). These calculations were all directed to model interstitial solute diffusion in metals. Murch (1980a) has examined the equivalent of $V \cdot W$ in the simple cubic lattice with nn exclusion. Calculations have also been made for V and W in the square-planar lattice gas with nn repulsion (Murch, 1982a) and with next nearest neighbor (nnn) attraction (Murch, 1981a). These calculations were directed to fast ion transport and adsorbate diffusion, respectively. All of the calculations above have made use of the transition probability as stated in Eq.

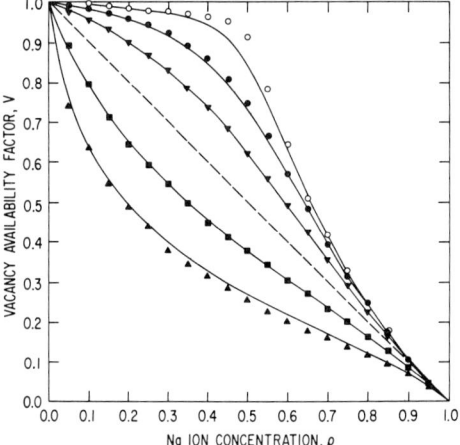

Fig. 1. The vacancy availability factor in the honeycomb lattice gas as a function of ion concentration, illustrating the effect of the strength of nn interactions ϕ_{nn}. The symbols ▲, ■, ▼, ●, ○ correspond to kT/ϕ_{nn} values of -0.5, -1.0, 1.0, 0.5, and 0.3, respectively; dashed line noninteracting. [From Murch and Thorn (1977a).]

(4). Kutner et al. (1982) calculated V and W for the fcc lattice gas with attractive interactions using Eq. (5) as the transition probability when a vacancy is in the nn position. Results thus obtained for V are similar to those found earlier by Murch and Thorn (1977c) and are typified by those shown in Fig. 1, since V does not depend on the choice of the transition probability. The results of Kutner et al. (1982) for W are shown in Fig. 3 for several temperatures. Note how their choice of the transition probability W_i leads to a symmetric behavior of W (cf. Fig. 2). So far, we have discussed only unary systems. Let us now focus on binary systems as exemplified by the binary substitutional alloy.

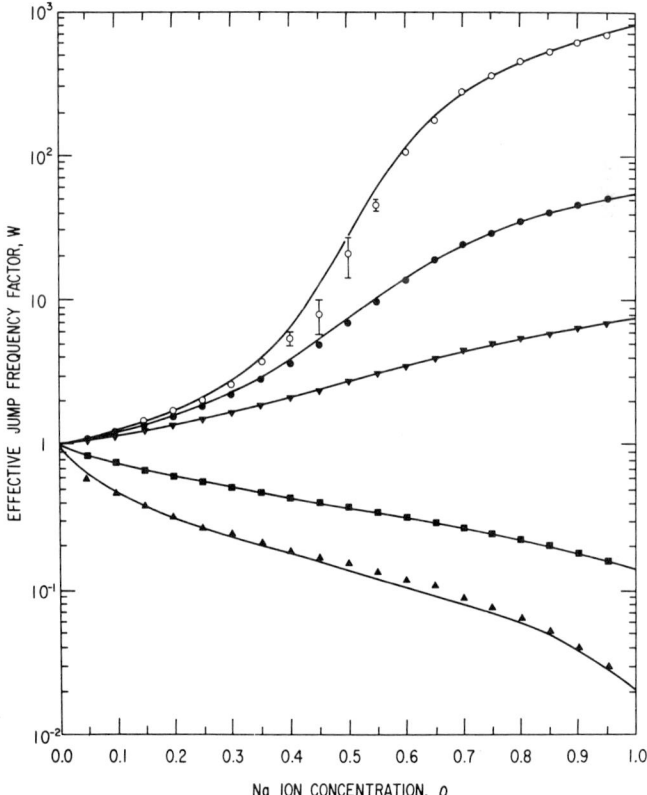

Fig. 2. The frequency factor W in the honeycomb lattice gas as a function of ion concentration: the effect of the strength of nn interactions ϕ_{nn}. The symbols ▲, ■, ▼, ●, ○ correspond to kT/ϕ_{nn} values of -0.5, -1.0, 1.0, 0.5, and 0.3, respectively. [From Murch and Thorn (1977a).]

For binary systems, one writes the equivalent of Eq. (3) for a component, say A,

$$\Gamma_A = Zv_A \exp(-E^m_{0,A}/kT)W_A p_{AV} \qquad (7)$$

where quantities of similar symbol have analogous meanings to those in Eq. (3) and p_{AV} is the probability of finding a vacancy next to an atom of type A: This is, of course, the binary analog of V. The first choice for W_A was that proposed by Flinn and McManus (1961) in their study of ordering kinetics in the bcc alloy. They proposed that

$$W_{A,i} = \exp(\Delta Z_{AB,i}\phi_{AB}/kT)/[1 + \exp(\Delta Z_{AB,i}\phi_{AB}/kT)] \qquad (8)$$

where ΔZ_{AB} is the number of unlike nn bonds that would be gained if the interchange were to occur between a vacancy and an atom at site i, and ϕ_{AB} is the energy of an unlike nearest-neighbor bond. Later, in a study of diffusion in an ordered alloy, Young and Elcock (1966) suggested using the well-known Metropolis et al. (1953) transition probability

$$W_{A,i} = \exp(\Delta Z_{AB,i}\phi_{AB}/kT), \qquad \Delta Z_{AB} < 0 \qquad (9a)$$
$$= 1, \qquad \Delta Z_{AB} > 0 \qquad (9b)$$

Neither of these transition probabilities provides a particularly realistic description of the diffusion process. In their analytical calculations based on the path probability method, Kikuchi and Sato (1969, 1970, 1972) suggested

$$W_{A,i} = \exp[(Z_{AA,i}\phi_{AA} + Z_{AB,i}\phi_{AB})/kT] \qquad (10)$$

where $Z_{AA,i}$ is the number of like bonds at the initial site i (similarly for

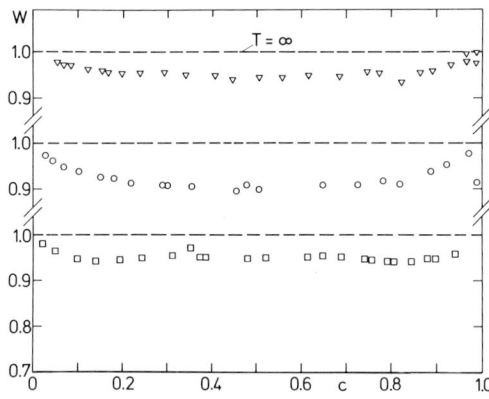

Fig. 3. The frequency factor W in the fcc lattice gas as a function of ion concentration: the effect of the strength of ϕ_{nn}. The symbols $\triangledown$, $\bigcirc$, $\square$ correspond to kT/ϕ_{nn} values of -5, -3, and -2.5, respectively. [From Kutner et al. (1982).]

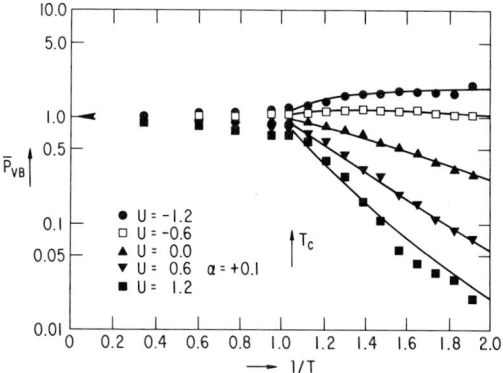

Fig. 4. The normalized availability factor $\bar{p}_{VB}$ in the bcc alloy lattice $A_{0.6}B_{0.4}$ as a function of reciprocal temperature: the effect of $U = (\phi_{AA} - \phi_{BB})/(\phi_{AA} + \phi_{BB} - 2\phi_{AB})$. [From Bakker et al. (1976); the solid lines are from Kikuchi and Sato (1969, 1970, 1972).]

$Z_{AB,i}$), and ϕ_{AA} the energy of an AA bond. Equation (10) corresponds to a potential that drops away rapidly with distance. This was later invoked in a Monte Carlo calculation by Bakker et al. (1976), see below.

Compared with the unary system, only a small number of Monte Carlo calculations of $W_{A(B)}$ and $p_{A(B)V}$ have been made on the binary system. Using Eq. (10), Bakker et al. (1976) made an extensive study of the bcc system at and in the vicinity of the stoichiometric composition with nn interactions above and below the order/disorder temperature. Some of their results on the temperature axis for $\bar{p}_{BV}$ (p_{BV} normalized by dividing by C_B) and $\overline{W}_B$ (W_B normalized by multiplying by $\exp(7\phi_{AB}/kT)$) are given in Figs. 4 and 5.

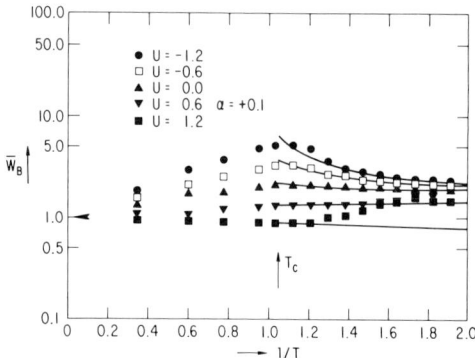

Fig. 5. The normalized frequency factor $\overline{W}_B$ in the bcc alloy lattice $A_{0.6}B_{0.4}$ as a function of reciprocal temperature: the effect of $U = (\phi_{AA} - \phi_{BB})/(\phi_{AA} + \phi_{BB} - 2\phi_{AB})$. [From Bakker et al. (1976); the solid lines are from Kikuchi and Sato (1969, 1970, 1972).]

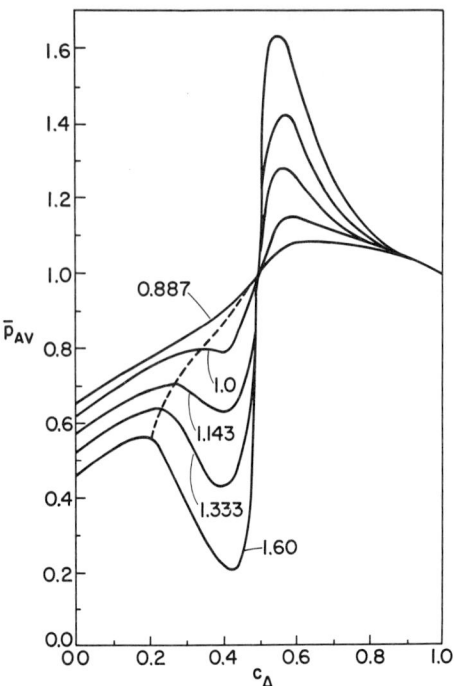

Fig. 6. The normalized vacancy availability factor $\bar{p}_{AV}$ in the sc alloy AB as a function of composition: the effect of the strength of interactions. The curves are labeled by $(\phi_{AA} + \phi_{BB} - 2\phi_{AB})/kT$. [From Murch (1982c).]

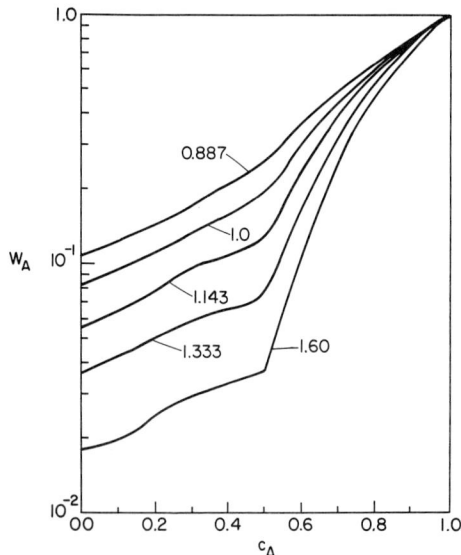

Fig. 7. The frequency factor W_A in the sc alloy AB as a function of composition: the effect of the strength of interactions. The curves are labeled by $(\phi_{AA} + \phi_{BB} - 2\phi_{AB})/kT$. [From Murch (1982c).]

7. SIMULATION OF DIFFUSION KINETICS

In general, reasonable agreement was found between these results and the results calculated from the path probability method of Kikuchi and Sato (1969, 1970, 1972). We refer also to Bakker's contribution in Chapter 4.

Murch (1982b,c) has studied the simple cubic system above and below the order/disorder boundary across the entire composition range. Again, the transition probability of Eq. (10) was used. Some of the results obtained on the composition axis for $\bar{p}_{BV}$ and W_B are shown in Figs. 6 and 7. The general behavior of these quantities follows what is intuitively expected as one enters and leaves the ordered phase, centered at $C_A = 0.5$.

B. THE TRACER CORRELATION FACTOR

1. The Einstein Equation Method

First let us write down the Einstein (1905) equation for the tracer diffusion coefficient D^* in one direction,

$$D^* = \langle X^2 \rangle / 2t \tag{11}$$

where $\langle X^2 \rangle$ is the mean-square displacement of an atom after time t in the x direction. The standard expansion of Eq. (11) is (Le Claire, 1970)

$$D^* = n\lambda^2 f / 2t, \qquad n \to \infty \tag{12}$$

where n is the number of displacements and λ is to be interpreted as the component of the jump distance in the x direction. Equation (12) in combination with Eq. (11) leads immediately to

$$f = \langle X^2 \rangle / n\lambda^2, \qquad n \to \infty \tag{13}$$

or, more generally, it can easily be shown that

$$f = \langle R^2 \rangle \bigg/ \sum_{\alpha=1}^{N} n_\alpha \lambda_\alpha^2, \qquad n \to \infty \tag{14}$$

where $\langle R^2 \rangle$ is the mean-square displacement in three dimensions and N the number of jump types.

Equations (13) or (14) immediately suggest a simple means of calculating f. By programming the random walk of a defect, say a vacancy, with machine-generated random numbers, one can record the individual displacements of a large number of atoms each represented as an element in a three dimensional matrix. After a large number of defect jumps, the *net* displacement of the atoms can be determined and one immediately forms Eq. (13) or (14). Since the array is usually made periodic, i.e., it is surrounded by images of itself, in the course of a run many atoms will obviously cross boundaries.

Nevertheless, their displacements are recorded as if no boundaries existed. Note that n is treated as an *average* number of atomic jumps.

This method of obtaining f was first used by Bennett and Alder (1971) in, incidentally, the first published Monte Carlo calculation of any kind of f. It has remained the most popular method, probably because of the obvious straightforwardness of the calculation. Typically, one uses an array of about 10^4 atoms. Obviously, one cannot realize the ideal infinite number of atomic jumps, but it has been found that with n taking an averaged value somewhere in the vicinity of 30 jumps per atom, one can routinely obtain f to within about a quarter of a percent of the exact analytical values for the common lattices (Manning, 1976; Murch, 1984a,b). One cannot achieve much better precision than this without substantially increasing the size of the array. A systematic study of f as a function of array size and number of jumps has not been undertaken, and would seem to be called for.

Let us now look at the applications of this method, first of all, in unary systems. In the course of their study of persistency effects in hard-sphere gases, Bennett and Alder (1971) calculated f for the vacancy and divacancy diffusion mechanisms in the fcc lattice. DeBruin and Murch (1973) calculated f for the vacancy mechanism in the fcc and sc lattices as well as a noncollinear anion interstitialcy mechanism in the fluorite lattice. Of particular interest, however, there was the calculation of f for a postulated mechanism for the Willis (1964) 2:2:2 defect complex in UO_{2+x}. The surprising result in this case was that f took a value of 1.587. This result has now been adequately explained from an analytical point of view (Murch et al. (1984). DeBruin and Murch (1973) also examined the effects of interactions among the defects. First, they investigated the functional dependence of f on defect concentration for self-blocking by the atoms. Next, they investigated the effect on f of nearest neighbor exclusion among defects. To do this, they first brought the array to thermal equilibrium using the well-known method of Metropolis et al. (1953), before proceeding to simulate diffusion. Invariably, the effect of such "hard-sphere type" interactions was to diminish f with increasing defect concentration. This is a manifestation of an effect which later came to be called the "physical correlation effect." Murch (1975a) further investigated the effect of such interactions on f for a noncollinear anion interstitialcy mechanism in the fluorite lattice. Murch (1975b) also studied f as a function of vacancy concentration in the self-blocking case in the fcc lattice.

Of particular interest in the theory of fast ionic conductors is the diffusive behavior of the unary lattice gas with nearest neighbor *soft* ion–ion interactions. Murch and Thorn (1977a) used Eq. (4) as their transition probability, but other choices of W_i are possible (see Section II.A) and some of these are clearly more appropriate to fast ionic conductors (Singer and

Peschel, 1980). Murch and Thorn studied the honeycomb lattice gas as a model for β''-alumina. To bring the system to thermal equilibrium they used the method of Metropolis *et al.* (1953) in the petit canonical ensemble. Because of the slow relaxation, a much faster method, used in later calculations of this type, uses the grand canonical ensemble for this part of the calculation. Some results for f for interionic repulsion are shown in Fig. 8. The prominent feature here is the existence of a minimum in the vicinity of 50% occupation. This minimum is not reproduced in the analytical calculations of Sato and Kikuchi (1971) (see Fig. 8.) The minimum is a consequence of the so-called physical correlation effect, wherein many tracer jumps are reversed in order to preserve local order. We should note that the depth of the minimum, perhaps even its existence, depends on the *choice* of the transition probability. Nor should such strong physical correlation effects be expected necessarily in a real fast ionic conductor unless the interionic potential drops rapidly with distance, as implicity assumed here (see Singer and Peschel, 1980).

Other calculations of f in the unary lattice gas along very similar lines include studies of (1) the simple cubic lattice gas with nn repulsion or attraction (Murch and Thorn, 1977d), (2) the fcc lattice gas with nn attraction (Murch and Thorn, 1977c), (3) the simple cubic lattice gas with nn exclusion

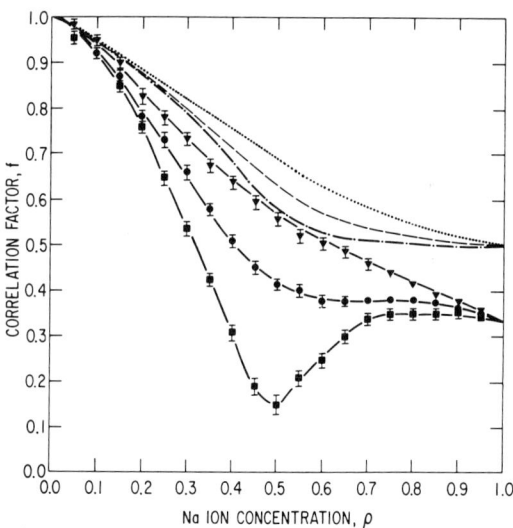

Fig. 8. The tracer correlation factor f in the honeycomb lattice gas as a function of ion concentration: the effect of the strength of nn ion–ion repulsion ϕ_{nn}. The symbols ●, ▲, ■ correspond to $k/T\phi_{nn}$ values of 0.5, 1.0, and 0.3, respectively. The curves without symbols are the analytical results of Sato and Kikuchi (1970). [From Murch and Thorn (1977a).]

(Murch, 1980a), (4) the square-planar lattice gas with nn repulsion at 50% occupation (Murch, 1982a), (5) the square-planar lattice gas with nnn exclusion and/or nn exclusion, and nn repulsion and nnn attraction (Murch, 1981b,c), (6) the fcc lattice gas with nn attraction (Kutner et al., 1982), and (7) the square-lattice gas with nn and nnn repulsion (Sadiq and Binder, 1983).

A slightly more complex addition to such lattice gas calculations is the inclusion of a difference in site potential energy a priori. Two calculations have been performed here. The first (Murch and Thorn, 1977b) was directed to the honeycomb lattice gas with nearest-neighbor repulsion and two types of site which were considered to exist alternately. The transition probability (Eq. 4) for the site of lower energy included another factor, $\exp(-w/kT)$, where w is the difference in site energy. The results are similar to those in Fig. 8. A second calculation, this time for the simple cubic lattice gas (Murch, 1982e) with self blocking, concentrated only on the effect of w/kT on the correlation factor. The results obtained again have the same form of Fig. 8.

A somewhat more complicated application has been to binary systems. In the spirit of the above the lattice gas with two atomic components still on a defective lattice has been studied (Sugimoto and Fukai, 1979). These authors investigated tracer correlation in the simple cubic lattice gas with nearest-neighbor attractions (independent of type) as a model for a ternary metal hydride such as NbH_xD_y. Hydrogen or deuterium differed in their ability to jump only in the term $\exp(-E_0^m/kT)$ [see Eq. (3)]. Some results are shown in Fig. 9 as a function of energy difference $\Delta E_0^m = E_0^m(D) - E_0^m(H)$.

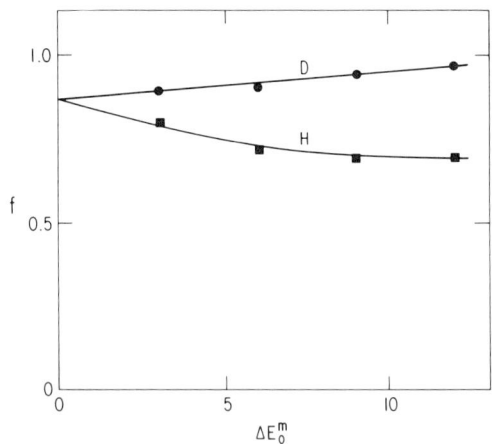

Fig. 9. The tracer correlation factor f in the sc two-component [deuterium (D) and hydrogen (H)], lattice gas at 40% occupation as a function of saddle point energy difference ΔE_0^m. Results were collected at $\phi_{nn}/kT = -0.5$. [From Sugimoto and Fukai (1979).]

7. SIMULATION OF DIFFUSION KINETICS

Sugimoto and Fukai concluded that a tagged H atom, which is surrounded by slower-moving D atoms, tends to be more correlated to its previous jump than if the other atoms are of the same type. A slower-moving D atom surrounded by faster-moving H atoms tends to have less correlation between its jumps. These conclusions are similar to the usual ones put forward to explain impurity diffusion (see, for example, Manning, 1968). More extensive studies in the fcc lattice gas, where the tracer was assumed to have a different jump rate from the host, have been given by Kutner and Kehr (1983).

Other applications of the Einstein equation method in binary systems include a study of tracer correlation in the five-frequency model for impurity diffusion in the fcc lattice (Murch and Thorn, 1978a). The results were in excellent agreement with Manning's (1964) analysis. Other studies have been directed to random systems consisting of two components and an exploring particle. The first of these is a study by Brandt (1975), who examined f for a particle that explored a random system consisting of varying fractions of "open," i.e., available, to "closed," i.e., unavailable, sites. He found that f, which he termed an "effective diffusivity," depended on the number of jumps and also exhibited an unphysical long tail below the percolation threshold for a network of infinitely connected "open" sites. The tail is due presumably to "local" diffusion. Similar behavior was found by Murch and Rothman (1981) when they calculated the vacancy or atomic tracer correlation factors in the alloy analog of Brandt's lattice, namely, a binary random alloy with one component immobile and with the vacancy mechanism operative. The vacancy is now the "exploring particle." Fortunately, the entire problem can be inverted so that the vacancy is eliminated and the atomic *exchange* mechanism between like (mobile) atoms is simulated. Many more configurations are thus effectively sampled. The correlation factor in this case for the mobile component is equal to the vacancy correlation factor f_v in the alloy example. This in turn can be related to the atomic tracer correlation factor f_i since (Manning, 1971)

$$f_i = f_0 f_v \tag{15}$$

where f_0 is the tracer correlation factor in a pure lattice. The results for f_v (and f_i) revealed the percolation threshold fairly well. An alternative method for calculating f_v in the random alloy from the drift of atoms in an external force is discussed in Section III.

Bocquet (1981) has used the present method to calculate f in the fcc random alloy with dissociating dumbell-like interstitials. He found good agreement between his results and those from his own mean-field calculation.

A calculation of f has been made with this method in the simple cubic binary alloy with order (Murch, 1982b,c). Murch chose Eq. 10 as the transition probability (see Section II.A for a discussion of some of the various

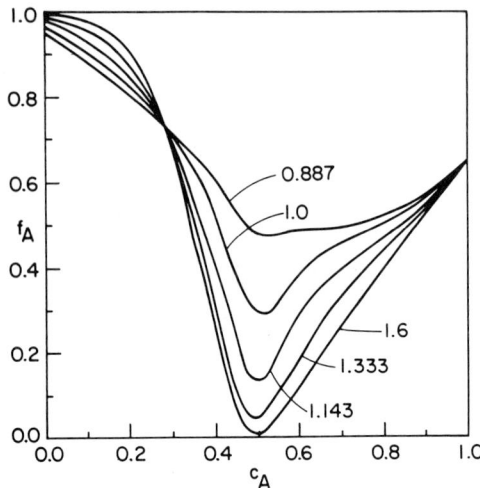

Fig. 10. The tracer correlation factor f_A in the sc alloy AB as a function of composition: the effect of the strength of interactions. The curves are labeled by $(\phi_{AA} + \phi_{BB} - 2\phi_{AB})/kT$. [From Murch (1982c).]

choices available). As in the unary system, in these calculations the familiar Metropolis *et al.* (1953) scheme in the grand canonical ensemble for bringing the system to equilibrium was used. A switch was then made to the petit canonical ensemble for the diffusion part of the calculation. In addition, at this point a single vacancy was introduced. Some results for f of the A component are shown in Fig. 10. Again the "physical correlation effect" in the vicinity of a concentration of 0.5, as discussed above with respect to the unary system, is very evident.

2. *The Estimator Method*

A variation of the Einstein equation method (see Section II.B.2) has been suggested by Stolwijk and Bakker (1977). These authors showed that when f is written as Eq. 14, the calculated f is in fact an unbiased estimate of the tracer correlation factor even when the average number of jumps per atom is very small, even less than one! The proof only applies (1) to systems where the defect concentration is very low, and (2) to pure systems and ideal systems such as random alloys. We make the formal distinction of defining this as a separate method from that given in Section II.B.1 when calculations put n equal to about 1, in contrast to about 30+ in the other method. To obtain adequate precision one normally must repeat the calculation, say 50 times, and average the results.

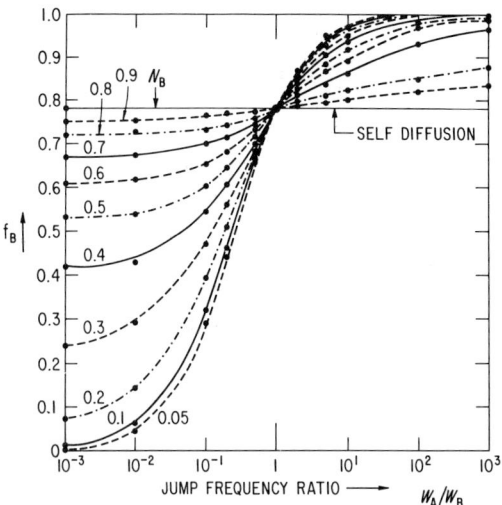

Fig. 11. The tracer correlation factor f_B in the random fcc alloy AB as a function of jump frequency ratio w_A/w_B: the effect of composition. [From de Bruin et al. (1974).]

The estimator method has been used on a number of occasions. De Bruin et al. (1974) used it in their study of correlation factors of the atomic components and the vacancy in fcc and bcc random alloys. In these calculations, the transition probabilities are predetermined by the specified atomic jump frequency ratio. Results for the frequency dependence of the B component in the fcc alloy are shown in Fig. 11. There is good agreement with the results of Manning's (1967, 1971) analysis, which are also shown in Fig. 11. Bakker et al. (1976) also used the estimator method in their study of tracer diffusion in the bcc ordered alloy with nearest neighbor interactions at compositions in the vicinity of stoichiometry. (See also II.A and Chapter 4.) They used Eq. (10) as their transition probability. Results for the temperature dependence of the correlation factor for the B component at various values of $U = (\phi_{AA} - \phi_{BB})/(\phi_{AA} + \phi_{BB} - 2\phi_{AB})$ are shown in Fig. 12. There is a large discrepancy between the Monte Carlo results and the analytical results of Kikuchi and Sato (1969, 1970, 1972) for this model. These discrepancies have now mostly been resolved in an improved path probability method calculation Sato (1980), but new results have not been published. However, the discrepancies may in fact be larger than that indicated in Fig. 12. In a preliminary study of the sc binary ordered alloy, this author found that the results for f from the estimator method are systematically higher in the long-range ordered alloy compared with those from the Einstein equation (see Section II.B.1). The results from the latter method were consistent with the vacancy-wind factor in the modified Darken equation, being equal to unity when both components have equal tracer diffusion coefficients (Murch, 1982f). This would seem to imply that the difficulty lies with the estimator method. Results for f from the two methods agree closely, however, in the disordered phase and in the random alloy.

De Bruin et al. (1977) have also used the estimator method in their detailed calculation of tracer correlation factors in the simple cubic random alloy. The results differed slightly from those obtained using the simulated matrix method (see Section II.B.3). This difference, which was greatest by far when the atomic jump frequencies differed by $\approx 10^3$, was interpreted as being indicative of the breakdown of the assumption that the jump vector has at least two-fold symmetry in the random alloy. This particular assumption is basic to the simulated matrix method. Later, Murch and Rothman (1981) studied the random alloy with one component immobile and obtained the tracer correlation factors using a method, unique to the random alloy, based on the calculation of the drifts in an external field (see Section III). They found good agreement with Manning's (1967, 1971) analysis, itself based on the two-fold symmetry assumption. Murch and Rothman (1981) suggested that the effect seen by de Bruin et al. might be due to inadequate sampling of the possible atomic networks when one component is virtually immobile.

Rothman et al. (1980) have used the estimator method in the calculation of tracer correlation factors in the random ternary alloy. This alloy served as a model for interpreting their tracer diffusion data in austenitic stainless

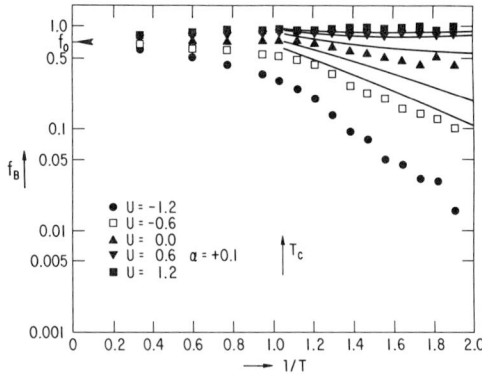

Fig. 12. The tracer correlation factor f_B in the bcc alloy lattice $A_{0.6}B_{0.4}$ as a function of reciprocal temperature: the effect of $U = (\phi_{AA} - \phi_{BB})/(\phi_{AA} + \phi_{BB} - 2\phi_{AB})$. [From Bakker et al. (1976); the solid lines are from Kikuchi and Sato (1969, 1970, 1972).]

steels. Lindström and Lindström (1979a, 1980, 1983) have used the estimator method in the calculation of tracer correlation factors for ions and vacancies when diffusion proceeds via an undissociating vacancy-pair mechanism in the NaCl lattice. The results were in good agreement with the analytical calculations of Compaan and Haven (1956) and Howard (1966). Lindström and Lindström (1979b) also used another method, called the moment method, in which f is calculated for *each* atom from Eq. (14), but with n being precisely the number of atomic jumps made by each atom rather than the average. Application to the fcc random alloy revealed that the moment method is far less efficient than the estimator method.

3. The Simulated Matrix Method

De Bruin et al. (1977) devised a Monte Carlo method of calculating f from the well-known equation (Le Claire, 1970) in cubic solids

$$f = (1 + t)/(1 - t) \tag{16}$$

where $t = P_+ - P_-$ and $P_\pm$ is the probability that the next tracer jump is in the same $(+)$ or opposite $(-)$ direction as its precursor. Evidently, it is the net probability that the tracer will return to its original position. The principal assumption behind Eq. (16) is the requirement that the basic jump vector is an axis of at least two-fold symmetry. This is equivalent to requiring that all jumps be of the same type.

Actual implementation of this method was inspired by the analytical matrix calculation of f. A simple cubic array of about 10^4 atoms was used with the vacancy occupying the reference origin. A tracer was introduced at $(-1, 0, 0)$, thereby suggesting the situation immediately after a tracer has jumped in the negative direction. The vacancy walk was programmed in the usual way with machine-generated random numbers, but if the vacancy jumped to $(-1, 0, 0)$, the run was terminated. If the vacancy jumped past the array boundary the run was also terminated. The quantity t was eventually calculated from

$$t = (n_+ - n_-)/N \tag{17}$$

where n_- is the number of tracer jumps to the origin, and n_+ the number to $(-2, 0, 0)$, and N the number of runs (typically 50).

De Bruin et al. (1977) used this method to calculate tracer correlation factors in the simple cubic random binary alloy. The results differed somewhat from those obtained with the estimator method (see Section II.B.2). This discrepancy was interpreted at the time as a breakdown of the assumption of two-fold symmetry of the jump vector but more recent work has cast some doubt on this (Murch and Rothman, 1981; see also the discussion in Section II.B.2).

A variant of the method has since been given by Wegener and Frischat (1982) using concepts dating back to Bardeen and Herring (1952). Wegener and Frischat claim that the method is simple and faster than the popular Einstein equation approach, but no comparative studies have been made. They applied the method to several well-known lattices and a series of complex crystalline forms of silica. Since Eq. 16 does not always apply (Kutner, 1983), caution is required before adopting this method.

4. Fick's First Law Method

For tracer diffusion into a chemically homogeneous host Fick's First Law is stated as

$$J^* = -D^*(\partial C^*/\partial x) \tag{18}$$

where J^* is the tracer flux, D^* the tracer diffusion coefficient, and $\partial C^*/\partial x$ the tracer concentration gradient. This equation has not been generally useful in experimental solid state diffusion studies because of the obvious difficulty in measuring the solid-state tracer flux. It does not present a difficulty in computer simulations, however, since we can readily probe for this quantity. On combining Eq. (18) with the well-known expansion of D^* from the Einstein equation [see Eq. (12)], we find that f is given by

$$f = -\frac{2J^*t}{n\lambda^2} \frac{\partial x}{\partial C^*} \tag{19}$$

This equation can, in principle, be used over *very short* times in a computer analog of a typical solid-state tracer diffusion experiment. That is, a source of tracers in the computer array is permitted to diffuse as a consequence of a programmed random walk of a defect, say, a vacancy. Guy et al. (1977) and Guy (1978) used this method to calculate f for the vacancy mechanism in the square planar lattice. They found that $f = 0.52$ with a standard deviation of 0.30 (the exact value is 0.47). This level of statistical uncertainty is unacceptable in a calculation of f. The reader should note that these calculations were interpreted incorrectly as a result of the false premise that the atomic and vacancy fluxes are unequal [see the comments by Manning (1981)].

A more appropriate method for using Eq. (19) is to set up steady-state conditions for the tracer flux (Murch and Thorn, 1979b). Accordingly, an inexhaustible reservoir of tracers is maintained at one end ($x = 0$) of the computer array, while at the other end, $x = 1$, the tracers are removed immediately to a sink. The array is made periodic in the usual way in the y and z directions. A single defect, say a vacancy, is introduced and its random walk is programmed in the usual way with machine-generated numbers. After simulating diffusion for time t, one can calculate J^*t, i.e., the *net* number of atoms crossing a given plane normal to the x direction or, indeed, averaged over all the available planes normal to the x direction. Since the other quantities in Eq. (19) are immediately available, one has an estimate of f. Results for some calculations of f using this and other methods are presented in Table I. Murch and Thorn (1979b) showed that for the cases where an exact result is shown, this method gives results about as accurately as the Einstein equation method (see Section II.B.1) for the same conditions (array size, number of defect jumps, computer time, etc.). The method would seem to have a future in the calculation of f in complicated structures or complex diffusion mechanisms where the Einstein equation method can be rather tedious to use.

TABLE I

SOME CALCULATED TRACER CORRELATION FACTORS AT INFINITELY LOW DEFECT CONCENTRATIONS

Lattice	Mechanism	Exact values (where possible) (Le Claire, 1970)	Simulated estimate	Reference
fcc	Vacancy	0.78145...	0.7813 ± 0.0011	Bennett and Alder (1971)
			0.7815 ± 0.0006	Murch and Thorn (1979b)
			0.7812 ± 0.0010	Murch and Thorn (1979b)
			0.782 ± 0.002	de Bruin and Murch (1973)
sc	Vacancy	0.6531...	0.6539 ± 0.001	Murch and Thorn (1979b)
			0.6526 ± 0.0012	Murch and Thorn (1979b)
			0.653 ± 0.002	de Bruin and Murch (1973)
bcc	Vacancy	0.72719...	0.7324	Wolf et al. (1977)
			0.7268 ± 0.0008	Murch and Thorn (1979b)
			0.7266 ± 0.0009	Murch and Thorn (1979b)
sp	Vacancy	0.46694...	0.4658 ± 0.001	Murch and Thorn (1979b)
			0.4671 ± 0.001	Murch and Thorn (1979b)
			0.52 + 0.3	Guy et al. (1977)
Fluorite	Anion noncollinear interstitialcy	0.9855...	0.987 ± 0.002	de Bruin and Murch (1973)
fcc	Divacancy	0.475 ± 0.024	0.458 ± 0.001	Bennett and Alder (1971)
			0.4579 ± 0.0005	Murch (1984b)

5. Fick's Second Law Method

For tracer diffusion into a chemically homogeneous host, Fick's Second Law is stated as

$$\frac{\partial C^*}{\partial t} = D^* \frac{\partial^2 C^*}{\partial x^2} \tag{20}$$

This equation has found extensive use in experimental solid state diffusion studies (see, for example, Chapter 1). To extract D^* from Eq. (20), one sets up initial and boundary conditions and solves Eq. (20). A well-known solution which immediately suggests a conceptually simple computer experiment is one corresponding to diffusion from a single-plane source into an infinite medium (Crank, 1975)

$$C^*(x, t) = M/2\pi D^* t)^{1/2} \exp(-x^2/4D^* t) \tag{21}$$

where M is the amount of diffusing substance. Expressing Eq. (21) in terms of f with the aid of Eq. (12), one finds

$$C^*(x, t) = M/(2\pi n \lambda^2 f)^{1/2} \exp[-x^2/(2n\lambda^2 f)] \tag{22}$$

To implement this method one introduces a plane of tracers (distinguished by a different digit) into a periodic array representing the crystal. A single defect, say a vacancy, is introduced and its random walk is programmed in the usual way by means of machine-generated random numbers. After some time t, one probes the array to determine the tracer distribution $C^*(x)$ in a way analogous to sectioning in an experiment. With this information in hand, f can be determined immediately from Eq. (22). The principal difficulty with the method is the large uncertainty in $C^*(x)$ simply because one cannot make the tracer source large enough. If the x dimension of the array is compressed in order to make the tracer source larger, one runs into the new problem that the "infinite" crystal condition is violated. One way over these problems, which is stated here for the first time, is to make every plane in the x direction a separate and independent tracer source by specifying different digits for each plane. After some time t one scans the arrays and builds up an average tracer concentration profile.

A closely related method makes use of the following solution to Eq. (2) for diffusion from a tracer source of width $2h$

$$C^*(x, t) = \frac{C_0^*}{2} \mathrm{erf}\left(\frac{h - x}{2(D^* t)^{1/2}}\right) + \mathrm{erf}\left(\frac{h + x}{2(D^* t)^{1/2}}\right) \tag{23}$$

or, with the aid of Eq. (12),

$$C^*(x, t) = \frac{C_0^*}{2} \mathrm{erf}\left(\frac{h - x}{(2n\lambda^2 f)^{1/2}}\right) + \mathrm{erf}\left(\frac{h + x}{(2n\lambda^2 f)^{1/2}}\right) \tag{24}$$

where C_0^* is the concentration at $t = 0$. This has been implemented by De Bruin and Murch (1973) by specifying tracer sources of width $2h$ in the computer array. In order to gain access to f the computed tracer profile can be graphically fitted to standard calculated profiles. Nonlinear least squares routines that employ user-supplied functions can also be useful here.

A general drawback of these methods, which are based on scanning a tracer concentration gradient, is the fact that the tracer distribution is affected directly by the atom distribution. The latter fluctuates considerably from plane to plane for such problems as binary alloys with order and lattice gases and this will be transmitted to the tracer gradient.

The final strategy for making use of Fick's Second Law should not really be considered in its own right but we mention it here for completeness. In the implementation of the Einstein equation method (see Section II.B.1), one calculates the mean square displacement R^2 (Eq. 14) for each atom i from the various component displacements in the three principal directions, i.e.,

$$R_i^2 = X_i^2 + Y_i^2 + Z_i^2 \tag{25}$$

Now to obtain the information for say X_i^2 one must have a file containing the current x coordinates for every atom referenced to $x = 0$ at $t = 0$. That is

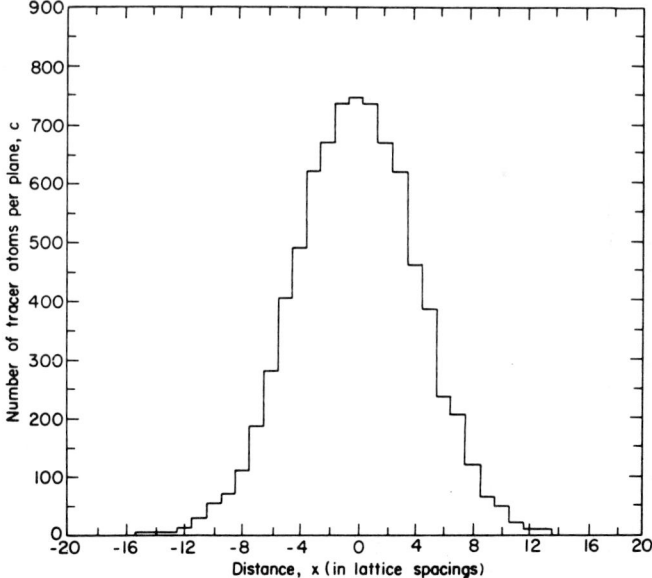

Fig. 13. A computer generated tracer concentration profile which has evolved from a δ-distribution of tracers at $t = 0$. [From Murch (1979a).]

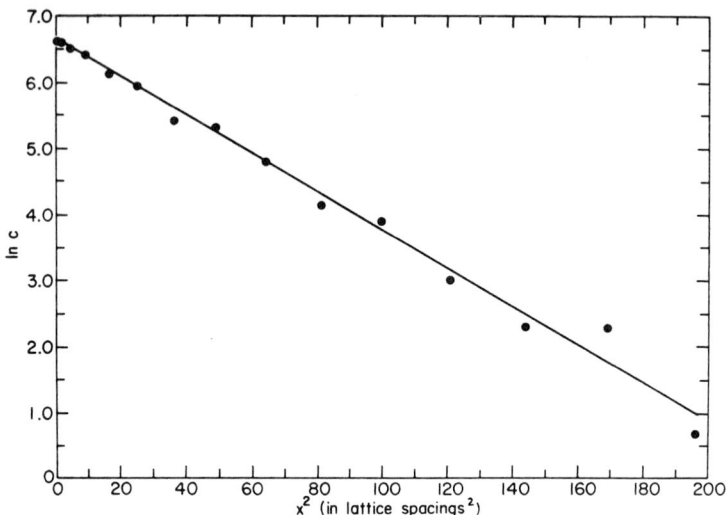

Fig. 14. An attempt to fit Eq. (22) to the tracer concentration profile presented in Fig. 13. [From Murch (1979a).]

to say every atom starts off from its own origin. With this information one can develop a quasi-tracer concentration profile by counting the number of atoms that have migrated, say, -20 lattice spacings, the number that have migrated -19, etc., up to $+20$. The result amounts to the equivalent of a tracer concentration profile that has evolved from a very thin tracer layer at $x = 0$ and $t = 0$. An example of such a profile is shown in Fig. 13. One can use Eq. 22 in order to extract f (see Fig. 14). Obviously, it would be preferable to calculate f from the Einstein equation, Eq. (14), since one already has the required information. Nonetheless, this procedure has pedagogical value, since it amounts to a Monte Carlo random-walk solution to the diffusion equation, Eq. (20) for the very thin tracer source diffusing into an infinite medium.[2] And, it shows the equivalence of the fickian and einsteinian definitions of D^*. In this context, a computer program has been made available (Murch, 1979a).

6. Waiting Time Distribution Method

Recently Kehr et al. (1981) have used the concept of waiting time distributions in a Monte Carlo calculations of f. Waiting time distributions

---

[2] This procedure has led to the development of a general Monte Carlo method of solution of the diffusion equation for such problems as diffusion in the presence of dislocations and diffusion-limited evaporation (Murch, 1984c).

7. SIMULATION OF DIFFUSION KINETICS

were developed originally by Montroll and Weiss (1965) in their analysis of continuous random walks.

Let $\psi(t)$ be the waiting time distribution of a particle that has made a jump at $t = 0$. Then $\psi(t)dt$ is the joint probability that the particle has made no jump until time t and performs a jump in the interval between t and $t + dt$. For tracer diffusion one considers $\psi_b(t)$ for reverse jumps and $\psi_f(t)$ for forward jumps. Only $\psi_b(t)$ is expected to exhibit the usual tracer correlation effects. It should be noted that waiting time distributions for forward jumps can often be different for various orientations of the jump with respect to the previous jump. Kehr et al. (1981) made the assumption that memory effects to the previous jump only were important. This is the usual assumption in tracer correlation theory (Le Claire, 1970). They were able to derive a generalized frequency-dependent tracer diffusion coefficient by way of the self-correlation function. The usual tracer correlation factor was given in the static condition by

$$f = \frac{1 - \tilde{\delta}(0)}{1 + \tilde{\delta}(0)} \qquad (26)$$

where $\tilde{\delta}(iw)$ is given in general by a weighted combination (dependent on the lattice) of the $\tilde{\psi}_i(iw)$, which are the laplacian transforms of the $\psi_i(t)$ (the subscript i denotes a direction). Note that Eq. (26) is of the form (Le Claire, 1970)

$$f = \frac{1 + \langle \cos \theta \rangle}{1 - \langle \cos \theta \rangle} \qquad (27)$$

where $\langle \cos \theta \rangle$ is the average of the cosine of the angle between two consecutive jump vectors. This equivalence to Eq. (26) is to be expected, since Eq. (27) also carries the assumption that only a memory between consecutive jumps of a tracer is considered.

The simulation itself follows the usual practice of setting up an array with periodic boundaries and programming the random walk of introduced defects by means of machine-generated random numbers. Kehr et al. (1981) computed waiting time distributions for different directions, i.e., the $\psi_i(t)$, integrated them to obtain $\tilde{\psi}_i(t)$, weighted them, and calculated $\tilde{\delta}(0)$ and finally f from Eq. (26). This method was implemented in the study of f as a function of occupation in the fcc lattice gas with self-blocking by the particles. The results obtained were in quite good agreement with those obtained using the Einstein equation method (see Section II.B.1).

Kehr et al. (1981) also made use of a closely related method that differs only in the calculation of the $\psi_i(t)$. They introduced the expression

$$\psi_i(t) = \Gamma_i(t)\Psi(t) \qquad (28)$$

where $\Gamma_i(t)$ are time-dependent jump rates for jumps in specified directions, which are defined under the condition that no jump of the tracer has occured from $t = 0$ until time t, and $\Psi(t)$ is the sojourn probability that a tracer which arrived at a site at $t = 0$ has made no jump until time t. The Γ_i each were expressed approximately in terms of the jump rate Γ. Equation (28) serves to gain access to the $\psi_i(t)$, which were then integrated to give $\tilde{\psi}_i(t)$, $\delta(0)$, and, finally, f. This approximate method was again implemented in the study of the fcc lattice gas described above. The results were in good agreement with those from the more direct calculation of $\psi_i(t)$ and from the Einstein equation. Further applications have been to the fcc lattice gas with nearest-neighbor repulsion, with particular emphasis on the ordered region (Kutner et al., 1984) and to the lattice gas, where the tracer has a different jump frequency from the host (Kehr, 1983). Kutner's (1983) work on the honeycomb lattice gas (where correlation effects are very prominent) showed that it is necessary to include correlations over multiple steps to obtain agreement with f obtained directly. That is to say, Eq. 26 is not general.

7. The Encounter Model Method

The essence of the encounter model (see, e.g., Wolf 1979a,b) lies in the distinction between those successive jumps of a tracer induced by the same point defect and those jumps that are a consequence of different point defects. Specifically, an encounter of a tracer with, say, a vacancy begins with the initial random exchange between the tracer and vacancy, and ends when a different vacancy interacts with that tracer.

One formally defines an encounter of the tracer with a point defect as the correlated rearrangement, in space and time, of the tracer and its neighbors as a result of an initial random exchange between the tracer and point defect. Three types of correlation are usually identified (Wolf, 1979b). These are (1) spatial correlations that described correlations between directions of successive jump vectors of the tracer (the familiar tracer correlation factor f is a result of this effect), (2) temporal correlations that describe the bunching of tracer jumps into groups of correlated jumps as a result of interaction with the same point defect, and (3) pair correlations that describe the fact that jumps of atoms in the vicinity of the tracer are correlated in both direction and time with the jumps of the tracer itself.

We shall deal here only with spatial correlations. Temporal and pair correlations arise in NMR and Mössbauer diffusion experiments, but space prevents us from dealing with these in this review. Wolf (1974a,b; Wolf and Differt, 1977) and Wolf et al. (1977) have made extensive Monte Carlo calculations in these other cases. The interested reader is directed to these papers for details.

7. SIMULATION OF DIFFUSION KINETICS

Let $P_{enc}(\mathbf{r}_m)$ represent the probability that as a result of a single encounter, the tracer has moved from the origin $\mathbf{r}_m = 0$ to any lattice site $\mathbf{r}_m$, which includes the origin. The mean squared displacement $\langle R_{enc}^2 \rangle$ of the tracer atom due to a typical encounter then reads

$$\langle R_{enc}^2 \rangle = \sum_{\mathbf{r}_m} P_{enc}(\mathbf{r}_m) |\mathbf{r}_m^2| \tag{28a}$$

If we now introduce Z_{enc}, the number of tracer jumps per encounter, f can be expressed as [cf. Eq. (14)],

$$f = \langle R_{enc}^2 \rangle / Z_{enc} a^2 \tag{29}$$

where a is the jump distance. Values of $\langle R_{enc}^2 \rangle$, Z_{enc}, and, therefore, f can be computed in a Monte Carlo simulation in the following way (Wolf and Differt, 1977; Wolf et al. 1977). The simulation does not make use of a periodic array like most of the other methods. Instead, one operates on a sphere of sites centered on the origin of the coordinate system. One initiates the simulation by placing a vacancy at the origin, while the tracer atom sits at a nearest-neighbor position. This presumes a tracer atom–vacancy exchange prior to $t = 0$. The random walk of the vacancy is programmed in the usual way with machine-generated random numbers. After about 600 jumps, the cycle is terminated and Z_{enc} and $\langle R_{enc}^2 \rangle$ are calculated. The whole process is performed about 1000 times in order to obtain good statistics to describe an "average" encounter. Further details of the simulation can be found in the papers by Wolf and Differt (1977) and Wolf et al. (1977) where the structure of the basic program is described in great detail. A machine-independent listing of the program has also been made available, and the interested reader is directed to the papers just mentioned.

The encounter model for tracer correlation was originally tested in Monte Carlo form with the calculation of the tracer correlation factor at very low defect concentrations for the vacancy mechanism in the well-known cubic lattices (Wolf et al. 1977) and the interstitialcy mechanism in the anion sublattice of the fluorite structure (Wolf et al., 1977). The results were in very good agreement with exact analytical values (Le Claire, 1970). More recently, the method has been extended to several cases in which the defect concentration is high. Wolf (1979a) proposed a comprehensive model for Na β-alumina that followed the observed site occupancies and included the charge compensating oxygen interstitials. The essence of the model is a noninteracting (self-blocking) lattice gas with an *ad hoc* interstitialcy mechanism superimposed on a lattice with randomly distributed pinning points or associated regions. Wolf used the encounter model method to calculate f as a function of interstitialcy concentration. Wolf (1979b) also studied various cubic lattice gases with self-blocking and calculated f as a function of vacancy concentration. However, those results are in rather poor agreement with accurate Monte Carlo calculations that make use of the Einstein equation method (see Section II.B.1 and Murch, 1984a). This suggests some difficulties in the definition of an encounter in such a situation. Apart from these studies of models for highly defective solids, Peterson et al. (1980) have used the encounter model method to calculate f for nine possible mechanisms for iron diffusion in Fe_3O_4. Correlation factors for four of these mechanisms are known exactly and in these cases the agreement was within the statistical uncertainty.

8. The Isotope Effect Method

The isotope effect method was first suggested by Murch (1984a). It makes use of a numerical analog of the experimental method of measuring diffusion isotope effects. One *assumes* that f can be written in the impurity form (for vacancy diffusion)

$$f = u/(u + w) \tag{30}$$

where u involves atom/vacancy exchange frequencies not involving the tracer (impurity), and w involves tracer (impurity)–vacancy exchange frequencies. In combination with the Einstein equation for a tracer of type α,

$$D_\alpha = A w_\alpha f_\alpha \tag{31}$$

where A contains geometrical terms and defect concentrations that do not depend on the exchange frequency w_α, it is easy to show (for tracers of type α and β) that (Schoen, 1958)

$$(D_\alpha - D_\beta)/D_\beta = f_\alpha(w_\alpha - w_\beta)/w_\beta \tag{32}$$

For assigned values of w_α and w_β, the LHS of Eq. 32 can be determined by way of the finite source solution to Fick's second law (Eq. 21) for both tracers (see Section II.B.5), that is to say

$$\ln(C_\alpha/C_\beta) = \text{const} - \ln C_\alpha[(D_\alpha - D_\beta)/D_\beta] \tag{33}$$

Accordingly, a plot of $\ln(C_\alpha/C_\beta)$ versus $\ln C_\alpha$ yields $(D_\alpha - D_\beta)/D_\beta$ and, with Eq. 32, f_α, which, for all practical purposes, is equal to f. Since everything depends on the validity of Eq. 30, then the method is mainly of use for testing this assumption for, as an example, high defect content solids or concentrated alloys as represented by the discrete lattice model. Application has been made to the simple cubic lattice gas with self blocking (Murch, 1984a). It was found that Eq. 30 was verified. The drawback with the method is the enormous amount of computer time required to get good statistics.

III. Ionic Conductivity

In pure ionic systems, where the defect concentration happens to be very low, the dc ionic conductivity $\sigma(0)$ is given simply in the hopping model formalism by [cf. Eq. (1)]

$$\sigma(0) = C_q q^2 a_q^2 \Gamma_q / \gamma k T \tag{34}$$

where C_q is the number of charge carriers per unit volume, q the charge, and the other quantities assume the same meanings as in Eq. (1), while the subscript q refers to a charge carrier. Clearly, no explicit calculation of $\sigma(0)$ is required beyond one for Γ.

The expression for the dc ionic conductivity is much more complicated, however, in ionic systems containing impurities or containing a high concentration of defects. In the first case, Manning (1968) showed that the contribution to the dc conductivity by the impurity $\sigma_i(0)$ is given by

$$\sigma_i(0) = C_{q,i} q_i^2 a_{q,i}^2 \Gamma_{q,i} \left[f_i \left(1 + \frac{2q_s}{q_i} \langle n_p \rangle \right) \right] / \gamma k T \tag{35}$$

where q_i and q_s are the charges on the impurity and the appropriate host ion, respectively, f_i is the correlation factor of the impurity, and $\langle n_p \rangle$ is a complex kinetic quantity. The latter factor depends on the jump frequencies near the impurity and is the result of the "vacancy-wind effect" or, more generally the "defect-wind effect," which distorts the equilibrium distribution of defects thereby changing the local disposition of defects with respect to a tracer atom (here the impurity) (Manning, 1968). Exact expressions for $\langle n_p \rangle$ and f_i exist in cubic lattices under the conditions of no impurity–vacancy binding and vanishingly small impurity content. Excellent approximate expressions for $\langle n_p \rangle$ and f_i also exist for such well-known impurity models as the five-frequency model in the fcc lattice. When the impurity has the same jump frequency (and the same jump distance) as the host and there is no impurity–vacancy binding, the term in parenthesis in Eq. 35 equals f_i^{-1} and the "correlation" in the conductivity vanishes. Under certain conditions it is possible (1) for the factor in square brackets in Eq. 35 to be negative, i.e., for the drift of the impurities to be against the direction of the field or (2) for this factor to be >1. These effects are discussed by Manning (1968). In the general case, where impurities are present at high concentrations, analytical expressions for $\langle n_p \rangle$ and f_i exist only for the random solution (alloy) model (Manning, 1968). Computer simulation based on the Monte Carlo method is, therefore, a natural method to use in the general case.

The second case for which the dc ionic conductivity assumes a more complex form than Eq. (34), is that of ionic conductors with high defect concentrations. As we pointed out in Section I, the most natural model here, which is consistent with the possibility of using the Monte Carlo method of this chapter, is the discrete lattice gas model. In this model, the pioneering analytical calculations of Sato and Kikuchi (1971), by using the path probability method, showed that the dc ionic conductivity takes the following form when interactions more extended than self blocking are present:

$$\sigma(0) = C_q q^2 a_q^2 \Gamma_q f_\mathrm{I}/\gamma kT \tag{36}$$

where f_I is the so-called physical correlation factor. This factor, which lies as $0 \le f_\mathrm{I} \le 1$, is a drift factor rather than a correlation factor in the strict Bardeen and Herring (1952) sense, as unfortunately suggested by the symbol f_I. It denotes a deviation from the equilibrium distribution at zero field when the unary lattice gas is placed in a static electric field. At very high field frequencies one presumes that the ions do not readjust quickly enough to the field polarity and the distribution remains the same as the distribution in zero field. By implication, $f_\mathrm{I} \equiv 1$ in the expression for $\sigma(\infty)$[3] and one may define f_I as

$$f_\mathrm{I} = \sigma(0)/\sigma(\infty) \tag{37}$$

The effect producing a nonunity value for f_I seems to be so closely related to the vacancy-wind effect that one cannot help thinking that they are manifestations of some more general phenomenon (see Murch, 1982g). Computer simulation based on the Monte Carlo method is again the most natural method to use in order to obtain exact results for f_I.

One general Monte Carlo method has been developed to deal with both of these cases: it forms the subject of the next section.

The Drift Mobility

The External Field Method

Only one Monte Carlo method has been designed to simulate ionic conductivity, and this seems to have been reported first by Murch and Thorn (1977e) and, independently, a little later by Richards (1977). One actually calculates not the ionic conductivity but the mobility u, which is related to $\sigma(0)$ by

$$\sigma(0) = C_q q u \tag{38a}$$

The mobility is usually partitioned as

$$u = \langle X \rangle / Et \tag{38b}$$

where $\langle X \rangle$ is the average drift at the charge carriers in time t under a static field of strength E.

---

[3] The symbol ∞ implies here a field frequency comparable with the jump frequency.

The concept behind the calculation of the key quantity $\langle X \rangle$ is very simple. Starting with a periodic array representing the lattice (see, for example, Section II.B.I) one imposes the electric field by introducing the quantity $\exp(-Eqa/kT)$ as a *further* conditional in, say, the $-x$ direction at each defect jump. This has the effect of drifting the charge carriers preferentially in the $+x$ direction (see Fig. 15). Combining Eq. (35) with Eqs. (38a) and (38b) for the case of impurity ionic conductivity, we have

$$f_i \left(1 + \frac{2q_s}{q_i} \langle n_p \rangle \right) = \frac{2kT \langle X \rangle}{Eqa^2 n_x} \tag{39}$$

where n_x is the number of jumps in the x direction in time t. And for defective solids modeled with the unary lattice gas we have, of course,

$$f_1 = \frac{2kT \langle X \rangle}{Eqa^2 n_x} \tag{40}$$

Thus we can gain access to the quantities in Eqs. (35) and (36), which give rise to the correlation in the ionic conductivity. For the rather strong fields commonly used in the computer simulations the rigorous expression $[1 - \exp(-Eqa/kT)]^{-1}$ replaces kT/Eqa in Eqs. (39) and (40) (Murch, 1979b).

A number of Monte Carlo calculations have now been made of these quantities. Murch and Thorn (1979c) calculated $f_i(1 + 2\langle n_p \rangle)$ for a selection of frequencies in the five-frequency model for impurity diffusion in the fcc

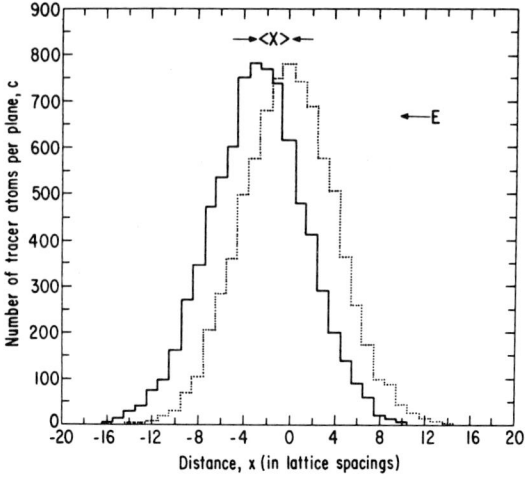

Fig. 15. Computer generated tracer concentration profiles infield and out-of-field having evolved from a δ-distribution of tracers at $t = 0$. [From Murch (1979b).]

lattice. The results were in excellent agreement with the analytical expression derived by Manning (1965). Murch and Thorn were able to demonstrate the unusual result predicted by Manning that under certain conditions $f_i(1 + 2\langle n_p \rangle)$ can be negative, that is, the flow of impurities can be upfield. Related Monte Carlo calculations have also been made in concentrated systems. In such systems, it is rather pointless to introduce more jump frequencies into the model in order to specify the large number of environments now encountered by a tracer. The reason is that one is now confronted with little more than an exercise in curve fitting with a large number of adjustable parameters (the jump frequencies) when applying the model to a real system. There are two approaches, both approximate in nature, that have been formulated to overcome this problem. The first is the postulate of a random distribution of both atomic components, e.g., the "random alloy." The jump frequencies are specified purely by the nature of the components themselves and not their environment. In the second approach, one specifies the jump frequencies indirectly through the interaction energies. Both approaches have been studied with the present Monte Carlo method. In an example of the first approach, Murch and Rothman (1981) studied the fcc random alloy. They showed that it was possible to relate the drift of component i to the partial vacancy correlation factor associated with that

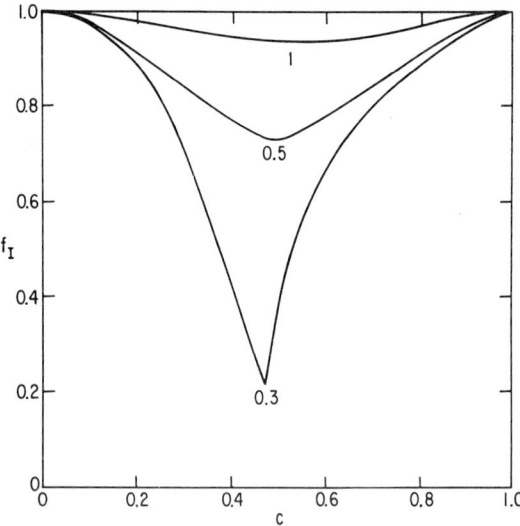

Fig. 16. The physical correlation factor f_I in the honeycomb lattice gas as a function of ion concentration: the effect of the strength of the nn ion–ion repulsion ϕ_{nn}. The curves are labeled by ϕ_{nn}/kT. [From Murch and Thorn (1977e).]

component. With this link, the method provided a particularly convenient means of studying correlation effects in the percolating random alloy, where one atomic component is immobile. Recently, further calculations have been directed to the simple cubic random binary alloy (Murch, 1982d) and, as an example of the second approach, to the simple cubic binary alloy with order (Murch, 1982f). Since the aim in these calculations was the understanding of vacancy-wind effects in chemical diffusion rather than ionic conductivity, we have delayed discussing these results until Section IV.6.

For highly defective solids, which are represented by unary lattice gas systems, there have been many Monte Carlo calculations of f_I. First, Murch and Thorn (1977e) studied the honeycomb lattice gas with nn repulsion and with all sites equivalent a priori. They employed the transition probability given in Eq. (4). In this model, the earlier analytical calculations of Sato and Kikuchi (1971), where the path probability method was used, gave the trivial result that $f_I \equiv 1$ at all temperatures and concentrations. The Monte Carlo results, on the other hand, gave the behavior shown in Fig. 16, which can hardly be described as trivial.[4] The characteristic minimum is partly a consequence of the form of the transition probability (Murch, 1982g) and also occurs in f, see Fig. 8 and Murch and Thorn (1977a). Again it is a consequence of the "physical correlation effect" (see Section II.B.1). Calculations along rather similar lines have also been performed for the simple cubic lattice gas with nn repulsion or attraction (Murch and Thorn, 1978b), the simple cubic lattice gas with nn blocking (Murch, 1980a), the square planar lattice gas with nnn and/or nn blocking (Murch, 1981b,c), the square planar lattice gas with nn repulsion and nnn attraction (Murch, 1981a), and finally the square planar lattice gas with nn repulsion at 50% occupation (Murch, 1982a).

A slightly more complex addition to such lattice gas calculations is the inclusion of a difference in site potential energy a priori. The first Monte Carlo calculation here was that undertaken originally by Richards (1977) for the 1D lattice where alternate sites are inequivalent and with self-blocking. The effect of site inequivalence is simply to reinforce the phenomenon causing the minimum in the nn repulsion case (see, e.g., Fig. 16). About the same time, a calculation by Murch and Thorn (1978a) was directed to the honeycomb lattice with nn repulsion with alternate sites inequivalent. This lattice gas, actually intended to be a model for Na β-alumina, was also studied analytically by Sato and Kikuchi (1971) with the path probability method. In contrast with their result for the lattice gas with all sites equivalent (see above), the agreement between their results and the Monte Carlo

---

[4] More recent analytical calculations give results that are similar to the Monte Carlo ones (Sato, 1980).

results was good. More recently, the simple cubic lattice gas with self blocking and with alternate sites inequivalent has been studied by Murch (1982e).

In the unary system, a further quantity of interest is the Haven ratio H_R (Le Claire, 1970; Murch, 1982g), which occurs in the relation connecting $\sigma(0)$ and the tracer diffusion coefficient

$$\sigma(0) = C_q q^2 D^*/H_R kT \tag{41}$$

For the lattice gas with single particle hopping, H_R is given by f/f_1. As we have seen, both f (Section II.B) and f_1 (this section) can be conveniently calculated by Monte Carlo methods and this provides us with the opportunity to form H_R and study its behavior. The Haven ratio has been the subject of a review and we refer the reader to that treatise (Murch, 1982g).

IV. Chemical Diffusion

THE CHEMICAL DIFFUSION COEFFICIENT

Chemical diffusion is the process by which a solid accommodates a change in chemical composition. The quantity of interest is the intrinsic diffusion coefficient D_i^I of component i, which is defined precisely by

$$J_i = -D_i^I \frac{\partial C_i}{\partial x} \tag{42}$$

where J_i is the flux of component i and $\partial C_i/\partial x$ is the concentration gradient of i. In the unary system, the intrinsic diffusion coefficient happens to be synonymous with the "chemical" diffusion coefficient $\tilde{D}$ (Manning, 1968). Sometimes D_i^I is also called the collective diffusion coefficient.

Most of the theory of chemical diffusion has been directed to systems that are electronic conductors or possess metalliclike conductivity. In such cases, the rate of change of composition is dependent only upon the rate of atomic migration. The general problem of chemical diffusion (formally ambipolar diffusion) in systems which possess mixed (ionic and electronic) conduction in comparable amounts has not been addressed from a fundamental and rigorous point of view.

It is possible to identify two classes of material in which chemical diffusion is of active interest with respect to an understanding at the atomic level. The first class consists of binary substitutional alloys. In this case, the intrinsic diffusion coefficient of component i can be written in the hopping model formalism as [Murch, 1982h); cf. Eq. (1)].

$$D_i^I = \frac{a^2 \Gamma_i}{\gamma} \left[\frac{1}{kT} \left(\frac{\partial \mu_i}{\partial \ln c_i} \right) \right] S_i \tag{43}$$

where μ_i is the chemical potential of component i and c_i is the mole fraction of i. The quantity in square brackets is the socalled "thermodynamic factor" and S_i is a factor arising from the vacancy-wind effect (Manning, 1968). This latter factor is of particular interest. An analytical expression exists for S_i in the case of the random alloy (Manning, 1968). Various other relations have been suggested for the general case (see, for example, Dayananda, 1981). Computer simulation based on the Monte Carlo method is a natural method to use in this case.

The second class of materials for which chemical diffusion is currently of interest consists of those solids that are termed highly defective *and* that have high electronic conductivity or metallic-like conductivity. Examples include most nonstoichiometric compounds, interstitial solid solutions, and intercalation compounds. In addition, the mobile surface layer in localized adsorption problems can also be considered as a two-dimensional defective material. The most natural model here is the discrete unary lattice gas model. In this case, the chemical diffusion coefficient takes the form [cf. Eqs. (1) and (43)]

$$\tilde{D} = \frac{a^2 \Gamma}{\gamma} \left[\frac{1}{kT} \left(\frac{\partial \mu}{\partial \ln c} \right) \right] f_{\mathrm{I}} \qquad (44)$$

where μ is the lattice gas chemical potential and f_{I} is the so-called physical correlation factor (see Section III.A). In the context of chemical diffusion, f_{I} apparently represents a further distortion of the distribution in the chemical potential driving force (Murch, 1982g). Probably, it should be interpreted as a vacancy-wind effect (Murch, 1982g). As in the case of the calculation of the dc ionic conductivity in fast ionic conductors, the Monte Carlo method is an excellent one with which to study f_{I}.

There are six techniques that have been devised to calculate quantities pertaining to chemical diffusion. The first five methods that have been devised calculate $\tilde{D}$ directly, in particular in the unary lattice gas system; no related calculations have yet been made for alloy systems. The sixth method, in contrast to the others, permits the calculation of the components entering into $\tilde{D}$ in the unary system and $D_i^!$ in the binary system. This has permitted a good deal of insight into the behavior of these diffusion coefficients and this seems rather desirable.

1. *The Time Autocorrelation Function Method*

This method, which was developed by Reed and Ehrlich (1981), makes use of the fact that the chemical diffusivity in the unary lattice gas system can be related to the time autocorrelation function $\rho(\tau)$ in a small region (arbitrarily assumed to be two-dimensional). Thus one writes $\rho(\tau)$ as

$$\rho(\tau) = \langle \delta n(\tau) \, \delta n(0) \rangle / \langle [\delta n(0)]^2 \rangle \tag{45}$$

where $n(\tau)$ is the number of particles in the small region a time interval τ after the previous calculation, and $\delta n(\tau)$ is the fluctuation in the number at time τ, i.e.,

$$\delta n(\tau) = n(\tau) - \langle n \rangle \tag{46}$$

where $\langle n \rangle$ is the ensemble average number of particles in the small region. Reed and Ehrlich (1981) were able to show that $\rho(\tau)$ is given by

$$\rho(\tau) = 1 - P/(1 - A/A'') \tag{47}$$

where A is the size of the small region and A'' is the size of a plane which has a large radius, compared with $(\tilde{D}t)^{1/2}$, and which contains the small region. The factor $1 - P$ is the conditional probability that material present in A at time $\tau = 0$ is also present at time τ. It is given by

$$1 - P = \frac{1}{N_s} \sum_{y_2} \sum_{y_1} \sum_{x_2} \sum_{x_1} W_{x_1,x_2}(\tau) W_{y_1,y_2}(\tau) \tag{48}$$

where N_s is the number of sites in A, and $W_{x_1,x_2}(\tau)$, the likelihood that material will be at x_2 at time τ having started at x_1, is given by

$$W_{x_1,x_2}(\tau) = \exp(-2\tilde{D}\tau/a^2) I_{x_2-x_1}(2\tilde{D}\tau/a^2) \tag{49}$$

where $I_j(\eta)$ is the modified Bessel function of order j.

The major drawback of this method is the assumption that the decay of concentration fluctuations is describable in terms of a chemical diffusion coefficient that is independent of concentration in the range possible in the computer experiment. In general, the chemical diffusion coefficient is a function of concentration, sometimes a sensitive one. However, it is not at all clear how serious a limitation this imposes on the method. Reed and Ehrlich (1981) implemented the method in a detailed study of the square planar lattice gas as a model for adsorbate diffusion. They used Eq. (4) as the transition probability. They studied the effect on $\tilde{D}$ of (1) self blocking, (2) nn repulsive interactions and (3) nnn attraction with nn repulsion. Figure 17 shows areas A and A'' for the nn repulsion lattice gas. The results for $\tilde{D}$ were in fair agreement with results obtained with the method based on Fick's second law (see also Section IV.3). However, in one case where the lattice gas exhibits long-range order [see case (3) above], Reed and Ehrlich did not find the expected peak in $\tilde{D}$ as generated by the thermodynamic factor (see Section IV.6). The cause of this deficiency is not known precisely.

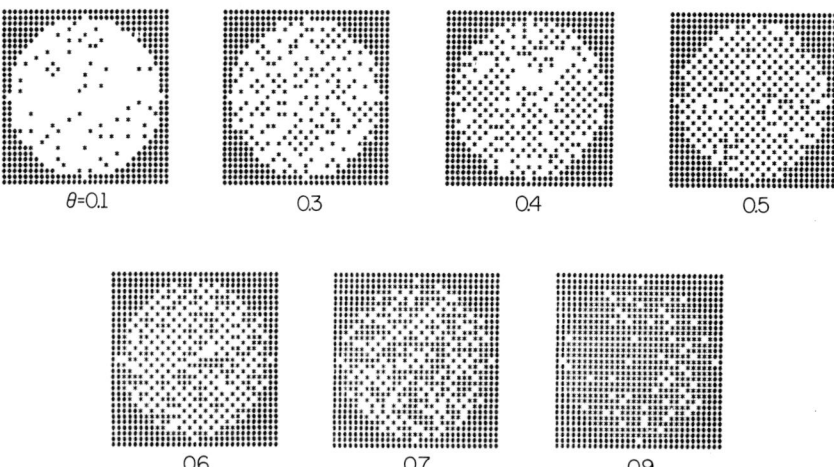

Fig. 17. Diagrams of areas A and A" for the square planar lattice gas with the ratio of nn repulsion to nnn $(\phi_{nn}/\phi_{nnn}) = -0.465$, and at a temperature of $\phi_{nn}/kT = 1.5$. The c (2 × 2) regions are evident for $0.4 \leq \theta \leq 0.6$. [From Reed and Ehrlich (1981).]

2. Fick's First Law Method

This method, which was developed by Murch (1980c), makes direct use of Fick's first law, Eq. (42) in order to calculate $\tilde{D}$ in the unary lattice gas. The method is implemented under steady-state conditions in a manner quite reminiscent of the way in which tracer correlation factors are calculated using a steady-state method and Fick's first law (see Section II.B.4). The method owes something also to the study by Gordon (1968) in which a steady-state flux was maintained across a lattice gas "membrane" of 32 sites. The primary interest in the earlier paper was the investigation of phase transition in a small system, and no diffusion parameters were assessed.

One starts with a periodic lattice gas array nominally in the petit canonical ensemble as depicted in Fig. 18. Two source/sink planes in the crystal are always maintained at differeing chemical potentials (and therefore differing concentrations) in a grand canonical ensemble sense. Upon simulating diffusion in the usual way (see Section II.B.1), one generates a steady-state particle flux between these planes both directly and via the periodic boundary. Since it can take quite a long time to reach steady-state conditions from an arbitrary starting configuration, the atoms can be distributed in a way which anticipates the intuitively expected steady-state distribution. Once at steady-state, the system is permitted to evolve for time t, during which

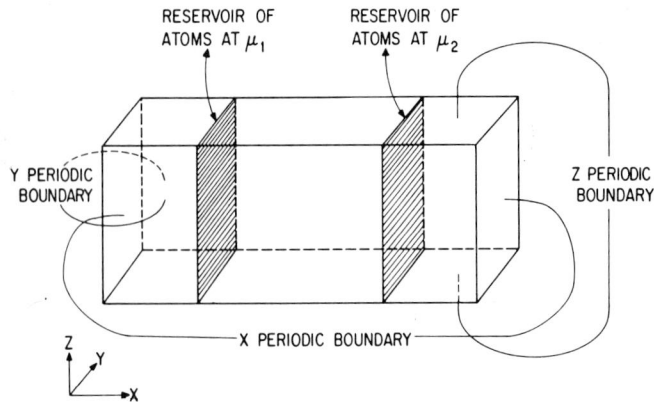

Fig. 18. Schematic diagram of the computer array in which steady state conditions are maintained. [From Murch (1980c).]

time average concentration profiles and the net flux of atoms passing across an average plane is calculated; $\tilde{D}$ is calculated simply from Eq. (42).

This method has been used once for the simple cubic lattice gas with nn attraction and nn repulsion between the diffusing particles (Murch, 1980c). The transition probability given in Eq. (4) was employed. High precision was found in the data, but the calculation may nonetheless have been rather forgiving, since temperatures were not examined below the order/disorder boundary (repulsive interactions). With this recognized, the disadvantages of this method are few, these being the requirement of knowledge of the concentration dependence of μ/kT and the difficulty in ascertaining whether the system is in the linear-response region.

3. Fick's Second Law Method

This method, which was first used by Bowker and King (1978a,b), makes use of the well-known Boltzmann–Matano analysis in order to compute the chemical diffusion coefficient from simulated concentration profiles. Like all of the other methods of the first five, it could be used to study the binary alloy system, but so far has been restricted to the unary lattice gas. One starts with Fick's second law

$$\frac{\partial C}{\partial t} = \frac{\partial}{\partial x} \tilde{D}(C) \frac{\partial C}{\partial x} \qquad (50)$$

The general solution for the concentration-dependent diffusion coefficient

7. SIMULATION OF DIFFUSION KINETICS

can be expressed as

$$\tilde{D}(C) = -\frac{1}{2t}\left(\frac{\partial x}{\partial C}\right)_c \int_{C_1}^{C} x \, dC \qquad (51)$$

where

$$\int_{C_1}^{C_0} x \, dC = 0$$

for the initial conditions

$$C = C_0, \qquad x < 0, \quad t = 0$$

and

$$C = C_1, \qquad x > 0, \quad t = 0$$

Bowker and King (1978a,b) started with a lattice gas consisting of two compartments. One compartment was entirely filled ($C_0 = 1.0$) while the other compartment was either entirely empty ($C_1 = 0.0$) or half-full ($C_1 = 0.5$). Diffusion was simulated in the usual way (see Section II.B.1). The transition probability used was that given in Eq. (4). After some time t, the array was scanned row by row so that the concentration profile could be established. The diffusivity $\tilde{D}$ was found numerically on application of Eq. (51).

Bowker and King (1978a) applied the method first to several one- and two-dimensional (square planar) lattice gases with self blocking, and, in the latter case, with nn interactions. These were of interest to the adsorption system O on W (110). Only temperatures above the critical and order/disorder temperatures were studied. In a later study, Bowker and King (1978b) examined $\tilde{D}$ for a lattice gas with nn repulsion and nnn attraction at a temperature that intersected the c(2 × 2) long-range ordered phase. They found a peak in $\tilde{D}(c)$ (see Fig. 19) that was roughly reminiscent of the peak found in experimental data for the O on W (110) system (Chen and Gomer, 1979; Butz and Wagner, 1977). A more precise study (Murch, 1981a) using the external field method (see Section IV.6) revealed substantial differences in the placement of data points, thereby suggesting problems with the evolution of the concentration profile. The method has also been used recently by Reed and Ehrlich (1981) as a reference calculation in their study of the lattice gas with nn interactions (see Section IV.1).

The main disadvantage of this method is the fact that the data seem to be subject to a rather high level of uncertainty. This may be due to the neglect of equilibration of the compartments prior to the simulation of diffusion and the fact that reflecting boundaries were used. Obviously these can be easily rectified.

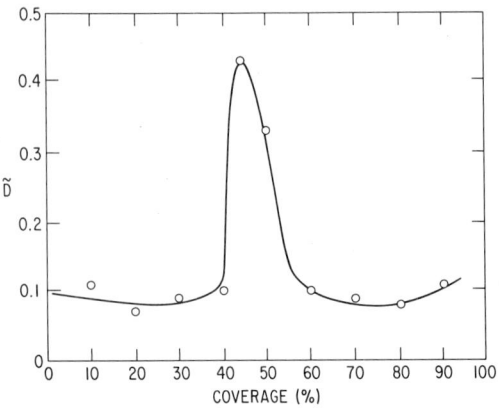

Fig. 19. The chemical diffusion coefficient $\tilde{D}$ in the square planar lattice gas as a function of occupation. The ratio of the strength of nn repulsion to nnn attraction is $\phi_{nn}/\phi_{nnn} = -0.465$, while the temperature is $\phi_{nn}/kT = 1.5$. [From Bowker and King (1978b).]

4. Time-Displaced Correlation Function Method

This method, which was developed by Sadiq (1974), makes use of the well-known Kubo (1957) expression for the chemical diffusion coefficient in the unary system

$$\tilde{D} = \frac{1}{dV\,kT} \frac{\partial \mu}{\partial C} \int_0^\infty dt \, \langle J(t) \cdot J(0) \rangle \qquad (52)$$

where V is the volume, d the dimension, and $J(t)$ the total particle flux. The method is implemented in a Monte Carlo simulation by programming the diffusion simulation in the usual way for the lattice gas (see Section II.8.1). At equilibrium, the time-dispaced microscopic fluxes are determined (averaged) in order to obtain $\tilde{D}$ directly. Sadiq (1974) used the method to determine the collective spin diffusion coefficient for an Ising magnet. His transition probability conserved the total magnetization (as in the lattice gas analog) *and* the total energy of the system. The results showed a fairly high level of statistical uncertainty, but this may be the consequence of a small system and insufficiently long runs.

Another alternative might be to impose a specific direction of particle flow as a consequence of a chemical potential gradient either in a steady state fashion or, over short times in a system that is relaxing. Again, Eq. (52) could be used but the calculation would reduce, at least in steady state form, to the Fick's first law method already described in Section IV.2.

5. The Relaxation Method

This method, which was developed by Kutner et al. (1982) makes use of the linear response of a lattice gas to a wave-vector imposed chemical potential. One starts with an arbitrary starting configuration of the lattice gas. A wave-vector-dependent chemical potential is superimposed on the external chemical potential in the usual grand canonical equilibration method (see, for example, Murch and Thorn, 1978d). The external chemical potential μ_{ext} is given the form

$$\mu_{ext} = \mu + \delta\mu \cos(q \cdot r_m) \tag{53}$$

acting at site m. This leads, of course, to a modulated structure in concentration of the same functional form provided that $\delta\mu$ is small enough that linear response applies. At this point, defined as $t = 0$, $\delta\mu$ is put equal to zero and petit ensemble conditions are enforced. Diffusion is simulated according to the usual recipe and an assigned transition probability, e.g., Eq. 5. We now wish to watch the relaxation or decay of the modulated structure. It is easy to show that the decay takes the form of a first-order relaxation

$$\delta C(q, t)/\delta C(q, 0) = \exp(-\tilde{D}q^2 t), \quad q \to 0 \tag{54}$$

and $\tilde{D}$ amounts to a type of rate constant for the first-order "self" reaction.

In practice, many equilibrium configurations of the grand canonical simulation are stored. These are used in short petit canonical simulations consecutively. The quantity $\delta C(q, t)$ is intended to mean an average of these runs.

Kutner et al. (1982) applied this method to the fcc lattice gas with nearest-neighbor attractive interactions with Eq. 5 as the transition probability. Some typical results for $\tilde{D}$ above the critical point are shown in Fig. 20. The minimum is the preprecursor to "critical slowing down" as the diphasic region is approached on the temperature axis. Another application, this time to the square lattice with nearest and next-nearest neighbor interactions, has been made by Sadiq and Binder (1983). Typical wavelengths for the chemical potentials in the calculations were $\lambda = (2\pi/q) = 16$ or 8 lattice spacings. Apart from the critical point, where the response must be nonlinear, the method appears to be quite satisfactory and rather simple to implement. Its only disadvantage probably is in the choice of the amplitude $\delta\mu$. If it is too small the modulations become lost in the noise of the usual concentration fluctuations; too large, and linear response will not apply. Presumably the optimum choice can be determined by the fit to Eq. (54).

A general disadvantage of all the methods discussed so far, in which $\tilde{D}$ is calculated directly, is precisely that fact. No information on the structure

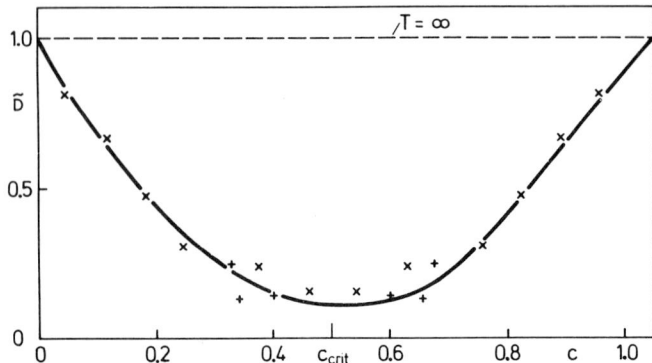

Fig. 20. The chemical diffusion coefficient $\tilde{D}$ in the fcc lattice gas with nn attraction at a temperature of $kT/\phi_{nn} = -2.47$. [From Kutner et al. (1982).]

of $\tilde{D}$ can be ascertained. The final method, discussion in Section IV.6, overcomes this disadvantage at the expense, however, of more labor.

6. *The External Field Method*

Of all the methods discussed for obtaining the chemical diffusion coefficient, only this one permits its detailed structure to be exposed. The background to the method has been the application of the phenomenological equations of nonequilibrium thermodynamics, which led to the realization that the terms S_i and f_1 in chemical diffusion [Eqs. (43) and (44)] can be extracted from simulations performed in a constant external field like the ionic conductivity simulations of Section III (Murch and Thorn, 1979d; Murch, 1982h) In particular, the quantity S_i in Eq. (43) can be expressed as

$$S_i = (s_i^0 - c_i s_i')/(1 - c_i) \tag{55}$$

with (cf. Eq. 39)

$$s_i^0 = 2kT\langle X_i^0 \rangle/q_i E n_i a^2, \quad q_i \neq q_j = 0, \tag{56a}$$

$$s_i' = 2kT\langle X_i' \rangle/q_i E n_i a^2, \quad q_i = q_j \tag{56b}$$

Similarly, f_1, appearing in the expression for $\tilde{D}$ in the unary system (Eq. 44), can be calculated from Eq. (40). The jump frequency Γ can be obtained using the methods described in Section II.A. Finally, the thermodynamic factor in Eqs. (43) and (44) can be determined by first calculating the μ/kT versus concentration isotherms, either trivially by means of the grand canonical ensemble or the particle insertion method in the petit canonical ensemble (Murch and Thorn, 1978d), and then differentiating by numerical means. In

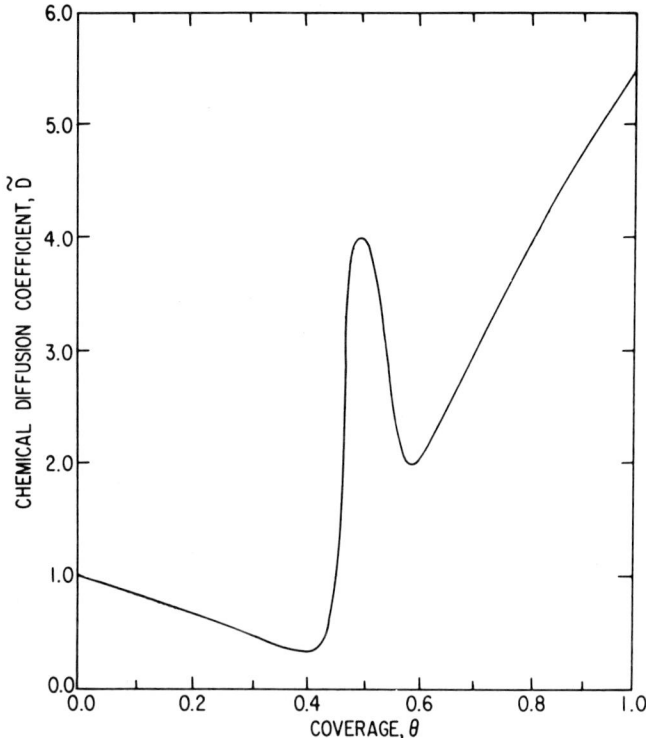

Fig. 21. The chemical diffusion coefficient $\tilde{D}$ in the square planar lattice gas as a function of occupation. The ratio of the strength of nn repulsion to nnn attraction is $\phi_{nn}/\phi_{nnn} = -0.465$, while the temperature is $\phi_{nn}/kT = 1.50$. [From Murch (1981a).]

this way $\tilde{D}$ or $D_i^!$ can be built up. Reed and Ehrlich (1981) have also attempted to calculate $\tilde{D}$ in the unary system in the same way, but, unfortunately, they have ignored the presence of f_1.

There have been several applications of this method. First, Murch (1980c) investigated the simple cubic lattice gas with nn attraction or repulsion and the transition probability given by Eq. (4). Results for $\tilde{D}$ built up from its components were in excellent agreement with those obtained by using a direct method based on Fick's first law (see Section IV.2). More extensive results have been presented more recently in a review article on chemical diffusion in defective solids (Murch, 1982h). Murch (1981a) also used the method to determine the cause of the peak in $\tilde{D}$ found in a Monte Carlo study by Bowker and King (1978b) of the square planar lattice gas with nn repulsion and nnn attraction (see Section IV.3 and Fig. 19). Murch found the rather different result as shown in Fig. 21, and showed unequivocally that the peak in $\tilde{D}$ is a consequence of a peak in the thermodynamic factor in the ordered region.

Another quantity of interest in the unary system is the Haven ratio H_R, which not only occurs in the relation connecting the dc ionic conductivity to the tracer diffusivity, Eq. (41), but also occurs in the Darken equation $\tilde{D}$ to the tracer diffusivity (Murch, 1982h), i.e.,

$$\tilde{D} = D^* H_R^{-1} \left[\frac{1}{kT} \left(\frac{\partial \mu}{\partial \ln c} \right) \right] \tag{57}$$

where H_R is given by f/f_1. This quantity has been discussed briefly in Section III and has recently been the subject of a review by Murch (1982g).

Murch (1982b,c,d,f) has published a series of papers dealing with the diffusion kinetics of the binary alloy. One of the objectives of that study is the behavior of D_i^I. Although the calculation of the thermodynamic factor, and, therefore, D_i^I itself, have not yet been completed, results for s_i' and s_i^0 have been collected: The former is shown in Fig. 22. The behavior of s_i^0 is quite similar. The behavior of the quantities is reminiscent of the behavior of f and f_1 in the unary system [Figs. (8) and (16)] and the tracer correlation

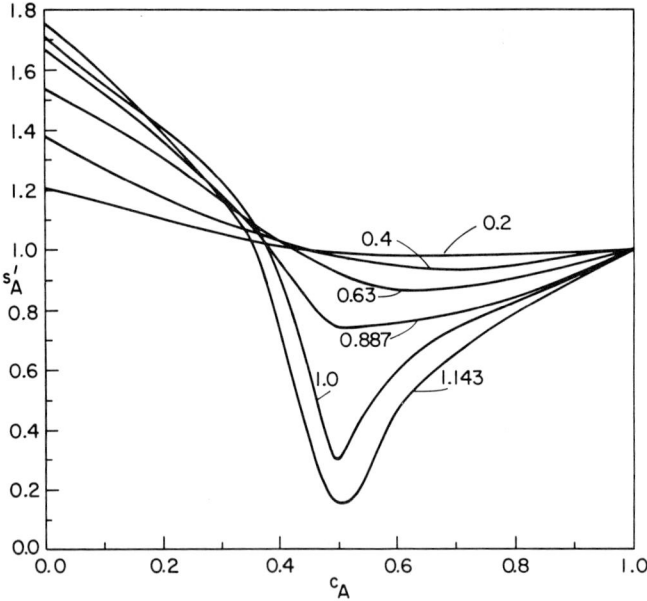

Fig. 22. The drift factor s_A' in the sc alloy AB as a function of composition: the effect of the strength of interactions. The curves are labeled by $(\phi_{AA} + \phi_{BB} - 2\phi_{AB})/kT$. [From Murch (1982f).]

factor in the alloy system (Fig. 10). The characteristic minimum is a manifestation of the physical correlation effect which is a consequence, to some extent, of the form of the transition probability, Eqs. (4) and (10), where the initial state determines the rate. Other transition probabilities such as those used in phase separation kinetics (Binder and Kalos, (1979)) would give quite a different behavior.

Another quantity of interest in the binary alloy system is the vacancy-wind effect factor r_i, which occurs in the Darken relation connecting D_i^I with the tracer diffusivity D_i^* (Stark, 1976; Murch, 1982d)

$$D_i^I = D_i^* \left[\frac{1}{kT} \left(\frac{\partial \mu_i}{\partial \ln c_i} \right) \right] r_i \tag{58}$$

An exact expression exists for r_i only for the random alloy (Manning, 1968). From Eqs. (55) and (43) it follows that r_i is given by

$$r_i = (s_i^0 - c_i s_i')/(1 - c_i) f_i \tag{59}$$

This quantity was calculated by Murch (1982f) directly and by using Manning's (1968) expression, which involves the component tracer diffusivities. Excellent agreement was found over a wide range of temperatures including the fully ordered alloy. This implies that Manning's expression is an excellent approximation for the alloy with order.

V. Conclusions

The Monte Carlo technique has proved itself to be of great value in understanding the diffusion kinetics in solids. This chapter has been directed mainly to the various methods that are possible with this technique. Further development will undoubtedly be limited only by available computing power and the ingenuity of its practitioners. It would seem appropriate here to make some recommendations concerning which of the present methods is simplest to implement for future applications. It must be admitted that for the straightforward calculation of the tracer correlation factor it is difficult to improve on the Einstein equation method (see Section II.B.1). The other methods are of interest only in the understanding of the physics of the correlation process. For ionic conductivity there is at present only one method, but see below. For chemical diffusion the choice is probably to be made among the Fick's first law method (Section IV.2), the relaxation method (Section IV.5), and the external field method (Section IV.6). The latter seems perhaps the easiest to implement in binary and ternary, etc., systems. A new method for ionic conductivity and chemical diffusion of possible major significance is that suggested by Allnatt (1982). He gives einsteinian expressions for the matrix of Onsager phenomenological coefficients, which are in a form suitable for Monte Carlo calculations *in a homogeneous system.* However, results for the fcc random binary alloy have been rather disappointing, particularly in the case of the cross-phenomenological coefficients (Allnatt, 1983; Allnatt and Allnatt, 1984).

ACKNOWLEDGMENTS

I wish to thank Dr. Hans Bakker and Prof. A. S. Nowick for critically reading the manuscript. Thanks are also due to Dr. K. W. Kehr, and Profs. A. R. Allnatt and R. L. Lindström for bringing some of their recent work to may attention. I wish to thank Katherine Archambeault for her careful typing of the manuscript.

REFERENCES

Allnatt, A. R. (1982). *J. Phys. C* **15**, 5603.
Allnatt, A. R., and Allnatt, E. L. (1984). To be published.
Arsenault, R. J., Beeler, J. R., and Simmons, J. A., eds. (1976). "Computer Simulation for Materials Applications," *Nucl. Metall.* **20**
Bakker, H., Stolwijk, N. A., van der Mey, L., and Zuurendonk, T. J. (1976). *Nucl. Metall.* **20,** 96.
Bardeen, J., and Herring, C. (1952). "Imperfections in Nearly Perfect Crystals," (W. Shockley, ed.). Wiley, New York.
Barker, W. W., and Knop, U. (1971). *Proc. Br. Ceram. Soc.* **19**, 15.
Beeler, J. R. (1964). *Phys. Rev. A* **134**, 1396.
Beeler, J. R., and Delaney, J. A. (1963). *Phys. Rev.* **130**, 962.
Bennett, C. H. (1975). *In* "Diffusion in Solids: Recent Developments" (A. S. Nowick and J. J. Burton, eds.), p. 73. Academic Press, New York.
Bennett, C. H., and Alder, B. J. (1971). *J. Phys. Chem. Solids* **32**, 2111.
Binder, K., ed. (1979). "Monte Carlo Methods in Statistical Physics." Springer-Verlag, Berlin.
Binder, K., and Kalos, M. H. (1979). "Monte Carlo Methods in Statistical Physics" (K. Binder, ed.), p. 225. Springer-Verlag, Berlin.
Bocquet, J.-L. (1980). Rep. CEA-R-5112, Centre d'Etudes Nucléaires de Saclay, France.
Bowker, M., and King, D. A. (1978a). *Surf. Sci.* **71**, 783.
Bowker, M., and King, D. A. (1978b). *Surf. Sci.* **72**, 208.
Brandt, W. W. (1975). *J. Chem. Phys.* **63**, 5162.
Butz, R., and Wagner, G. (1977). *Surf. Sci.* **63**, 448.
Chen, J. R., and Gomer, R. (1979). *Surf. Sci.* **79**, 413.
Compaan, K., and Haven, Y. (1956). *Trans. Faraday Soc.* **52**, 786.
Crank, J. (1975). "The Mathematics of Diffusion." Oxford Univ. Press (Clarendon), London and New York.
Dayananda, M. A. (1981). *Acta Metall.* **29**, 1151.
de Bruin, H. J., Bakker, H., and van der Mey, L. P. (1977). *Phys. Status Solidi B* **82**, 581.
de Bruin, H. J., and Murch, G. E. (1973). *Philos. Mag.* **27**, 1475.

de Bruin, H. J., Murch, G. E., Bakker, H., and van der Mey, L. P. (1974). *Thin Solid Films* **25**, 47.
Einstein, A. (1905). *Ann. Phys.* **17**, 549.
Flinn, P. A., and McManus, G. M. (1961). *Phys. Rev.* **124**, 54.
Franklin, W. M. (1975). *In* "Diffusion in Solids: Recent Developments" (A. S. Nowick and J. J. Burton, eds.), p. 1. Academic Press, New York.
Gordon, R. (1968). *J. Chem. Phys.* **49**, 570.
Guy, A. G. (1978). *Trans. Jpn. Inst. Met.* **19**, 483.
Guy, A. G., Cooper, W. D., and Poole, R. L. (1977). *Mater. Sci.* **3**, 103.
Howard, R. E. (1966). *Phys. Rev.* B **114**, 650.
Kehr, K. W. (1983). *J. Stat. Phys.* **30**, 509.
Kehr, K. W., Kutner, R., and Binder, K. (1981). *Phys. Rev.* B **23**, 4931.
Kikuchi, R., and Sato, H. (1969). *J. Chem. Phys.* **51**, 161.
Kikuchi, R., and Sato, H. (1970). *J. Chem. Phys.* **53**, 2702.
Kikuchi, R., and Sato, H. (1972). *J. Chem. Phys.* **57**, 4962.
King, G. W. (1951). *Ind. Eng. Chem.* **43**, 2475.
Kubo, R. (1957). *J. Phys. Soc. Jpn.* **12**, 570.
Kutner, R., and Kehr, K. W., 1983. *Philos. Mag.* [*Part*] A **48**, 199.
Kutner, R., 1984. *Phys. Rev. B.* To be published.
Kutner, R., Binder, K., and Kehr, K. W. (1982). *Phys. Rev.* B **26**, 2967.
Kutner, R., Binder, K., and Kehr, K. W. (1984). *Phys. Rev. B*, in press.
Le Claire, A. D. (1970). *In* "Physical Chemistry: An Advanced Treatise," Vol. 10 (H. Eyring, D. Henderson, and W. Jost, eds.), p. 261. Academic Press, New York.
Lindström, R.-L., and Lindström, R. (1979a). Rep. No. 98-A, Inst. for Appl. Math., Univ. of Turku, Finland.
Lindström, R.-L., and Lindström, R. (1979b). Rep. No. 103-A, Inst. for Appl. Math., Univ. of Turku, Finland.
Lindström, R.-L., and Lindström, R. (1980). *Proc. Ann. Conf. of Finnish Phys. Soc., Haikko, Finland.*
Lindström, R.-L., and Lindström, R. (1983). *Philos. Mag.* [*Part*] A **47**, 627.
Manning, J. R. (1964). *Phys. Rev.* A **136**, 1758.
Manning, J. R. (1965). *Phys. Rev.* A **139**, 2027.
Manning, J. R. (1967). *Acta Metall.* **15**, 817.
Manning, J. R. (1968). "Diffusion Kinetics for Atoms in Crystals," Van Nostrand-Reinhold, Princeton, New Jersey.
Manning, J. R. (1971). *Phys. Rev.* B **4**, 1111.
Manning, J. R. (1976). *Nucl. Metall.* **20**, 109.
Manning, J. R. (1981). *Scr. Metall.* **15**, 1.
Mehrer, H. (1969a). *Z. Naturforsch.* **24a**, 358.
Mehrer, H. (1969b). *Z. Naturforsch.* **24a**, 367.
Metropolis, N., Rosenbluth, A. W., Rosenbluth, M. N., Teller, A. H., and Teller, E. (1953). *J. Chem. Phys.* **21**, 1087.
Montroll, E. W., and Weiss, G. W. (1965). *J. Math. Phys.* **6**, 167.
Murch, G. E. (1975a). *Philos. Mag.* **32**, 1129.
Murch, G. E. (1975b). *J. Nucl. Mater.* **57**, 239.
Murch, G. E. (1979a). *Am. J. Phys.* **47**, 78.
Murch, G. E. (1979b). *Am. J. Phys.* **47**, 958.
Murch, G. E. (1980a). *Philos. Mag.* [*Part*] A **41**, 701.
Murch, G. E. (1980b). "Atomic Diffusion Theory in Highly Defective Solids," Trans Tech, Aedermannsdorf, Switzerland.

Murch, G. E. (1980c). *Philos. Mag.* [*Part*] *A* **41**, 157.
Murch, G. E. (1981a). *Philos. Mag.* [*Part*] *A* **43**, 871.
Murch, G. E. (1981b). *In* "Fast Ionic Transport in Solids" (J. B. Bates and G. C. Farrington, eds.). North-Holland, Amsterdam.
Murch, G. E. (1981c). *Philos. Mag.* [*Part*] *A* **44**, 699.
Murch, G. E. (1982a). *Philos. Mag.* [*Part*] *A* **45**, 929.
Murch, G. E. (1982b). *Philos. Mag.* [*Part*] *A* **45**, 941.
Murch, G. E. (1982c). *Philos. Mag.* [*Part*] *A* **46**, 565.
Murch, G. E. (1982d). *Philos. Mag.* [*Part*] *A* **46**, 151.
Murch, G. E. (1982e). *J. Phys. Chem. Solids* **43**, 243.
Murch, G. E. (1982f). *Philos. Mag.* [*Part*] *A* **46**, 575.
Murch, G. E. (1982g). *Solid State Ionics* **7**, 177.
Murch, G. E. (1982h). *Solid State Ionics* **6**, 295.
Murch, G. E. (1984a). *Philos. Mag.* [*Part*] *A*. **49**, 21.
Murch, G. E. (1984b). *J. Phys. Chem. Solids*. In press.
Murch, G. E. (1984c). *High Temp. Sci.*, in press.
Murch, G. E., and Rothman, S. J. (1981). *Philos. Mag.* [*Part*] *A* **43**, 229.
Murch, G. E., and Thorn, R. J. (1977a). *Philos. Mag.* **35**, 493.
Murch, G. E., and Thorn, R. J. (1977b). *Philos. Mag.* **36**, 517.
Murch, G. E., and Thorn, R. J. (1977c). *Philos. Mag.* **35**, 1441.
Murch, G. E., and Thorn, R. J. (1977d). *J. Phys. Chem. Solids* **36**, 789.
Murch, G. E., and Thorn, R. J. (1977e). *Philos. Mag.* **36**, 529.
Murch, G. E., and Thorn, R. J. (1978a). *Philos. Mag.* **37**, 85.
Murch, G. E., and Thorn, R. J. (1978b). *Philos. Mag.* **38**, 125.
Murch, G. E., and Thorn, R. J. (1978c). *J. Phys. Chem. Solids* **39**, 1301.
Murch, G. E., and Thorn, R. J. (1978d). *J. Comput. Phys.* **29**, 237.
Murch, G. E., and Thorn, R. J. (1979a). *J. Phys. Chem. Solids* **40**, 389.
Murch, G. E., and Thorn, R. J. (1979b). *Philos. Mag.* [*Part*] *A* **39**, 673.
Murch, G. E., and Thorn, R. J. (1979c). *Philos. Mag.* [*Part*] *A* **39**, 259.
Murch, G. E., and Thorn, R. J. (1979d). *Philos. Mag.* [*Part*] *A* **40**, 477.
Murch, G. E., Le Claire, A. D., and de Bruin, H. J. (1984). To be published.
Murch, G. E., Rolls, J. M., and de Bruin, H. J. (1974). *Philos. Mag.* **29**, 337.
Peterson, N. L. (1975). *In* "Diffusion in Solids: Recent Developments" (A. S. Nowick and J. J. Burton, eds.), p. 115. Academic Press, New York.
Peterson, N. L., Chen, W. K., and Wolf, D. (1980). *J. Phys. Chem. Solids* **41**, 709.
Reed, D. A., and Ehrlich, G. (1981). *Surf. Sci.* **105**, 603.
Richards, P. M. (1977). *Phys. Rev. B* **16**, 1393.
Rothman, S. J., Nowicki, L. J., and Murch, G. E. (1980). *J. Phys. F* **10**, 383.
Sadiq, A. (1974). *Phys. Rev. B* **9**, 2299.
Sadiq, A., and Binder, K. (1983). *Surf. Sci.* To be published.
Sato, H. (1980). Private communication.
Sato, H., and Kikuchi, R. (1971). *J. Chem. Phys.* **55**, 677, 702.
Scholz, A. (1966). *Phys. Status Solidi* **14**, 169.
Singer, H., and Peschel, I. (1980). *Z. Phys. B* **39**, 333.
Stark, J. P. (1976). "Solid State Diffusion." Wiley, New York.
Stolwijk, N. A., and Bakker, H. (1977). *Phys. Status Solidi B* **79**, K1.
Sugimoto, H., and Fukai, Y. (1979). *Proc. Int. Conf. on Hydrogen in Metals, Minakami Spa, Japan*, November, 1979.
Valleau, J. P., and Whittington, S. G. (1976). *Mod. Theor. Chem.* **5**, 137.

Wegener, W., and Frischat, G. H. (1982). *J. Phys. Chem. Solids* **43**, 189.
Willis, B. T. M. (1964). *Proc. Br. Ceram. Soc.* **1**, 9.
Wolf, D. (1971). *Z. Naturforsch. A* **26**, 1816.
Wolf, D. (1974a). *Phys. Rev. B* **10**, 2710.
Wolf, D. (1974b). *Phys. Rev. B* **10**, 2724.
Wolf, D. (1979a). *J. Phys. Chem. Solids* **40**, 757.
Wolf, D. (1979b). *J. Phys. Chem. Solids* **41**, 1053.
Wolf, D., and Differt, K. (1977). *Comput. Phys. Commun.* **13**, 167.
Wolf, D., Differt, K., and Mehrer, H. (1977). *Comput. Phys. Commun.* **13**, 183.
Wolf, D., Figueroa, D. R., and Strange, J. H. (1977). *Phys. Rev. B* **15**, 2545.
Young, W. M., and Elcock, E. W. (1966). *Proc. Phys. Soc.* **89**, 735.

DIFFUSION IN CRYSTALLINE SOLIDS

8

Defect Calculations beyond the Harmonic Model

GIANNI JACUCCI

CENTER OF STUDIES OF THE NATIONAL RESEARCH COUNCIL AND
DEPARTMENT OF PHYSICS, UNIVERSITY OF TRENTO, POVO, ITALY
AND
DEPARTMENT OF PHYSICS AND MATERIALS RESEARCH LABORATORY
UNIVERSITY OF ILLINOIS AT URBANA–CHAMPAIGN
URBANA, ILLINOIS

	Prefatory Remarks	430
I.	Introduction	431
	A. Plan of the Chapter	431
	B. Scope of Machine Calculations on Point Defects	433
	C. Point Defects in Machine Calculations	435
II.	Lattice Dynamics	436
	A. Brute Force versus Analytical Approximations	436
	B. LD Formulas	438
	C. Finding Extrema	440
III.	Vacancy Formation in fcc Lennard–Jones Crystals	442
	A. LD for the Perfect LJ Crystal	442
	B. LD for the LJ Crystal Containing a Lattice Vacancy	446
	C. Test of Common Approximations	449
	D. Comparison with MC Results	451
	E. Relevance for LD Calculations in Metals	452
IV.	Vacancy Migration	453
	A. Rate Theory Formulas	453
	B. Anharmonic Contributions to Thermodynamics	454
	C. Dynamical Corrections Due to System Memory	455
V.	Analytical Treatment of Anharmonic Jump Frequency	460
	A. The Saddle Surface S	462
	B. Taylor Expansion Form for S	462
	C. A Numerical Procedure for Extracting the Coefficients F_{ij}	467
	D. Dynamical Corrections to Rate Theory Due to a Curved S	468
	E. Numerical Results and Comparison with MD	470
VI.	Conclusions	473
	References	474

Copyright © 1984 by Academic Press, Inc.
All rights of reproduction in any form reserved.
ISBN 0-12-522662-4

Prefatory Remarks

Advances in the theory of atomic diffusion in crystals due to thermal generation and migration of point defects are currently sought in the two distinct direction of (i) an accurate determination of the interaction potential energy of the crystal as a function of atomic positions, and (ii) a precise calculation of the consequences of the N-body potential energy surface on thermodynamic and dynamic properties expressed as thermal averages.

Let us loosely characterize the present status of the theory. Regarding point (i), carefully determined potential functions have become available, often resulting from sophisticated ab initio calculations, if only for the simpler ionic, rare gas, and metallic crystals. Many body effects have also been introduced via polarization, screening, and the like.

The theoretical task is far from exhausted, however. An accurate and manageable description of the crystal energy is awaiting further progress for large and important classes of materials. This is the case most noticeably of transition metals and covalent-bonded solids. For these and other systems work is actively being done with pseudopotential theories (often pushing the expansion to higher order), self-consistent calculations, and other methods. Even for the simpler solids the available theoretical calculations are not final. Choice of theoretical parameters contained in the ab initio potentials, of the form for the dielectric function, of the description of the polarizability, and so on, are still a matter of debate. Because of the resulting variation of theoretical predictions, and because of further sources of indetermination in practical implementations, like inadequate treatments of the long-range part of the interaction, comparison with experiment is often aleatory and not conclusive.

Regarding point (ii), exact treatments of structural relaxation on the one hand, and of vibrational energy and entropy on the other, have made possible the numerical solution of the problem in the harmonic approximation and assuming memory randomization, using long-developed theoretical treatments based on equilibrium statistical mechanics and transition state theory. Within these limits, important factors in solid state diffusion, like defect concentration, jump frequency, and the isotope effect, can all be evaluated and compared with experiment. This achievement is made possible by the capability of modern electronic computers to handle large dynamical matrices in a reasonably short time.

However, the statistical mechanics problem should be approached in a manner amenable to solution without the assumptions and approximations mentioned above. The effect of anharmonicity on canonical averages as well as the role of crystal memory in such dynamical effects as multiple jumps and immediate return jumps remain largely open problems. As a consequence, such fundamental questions as the origin of curvatures of Arrhenius plots,

8. DEFECT CALCULATIONS BEYOND THE HARMONIC MODEL

high-temperature deviations from rate theory predictions, and anomalous isotope effects cannot be resolved. The conditions for a change of this situation have been brought about in the last decade by computer simulation techniques. These methods and their applications to point defects were discussed in a beautiful chapter by Bennett (1975). Bennett (1976a) also outlined a complete strategy based on them to calculate the features of point defect diffusion in simple model substances.

As a consequence, we should now be in a position to confront the task of precisely assessing the shortcomings of the harmonic model and of testing the adequacy of analytical treatments of anharmonicity in solid state diffusion.

I. Introduction

A. Plan of the Chapter

This chapter is devoted to a description of recent developments regarding the statistical mechanical treatment, using the computer, of point defect diffusion. In recent years Monte Carlo (MC) calculations, i.e., the numerical evaluation of the configurational integrals involved in thermal averages by using sampling techniques based on computer-generated random numbers [see, for example, Binder (1979) and references therein], and molecular dynamics (MD) calculations, i.e., the numerical solution of the classical equations of motion describing the time evolution of a system of interacting particles [see, for example, Kushick and Berne (1977) and references therein], have been carried out in this field for a few model crystals along with accurate lattice dynamics (LD) calculations, i.e. the numerical evaluation of the thermodynamic properties of a model crystal involving the knowledge of the crystal static energy in the equilibrium configuration along with its derivatives with respect to particle displacements [see, for example, Maradudin et al. (1971) and references therein]. Yet they have been far less extensive and have covered far less ground that one might have hoped ten years ago. This may be due to the fact that these calculations require both a great deal of computer time, and the use of rather sophisticated techniques to extract the interesting result, i.e., to isolate nonharmonic effects and dynamical memory effects. Account will be given here of what can be learned from such studies by comparing the results to LD calculations. Methods will be outlined along with examples of their application. The necessity for increased emphasis on the detailed analytical investigation of the N-body potential energy surface to include anharmonic features should become apparent from this presentation.

The precise nature of the assumed "force law" between the particles is not of paramount importance. Of course, care is taken to choose model potentials likely to embody the main qualitative features of the systems in hand, rare gas crystals, alkali metals, salts, or other. But the main thrust is always on the assessment of the precise consequences of an assumed force law on the statistical mechanical behavior of the model. This is because it is quite unknown at the outset what the findings will be, and because these are expected to be qualitatively similar for large classes of systems.

Because a presentation of this character is best developed from a single point of view, and even more because that work is better known to the present author, he shall mostly describe work in which he was directly involved. Part of it is yet to be published. Some of it is still under way. The resulting picture will be necessarily incomplete.

A major decision in preparing the chapter has been the inclusion of a rather extensive treatment of a specific investigation, i.e., lattice dynamics of an fcc Lennard–Jones model crystal containing a vacancy. This was done to create the occasion for pointing out the conceptual and methodological intricacies of defect thermodynamics. Vacancy formation at constant lattice parameter, constant atomic volume, and constant pressure will be compared showing the relation between LD predictions, both at zero and finite temperature, for the different circumstances. The possible size of errors involved in using approximate expressions to get one from the other, e.g., constant pressure results from constant volume data, or worse, in confusing one of these with the constant lattice parameter data, will be thus indicated. System size and long-range potential cutoff dependence of the results will be mentioned. Accurate treatment of these technical aspects of the calculation is important. Furthermore, classical and quantum mechanical LD results will be compared for the case of the Lennard–Jones model of argon.

The role of anharmonicity in the Lennard–Jones fcc crystal model has been studied by Monte Carlo (Holt et al., 1970). Methods permitting the evaluation of free energy differences were successfully applied to the calculation of the free energy of formation of a lattice vacancy (Squire and Hoover, 1969; Jacucci and Ronchetti, 1980). A comparison of these thermodynamic results, including a recent calculation for the vacancy formation volume, with corresponding LD data (De Lorenzi et al., 1984c), will indicate clearly the relevance of "explicit" anharmonic contributions, i.e., additional to those included in the quasiharmonic treatment via thermal expansion of the lattice.

Vacancy migration properties will also be discussed. Monte Carlo investigations of the free energy on the saddle plane are available (Bennett, 1975) and will be compared to recent accurate LD data (De Lorenzi et al., 1984c).

Anharmonic effects on the jump event have been recently investigated by approximating the saddle surface S, using the Taylor expansion of the N-body potential energy surface truncated to third order (Jacucci et al., 1984). Dynamical corrections to rate theory have thus been evaluated for the first time on realistic model systems. The temperature-dependent fraction of immediate return jumps and correction to the isotope effect factor are obtained to lowest order and compared to computer simulation studies.

These investigations provide a thorough testing of the conceptual foundations of rate theory and a demonstration of its wide applicability to diffusion in solids.

B. Scope of Machine Calculations on Point Defects

Consider a crystal consisting of N interacting atoms disposed either in a regular lattice or in a lattice containing a defect. Place it in contact with a heat bath at a given temperature and pressure. The exact evaluation of the energy, entropy, and volume of this system from knowledge of the interaction potential is a formidable computational task even for pair additive forces and $N \approx 100$.

Computer simulation techniques, or machine calculations, like MC and MD have been used extensively to this end for liquids and, more recently, high-temperature solids. These sampling techniques are needed to evaluate the $3N$-dimensional configurational integrals related to the desired thermodynamic quantities in classical statistical mechanics. No sufficiently accurate analytical treatment is available for such highly disordered systems.

A quite different case are crystals below the Debye temperature, whether consisting of regular, or defected, lattices. Here the disorder can be subjected to analytical treatment. Terms beyond quadratic in the Taylor expansion of the potential energy in powers at the atomic displacements about static equilibrium configurations can be neglected, the quadratic form diagonalized, and the integrals over normal coordinates analytically carried out to obtain precise expressions of the energy, entropy and pressure as a function of temperature and volume. The computational task is reduced to the determination of the crystal static energy ϕ and of the vibrational eigen-frequencies ω_α in the relaxed equilibrium configuration, as functions of the crystal volume. This task still requires the solution of an N-body problem. However, it can be carried out on modern machines for N up to ~ 100 using standard computer routines performing function minimization and matrix diagonalization. The results are of course affected by zero variance, these procedures being analytical, and not statistical, in character. These calculations are called lattice dynamics.

At temperatures at which atomic diffusion is important, terms beyond quadratic in the expansion of the crystal energy are not negligible. As a result, anharmonic contributions to the thermodynamics of lattice defects are to be expected. Of course, machine calculations such as MC and MD are exact in this respect. They properly include in the results the effect of all features of a given interaction potential. The reason that existing calculations of this kind are so scarce is that point defects contribute only terms of the order of $1/N$ to the total crystal properties. As a consequence, the sought result is buried in the statistical noise in standard calculations. One must revert to sophisticated, if available, difference techniques. Apart from its use as a means of sampling configuration space to perform thermal averages, the MD method is the primary tool to investigate in full the dynamics of a classical many-particle system. Trajectories of the representative point of the system in $6N$-dimensional phase space can be generated and analyzed. Jump frequencies can be measured in a simple way, at least when the residence time of the defect is not much longer than a picosecond. The actual path taken in the jump process can be monitored. All questions connected with crystal memory and persistence in the direction of motion of the defect, multiple jumps, and anharmonic isotope effects are open to direct observation.

In short, there exist two essentially different ways to exploit the knowledge of the N-body potential energy surface ϕ. One was just mentioned, i.e., sampling configurations from a thermal ensemble using the potential energy as statistical weight in MC or MD calculations to evaluate thermal averages related to free energies, and jump frequencies. Alternatively, one can investigate the shape of the surface ϕ directly and make use of the information thus gained in the framework of an analytical theory. This is the strategy used in LD. From knowledge of ϕ and ω_α at special points, thermodynamic parameters of point defects are calculated with the quasiharmonic theory of lattice vibrations, and jump frequencies are derived with the help of transition state theory.

One can take the latter point of view and extend the Taylor expansion approach beyond harmonicity. While the harmonic treatment truncates the Taylor expansion of the energy surface to second order, the third-order terms can be calculated as a next step. Lowest-order anharmonic contributions to the free energy can be evaluated in this way, as well as lowest-order dynamical corrections to rate theory in the form of the occurrence of immediate return jumps or deviations from harmonic isotope effects.

Whether the evaluation of local properties of the N-body surface at special points (minima, saddle points, etc.) limited to third derivatives of the crystal potential energy ϕ with respect to atomic displacements is an exhausting undertaking, or even a relevant one, cannot be said at the start. Whether strongly anharmonic hills and canyons dominate the shape of the surface on a distance scale that is small compared with atomic displacements characteristic of diffusion paths, thus hindering the possibility of projecting the behavior of the surface looking out from special points and using only the first few terms in the Taylor expansion, is a matter to be settled by extensive testing. In view of the considerable success met by the harmonic approximation, it seems certainly highly desirable and probably a rewarding endeavor to investigate the role of the next order in the expansion and to consistently extend the theoretical description. Such an investigation represents a formidable computational problem, but it will probably bring about a most important advance in the fundamental theory of point defect diffusion.

C. POINT DEFECTS IN MACHINE CALCULATIONS

Well-localized point defects in otherwise perfect crystals are especially suited for investigation by using machine calculation methods on a crystal comprising a few hundred particles with periodic boundary conditions. Short-range interactions between point defect pairs can also be handled in this way. Information thus obtained is sufficient to describe the thermodynamics of point defects in crystals in the low concentration limit. At high concentration of lattice defects, e.g., in superionic conductors, deviations from isolated point defect properties should be expected. The time-honored Born–von Kármán cyclic conditions are still a good way to minimize the effect of surfaces or interfaces. The limit of large system size has, of course, to be considered.

In computer simulation studies certain types of defects can be spontaneously generated by thermal fluctuations. This is the case of Frenkel pairs in, e.g., CaF_2 (Jacucci and Rahman, 1978). However, because of the absence of sources and sinks for point defects (e.g., dislocations, grain boundaries, surfaces), in most such studies unpaired point defects like vacancies are not generated or annihilated. Furthermore, the free energy required to generate Frenkel pairs is too high in many systems, e.g., monoatomic solids. As a consequence, dynamical simulations of the migration of an isolated vacancy or interstitial in otherwise perfect crystals are easily carried out by artifically introducing a single defect into the lattice from the start. The situation is quite different in superionic conductors and on crystal surfaces. The generation of point defects on surfaces costs much less free energy than in the bulk. High concentrations of adatoms and surface vacancies are commonly attained above one half of the bulk melting temperature. As a consequence, point defect pairs spontaneously appear within a few picoseconds on the top of the few-hundred square angstrom surface being simulated (De Lorenzi et al., 1982). The regular lattice structure of the surface is readily spoiled and the very description of the system in terms of point defects breaks down, as it happens, e.g., in bulk AgI or certain oxides.

A last remark is in order about machine calculations. The techniques employed in them have become rather involved and many practical details influence the final result. In a sense they have lost the character of directness and self evidence that theoretical work often has. They are really experiments, albeit numerical and on model systems, and should be regarded and treated as such. In reporting them it is essential to state precisely the nature and order of all the steps taken. All relevant technical details affecting procedures, routines and computer usage are essential parts of the experiment. Failure to record and convey to others all relevant aspects included in the calculation results in the impossibility to repeat the experiment recovering the same result. This general remark is expecially true for defect calculations, because defect properties are of order $1/N$. In order to assess whether a theoretical prediction of defect properties is really sound, it is essential that the model and techniques employed be consistently and extensively tested, so that all aspects of them be well understood by the authors as well as by others. This is seldom done. The one bad feature of this state of affairs is that often nobody knows how to assess the degree of confidence of numbers provided by calculations and being compared to experimental results!

II. Lattice Dynamics

A. Brute Force versus Analytical Approximations

Until recently, coherent use of well-developed statistical mechanical concepts was hindered by computational limitations. Use of drastic approximations both in the models examined and in the technical procedures employed biased the resulting theoretical predictions in an essentially unknown manner. Starting about 25 years ago, a great deal of effort went into the development of valuable analytical schemes for calculating defect properties with an acceptable computational burden on available computers. Careful theoretical backing was provided for the various approximations involved. However, the precise determination of point defect properties has turned out to be an extremely difficult task that is very often out of the reach of simplified schemes.

It should be clearly stated that the older schemes are no substitute for current codes. The latter are specifically suited for handling disordered systems and investigating the N-body potential energy surface exactly. The earlier schemes are still of value insofar as a careful comparison of their predictions with the results of modern accurate calculations represents a test of the adequacy of well-identified theoretical approximations. But, unfortunately, their use for such careful comparisons is rare. In our opinion the use of old schemes has somewhat outlived their research value. The effort needed to precisely identify their limitations has outgrown any useful knowledge they provide.

To some, this transformation of the theoretical work in the direction of "brute force" methods may not be palatable. However, a comment similar to

8. DEFECT CALCULATIONS BEYOND THE HARMONIC MODEL

ours was made almost 15 years ago by Maradudin et al. (1971, p. 196), where a complete account can be found of the approximate methods that were introduced to study frequency distributions and related properties such as zero-point energies and Debye thetas in crystalline solids:

> In the final analysis, however, the advent of high speed computers may render obsolete approximate methods for the calculation of θ_D [and the frequency spectra, etc.]... it is only necessary to solve for the eigenvalues of the dynamical matrix for a very large number of direction in q-space; [frequency spectra, etc.] can then be evaluated numerically to almost any desired degree of accuracy.

Here are some examples of highly idealized models and approximate procedures that were employed for defect studies in order to keep the computational task within the capability of available machines.

(i) Regarding the interaction potential:
    (a) limitation to first near neighbors only, or to the first few shells,
    (b) approximation of the potential and forces at distorted lattice configurations using truncated Taylor expansion of the potential function from the regular lattice configuration.

(ii) Regarding the determination of relaxed configurations:
    (a) limitation of the relaxation to the shell of first neighbors of the point defect, or to the first few shells,
    (b) approximation of the relaxed displacements as a truncated expansion in powers of the forces resulting from creation of the defect; a linear response scheme was used more often than not.

(iii) Regarding the thermodynamic framework:
    (a) formation of the defect at constant lattice spacing (without further discussion of the effect of imposing this condition, rather than constant pressure on the prediction of defect parameters),
    (b) neglect of thermal expansion.
    (c) neglect of finite temperature effects related to the crystal entropy and free energy when considering volume thermal expansions,
    (d) neglect of the contribution of atomic vibrations to the pressure and to the compressibility.

It is only fair to repeat that not all restrictions are removed in up-to-date calculations. Apart from the fact already mentioned that fully anharmonic machine calculations are still very rare, and that the treatment of many-body forces is mostly limited to schemes including polarizability, it should be emphasized that the panorama of properLD calculations is far from complete even for the simpler model crystals, i.e., rare gases, alkali halides, and alkali metals. The situation is quite the opposite: The picture is so fragmented that it would be quite embarrassing to have to produce a detailed evaluation of it.

The author prefers to take up an example and to present it in some detail for the purpose of illustrating what can be learned from careful LD calculations. The scope of the calculation remains that of collecting in a consistent way information of a qualitative nature on defect properties, in view of the fact that the relevance of the models studied to the natural substances they are intended to model is still to be satisfactorily assessed. However, theoretical calculations of defect properties of these model substances now can, and should, be exact from the point of view of statistical mechanics, be it for the simplified quasi-harmonic hamiltonian of LD.

B. LD Formulas

Let us lay down the expressions of relevant thermodynamic quantities in LD. The free energy F of a vibrating crystal is expressed in terms of the potential energy ϕ_0 and the normal mode frequencies ω_α calculated at the relevant minimum $\mathbf{R}_0$ in $3N$-dimensional configurational space as

$$(F)_Q = \phi_0 + \frac{1}{2} \sum_\alpha^{3N-3} \hbar\omega_\alpha + k_B T \sum_\alpha^{3N-3} \ln\left[1 - \exp\left(\frac{-\hbar\omega_\alpha}{k_B T}\right)\right] \quad (1a)$$

which reduces at high temperatures (i.e., $\hbar\omega/k_B T \ll 1$) to the classical approximation

$$F = \phi_0 + k_B T \sum_\alpha^{3N-3} \ln \frac{\hbar\omega_\alpha}{k_B T} \quad (1b)$$

Having periodic boundary conditions in mind, one sees that the three translational degrees of freedom of the center of mass have been excluded from the sums. For pair potentials the potential energy is written as

$$\phi_0 = \sum_{i<j=1}^N v(r_{ij}) + \phi_c \quad (2)$$

with the term ϕ_c containing the correction due to long-range cutoff of pair contributions excluded from the sum for $r_{ij} > r_c$. The normal mode frequencies ω_α are simply related to the eigenvalues λ_α of the dynamical matrix $\|\phi_{lm}\|$:

$$\omega_\alpha = \sqrt{\lambda_\alpha/m} \quad (3)$$

m being the mass of the particles, here taken to be the same for all. Other relevant quantities are the internal energy U, the entropy $S = (U - F)/T$,

8. DEFECT CALCULATIONS BEYOND THE HARMONIC MODEL

and the pressure $P = -(\partial F/\partial V)_T$:

$$(U)_Q = \phi_0 + \frac{1}{2} \sum_\alpha^{3N-3} \hbar\omega_\alpha + \sum_\alpha^{3N-3} \frac{\hbar\omega_\alpha}{\exp(\hbar\omega_\alpha/k_B T) - 1} \qquad (4a)$$

$$U = \phi_0 + 3Nk_B T \qquad (4b)$$

$$(S)_Q = -k_B \left\{ \sum^{3N-3} \ln\left(1 - \exp - \frac{\hbar\omega_\alpha}{k_B T}\right) + \sum^{3N-3} \frac{\hbar\omega_\alpha/k_B T}{1 - \exp(\hbar\omega_\alpha/k_B T)} \right\} \qquad (5a)$$

$$S = -k_B \left[\sum^{3N-3} \ln\left(\frac{\hbar\omega_\alpha}{k_B T}\right) - 1 \right] \qquad (5b)$$

$$P = P_\phi + P_\omega(T) \quad \text{with} \quad P_\phi = -\frac{\partial \phi_0}{\partial V} \quad \text{and} \quad P_\omega(T) = -\left(\frac{\partial (F - \phi_0)}{\partial V}\right)_T \qquad (6)$$

When allowing for changes of the crystal volume V the parameters of the quasi-harmonic description, i.e., ϕ_0 and the ω_α, must change. Otherwise, the featureless, exactly harmonic description will be recovered, where there is no internal pressure. Given the interaction potential law, ϕ_0 and ω_α will be determined for the different equilibrium configurations corresponding to different values of the crystal volume V. And, quite naturally, they will depend on V. The internal pressure is then fixed by Eq. (6).

Thermal volume expansion is automatically included in this description. This is because P from Eq. (6) depends on the temperature at a given V. Fixing the value of the external pressure P_{ex} and letting T vary produces the LD equation of state for the crystal, i.e., the volume V for which P from Eq. (6) equals P_{ex} at a chosen T. The thermal expansion of the crystal volume being one of the effects of anharmonicity of lattice vibrations, LD is also called a quasi-harmonic treatment. It cannot be overemphasized that a thermodynamically consistent implementation of LD requires that when thermal volume expansion is considered and ϕ_0 recalculated at different values of V corresponding to values of T derived from some equation of state (not necessarily the LD one) the variations of ω_α should also be considered, together with the resulting variations of S and P (and eventually $(U)_Q$).

We see from Eq. (6) that P is the sum of two terms, one coming from variations of ϕ with V, the other from variations of ω_α with V. The first term is easily treatable analytically, giving the virial expression

$$P_\phi = -\left(\frac{N}{3V}\right) \sum_{i<j}^{3N-3} \frac{\partial v(r_{ij})}{\partial r_{ij}} r_{ij} + P_c \qquad (7)$$

where P_c indicates the correction due to long-range cutoff. On the contrary,

P_ω is best evaluated numerically by using incremental ratios built from recalculations of the ω_α at two closely spaced values of V.

It is important to realize that the virial formula in Eq. (7) differs from the usual formula based on the thermal average of the virial sum precisely by the term $P_\omega(T)$:

$$P = -\left(\frac{\partial F}{\partial V}\right)_T = -\left(\frac{N}{3V}\right)\left\langle \sum_{i<j}^{3N-3} \frac{\partial v(r_{ij})}{\partial r_{ij}} r_{ij} \right\rangle_T + P_c = P_\phi + P_\omega(T) \quad (8)$$

As a result the substitution of the thermal average of the virial sum with the value of the sum calculated in the most probable configuration, i.e., the equilibrium configuration $\mathbf{R}_0$, may lead to disastrous results, because it essentially consists in using P_ϕ alone disregarding the volume dependence of the ω_α. This is only correct at $T = 0$ and in the classical approximation.

Laboratory experiments are more commonly done at constant pressure. Machine calculations are more commonly done at constant volume, i.e., fixing the basic box of the cyclic boundary conditions, although constant pressure MC and MD schemes have also been used. Fixing the volume is a more or less imposed choice in LD. In the static lattice, with thermal vibrations frozen out, the pressure as read by the virial formula does not contain $P_\omega(T)$. This term, of course, can be evaluated separately and added on to get P from Eq. (6). However, a numerical iteration procedure is necessary to precisely determine V at given P_{ex} and T. This discussion should be generalized, if appropriate in view of the anisotropy of the solid, to constant stress or constant strain conditions. The formalism is easily extended, accordingly, by the introduction of the components of the strain tensor as variables on which ϕ_0 and ω_α depend.

C. Finding Extrema

We have illustrated how LD calculations are based on the determination of the static equilibrium configuration $\mathbf{R}_0$ of the crystal. $\mathbf{R}_0$ is defined at fixed volume (or strain). The atomic positions correspond to a fully relaxed configuration, i.e., all forces are zero and $\nabla\phi(\mathbf{R}_0) = 0$. In regular crystals having a high degree of symmetry (e.g., sc, fcc, bcc), the configuration $\mathbf{R}_0$ is obvious. In defect formation calculations $\mathbf{R}_0$ denotes also the relaxed configuration of the crystal containing the defect. In defect migration studies, transition state theory requires use of the thermodynamic quantities given by Eqs. (1)–(7) evaluated at the saddle point $\mathbf{R}_s$ for the $3N - 1$ dimensional space normal to the eigenvector with negative eigenvalue, i.e., obtained by letting the reaction coordinate ξ equal zero, in order to estimate the jump frequency. In both cases, one needs to locate an extremum of ϕ, $\mathbf{R}_0$, or $\mathbf{R}_s$ ($\mathbf{R}_0$ being also a minimum). The location of one of these points is identified

8. DEFECT CALCULATIONS BEYOND THE HARMONIC MODEL

with the help of one of many function minimization routines applied to $|\nabla\phi|^2$, or even to ϕ itself for $\mathbf{R}_0$, starting the search from estimated configurations (Bennett, 1977). The chosen procedure has no relevance to the final result if one makes certain that the right extremum has been exactly determined within the numerical precision available. This is easily checked because, fortunately, the property of an extremum is a local property where $\nabla\phi(\mathbf{R}_0) = 0$. The neighborhood of $\mathbf{R}_0$ is first attained with routines that converge rather slowly but have a large radius of convergence. Then a few iterations of the extremely fast first-order Taylor expansion of the gradient bring the system exactly at $\mathbf{R}_0$ within the precision of the computer:

$$\Delta\phi(\mathbf{R}) \cdot (\mathbf{R} - \mathbf{R}_0) = \nabla\phi(\mathbf{R}) \tag{9}$$

This, of course, requires matrix inversion, but the time involved is negligible with respect to that for the matrix diagonalization to follow.

Lattice dynamics, as with other techniques employed to study the N-body potential energy surface ϕ, is based on local differential properties of the surface. Therefore, it is essential that ϕ be continuous and differentiable, at least in a reasonably large domain including the special points $\mathbf{R}$ investigated. When ϕ is given as a sum of pair interactions that do not go smoothly to zero, but are cutoff at r_c, care must be used in the investigation of ϕ. Whenever two nearby configurations are compared, such as the evaluation of incremental ratios in computing P_ω, the list of interacting neighbors must be employed and kept unchanged. If the list of neighbors is kept constant during the search of $\mathbf{R}_0$, however, one last iteration of Eq. (9) must be tried as a check with the list recalculated in $\mathbf{R}_0$. Because the shape of the surface ϕ far from $\mathbf{R}_0$, as well as all thermodynamic consequences thereof, are solely predicted from shapes and curvatures of ϕ at $\mathbf{R}_0$, great care is required in dealing with such technical points in order to avoid gross miscalculations. Whether the curvatures of ϕ at $\mathbf{R}_0$ and $\mathbf{R}_s$, and eventually their spatial derivatives involving ϕ_{lmn}, are indeed sufficient information to accurately determine defect free energies at a given temperature is of course a different issue. But in order to properly test this very important point one has to be exceedingly careful in treating the N-body potential energy surface. In MC and MD calculations this problem does not exist because the representative point directly samples the various relevant regions in configuration space, and ϕ is properly recalculated at every point. Any residual roughness of ϕ due to cutoff effects is buried in the statistical noise.

Often model pair potentials are not very short range, and it is not possible to place the cutoff within the basic box to which periodic conditions apply. Use of larger boxes is hindered by computer limitations on N. The solution is to let each particle interact with all the images of the other particles within the appropriate cutoff distance r_c. Many shells of boxes surrounding the basic one may be included, for in LD calculations the time needed to sum up pair interaction is not the main limitation as in MC or MD. Again, all should be controlled by repeating the calculation for different values of N. In any way, use of short cutoff distances may result in much larger deviations from the sought result.

As a practical matter we remark that the classical approximation to the free energy equation [Eq. (1b)] can be obtained without performing the diagonalization of the dynamical matrix $\|\phi_{lm}\|$. Its determinant $D = \prod_\alpha \lambda_\alpha$ is invariant under coordinate transformations, and can be computed for the cartesian coordinate form of the matrix. For equal masses we have

$$\exp\left(\sum_\alpha^{3N-3} \ln \frac{\hbar \omega_\alpha}{k_B T}\right) = \prod_\alpha^{3N-3} \frac{\hbar \omega_\alpha}{k_B T} = \left(\frac{\hbar}{k_B T m^{1/2}}\right)^{3N-3} \sqrt{D}$$

In the evaluation of D care must be taken to deal with translational invariance and the three null eigenvalues thereof. The simplest way is to change these eigenvalues from zero to unity, leaving the rest unchanged, by adding to the hamiltonian harmonic potentials of appropriate force constants relative to displacements of the center of mass along the three cartesian coordinates. The sought value of D consisting in the product of the $3N - 3$ relevant eigenvalues is then obtained from the matrix $\|\phi_{lm}\|$, modified accordingly. If the masses are not all equal, the situation is, of course, more complicated [see, for example, Vineyard (1957), Jacobs et al. (1982)]. Otherwise, proper diagonalization is only needed for the exact quantum mechanical formulas [although the first quantum correction to Eq. (1b) is expressed in terms of the mean sqaure value of ω_α or the trace of $\|\phi_{lm}\|$, also invariant] and for the saddle point calculations, to determine the directions spanning the saddle plane, normal to the reaction coordinate, and the isotope effect factor. For many applications it is sufficient to determine the eigenvector corresponding to the lowest (negative) eigenvalue, i.e., to the reaction coordinate.

III. Vacancy Formation in fcc Lennard–Jones Crystals

A. LD for the Perfect LJ Crystal

We shall discuss in some detail a recent LD study (De Lorenzi et al., 1984c) of vacancy formation and migration in a simple model for rare gas crystals: the Lennard–Jones (6–12) pair potential

$$v(r) = 4\varepsilon[(\sigma/r)^{12} - (\sigma/r)^6]$$

Within the framework of classical mechanics, thermodynamic quantities like F and P scale with σ and ε in this model. As a consequence, results in reduced units like $F^* = F/\varepsilon$ and $P^* = P\sigma^3/\varepsilon$, obtained in calculations performed at the state point ($V^* = V/\sigma^3$; $T^* = k_B T/\varepsilon$), can be referred to argon, e.g., by using appropriate values for σ and ε:

$$F = F^*\varepsilon_{Ar}, \qquad P = P^*\varepsilon_{Ar}/\sigma_{Ar}^3$$

Quantum mechanical expressions, however, do not scale. Fixing the value of Planck's constant $\hbar$ in reduced units $\sigma(m\varepsilon)^{1/2}$ is equivalent to picking a specific atomic element. This is consistent with the common fact that the magnitude of quantum effects is different for crystals of the various rare gas elements at corresponding state points. We must therefore specialize the quantum mechanical results to one element. Argon is chosen, and for simplicity all results will refer to it. The values employed for the parameters are $\sigma = 3.405$ Å and $\varepsilon/k_B = 119.8$ K.

Table I contains thermodynamic data for the perfect crystal in the fcc lattice structure at four different state points. Computations were done using a cutoff value of $r_c = 4.5\sigma$. Long-range corrections ϕ_c to U and P_c to P were added to the results. The basic box contains $N = 108$ atoms. All sums thus contain $3N - 3 = 321$ terms and the center of mass is considered fixed. A different treatment of the center of mass contribution to the thermodynamic properties would make a difference of order $1/N$ to the quantities per particle.

The most striking features of Table I refer to the pressure data. The four state points have been chosen on the equation of state of crystalline argon at

TABLE I

Lattice Dynamics Results for an FCC Lennard–Jones Crystal with $N = 108$ Atoms and Cutoff Distance $r_c = 4.5\sigma^a$

$k_B T$	0	0.3339 (~40 K)	0.4923 (~60 K)	0.6678 (~80 K)
V/N	0.9535	0.9734	0.9974	1.0323
P	−2.5510	−0.4024	0.1277	0.7778
P_Q	−0.1873	0.1285	0.4432	0.9662
P_ϕ	−2.5510	−3.5733	−4.5514	−5.5804
P_c	−0.2022	−0.1940	−0.1848	−0.1725
U/N	−8.5551	−7.4940	−6.9206	−6.2190
U_Q/N	−7.8250	−7.3405	−6.8377	−6.1802
ϕ/N	−8.5551	−8.4957	−8.4002	−8.2224
ϕ_c/N	−0.0964	−0.0944	−0.0922	−0.0891
S/Nk_B	—	2.1449	3.5314	4.7643
S_Q/Nk_B	—	2.3838	3.6289	4.8070
K_T	0.0184	0.0219	0.0278	0.0418
$(K_T)_Q$	0.0161	0.0204	0.0263	0.0398

[a] The value of $\hbar$ is $0.029569\sigma\sqrt{m\varepsilon}$. Energy is in units of ε and pressure in units of ε/σ^3; K_T is the isothermal compressibility.

vapor pressure, i.e., $P \ll 1$ in units of ε/σ^3. Therefore LD values of $|P| \ll 1$ would indicate that

(i) The LJ model provides a satisfactory description of the equation of state.

(ii) The values of σ and ε employed in converting V and T to reduced units have been chosen satisfactorily.

(iii) Quantum effects are either small (both $|P|$ and $|P_Q| \ll 1$) or satisfactorily taken into account (only $|P_Q| \ll 1$).

(iv) The residual anharmonicity of lattice vibrations, after the volume dependence of the ω_α has been taken into account, contributes only a small term to the free energy of the crystal; this contribution should be small at low T and vanish for $T = 0$ in classical mechanics.

In fact, Table I shows that

(a) P_Q is quite close to zero at small and intermediate temperatures,
(b) P is close to zero only at intermediate temperatures, and
(c) P_ϕ is large and negative everywhere, but much more so at high T.

Observations (a) and (b) indicate that points (i) and (ii) are satisfied, i.e., the model is not bad. Furthermore, quantum effects are rather large in argon. They are never really negligible in the pressure, the deviation from the classical value exceeding $0.5\varepsilon/\sigma^3$ below ~ 40 K.

Observation (c) indicates that the contribution of thermal vibrations to the pressure have a huge effect on the equation of state. We may expect in general that LD calculations neglecting P_ω can meet with disaster at high T.

When the volume dependence of ϕ_0 and the ω_α are correctly included in the calculation of the pressure by adding P_ϕ and P_ω, much of the anharmonicity of ϕ is accounted for in these quasi-harmonic calculations. This can be argued from the fact that the LD equation of state does not depart too much from that of the real crystal. In particular, P is less than one-fifth of P_ϕ for the state point of argon close to melting.

The residual anharmonicity quoted in (iv) is, of course, not taken into account by LD. This fact probably causes the positive deviations of P_Q and P from zero that grow towards high T. In fact, these deviations can be taken as a measure of the residual anharmonicity. A direct comparison of LD and MC data on the same model would definitely confirm this interpretation. The comparison indeed exists, and it had been carefully carried out by Holt et al. (1970). We note that LD results from Holt et al. (1970) and this work coincide to the reported precision when account is taken of the slightly different choice of state points.

From the results reported by Holt et al. (1970) we see that the effect of residual anharmonicity is to lower the value of P in MC with respect to P in

LD by about -0.17, -0.45, and -0.91 in units of ε/σ^3, at $T \sim 40$, 60, and 80 K, respectively. The deviations of $(U - \phi_0)/Nk_BT$ from the classical value of 3 in LD are also negative in MC. The measured values of this quantity at the same state point as previously specified are: 2.89, 2.85, and 2.79. The effect on the values of U/N listed in Table I is to lower them from -7.49, -6.92, and -6.22 to -7.53, -6.99, and -6.36 in units of ε, the effect of residual anharmonicity on U_Q probably being not too different. If we correct the value of P_Q in Table I with the deviation $P_{MC} - P_{LD}$ measured by Holt et al. (1970), we get the gratifyingly small values of -0.04, $+0.01$, and $+0.06$ at $T \sim 40$, 60, and 80 K, i.e., a good description of the argon crystal.

The LD values for the crystal entropy per particle are also a fair estimate of the properties of natural argon. The values in Table I can be transformed into excess entropies S_e/Nk_B with respect to the classical ideal gas at the same density and temperature. The values obtained for $T \sim 60$ and 80 K are -5.62 and -4.89, i.e., about 0.19 and 0.35 high with respect to experiment for Ar, and 0.29 and 0.43 high for Kr (Squire and Hoover, 1969). Note that these LD values do not include the correction term $-\ln(2\pi N)^{1/2} \sim -0.030$ from Stirling's formula. Furthermore, they are probably affected by small crystal corrections of order $\ln N/N$.

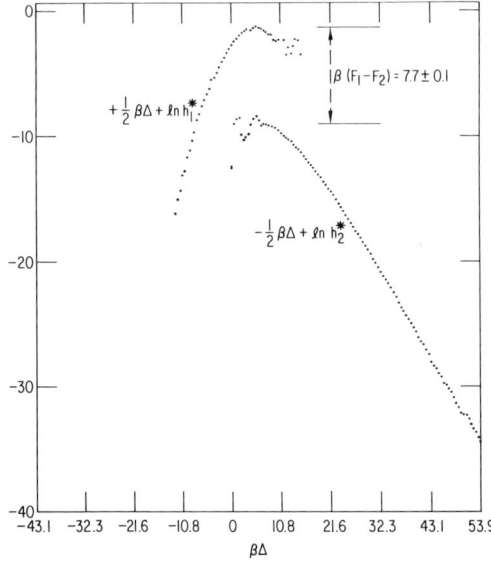

Fig. 1. Use of the graphical method for evaluation of free energy differences in the case of little or no overlap between the energy difference distributions $h_1^*(\Delta)$ and $h_2^*(\Delta)$. Data are taken from the comparison of an LJ crystal of fcc structure to the corresponding quasi-harmonic model. From the graph we get $\beta \Delta F$ per particle $= 0.143 \pm 0.002$ at $T = 0.5$ and $\rho = 1.0$. [From Rahman and Jacucci (1984).]

Whether the observed deviations from experiment are due to the neglect by LD of anharmonic terms can be checked using MC data. Rahman and Jacucci (1984) have measured the free energy difference between the fcc LJ crystal, at $T \sim 60$ K for Ar, and the corresponding harmonic solid (i.e., the LD approximation). Monte Carlo calculations based on the acceptance ratio method (Bennett, 1976b) and graphical display (see Fig. 1) gave $(F_{LJ} - F_{LD})/Nk_BT = 0.143 \pm 0.002$ for a 54 particle system. Previously quoted MC energy values gave $(U_{LJ} - U_{LD})/Nk_BT = -0.15$ at this temperature. As a result, we have $(S_{LJ} - S_{LD})/Nk_B = -0.29$, or $S_{LJ}/Nk_B = -5.62 - 0.29 = -5.91$, which compares favorably with the experimental values of -5.81 (Ar) and -5.91 (Kr). (Note that anharmonic contributions lower the energy, but lower the entropy twice as much, so that the effect on the free energy is a net increase.)

B. LD FOR THE LJ CRYSTAL CONTAINING A LATTICE VACANCY

Results for vacancy formation may be obtained by comparing properties relative to the regular lattice crystal C_0 containing N atoms with those relative to a crystal C_v containing $N - 1$ atoms, and one vacant site, both systems being subjected to periodic boundary conditions. Extensive properties of the regular lattice must be reduced by the factor $(N - 1)/N$ before taking differences, e.g.,

$$\Delta U = U(N - 1, v) - U(N, 0)(N - 1)/N$$

$$\Delta S = S(N - 1, v) - S(N, 0)(N - 1)/N$$

Because an extra bulk atom results upon forming the vacancy in a lattice. In fact, the Nth atom is missing from the system C_v containing the vacancy considered above, and it is not practical to introduce it anywhere in the lattice of the periodic box. Its properties can be taken to be identical to those of any one atom in the regular lattice C_0. Hence the factor $(N - 1)/N$ is to subtract its contribution from the extensive properties of C_0.

The vacancy formation process can be investigated in three different thermodynamic conditions: constant lattice (CL) parameter, constant atomic volume (CΩ), and constant external pressure (CP). In the CΩ calculation the lattice of C_v is squeezed so that its total volume V be reduced by the factor $(N - 1)/N$ to leave the atomic volume $\Omega = V/N$ unaltered. Of course, this procedure produces different values of the lattice parameter of C_v for different values of N. As a consequence, N-dependent correction terms may be expected to be substantial. In the CP calculation, the lattice parameter of C_v is varied until the reading of the pressure coincides with that of C_0. The resulting variation of the lattice parameter will be found to be smaller than in

CΩ calculations, for a given value of N. Relevant vacancy parameters are identified as limiting values obtained with these procedures for $N \to \infty$.

The values listed in Table II were obtained along these lines with $N = 108$ (De Lorenzi et al., 1984c). The cutoff distance of the LJ potential was $r_c = 4.5\sigma$. Long-range corrections are not included in the values shown in the table, i.e., uncorrected values for the pressure $P - P_c$ have been employed in evaluating ΔH and ΔG, CP calculations refer to equal values of the uncorrected pressures, and so on. As a result, the data reported can be regarded as referring to a consistently treated interaction potential model, i.e., LJ + cutoff at the specified distance r_c. Thermodynamic checks can therefore be carried out precisely. In fact, long-range corrections due to cutoff are not as straightforward for C_v as for C_0, and a consistent treatment for the full LJ potential is difficult. Cutoff dependence of the results will not be considered here.

We shall now briefly discuss these results, perform a thermodynamic check of their internal consistency, and verify quantitatively whether certain approximations often encountered in evaluations of defect properties are indeed acceptable. The discussion will be based on the material presented in Table II. Table III is included to display the role of quantum effects in argon, which are found to be rather small. For a further discussion of these effects, as well as extrapolations to large N and cutoff corrections, see De Lorenzi et al. (1984c).

As a first observation, we note that the formation volume at constant

TABLE II

Variation of LD Thermodynamic Quantities of an LJ Crystal upon the Formation of a Lattice Vacancy[a]

$k_B T$	0	0.3339 ($\sim$40 K)	0.4932 ($\sim$60 K)	0.6678 ($\sim$80 K)
$(\Delta S)_L$	2.4493	2.3249	2.1451	1.7988
$(\Delta S)_\Omega$	-6.5043	-6.7791	-7.1558	-7.8234
$(\Delta S)_P$	3.0112	2.4658	2.1846	1.0130
$(\Delta U)_L$	8.4185	8.3810	8.2929	8.0894
$(\Delta U)_\Omega$	6.4502	5.3525	4.1880	2.7451
$(\Delta U)_P$	8.5583	8.4312	8.3111	7.6440
$(\Delta P)_L$	0.0309	0.0064	0.0014	-0.0182
$(\Delta P)_\Omega$	0.5460	0.4404	0.3464	0.2175
$(\Delta V)_P/\Omega$	1.0628	1.0155	1.0043	0.9184
$(\Delta F)_L$	8.4185	7.6047	7.2349	6.8882
$(\Delta F)_\Omega$	6.4502	7.6160	7.7172	7.9696
$(\Delta G)_P$	6.1781	7.4019	7.5468	7.8685
$(\Delta H)_P$	6.1781	8.2252	8.6242	8.5449

[a] All quantities are in reduced units; the values refer to the classical approximation.

TABLE III

Variations of Quantum LD Thermodynamic Quantities of an LJ Crystal upon the Formation of a Lattice Vacancy

$k_B T$	0	0.3339 (~40 K)	0.4932 (~60 K)	0.6678 (~80 K)
$(\Delta S)_L$	—	1.9835	1.9464	1.6533
$(\Delta S)_\Omega$	—	−5.6931	−6.7356	−7.6816
$(\Delta S)_P$	—	1.8743	1.7808	0.7126
$(\Delta V)_L$	7.9313	8.2018	8.1865	8.0202
$(\Delta U)_\Omega$	8.2061	6.1690	4.7028	3.0714
$(\Delta U)_P$	7.9315	8.1697	8.1168	7.5180
$(\Delta P)_L$	−0.0160	−0.0063	−0.0066	−0.0235
$(\Delta P)_\Omega$	0.5707	0.4582	0.3572	0.2237
$(\Delta V)_P/\Omega$	0.9716	0.9858	0.9810	0.8993
$(\Delta F)_L$	7.9313	7.5395	7.2265	6.9250
$(\Delta F)_\Omega$	8.2061	8.0699	8.0248	8.2012
$(\Delta G)_P$	7.9454	7.8533	7.8530	8.0992
$(\Delta H)_P$	7.9454	8.4792	8.7313	8.5751

pressure is predicted to be always within about 10% of unity. As a consequence, CL values are much closer to CP value than to CΩ ones. The obtained CΩ values are indeed remarkable: $(\Delta S)_\Omega$ is large and negative, and $(\Delta U)_\Omega$ decreases roughly linearly with temperature, so that in $\Delta F = \Delta U - T \Delta S$ the temperature variation of the energy and entropy terms cancels to a large extent. Constant Lattice parameter and CP values for ΔU are much less dependent on T, and ΔS is much smaller and positive in those cases; as a consequence, ΔF is again only moderately dependent on T.

Second, $(\Delta U)_L$ at $T = 0$ is not too different from $-U/N$ of Table I. As T increases, however, $-U/N$ decreases while $(\Delta U)_L$ remains remarkably constant. A similar comparison for $(\Delta F)_L$ and $-F/N$ shows that $-F/N$ increases with temperature while $(\Delta F)_L$ decreases. In conclusion, the formation energy of the vacancy is not accurately described by the opposite of the energy per particle, relative deviations of about one third being observed close to the melting temperature.

Third, $(\Delta F)_\Omega$ and $(\Delta G)_P$ are always only a few percent apart. This is a gratifying result and it constitutes the first indication that the thermodynamics of the vacancy formation process are correctly described. It is in fact a general result (Flynn, 1972) that in thermodynamic transformations involving volume changes one has

$$(\Delta F)_\Omega = (\Delta G)_P \qquad (10)$$

to order $\Delta V/V$, or $1/N$ in this case. This result conveys the information that energy and entropy changes in the transformations with and without lattice dilatation balance out. Therefore, only a thermodynamically consistent description can reproduce it.

C. TEST OF COMMON APPROXIMATIONS

The basic approximation to be encountered in this type of calculation is the static model, i.e., the neglect of entropy and related contribution, resulting from lack of knowledge of the ω_α. Therefore, the only quantity entering the description is ϕ_0. Quantum mechanics is, of course, out of the reach of this approach. However, since the contributions related to entropy are invariably multiplied by the factor T, this description becomes exact at $T = 0$ in classical mechanics, the expected errors growing with T. One simple check of the magnitude of these errors can be performed on the mentioned equality $(\Delta G)_P = (\Delta F)_\Omega$ (to order $1/N$), that can also be written as

$$(\Delta H)_P = (\Delta U)_\Omega + T[(\Delta S)_P - (\Delta S)_\Omega] \tag{11}$$

This relation is particularly useful whenever $C\Omega$ calculations are to be preferred, e.g., for metals described with the pseudopotential model (Jacucci and Taylor, 1979). The outcome of these calculations is $(\Delta U)_\Omega$, and $(\Delta H)_P$ is the sought result, being the slope of experimental Arrhenius plots. We see from Table I that the neglect of the entropy term in argon at high temperature brings disaster: $(\Delta U)_\Omega$ is only about one-third of $(\Delta H)_P$. It has become customary in these cases to evaluate separately the entropy term using a value for the thermal expansion coefficient α_P independently available (often from experiment) and the approximate relation (Flynn, 1972; Jacucci and Taylor, 1979)

$$T(\Delta S_P - \Delta S_\Omega) \cong T \Delta\Omega_P \alpha_P / K_T \tag{12}$$

based on the evaluation of the entropy change upon expansion of the regular lattice C_0. It is indeed found that the estimate of Eq. (12) provides the desired value of the entropy difference term within the LD model to order $1/N$. However, in the model argon being investigated MC values of $(T\Delta\Omega_P \alpha_P)/K_T \equiv P_\omega$ are some 15% lower than LD values at $T \sim 80$ K (see pressure comparisons for the perfect lattice reported above; the discrepancy would be even larger were it not for a partial cancellation of differences in α_P and K_T). As a result, the value of $T(\Delta S_P - \Delta S_\Omega)$ at high temperature predicted in this way is accordingly lower than what it would be using LD consistently.

Another feature of the static approximation is to neglect the entropy related term P_ω in the pressure. We already noted that the size of this term is by no means negligible for the external pressure of the regular crystal. Let us see what the value of $(\Delta P_\phi)_L$ is and compare it with the sum $(\Delta P_\phi)_L + (\Delta P_\omega)_L = (\Delta P)_L$. It is found (De Lorenzi et al., 1984c) that $(\Delta P_\phi)_L$ is 0.0567, 0.0828, and

0.1136 at $T \sim 40$, 60, and 80 K, instead of the values for $(\Delta P)_L$ of 0.0064, 0.0014, and -0.0182 of Table II. This means that the formation volume $(\Delta V)_P$ predicted using $(\Delta P_\phi)_L$, or $(\Delta P_\phi)_\Omega$, comes out to be much too large. In fact, in the static approximation it is found that $(\Delta V)_P = 1.24\Omega$ and 1.50Ω at $T \sim 60$ and 80 K instead of 1.00 and 0.92! Even the sign of the relaxation volume is wrong. This huge discrepancy of the static approximation to $(\Delta V)_P$ from the LD value is probably the most noticeable effect of neglecting P_ω and represents a very important discovery in the work of De Lorenzi et al. (1984c) described here. Although argon (and its models) is expected to be badly behaved in this respect, because of the relatively high value of its Grüneisen parameter, from now on the effect of P_ω on defect properties cannot be disregarded light heartedly.

The results of the static treatment are thermodynamically consistent only if read at $P_{ex} = P_\phi$ and $T = 0$, where they are also exact in the classical framework. Values corresponding to different lattice parameters will then refer to different external pressures at $T = 0$. Attempts to interpret the data in terms of different temperatures assuming thermal volume expansion are bound to meet with inconsistencies, because of the neglect of the contribution of S to F, or of $\partial S/\partial V$ to P, or both. For instance, Eq. (11) holds exactly at $T = 0$ in the classical static model, with $P \equiv P_\phi$ given by Eq. (7), for any given lattice parameter:

$$(\Delta U)_{P_\phi} + P_\phi (\Delta V)_{P_\phi} = (\Delta U)_\Omega \tag{13}$$

If one now desires to use the same data, obtained from knowledge of ϕ_0, ϕ_v, and their volume derivatives, to exploit or check Eq. (11) in full, i.e., at finite temperature T, one must alter both sides to include entropy contributions. In the left-hand side $P = P_\phi + P_\omega$ will replace P_ϕ throughout, and in the right-hand side the term $T(\Delta S_P - \Delta S_\Omega)$ will be added. Performing only one such modifications will, of course, unbalance Eq. (13). An approximate treatment consistently neglecting P_ω, but properly including S, produces $(\Delta F)_\Omega \simeq 8.06$ and $(\Delta G)_P \simeq -1.86$ at $T \simeq 80$ K in complete disregard of Eq. (10) and grave prejudice of the stability of the lattice held at constant external pressure!

The moral of these findings is that now is a good time to stop fooling the experimentalists. Predictions from unreliable procedures cannot be sold as reliable with the only excuse that they are the best one can do. The calculation of defect properties has turned out to be a very difficult task, and the use of approximations is seldom forgiving. Most of these approximations now can and should be dropped. Remaining sources of indeterminations must be spelled out and their effect analyzed carefully.

The only source of indeterminations of the LD treatment of the LJ model presented above sits in the residual anharmonic effects already mentioned

8. DEFECT CALCULATIONS BEYOND THE HARMONIC MODEL

for the case of the regular lattice. These will be treated below, apart from cutoff and finite size effects, for which the reader is referred to the original work.

D. COMPARISON WITH MC RESULTS

The free energy of formation of lattice vacancies in LJ crystals was first measured by Squire and Hoover (1969) using MC and a path integration method. The calculation was repeated by Jacucci and Ronchetti (1980) using the acceptance ratio method, and has been recently extended to include the evaluation of the formation volume (De Lorenzi et al., 1984c). In these calculations one measures the free energy Δf associated with reversibly removing a particle from a perfect crystal, keeping the lattice parameter constant. The free energy contribution G/N of reintroducing the particle in the bulk of the system must be accounted for separately. This is commonly done by using MC energies and pressures, and experimental excess entropies. Eventually, the excess free energy should also be measured by MC in the model system.[1]

The formation Gibbs free energy is obtained as (Squire and Hoover, 1969; De Lorenzi et al., 1984c).

$$(\Delta G)_P = \Delta f + G/N \tag{14}$$

The formation volume requires a finite difference evaluation of the derivative of Δf with respect to volume. A small statistical uncertainty on Δf must therefore be achieved. Furthermore, this delicate difference operation may be spoiled by problems connected with cutoff of the interaction potential. In the work of De Lorenzi et al. (1984c) LD results for various system sizes and cutoff distances were compared, providing an indirect check of the adequacy of the MC procedure used to measure $\Delta \Omega_P$ in a 108 particle system.

The LD value of the pressure is large and positive along the experimental equation of state of rare gas solids, while the MC pressure is fairly low. Since a term PV directly enters in G, and hence in $(\Delta G)_P$ through G/N, the comparison of the two sets of data (MC and LD) should be done disregarding its contribution. Larger discrepancies are unnecessarily obtained otherwise. The comparison is then done on the quantity $(\Delta G)_P - P\Omega$.

The LD value of this quantity exceeds the MC value by 6% at 60 K and 10% at 80 K (these deviations double if the term $P\Omega$ is not subtracted). Since the difference in the Helmholtz free energy per particle (~ 0.07 at 60 K, see above) is only one fifth of this, almost all the large explicit anharmonic contribution to (ΔG) comes from Δf, i.e., from the lattice relaxation around

---

[1] Lowercase f is used to stress the difference of the quantity Δf presently in hand, which refers to the comparison of two crystals, with and without defects, i.e., consisting of different numbers of atoms $N - 1$ and N, and ΔF defined previously, which referred to the same number of atoms.

the defect. The conclusion is easily checked by a direct comparison of Δf values. The effect of anharmonicity is to lower the free energy of formation, with an increase by a factor ~ 4 of the number of lattice vacancies at 80 K in Ar.

An interesting result of the work of De Lorenzi et al. (1984c) is that the MC value of the formation volume, $\Delta\Omega_P/\Omega = 1.21 \pm 0.05$ for $T \sim 60\ K \div 80\ K$, is intermediate between LD and LS values, LD being too low and LS too high, by over 20%.

E. Relevance for LD Calculations in Metals

The test of approximations performed above is particularly interesting with respect to the case of alkali metals, for which lattice statics calculations that use accurate interaction potentials exist (Jacucci and Taylor, 1979). The relevance of the entropic term of Eq. (11) is clearly recognized in that work, and its contribution evaluated using experimental values for compressibility and thermal expansion. The contribution P_ω of vibrations to the pressure was, however, neglected so that all calculated volumes are confined to the LS model. Despite the presently described evidence that the latter approximation is indeed very bad for rare gas crystals, LS results for alkali metals met with precise agreement with careful experimental data. For Na and K adding the migration energy, obtained by MD by Da Fano and Jacucci (1977), to the static formation energies one can derive the temperature dependence of the self diffusion coefficient. Strikingly good agreement found for both metals with the slope of the best-fit formula of Mundy's data (Mundy, 1971; Mundy et al., 1971) points to the fact that monovacancies can explain the curvature of the Arrhenius plot of Na.

TABLE IV

The Temperature Dependence of ΔH_P (Migration) for Al Calculated Using $V(R, \Omega)$ Truncated at 1.8 a (Lattice Spacing)

T	ΔU	$T(\Delta S_P - \Delta S_V)$	ΔH_P (migration)
75	0.45	0	0.45 (0.41)[a]
350	0.40	0.11	0.51
450	0.38	0.16	0.54
705	0.34	0.27	0.61
860	0.31	0.38	0.69 (0.60)[a]
960	0.29	0.48	0.77

[a] The values in parentheses were obtained by using $V(R, \Omega)$ properly summed.

8. DEFECT CALCULATIONS BEYOND THE HARMONIC MODEL

On the basis of this evidence, one is inclined to think that the approximation involved in neglecting P_ω in alkali metals will prove to be much better than in Ar. This is in accord with expectations on the role of anharmonic effects in these materials.

On the other hand, the other LS approximation of neglecting $T(\Delta S_P - \Delta S_V)$ in calculating finite temperature enthalpies from constant volume energies is always bad. A striking example is given for Al (Jacucci, et al., 1981) by calculations of the migration enthalpy (see Table IV). From the LJ study of this section we can infer that if the volume changes are evaluated using only P_ϕ, as in this case, the correction is probably overestimated.

IV. Vacancy Migration

A. RATE THEORY FORMULAS

In the limit of classical mechanics rate theory is used to calculate the vacancy jump frequency. Calling S the saddle surface, or "watershed," separating the regions τ and τ' in configuration space corresponding to the defect being located at two neighboring lattice sites, one writes the jump rate as (Vineyard, 1957)

$$\Gamma = \sqrt{\frac{k_B T}{2\pi}} \frac{\int_S \exp(-\phi/k_B T)\, dS}{\int_\tau \exp(-\phi/(k_B T))\, d\tau} \tag{15}$$

The integral in the denominator extends over one of the two equivalent regions. In using Eq. (15), it is assumed that no memory of previous jumps is kept, so that the equilibrium distribution is established before each jump, and that no recrossings occur, i.e., that the surface S is crossed only once during the jump event. In the quasi-harmonic approximation Γ is written as (Vineyard, 1957)

$$\Gamma_0 = \frac{1}{2\pi} \frac{\prod_\alpha^{3N-3} \omega_\alpha}{\prod_\alpha^{3N-4} \omega_{s\alpha}} \exp\left[-\frac{\phi_s - \phi_v}{k_B T}\right] \tag{16}$$

where again the center of mass motion has been excluded. The $\omega_{s\alpha}$ are the normal mode frequencies at the saddle point, and $\alpha = 3N - 3$ is the index of the imaginary frequency relative to the reaction coordinate, that has been left out from the product at the denominator. In this approximation S is replaced by the hyperplane S_0 tangent to S at the saddle point. We choose to rewrite Eq. (16) as

$$\Gamma_0 = (\bar{\omega}/2\pi)\exp(\Delta S/k_B)\exp(-\Delta U/k_B T) \tag{17}$$

with

$$(\bar{\omega})^{3N-3} = \prod_{\alpha}^{3N-3} \omega_\alpha, \quad \Delta S = k_B \sum_{\alpha}^{3N-4} \ln \frac{\bar{\omega}}{\omega_{s\alpha}} \quad \text{and} \quad \Delta U = \phi_s - \phi_v$$

Values of ΔU, ΔS, and $\bar{\omega}$ can be calculated by LD in the quasi-harmonic approximation, as for vacancy formation quantities. Results obtained by De Lorenzi et al. (1984c) are listed in Table V.

The calculations are done at constant Ω and at constant P. These conditions refer to the saddle point configuration with respect to the equilibrium one, i.e., they are chosen to be at values of P and Ω appropriate to the formation of the defect at constant pressure. Building $(\Delta G)_P$ and $(\Delta F)_\Omega$ from ΔU, ΔS, and $(\Delta U)_P$, Eq. (10) is seen to be again rather well-obeyed.

Results of Table V warrant a discussion similar to that presented for the formation data of Table II, that will not be repeated here. As in that case, the cutoff value was $r_c = 4.5\sigma$ and the data have not been corrected for long-range cutoff effects.

B. Anharmonic Contributions to Thermodynamics

The calculation of Γ in Eq. (15) is essentially a free energy difference problem. It can therefore be approached by using methods based on sampling of the type encountered earlier for the calculation of the free energy of formation of defects. In fact Bennett (1975) has invented one such method to

TABLE V

Variations of LD Thermodynamic Quantities of an LJ Crystal Containing a Vacancy upon Raising from Equilibrium to the Saddle for Vacancy Migration[a]

$k_B T$	0	0.3339 (~40 K)	0.4932 (~60 K)	0.6678 (~80 K)
$(\Delta S)_\Omega$	−1.1950	−1.2284	−1.2742	−1.3506
$(\Delta S)_P$	3.1612	3.1575	3.4404	3.9543
$(\Delta U)_\Omega$	5.2652	4.8035	4.3064	3.6314
$(\Delta U)_P$	6.3061	6.3192	6.4588	6.6048
$(\Delta P)_\Omega$	0.2402	0.2008	0.1669	0.1242
$(\Delta \Omega)_P/\Omega$	0.4897	0.4847	0.5098	0.5542
$(\Delta F)_\Omega$	5.2652	5.2137	4.9348	4.5333
$(\Delta G)_P$	5.2094	5.1666	4.9209	4.5078
$(\Delta H)_P$	5.2094	6.2209	6.6178	7.1485
$\bar{\omega}$	15.7509	14.8043	13.7263	12.3219

[a] From De Lorenzi et al. (1984). Definitions of ΔS and ΔU are in the text. Values of the geometrical mean $\bar{\omega}$ of the normal mode frequencies at equilibrium are also given (in reduced units $\sqrt{\varepsilon/m\sigma^2}$).

8. DEFECT CALCULATIONS BEYOND THE HARMONIC MODEL

evaluate Γ. For LJ at $T \sim 60$ and 80 K, Bennett finds $\ln[\Gamma(m\sigma^2/\varepsilon)^{1/2}] = -7.5 \pm 1.0$ and -5.0 ± 0.5. The LD estimates from Table IV are -9.2 and -6.1, showing a rather large discrepancy between the quasi-harmonic approximation and fully anharmonic rate theory predictions. Because in Bennett's work direct jump frequency measurements were also made and gave the value of -5.3 ± 0.2 for the logarithm of the jump frequency at $T \sim 80$ K, in agreement with his anharmonic rate theory prediction, we conclude that (i) rate theory is reliable, while (ii) LD evaluations at high temperature may not be. In this case LD predictions are low by a factor anywhere from 2 to 15.

It should be noted that previous evaluations of Γ_0 using lattice statics values of $(\Delta U)_\Omega$ and an effective attack frequency v^* in the place of the LD factor $(\bar{\omega}/2\pi)\exp(\Delta S)_\Omega$ accidentally yielded better agreement with anharmonic jump rates (Bennett, 1975). The LD value for $(\Delta S)_\Omega/k_B$ is larger than unity in modulus and negative in sign. This somewhat unexpected fact lowers the value of the LD prediction for Γ below previous estimates, thus destroying the apparent agreement between static predictions and dynamical results.

This is the first accurate comparison between harmonic and anharmonic rate theory predictions. That LD overestimates the barrier to atomic jumps may in fact be a general feature (Jacucci et al., 1981; De Lorenzi et al., 1982). The extension to metallic and ionic crystals is very desirable.

C. Dynamical Corrections Due to System Memory

Molecular dynamics studies of vacancy migration have revealed that memory effects are indeed present (Bennett and Alder, 1968; Bennett, 1975; Bennett, 1976a; Da Fano and Jacucci, 1976, 1977) as a tendency of the vacancy to persist in the direction of motion in successive jumps closely spaced in time. The most relevant consequence of this fact can be described as the occurrence of *multiple jump* processes in which the vacancy is displaced by more than one lattice spacing in one complex dynamical event. These events are clearly separated from regular single jumps in the distribution of time delays between successive jumps (see Fig. 2). In fact, the average delay allowed in multiple jumps is only about 10% of the Debye period. One may consider the second jump to be a dynamical consequence of the first. The distribution of time delays for longer delays closely follows an exponential decrease appropriate to randomization of the system memory between successive jump events. The contribution to diffusion of multiple jump events must be evaluated separately.

Double jumps have been observed in fcc and bcc lattices. They appear to be a general feature of solid-state diffusion, quite independent from the model

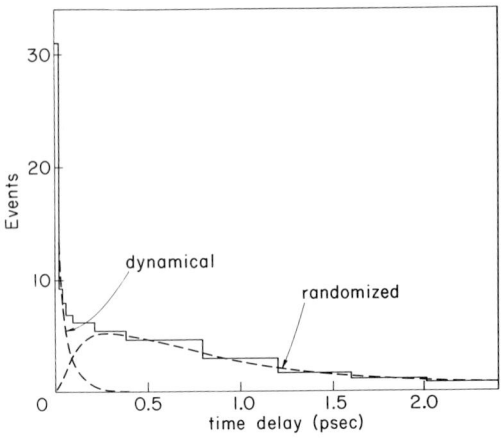

Fig. 2. Time delay between successive jumps of a vacancy; dashed lines indicate the separation of double events into the exponential long-time behavior of the randomized system and short-time behavior due to dynamical correlations. [From Da Fano and Jacucci (1976) and Flynn and Jacucci (1982).]

potential employed. The question is rather what fraction of double jumps occur at the melting point of crystals, than if they in fact exist. The activation energies (E_2) for double jumps are much higher than those (E_1) for single jumps, possibly higher than the double of it. Da Fano and Jacucci (1977) found the values (in electron volts) for E_1 and E_2 in Na, K, and Al to be 0.12, 0.08, 0.42 and 0.55, 0.40, and ~ 2, respectively, although it must be pointed out that multiple jumps involving more than two atoms were broken down into double jumps in the bookkeeping.

Because of their high activation energy double jumps do not contribute appreciably to atomic diffusion except in the close neighborhood of the melting point T_M. Even if they rise to 25% of the total jump events in Na, where they have been observed in greater number, when $T \approx 0.9\ T_M$ their relevance is much reduced. For instance, in K they drop from 17% at 330 K to 9% at 310 K. As a consequence, the double jump is not a possible candidate to explain the curvature of Arrhenius plots over large temperature intervals. Failure to clearly state this point in the original work has perhaps mislead some authors. Double jumps are most probably responsible (Bennett, 1976a; Da Fano and Jacucci; 1977) for the notorious "premelting phenomenon," e.g. in Na, i.e., an anomalous sudden increase in diffusivity accompanied, also in Na, by an anomalous decrease of the isotope effect factor.

The double jump must be regarded as a second diffusion channel available to the vacancy. In fact, there are many such channels for the various possible angles between the two atomic displacements, plus multiple jumps involving

8. DEFECT CALCULATIONS BEYOND THE HARMONIC MODEL

more than two atoms. In addition, vacancy jumps involving the displacement of only one atom to a second nearest-neighbor site have been also observed in Na (5% at T_M!) and in K.

Another observation of peculiar dynamical events in vacancy jumps (Bennett, 1975) refers to the fact that the system trajectory in configuration space may cross more than once, in a short time, the hyperplane S_0 tangent to the "watershed" S at the saddle point. Here S_0 is orthogonal to the direction of the eigenvector $\mathbf{\eta}_\xi$ having negative eigenvalue, and it is easily identified in the simulation by the vanishing of the reaction coordinate ξ. Because Eq. (16) gives an estimate of Γ_0 on the assumption that all crossings of S_0 are successful, these immediate recrossings are an additional source of error of the quasi-harmonic formula.

The problem of the recrossings is more general, however. Also, the watershed S in configuration space does not exactly separate trajectories falling in the two neighboring minima. This would happen only for purely viscous motion with no inertia, but not for newtonian motion. In the latter case, S can indeed be crossed more than once within a short time (Flynn, 1975) just because it is curved. Therefore, a more correct way of writing the jump frequency is

$$\Gamma = \Gamma'c \qquad (18)$$

with Γ' given by Eq. (15), and c a conversion factor representing the fraction of forward crossing of S for which the jump is successfully completed. Equation (18) is more general than hitherto implied. Here Γ' and c can, in fact, be defined with reference to any convenient hypersurface between τ and τ' (Bennett, 1977); S_0 can be used to this end. The watershed S has the only advantage to maximize c. In any case a correct estimate of Γ requires an evaluation of c.

In summary, evidence gathered by MD studies indicates that dynamical events at variance with the assumptions of randomization and single crossings are restricted to short times. On the basis of this observation, an analysis of the origin of dynamical correlation effects has suggested (Flynn and Jacucci, 1982) a way to correct the predictions of diffusion coefficients from rate theory. The basic idea is that since these dynamical effects are restricted to short times, they can be related to local geometrical properties of the potential energy surface.

The rate at which return jumps occur may be calculated directly as the outcome of trajectories almost parallel to S that cross S twice because S is curved. This happens if the momentum along the trajectory is high enough to ensure a radius of curvature of the trajectory larger then that of S (see Fig. 3). Similarly, the frequency of multiple jumps can be evaluated from the thermal expectation rate of trajectories that cut the relevant dividing hypersurfaces

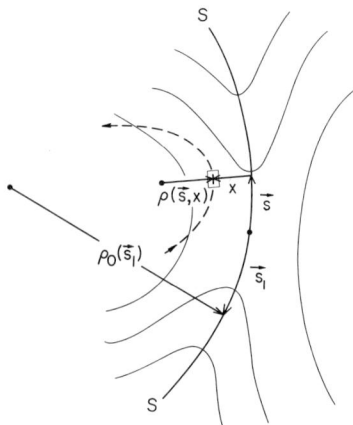

Fig. 3. Schematic representation of a curved saddle surface and the potential energy at neighboring points. The location of a volume element at x, $\mathbf{s}$ is indicated, together with a trajectory which is parallel to S inside the volume element. Return jumps occur when the radius of curvature $\rho(\mathbf{s}, x)$ of the trajectory is greater than the radius of curvature $\rho_0(\mathbf{s})$ of the saddle surface at the same value of $\mathbf{s}$, so that the trajectory cuts S twice. [From Flynn and Jacucci (1982).]

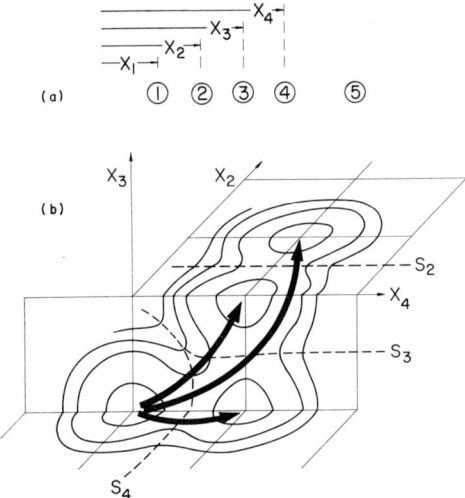

Fig. 4. (a) A row of atoms with a vacant site. (b) Schematic diagram of the potential energy contour for motion of atoms 2, 3, and 4, and the saddle surfaces S_4, S_3, and S_2 for successive jumps of atoms 4, 3, and 2 into the vacancy. Progressively longer heavy arrows indicate the jump of atom 4 from the initial configuration, the double jump of atoms 4 and 3, and the triple jump of atoms 4, 3, and 2 in a single dynamical event. It is possible, in principle, to calculate the rate at which multiple events occur and hence to correct errors in the rate theory predictions. [From Flynn and Jacucci (1982).]

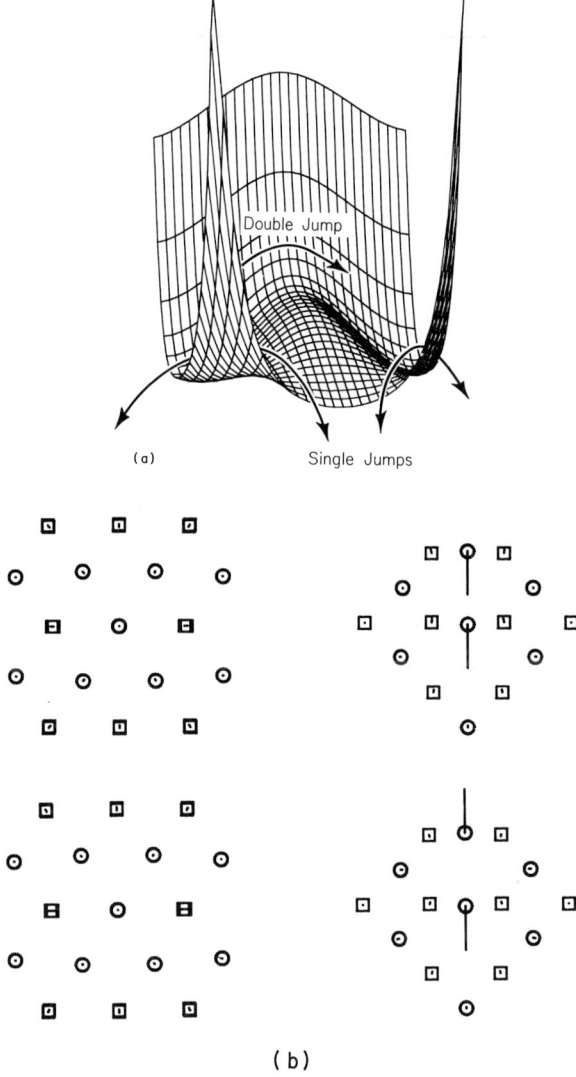

Fig. 5. Potential energy surface showing the path to double jumps of the vacancy in the fcc LJ crystal. (a) V plotted along two normal mode directions, ζ_1 and ζ_2, from the inflection point; ζ_1 (running from right to left) is the unstable mode corresponding to double jumps; ζ_2 is the decomposition mode for the jumping pair: it has zero frequency and exhibits a flexus of nonvanishing slope. Along this direction the potential falls from high values to the equilibrium configuration corresponding to the vacancy being in the middle. Usual saddle points for single jumps connect this configuration to the neighboring ones. The plot spans 1.4σ in ζ_1 and 0.6σ in ζ_2, resulting in a maximum variation of V of about 15ε. (b) Atomic displacements involved in modes ζ_1 and ζ_2 are projected in (i) the (110) plane normal to the atomic jump and (ii) the parallel (100) plane. Circles denote atoms in the central plane, squares denote atoms belonging to the two adjacent planes above and below. [From De Lorenzi *et al.* (1984d).]

in direct succession (see Figs. 4 and 5a,b). In both calculations local properties of ϕ (e.g., curvatures of S at the saddle point) are used to evaluate the first correction term in a T expansion of these anharmonic contributions. This analytic program can produce predictions useful also at high T if the Taylor series converges quickly.

The calculation of the isotope effect in tracer diffusion is also affected by dynamical contributions. The dependence of Γ on the mass of the jumping atom comes both from Γ' and c. If Γ' is defined with respect to the hyperplane S_0, the contribution of Γ' to the isotope effect is equal to the harmonic ΔK, and the only source of anharmonic contribution is c. Bennett (1975) has deviced a difference method to evaluate dc/dm by MD with reference to the planar S_0. Again, it has been proposed to use knowledge of the curvature of S at the saddle point to evaluate anharmonic contributions to the isotope effect to lowest order in a T expansion (Flynn, 1975).

It should be emphasized that a coherent program using modern calculations based on sampling, i.e., MC and MD, can be carried out to evaluate Γ' of Eq. (15) and c, as well as their mass dependence, at least with reference to S_0 (Bennett, 1976a). This last limitation is dictated by the necessity of having a simple criterion available to decide whether a given point is on the surface of interest. No such simple criterion, based on local properties of ϕ, is known to exist for S itself. Furthermore, MD can be employed to measure jump frequencies directly. However, all these calculations are still rare.

V. Analytical Treatment of Anharmonic Jump Frequency

If the analytical approach based on the Taylor expansion of ϕ to third order were successful, it would provide a theory of dynamical effects, the predictions of which would be valid for different materials and different temperature and pressure, given analytically in terms of potential energy and atomic masses. The structure of the theoretical results would make intuitively clear what factors cause rate theory to break down, and which physical systems are most likely to exhibit these effects. Therefore, analytical calculations of first order corrections to rate theory, including the fraction of return jumps, anharmonic isotope effect, and deviation of the flux Γ' through S or S_0 from Eq. (15) because of higher-order terms in the variation of the potential energy ϕ (see parts a, b, and c of Fig. 6), should be carried out numerically on realistic models. These first-order corrections could, in turn, be compared with available MD observations. This work is in progress (Jacucci et al., 1984; De Lorenzi et al., 1984a; De Lorenzi et al., 1984b) and a partial description of it will be anticipated here.

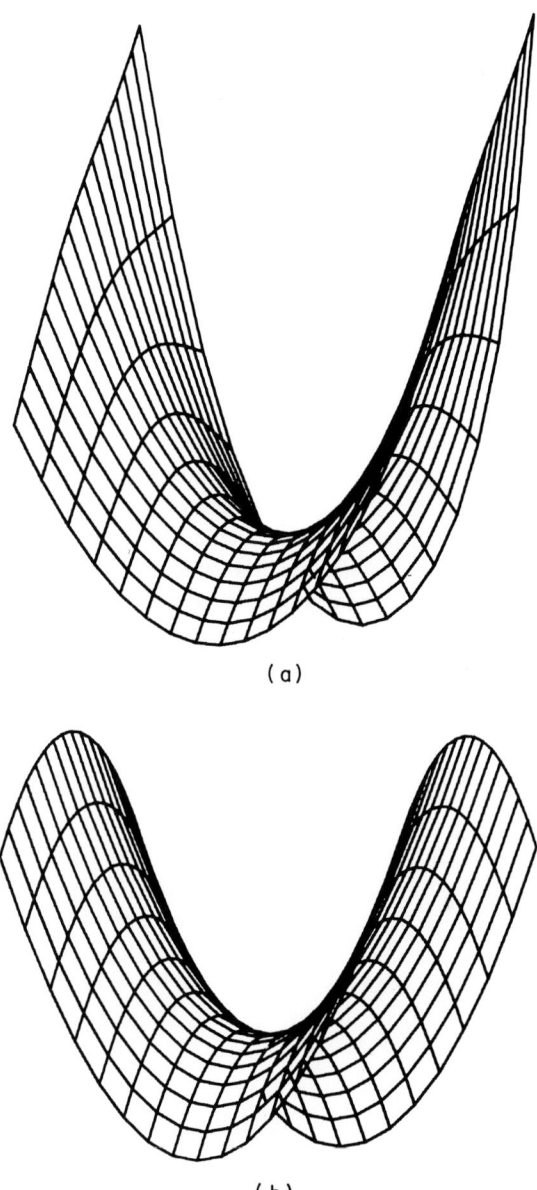

Fig. 6. Curved saddle for vacancy jumps in the fcc LJ crystal (a) compared to the osculating quadratic saddle and (b) plotted along the direction of the reaction coordinate z and along the direction of the saddle plane corresponding to maximum curvature of the saddle surface, v_1. The plot spans 0.4σ in both directions, resulting in a maximum variation of V of about 4ε for (a) and 3ε for (b). The atomic displacements involved in the reaction mode are shown in (c). [From De Lorenzi *et al.* (1984b).] (*Figure continues on next page.*)

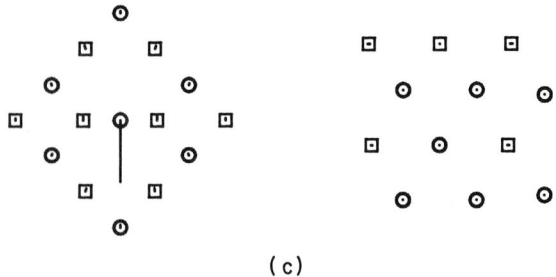

(c)

Fig. 6. (*Continued*)

A. The Saddle Surface S

Consider the potential energy function ϕ; the gradient $\nabla\phi \equiv (\partial\phi/\partial x_1 \cdots \partial\phi/\partial x_{3N})$ defines a vector field in configuration space. Constant ϕ contours consists of $3N - 1$ dimensional hypersurfaces, everywhere orthogonal to $\nabla\phi$. The vector field has singular points where $\nabla\phi = 0$. Among these we are particularly interested in stable minima, where all the eigenvalues of $\Delta\phi$ are positive. Consider two nearby local minima, along with all possible trajectories in configuration space linking the two. To each trajectory one can associate the maximum value taken by ϕ on it. The point identified by the lowest value of these maxima will be called the saddle point P_0. In P_0, $\nabla\phi = 0$ and all but one eigenvalue of $\Delta\phi$ are positive. We wish to define a suitable "watershed" hypersurface dividing the space around P_0 into two regions surrounding the two local minima. These regions have the property that any curve everywhere tangent to the vector field $\nabla\phi$ belongs entirely either to one or to the other region and terminates in the corresponding local minimum, *unless* it belongs entirely to the watershed and terminates in the saddle point P_0. The watershed hypersurface shall be named the saddle surface S.

The $3N - 1$ dimensional hypersurface S is everywhere orthogonal to constant ϕ contours, and its normal **n** obeys the equation

$$\mathbf{n} \cdot \nabla\phi = 0 \tag{19}$$

in all points.

Furthermore, S contains the saddle point P_0. Its normal $\mathbf{n}_0$ in P_0 coincides with the direction of the eigenvector of $\Delta\phi$ having negative eigenvalue. The hyperplane through P_0 and normal to $\mathbf{n}_0$ is therefore tangent to S in P_0. It is called the saddle plane S_0.

B. Taylor Expansion Form for S

We wish to find the Taylor expansion form for the equation of S about P_0. From now on we shall use the reference system with origin in P_0 having the principal axis parallel to the eigenvectors of $\Delta\phi$ at P_0. Call z the coordinate

8. DEFECT CALCULATIONS BEYOND THE HARMONIC MODEL

in the direction of the unstable normal mode, or reaction coordinate. The saddle plane is described by the equation

$$z = 0 \tag{20}$$

The equation of an hyperplane through a point P with normal $\mathbf{n} \equiv (n_1, \ldots, n_i, \ldots, n_{3N-1}, n_z)$ is

$$z - z(P) = -\sum_{i=1}^{3N-1} \frac{n_i}{n_z}[x_i - x_i(P)] \tag{21}$$

We can verify that the saddle plane has a normal $\mathbf{n}_0$ of components, $n_i(P_0) = 0$, $n_z(P_0) = 1$, as it should have.

The equation for S is of the type

$$z = F(\{x_i\}) \tag{22}$$

with $F(P_0) = 0$ and $\partial F(P_0)/\partial x_i = -n_i(P_0)/n_z(P_0) = 0$. The first non-vanishing terms in a Taylor expansion of the function F in powers of the displacements x_i are of second order. To this order, S is approximated by the quadratic form

$$z = \frac{1}{2}\sum_{i,j=1}^{3N-1} F_{ij} x_i x_j$$

Now

$$\frac{\partial^2 F}{\partial x_j \partial x_i} = \frac{\partial}{\partial x_j}\left(\frac{\partial F}{\partial x_i}\right) = \frac{\partial}{\partial x_j}\left(-\frac{n_i}{n_z}\right) = \frac{1}{n_z}\left(\frac{n_i}{n_z}\frac{\partial n_z}{\partial x_j} - \frac{\partial n_i}{\partial x_j}\right)$$

so that

$$F_{ij} = \left(\frac{\partial^2 F}{\partial x_i \partial x_j}\right)_{P_0} = \left(\frac{\partial^2 F}{\partial x_j \partial x_i}\right)_{P_0} = -\left(\frac{\partial n_i}{\partial x_j}\right)_{P_0} = -\left(\frac{\partial n_j}{\partial x_i}\right)_{P_0} \tag{23}$$

Note that the coefficients F_{ij} are the opposite of the elements of the curvature matrix

$$\mathbf{c} \equiv \begin{bmatrix} \frac{\partial n_1}{\partial x_1} & \cdots & \frac{\partial n_1}{\partial x_N} \\ \vdots & & \vdots \\ \frac{\partial n_N}{\partial x_1} & \cdots & \frac{\partial n_N}{\partial x_N} \end{bmatrix} \tag{24}$$

calculated in P_0. They can be related to derivatives of ϕ using the fact that S if a solution of $\nabla\phi \cdot \mathbf{n} = 0$, or

$$n_z \phi_z + \sum_{i=1}^{3N-1} n_i \phi_i = 0 \tag{25}$$

where $\phi_i = \partial\phi/\partial x_i$ and $\phi_z = \partial\phi/\partial z$. Equation (25) holds for all points of S. Therefore, we can impose the condition that the derivative of the left-hand side of the equation along any direction in space contained in S_0 must vanish in P_0. Iterating the procedure twice, we find after some algebra that

$$\phi_{zij} + \frac{\partial n_i}{\partial x_j}\phi_{ii} + \frac{\partial n_j}{\partial x_i}\phi_{jj} = 0 \tag{26}$$

or, remembering Eq. (23),

$$F_{ij} = \phi_{zij}/(\phi_{ii} + \phi_{jj} - \phi_{zz}) \tag{27}$$

all quantities being evaluated in P_0.

Equation (27) is the main result of this section and it shall be used in Section C as the base for the description of S_0 with a second order Taylor expansion of ϕ about P_0. Needless to say, the procedure used to obtain Eq. (27) can be iterated any number of times to get higher-order coefficients in the Taylor expansion (De Lorenzi et al., 1984a).

An additional important feature of S being present in many different circumstances can be derived by symmetry arguments. Imagine that the saddle point geometry is equivalent looking from a positive or negative value of z. Then the derivatives, with respects to displacements along the unstable normal mode, of the eigenvalue ϕ_{ii} must be zero, viz.,

$$\phi_{zii} = \phi_{-zii} = 0 \tag{28}$$

No such requirement exists on ϕ_{zij} for $i \neq j$, since the symmetry $\phi_{z-ij} = \phi_{-zij} = -\phi_{zij}$ is sufficient for the two sides of S to be equivalent. This symmetry property is to be found in many important instances of solid state diffusion, e.g., vacancy diffusion in fcc crystals. In all these cases Eq. (22) becomes

$$z = \frac{1}{2}\sum_{i=1}^{3N-1}\sum_{j\neq i=1}^{3N-1} F_{ij}x_i x_j \tag{29}$$

because $F_{ii} = 0$. The matrix $\mathbf{c}(P_0)$ has diagonal elements that are all zero. It can be diagonalized by finding normal modes of the saddle surface quadratic form having eigenvalues $-\rho_i^{-1}$, i.e., the inverse of the radii of curvature. The equation for S becomes

$$z = -\frac{1}{2}\sum_{i=1}^{3N-1}\rho_i^{-1}\zeta_i^2 \tag{30}$$

Because of the invariance of the trace, the sum of the principal curvatures $-(\rho_i)^{-1}$ must vanish, so that $\sum_{i=1}^{3N-1}(\rho_i)^{-1} = 0$. The eigenvalues of $\mathbf{c}$ are in

8. DEFECT CALCULATIONS BEYOND THE HARMONIC MODEL

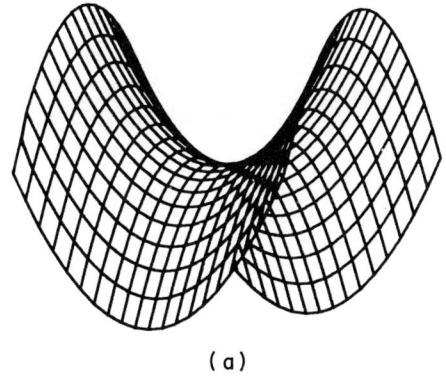

(a)

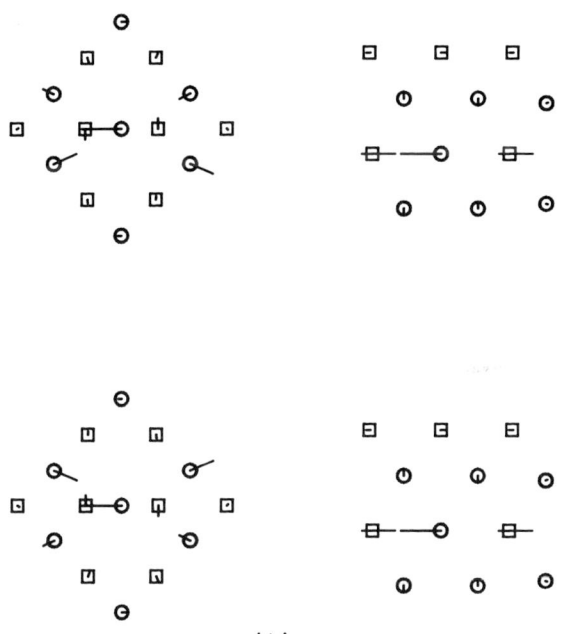

(b)

Fig. 7. (a) Plot of the *saddle-shaped* saddle surface for vacancy jumps in the fcc LJ crystal along the twin directions of maximum principal curvature (v_1); the plot spans 2σ along v_1^+ and v_1^-, and 1.1σ along z. (b) Projections of atomic displacements involved in v_1^+ and v_1^-. (c) Surface along the twin directions of principal curvature v_2 showing the second largest curvature; the plot spans 2σ along v_2^+ and v_2^-, and 0.75σ along z. (d) Projections of atomic displacements involved in v_2^+ and v_2^-. [From De Lorenzi *et al.* (1984b).] (*Figure continues on next page.*)

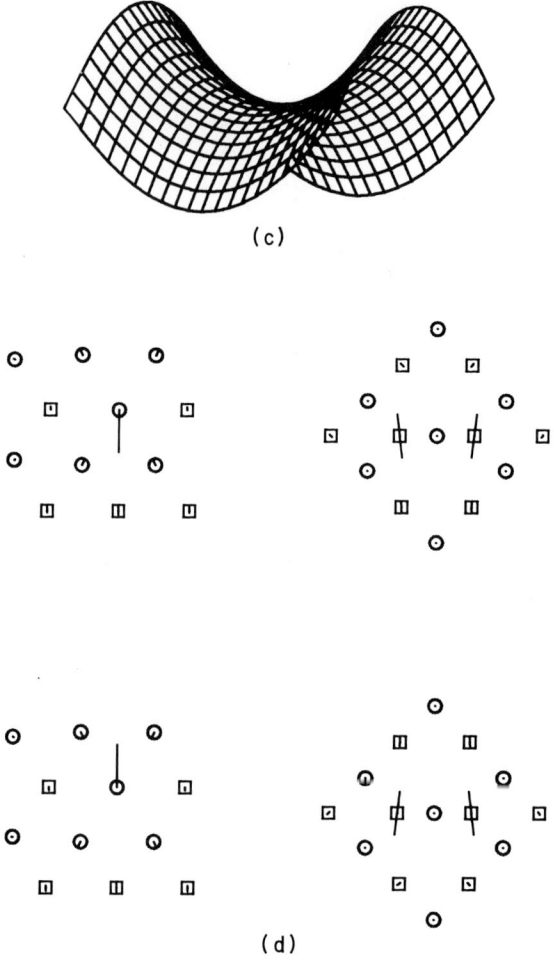

Fig. 7. (*Continued*)

fact seen to come in pairs having opposite sign. As a consequence the saddle surface S has itself the shape of a many-dimensional saddle (Figs. 7 and 8)!

In conclusion, a quadratic description of S in powers of the displacements from P_0 is simply done in terms of third-order derivatives of ϕ at P_0 of the type ϕ_{zij} and of the eigenvalues of the dynamical matrix also at P_0, the coefficients F_{ij} being given by Eq. (27). In practical cases, numerical knowledge of these coefficients will permit to take into account, to lowest order, corrections due to the curvatures of S, in diffusion processes. The extension to higher orders is, in principle, also possible.

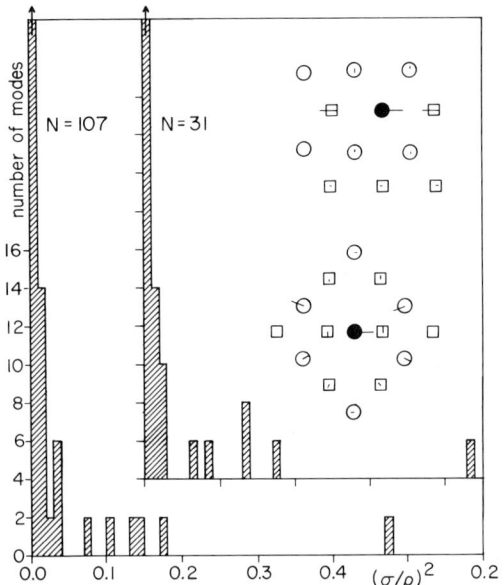

Fig. 8. Distributions of squared principal curvatures of the saddle surface for vacancy jumps in Lennard–Jones 31 and 107 atom cyclically repeating crystals. The atomic displacements in a direction of highest principal curvature are shown again. The twin directions of identical principal curvature are related by mirror reflection from the (110) plane through the central atom. [From Jacucci *et al.* (1984).]

C. A Numerical Procedure for Extracting the Coefficients F_{ij}

Given $\Delta\phi(P_0)$, its eigenvectors in cartesian coordinates, and the respective eigenvalues, and given the possibility of evaluating $\nabla\phi$ at any point P, a numerical procedure has been developed to obtain the normal coordinates derivatives ϕ_{zij} entering the coefficients F_{ij} (Jacucci *et al.*, 1984). This problem cannot be approached analytically in general for $N \approx 100$. Imagine having computed all third-order derivatives of ϕ in cartesian coordinates. The problem of transforming this matrix to normal coordinate derivatives is a formidable computational task. One must adopt a numerical procedure that yields ϕ_{zij} directly.

Starting from P_0 consider a displacement δ along the direction $\mathbf{x}_i$, and evaluate $\phi_z(x_i = \delta, x_j = z = 0, j \neq i) \equiv v_1$. Then consider a displacement 2δ also along $\mathbf{x}_i$, and evaluate $\phi_z(x_i = 2\delta, x_j = z = 0, j \neq i) \equiv v_2$. If δ is small enough, we shall pick up only the leading term in the power expansion of ϕ_z with respect to x_i. Calling n the order of this leading term, we write

$$\phi_z(x_i, x_j = z = 0, j \neq i) = \alpha x_i^n \quad \text{and} \quad v_2 = v_1(2)^n \quad (31)$$

We can get n simply as

$$n = \ln(v_2/v_1)/\ln 2 \tag{32}$$

If we find $n = 2$, then we can evaluate ϕ_{zii}, because near P_0 it must be

$$\phi_z(x_i, x_j = z = 0, j \neq i) = \tfrac{1}{2}\phi_{zii}x_i^2 \tag{33}$$

If we find $n = 3$, then ϕ_{zii} must be zero. Let us extend the discussion to include cross derivatives entering F_{ij} for $i \neq j$. For simplicity we shall consider the important case in which $\phi_{zii} = 0$. Consider a displacement δ from P_0 along $\mathbf{x}_i$, and at the same time a displacement δ along $\mathbf{x}_k$.

From displacements δ and 2δ we get v_1 and v_2 and from Eq. (32) n. If $n = 2$, we have

$$\phi_z(x_i = x_k, x_j = z = 0, j \neq i \neq k) = \tfrac{1}{2}(\phi_{zii}x_i^2 + 2\phi_{zik}x_ix_k + \phi_{zkk}x_k^2)$$
$$= \phi_{zik}x_ix_k \tag{34}$$

Again, if we find $n = 3$, then ϕ_{zik} is zero. In practice the procedure is repeated for $3N(3N - 1)/2$ times to fill the symmetrical ϕ_{zij} matrix, and hence $F_{ij} = -c_{ij}$. The stability of the results with respect to the parameter δ is found in typical applications to span several order of magnitude. Once the saddle point is well located within the precision of the computer, results for the various quantities of interest (listed below) are stable to one part in ten thousand (Jacucci et al., 1984).

D. Dynamical Corrections to Rate Theory Due to a Curved S

Knowledge of the coefficients F_{ij} permits the calculation of dynamical corrections to rate theory due to the curvature of the saddle surface S, as proposed by Flynn (1975) and Flynn and Jacucci (1982), and carried out by Jacucci et al. (1984). The procedure is to construct the "watershed" corresponding to the energy barrier to atomic jumps, calculate the gross flux across it by rate theory, and subtract dynamical return jumps (coming from both sides!) to get the net flux. These calculations involve thermal averaging over the appropriate surface, which can be done part analytically and part by MC once S is approximated by Taylor expansion including third-order terms in ϕ.

The expression of the return jump fraction R'/R through the saddle surface S is (De Lorenzi et al., 1984a)

$$\left(\frac{R'}{R}\right)_S = \frac{k_B T}{-\phi_{zz}} \sum_{i=1}^{3N-1} \frac{1}{\rho_i^2} = \frac{k_B T}{-\phi_{zz}} \sum_{i \neq j=1}^{3N-1} F_{ij}^2 \tag{35}$$

in terms of normal coordinate derivatives of ϕ, or in terms of principal radii of curvature of S. Similarly, the fraction of immediate return jumps through the saddle plane S_0 is found to be (De Lorenzi et al., 1984a)

$$\left(\frac{R'}{R}\right)_{S_0} = \frac{k_B T}{-4\phi_{zz}} \sum_{i \neq j=1}^{3N-1} \frac{\phi_{zij}^2}{\phi_{ii}\phi_{jj}} \tag{36}$$

In both calculations only U-turns are considered; S-shaped trajectories crossing three and more times are disregarded, although they may exist. Furthermore, since the derivation is based on local differential properties (up to second order), trajectories that change their mind before recrossing may be spuriously included. Both errors tend to overestimate the fraction of return jumps.

Another quantity of interest is the anharmonic correction to the gross rate Γ through the free energy related configurational integrals appearing in Eq. (15). Unfortunately, a complete evaluation of this correction term to lowest order in T demands the inclusion of fourth-order terms in the expansion of ϕ. This task is out of reach of today's computational capacity if attacked analytically, but can be approached by MC. Third-order terms of the form $\phi_{zij}zx_ix_k$ do not contribute to the flux through S_0, where $z = 0$, but contribute to S. Third-order terms like $\phi_{ijk}x_ix_jx_k$ do not contribute to thermal averages, on symmetry grounds. The contribution to the flux through S of the former third-order term is (De Lorenzi et al., 1984a)

$$\left(\frac{R''}{R}\right)_S = \frac{-k_B T}{4} \sum_{i \neq j=1}^{3N-1} \frac{\phi_{zij}^2}{\phi_{ii}\phi_{jj}(\phi_{ii} + \phi_{jj} - \phi_{zz})} \tag{37}$$

so that the net flux Γ_S through S in this approximation is

$$\Gamma_S = \Gamma_0(1 - (R'/R)_S + (R''/R)_S) \tag{38}$$

while the net flux Γ_{S_0} through S_0 in the same approximation is

$$\Gamma_{S_0} = \Gamma_0(1 - (R'/R)_{S_0}) \tag{39}$$

In these equations Γ_0 is the harmonic flux of Eq. (16).

Γ_S and Γ_{S_0} are inescapably different, as is easily verified by substitution. The topological argument that all trajectories going from a minimum of ϕ, say, A, to one of the neighbor minima B must cut any suitable dividing surface does not apply here. However, it should be realized that this result pertains to the highly idealized hamiltonian of potential energy ϕ_3 given by

$$\phi_3 = -\frac{1}{2}\phi_{zz}z^2 + \frac{1}{2}\sum_{i=1}^{3N-1}\phi_{ii}x_i^2 + \frac{1}{2}z\sum_{i \neq j=1}^{3N-1}\phi_{zij}x_ix_j \tag{40}$$

In fact, ϕ_3 bears no resemblance to ϕ away from P_0, and the minima of ϕ are not reproduced in ϕ_3. As a consequence, trajectories may exist that are solutions of the equation of motion under ϕ_3 and cross S but not S_0, whereas the corresponding trajectory through S does also cross S_0 under the action of ϕ. The ratio of the flux of the "missing" trajectories, over the total magnitude of the anharmonic contribution to the flux, is independent of the temperature, i.e., this problem is present in the lowest-order perturbation. A second source of discrepancy in Γ_S and Γ_{S_0}, as evaluated by the formulas above, consists in contributions from complicated trajectories undergoing several recrossings. This point will receive further attention below. Here we wish to underline that dynamical recrossings of S and S_0 have a quite different origin. Those of S are related to high momentum trajectories almost parallel to S. Those of S_0 are related to low-energy trajectories reaching S_0, but not the somewhat higher watershed S.

In conclusion, the net fluxes through S and S_0 can be evaluated by using ϕ_3. Γ_S is found to be somewhat larger than Γ_{S_0}, the discrepancy being a fixed fraction of contributions to order $k_B T$. The predictions for Γ_S are believed to be more reliable, and should be employed, e.g., in the evaluation of the isotope effect factor ΔK, in place of Γ_{S_0}. Numerical evaluation of the fluxes for specific cases, e.g., vacancy migration in model fcc crystals, can show the magnitude of these anharmonic corrections in typical solid-state diffusion instances and indicate whether terms of lowest order in $k_B T$ may be expected to be sufficient to accurately describe jump rates close to the melting point. A detailed comparison of the predicted fluxes with MD data would be of great help in drawing these conclusions. The remaining sources of anharmonic corrections, not included in this treatment, are (i) fourth-order terms in ϕ contributing to order $k_B T$ to Γ', and (ii) higher order terms in $k_B T$.

Before closing this section, we note that two anharmonic contributions affect the isotope effect factor ΔK (Flynn, 1975). One comes from the variation of the direction of the normal n to S, the other from the mass dependence of the fraction R'/R of return jumps. The first term is absent when calculating ΔK using the saddle plane, because n is always equal to n_0 on S_0. Furthermore, by using $(R'/R)_{S_0}$ from ϕ_3, the fraction of return jumps from S_0 is found to be mass-independent (De Lorenzi et al., 1984a).

In the case of the watershed S, the anharmonic expression of ΔK contains both contributions, neither of which vanishes. We see here effects of the observed difference of the two fluxes: $\Gamma_S \neq \Gamma_{S_0}$.

E. Numerical Results and Comparison with MD

Numerical results on model systems appropriate to Al, Ar, and Cu were obtained by Jacucci et al. (1984). The investigation included a careful check of cutoff and size dependences, as well as calculations for various densities and temperatures (De Lorenzi et al., 1984a).

8. DEFECT CALCULATIONS BEYOND THE HARMONIC MODEL

TABLE VI
Return Jump Rates Close to the Melting Point

Metal	Jump rate (%)		
	$N = 31$	$N = 107$	$N \to \infty$
Al	5.1	4.6	4.4
Ar	3.5	4.3	4.6
Cu	4.0	—	—

The main results can be summarized as follows:

(i) The mean curvature of the saddle surface is similar for the various potential functions used (Lennard–Jones and Morse) and comes from only a few directions in the saddle plane.

(ii) $(R'/R)_S$ is invariably of the order of 5% at melting, so that statistical theories of diffusion are indeed valid, and the first order in $k_B T$ should be quite sufficient (see Table VI).

(iii) The temperature-dependent correction $-K_T T_M$ to ΔK related to the variation of the direction of $\mathbf{n}$ is very small, less than 2% (see Table VII).

(iv) The correction term originating from the mass dependence of the return jumps is large and positive, and unambiguously predicts unphysical results upon mass substitution. This isotope effect "catastrophe" points to a basic flaw in the formulation of rate theory.

These results are of central importance for the theory of diffusion in solids. Their confirmation in MD studies is of great interest. Let us focus on

TABLE VII
Harmonic and Anharmonic ($\Delta K = K_0 - K_T T_M$) Isotope Effect Factors

Metal	N	K_0	$-K_T T_M$	ΔK
Al	31	0.9229	−0.0188	0.9041
	107	0.9133	−0.0194	0.8939
	255	0.9125	—	—
	$\to \infty$	0.9124	−0.0196	0.8929
Ar	31	0.9609	−0.0139	0.9470
	107	0.9774	−0.0167	0.9607
	$\to \infty$	0.9789	−0.0178	0.9611
Cu	31	0.9365	−0.0153	0.9212
	107	0.9363	—	—

the results for LJ at ~ 80 K: $(R'/R)_S \sim 0.05$ and $(R'/R)_{S_0} \approx 0.09$. Careful measurements of R'/R and ΔK were done by Bennett (1975) by using specially devised MD techniques. Both quantities were measured with respect to S_0, for the present case of vacancy migration in fcc LJ crystals.

The reported results are $R'/R = 10\%$ and $\Delta K = 0.89 \pm 0.05$. In the harmonic case these values are 0 and 0.98, so that anharmonic contributions are appreciable. The agreement of the two evaluations for $(R'/R)_{S_0}$ is particularly rewarding, in view of the fact that Bennett sees *all* return jumps, while the calculation of Jacucci *et al.* (1984) estimates only immediate return jumps related to the curvature of S. These, therefore, represent substantially all return jumps.

One further step is needed, however, to clear up the situation for the isotope effect. Complicated trajectories that cross the watershed three or more times, or start heading for a return jump and then change their mind, have not been accounted for, as mentioned above. Albeit very small, the number of these trajectories may be so strongly mass-dependent that the isotope effect cannot be accurately evaluated within the above analytical scheme, while the conclusions for Γ_S remain essentially unaltered in practical cases.

It is obvious that the possible cure of the errors in the bookkeeping of re-crossings rests in the implementation of an exact jump condition that eliminates immediate return jumps. The existence of invariant manifolds in phase space that cannot be crossed by trajectories provides a critical jump condition that can be used in this context. A $(6N - 2)$-dimensional surface can be defined in phase space, called the *center manifold*, made up of trajectories indefinitely trapped in the saddle neighborhood. These trajectories are, of course, highly unstable and, if perturbed, fall into one of the two wells. In the harmonic approximation, the equation of the center manifold is $z = 0$; $\dot{z} = 0$, and the trajectories on it correspond to many dimensional harmonic motion. Presence of the term $\frac{1}{2} z \sum_{i \neq j=1}^{3N-1} \phi_{zij} x_i x_j$ displaces the center manifold from the watershed, and the two surfaces are quite unrelated.

The center manifold can be used to calculate the flux in such a way that there are no immediate return jumps (Jacucci *et al.*, 1984), as shown in Fig. (9). It is found in this way that the anharmonic jump frequency Γ_{CM} is only slightly larger than Γ_S, as expected because of the overestimation of return jumps. Furthermore, no isotope effect catastrophe is met. Calculations using these ideas are being carried out (De Lorenzi *et al.*, 1984d).

In conclusion, accurate jump rates can be calculated by using the notion of anharmonic saddle and geometrical constructions like the watershed and the center manifold in phase space. An improved, and for certain aspects new, conceptual basis for jump theories is achieved, and the validity of statistical jump theories is demonstrated for diffusion in solids. The path is

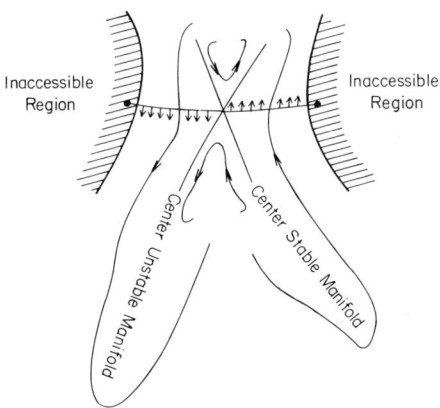

Fig. 9. Crossing surface in phase space with no immediate returns. The center *stable* and *unstable* manifolds cannot be crossed by trajectories. They meet on the *center manifold* and precisely separate trajectories crossing the dividing surface in opposite senses. Returns must go around these invariant manifolds and cannot be immediate. The flux toward one of the two sides is obtained by integrating over half of the dividing surface. The latter can be chosen rather arbitrarily as long as it is bounded by inaccessible high-energy regions. [From De Lorenzi *et al.* (1984).]

therefore open to precise future calculations of jump rates and of related thermodynamic parameters for model systems using realistic interatomic forces.

VI. Conclusions

Important results have been achieved. Comparison of Monte Carlo data with accurate quasi-harmonic lattice dynamics results for formation and migration of lattice vacancies demonstrates reliably, for the first time, that the *harmonic model overestimates* both contributions to the *free energy of activation* of diffusion in rare gases.

In addition, it has been analytically shown that the fraction of unsuccessful crossings of the "watershed" saddle surface in fcc crystals does not exceed 5% at melting. *Jump frequencies are therefore accurately given by rate theory* and may be corrected for dynamically inevitable immediate return jumps. This result is obtained by extending to third order the Taylor expansion of the potential energy in the analytical approach to rate theory.

The very success of the perturbation approach is possibly the most important new finding. It reopens the analytical path to the understanding of diffusion in solids, a path that appeared for some time bypassed by computer simulation. Reentering the stage, analytical mechanics seeks its revenge with elegance: The critical jump condition in phase space is effectively provided by invariant manifolds at the singular point, and the expression of these manifolds to first order in anharmonic perturbation is sufficient.

Further development and implementation of machine calculation methods, as they are coming within the reach of existing computers, for the investigation of point defect formation and migration in crystals will no doubt constitute one of the major areas of activity in this field during the coming years (see also Jacucci, 1984). A substantial advance in the understanding of point defect diffusion is to be expected as a result of using well handled statistical mechanics methods.

Challenging calculations to be confronted next, along the lines exposed in this chapter, are:

(i) formation free energies, enthalpies, and volumes using MC and MD in metals, starting from the alkalis for which excellent ab initio pair potentials exist, and in simple ionic solids;

(ii) jump rates for vacancy migration in bcc crystals; here the conformation of the saddle region should be more complex than in the fcc case, possibly featuring a metastable transition state;

(iii) multiple jump rates in various circumstances from knowledge of local properties of the N-body potential energy surface;

(iv) extension to highly disordered systems, e.g., superionic conductors (Flynn, 1979) and liquids (Stillinger and Weber, 1983), of method and concepts developed for point defect diffusion in crystalline solids.

Acknowledgments

The material included in this chapter comes to a great extent from work done in collaboration with Giulia De Lorenzi, Colin Peter Flynn, Marco Ronchetti, and Marco Toller. Their contribution of ideas and effort to the preparation of this chapter together with their consent to the use of material prior to publication is gratefully acknowledged. The author is also indebted to Charles H. Bennett and Aneesur Rahman for many useful discussions, to Graeme E. Murch for encouragement and advice in the preparation of the manuscript, and to Mary Ostendorf for very competent typing.

This work was supported in part by Gruppo Nazionale di Stuttura della Materia through the University of Trento and by National Science Foundation grant DMR80-20250 through the University of Illinois.

References

Bennett, C. H. (1975). In "Diffusion in Solids: Recent Developments" (A. S. Nowick and J. J. Burton, eds.), Chap. 2. Academic Press, New York.
Bennett, C. H. (1976a). *Colloq. Metall.* **19,** 65.
Bennett, C. H. (1976b). *J. Comput. Phys.* **22,** 245.

8. DEFECT CALCULATIONS BEYOND THE HARMONIC MODEL

Bennett, C. H. (1977). *In* "Algorithms for Chemical Computations" (R. E. Christoffersen, ed.), Chap. 4. American Physical Society, Washington, D.C.
Bennett, C. H., and Alder, B. J. (1971). *J. Phys. Chem. Solids* **32**, 2111.
Binder, K. (1979). *Top. Curr. Phys.* **7**, 1.
Da Fano, A., and Jacucci, G. (1976). *Colloq. Metall.* **19**, 687; *J. Nucl. Mater.* **69/70**, 549.
Da Fano, A., and Jacucci, G. (1977). *Phys. Rev. Lett.* **39**, 950.
De Lorenzi, G., Flynn, C. P., and Jacucci, G. (1984a). To be published.
De Lorenzi, G., Flynn, C. P., and Jacucci, G. (1948b). To be published.
De Lorenzi, G., Jacucci, G., and Pontikis, V. (1982). *Surf. Sci.* **116**, 391.
De Lorenzi, G., Jacucci, G., and Ronchetti, M. (1984c). To be published.
De Lorenzi, G. Flynn, C. P., Jacucci, G., and Toller, M. (1984d). To be published.
Flynn, C. P. (1972). "Point Defects and Diffusion." Oxford Univ. Press, London and New York.
Flynn, C. P. (1975). *Phys. Rdv. Lett.* **35**, 1721.
Flynn, C. P. (1979). *In* "Fast Ion Transport in Solids" (Vashishte, Mundy, and Shenoy, eds.), p. 559. Elsevier-North Holland, New York.
Flynn, C. P., and Jacucci, G. (1982). *Phys. Rev. B* **10**, 6225.
Holt, A. C., Hoover, W. G., Gray, S. G., and Shortle, D. R. (1970). *Physica* **49**, 61.
Jacobs, P. W. M., Nerenberg, M. A., and Govindarajan, J. (1983). *Lect. Notes Phys.* Vol. 166, 21.
Jacucci, G. (1984). *In* "Nontraditional Methods in Diffusion" (G. E. Murch, H. K. Birnbaum, and J. R. Cost, eds.), *Symp. Proc. TMS-AIME Fall Meet.*, Philadelphia, Pennsylvania, *October 3-6, 1983.*
Jacucci, G., and Rahman, A. (1978). *J. Chem. Phys.* **69**, 4117.
Jacucci, G., and Taylor, R. (1979). *J. Phys. F* **9**, 1489.
Jacucci, G., and Ronchetti, M. (1980). *Solid State Commun.* **33**, 35.
Jacucci, G., Taylor, R., Tenenbaum, A., and Van Doan, N. (1981). *J. Phys. F* **11**, 793.
Jacucci, G., Toller, M., De Lorenzi, G., and Flynn, C. P. (1984). *Phys. Rev. Lett.* **52**, 295.
Kushick, J., and Berne, B. J. (1977). *Mod. Theor. Chem.* **6**, 1.
Maradudin, A. A., Montroll, E. W., Weiss, G. H., and Ipatova, I. P. (1971). "Theory of Lattice Dynamics in the Harmonic Approximation," 2nd ed. Academic Press, New York.
Mundy, J. N. (1971). *Phys. Rev. B* **3**, 2431.
Mundy, J. N., Miller, T. E., and Porte, R. J. (1971). *Phys. Rev. B* **3**, 2445.
Rahman, A., and Jacucci, G. (1984). Submitted to *Nuovo Cimento D*.
Squire, D. R., and Hoover, W. G. (1969). *J. Chem. Phys.* **50**, 701.
Stillinger, F. H. (1983). Preprint, Bell Labs., Murray Hill, New Jersey.
Vineyard, G. H. (1957). *J. Phys. Chem. Solids* **3**, 121.

Index

A

Absorption coefficient, 29–30
β-AgMg, Mg in, 239
β-alumina, Na in, 384, 405, 411
β''-alumina, Na in, 384, 391
Aluminum
  self-diffusion (grain boundary), 363
  tilt boundary in, 335, 347, 362
Aluminum–zinc alloy, Zn in, 202
Anelastic relaxation, 158–183
Anharmonic effects, 454–473
Anisotropic diffusion, 7, 55
Anodizing, 33–36
Area (of sample), measurement of, 41
Argon, *see also* Lennard–Jones model
  equation of state, 443–446
β-AuCd, 243
Auger electron spectroscopy, 36, 38
Autoradiography, 41–42
β'-AuZn
  Au in, 240–243
  defect structure in, 240–243
  isotope effect in, 240, 243
  Zn in, 240–243

B

bcc alloys, 208–213
δ-Bi$_2$O$_3$
  dopants in, 183
  ionic conductivity in, 183
Boltzmann–Matano analysis, 416

C

Calcia-doped ceria
  anelastic and dielectric relaxations in, 161–162
  ionic conductivity in, 175–176
Calcia-doped thoria
  anelastic and dielectric relaxations in, 161–162
Calcia-stabilized zirconia (CSZ)
  ionic conductivity in, 161, 179, 183–184
  O in, 152, 183–184
  ordering in, 148, 149, 180, 184
Calcium fluoride, defect structure in, 435
Cerium dioxide, *see also* Calcia; Yttria-doped ceria
  effect of dopants on ionic conductivity in, 172–177
  O in, 149–152
Chemical diffusion, 2, 3, 11
Chemical diffusion coefficient, 412–423
Cluster variation method, 215
Cobalt–nickel alloy, Ni in, 201
β-CoGa
  Co in, 245, 246, 247
  defect structure, 245
  Ga in, 245, 246, 247
Complex impedance analysis, 165–166
Concentrated alloy, 189–253, *see also* Ordered alloy
Concentration profiles, *see* Penetration plots
Copper
  Ag in, 366
  tilt boundary in, 324–328, 339, 361, 370

Copper (*continued*)
  twist boundary in, 329
Copper–aluminum alloy, Cu in, 200–201
Copper–nickel alloy, 201
Copper–zinc alloy
  Cu in, 205
  isotope effect in, 205, 236
  Zn in, 201–202, 205
Correlation factor, *see* Tracer correlation factor
Counting equipment, 44–46
  cocktails, 46
  dead time, 43
Counting statistics, 43
Creep, 348–349
β-Cu_3Sb
  Cu in, 250
  defect structure of, 248–250
β-Cu_3Sn, 250
β-CuZn
  Cu in, 235–237
  Sb in, 235–236
  Zn in, 235–237

D

Debye–Hückel theory, 157–158, 170, 176
Defect cluster, 151, 158, 390
Deposition of radiotracers, 11–15
Depth measurement, 22, 25–26, 31, 40–41
Dielectric relaxation, 158–183
Diffusion, empirical rules, 193–197, 200–205
Diffusion equation, 3, 264, 400
  solutions of, 3–8, 266–274, 400
Diffusion-induced grain boundary migration (DIGM), 352–353
Dislocations
  analysis of diffusion along, 257–316
  climb of, 112–115
  estimates of density, 299–302, 310–311
  radius of "pipe," 262–263, 300–304, 310–311
  tails on penetration plots, *see* Penetration plots
Dissociative mechanism, 67–69, 122–135
Double jumps, 455
Dumbell mechanism, 393

E

Einstein equation, 389
Electrochemical sectioning, *see* Sectioning by chemical removal
Electromigration, effect of grain boundaries on, 351
Emitter-push effect, 110–115
Encounter model, 404–406
Erbia-stabilized hafnia, 148
Exchange experiments, 7
Exchange mechanism, 66, 393

F

Fast ionic conductors, 143–185, 382, 384, 390, 413, 435
fcc alloys, 200–208
FeAl, 247–248
Fe_3O_4, Fe in, 406
$Fe_{1-\delta}S$, Fe in, 383
Furnaces, 15–17

G

Germanium
  Al in, 89
  As in, 89
  Cu in, 89, 116, 135
  dislocation density in, 302
  Ga in, 89, 302–303
  group III elements in, 89
    effect of doping on, 108–113
  group V elements in, 89
    effect of doping on, 108–113
  In in, 89
  Li in, 89
  P in, 89
  Sb in, 89
  self-diffusion, 74–77, 89, 135–136
    effect of doping on, 82–84, 108
    isotope effects in, 84–86
    pressure effects on, 87
Gold
  tilt boundary in, 331–332, 340, 344
  twist boundary in, 346, 369, 371

INDEX

Grain boundaries in metals, 322–374, *see also* Grain boundary diffusion
  structure of, 324–360
  width of, 360
Grain boundary diffusion, 7, 52–53, 260, 348–374
  effect of pressure on, 353–354
Grain boundary energy, 346–347
Gruzin technique, *see* Residual activity technique

H

HADES calculations, *see* Molecular statics
Hafnium dioxide, 144, *see also* Erbia-stabilized hafnia
Harmonic approximation, 435, 455
Hart equation, 282
Haven ratio, 412, 422

I

Impurity diffusion, 393, 409–410
Integral activity technique, 30
Intermediate phases, 235–252
Intermetallic compounds, *see* Intermediate phases
Interstitialcy mechanism, 66–69
Interstitial mechanism, 66
Ionic conductivity, 152–184, 407–412
  effect of ordering on, 146–149, 180–182
Ionic thermocurrent (ITC), *see* Thermally stimulated depolarization currents
Iron
  edge dislocation in, 334
  tilt boundary in, 333, 340–341
  twist boundary in, 338–339
  Zn in, 366
Iron–cobalt alloy (bcc)
  Fe in, 209
  isotope effect in, 209
Iron–cobalt alloy (fcc)
  Co in, 206
  Fe in, 206
  isotope effect in, 205

Iron–nickel alloy (fcc), Ni in, 201
Iron–palladium alloy, 206
Iron–silicon alloy, Fe in, 209
Iron–vanadium alloy, 208
Ising model, 214, 418
Isoconcentration contours, 22, 41, *see also* Sectioning
Isotope effect, 193, 205, 354, 406, 430, 433, 460, 470–472
  in β'-AuZn, 240
  in Cu–Zn alloy, 205
  in Fe–Co alloy, 205–206, 209
  in Ge, 84–86
  measurement of, 44, 46–48
  in ordered alloy, 218–219, 238
  in Si, 84–86

K

Kick-out mechanism, 67–69, 117–136
Kirkendall effect, 353

L

Lattice dynamics, 431, 433, 436–455
Lattice gas models, 158, 214, 382–424
Lattice statics, 449, 452–455, *see also* Molecular statics
Lead–thallium alloy, 202
Lennard–Jones model, 432, 442–452, 461, 465–467

M

Microsectioning, 22, 30–42, 74
Mn$_3$Pt, Mn in, 252
Molecular dynamics, 334–336, 355–360, 431–435, 440–442, 452, 455–460, 470–472
Molecular statics, 151, 179, 324–327, 334, 338, 340, 344–347, 359–360, *see also* Lattice statics
Molybdenum–tungsten alloy, 210
Monte Carlo method, 200, 215, 219, 226–228, 231–237, 242, 274, 379–425, 431, 433, 440–442, 449, 451, 460, 469

Morse potential, 215, 239
Multiple jumps, 455

N

Nernst–Einstein equation, 302
β-NiAl
  Co in, 244
  defect structure, 243
  In in, 244
  Ni in, 243
Ni_3Al, Ni in, 251
Nickel, In in, 367
Nickel–chromium alloy (bcc), Cr in, 210
Nickel–chromium alloy (fcc), 207
β-NiGa, 245
β-Ni_3Sb
  defect structure of, 248–249
  Ni in, 249
Niobium
  D in, 392–393
  H in, 392–393
Nonstoichiometric compound, 2, 144, 146, 149–152, 381, 390, 413
Nuclear reactions, 74, 149

O

Optical pyrometry, 19
Ordered alloy, 213–252, 386, 393–395
Oxidation-enhanced diffusion (OED), 94, 98–105, 136
Oxidation-induced stacking faults (OSF), 94, 98–105, 112–113
Oxidation-retarded diffusion (ORD), 97–105
Oxides (of the fluorite structure), 143–185

P

Path probability method, 220–226, 231–234, 384, 389, 395, 411
Penetration plots, 4–7, 9, 22, 27, 31–32, 41, 43, 48–55, 280–304
Percolation, 393
Physical correlation factor, 153–154, 180, 408–413, 420–421

Plutonium dioxide, O in, 149
Point defects
  in grain boundaries, 338–343
  sinks for, 367–372
  sources of, 367–372
  vacancy formation energy, 442–453
  vacancy migration energy, 453–460
Potassium, self-diffusion, 456
Potassium bromide
  Br in, 309
  dislocation density in, 310–311
  dislocation radius in, 310, 311

R

Random alloy, 197–200, 205–212, 393, 395–397, 410–411, 423
Random walk theory, 197–200, 228–234, 381, 395
Rate theory, 432, 453–474
Residual activity technique, 29, 46, 77
Ring mechanism, 66, *see also* Six-jump cycle mechanism

S

Sample preparation, 10–15
Secondary ion mass spectrometry (SIMS), 39–40, 74, 81, 149, 184 185
Sectioning, *see also* Microsectioning
  alignment for, 22, 24, 26–29
  by chemical removal, 29, 33, 34, 35
  by grinding, 25–27, 75
  by lathe, 22–25
  by microtome, 28
Selenium, self-diffusion, 72
Semiconductors, 63–136
Si–Ge alloy, 87–88
Silicon
  Al in, 71, 90
  As in, 90, 112
  Au in, 91–92, 116, 125–136
  B in, 90, 112, 116
  C in, 90, 91
  Co in, 116
  Cu in, 90
  Fe in, 90, 116
  Ga in, 90, 116

INDEX

Ge in, 80–81, 90
  effect of doping on, 82–84
  isotope effects in, 84–86
  group III elements in, 71, 88–90, 94, 101, 112, 116
    anomalous diffusion of, 110–116
    effect of doping on, 108–110
    effect of oxidation on, 98–103
  group V elements in, 88–90, 94, 101, 112–116
    anomalous diffusion of, 110–116
    effects of doping on, 108–110
    effect of oxidation on, 98–103
  In in, 90
  isotope effects in, 84–86
  Li in, 90
  Ni in, 90, 116, 135–136
  O in, 90–91
  oxidation of, 79, 93–96
  oxidation-influenced diffusion in, 93–105
  Sb in, 90
  self-diffusion, 77–80, 90, 131, 136
    effect of doping on, 82–84, 108–109
    isotope effects in, 84–86
  P in, 90, 112–116
Silver
  tilt boundary in, 360
  self-diffusion (grain boundary), 360, 364
Silver–aluminum alloy, Ag in, 202–203
Silver–cadmium alloy, 201–202
Silver–gold alloy, 200–201
Silver–palladium alloy, Ag in, 202
SIMS, *see* Secondary ion mass spectrometry
Sintering, 348, 350
Six-jump cycle mechanism, 216–219, 221, 239–242, 244
Sodium, self-diffusion, 456
Sodium chloride
  dislocation density, 301
  dislocation radius, 301
  ionic conductivity, 302
  Na in, 300
  vacancy-pair mechanism in, 397
Sputtering, 36–39, 74
Superionic conductors, *see* Fast ionic conductors
Surface diffusion, 322
  adsorbate, 384, 417–418
Surface holdup, 50–51
Surface preparation, 10, 30–31

T

Tellurium, self-diffusion, 72
Temperature measurement, 18–20
Thermally stimulated depolarization currents (TSDC), 160, 163
Thermocouples, 15, 18–21
Thermodynamic factor, 412–413, 422–423
Thermomigration, effect of grain boundaries on, 351
Thorium dioxide
  O in, 149
  Th in, 146
Titanium, self-diffusion, 210–211
Titanium–manganese alloy, Ti in, 211
Titanium-niobium alloy, 212
Titanium–vanadium alloy, 211
Tracer correlation factor, 70, 85, 145, 192, 197–199, 205–207, 209, 212–213, 218–220, 228–234, 238, 246, 302, 353, 354–355, 381, 389–406
Tracer diffusion coefficient, 145–146, 192, 220, 221, 353, 354
  errors entering into, 8, 20, 40–44, 49–50
  measurement of, 1–55
Triple-defect mechanism, 246
Tungsten (110), O on, 417

U

(U, Pu)O$_2$, O in, 149
Uranium dioxide
  defect structure, 144, 151
  O in, 149–151, 383, 390
  U in, 146

V

Vacancy, *see* Point defects
Vacancy-wind effect, 407–412, 420–423

W

Welding diffusion couples, 10–11
Wigner–Seitz type cell, 263

Y

Yttria-doped ceria
  dielectric and anelastic relaxations in, 162–163
  ionic conductivity in, 166–180, 184
  O in, 151–152, 184

Yttria-stabilized zirconia (YSZ), ionic conductivity in, 161

Z

Zirconium dioxide, dopants in, 177, 180, 182, *see also* Calcia-stabilized zirconia; Yttria-stabilized zirconia

MATERIALS SCIENCE AND TECHNOLOGY

EDITORS

A. S. NOWICK
Henry Krumb School of Mines
Columbia University
New York, New York

G. G. LIBOWITZ
Solid State Chemistry Department
Materials Research Center
Allied Corporation
Morristown, New Jersey

A. S. Nowick and B. S. Berry, ANELASTIC RELAXATION IN CRYSTALLINE SOLIDS, 1972

E. A. Nesbitt and J. H. Wernick, RARE EARTH PERMANENT MAGNETS, 1973

W. E. Wallace, RARE EARTH INTERMETALLICS, 1973

J. C. Phillips, BONDS AND BANDS IN SEMICONDUCTORS, 1973

J. H. Richardson and R. V. Peterson (editors), SYSTEMATIC MATERIALS ANALYSIS, VOLUMES I, II, AND III, 1974; IV, 1978

A. J. Freeman and J. B. Darby, Jr. (editors), THE ACTINIDES: ELECTRONIC STRUCTURE AND RELATED PROPERTIES, VOLUMES I AND II, 1974

A. S. Nowick and J. J. Burton (editors), DIFFUSION IN SOLIDS: RECENT DEVELOPMENTS, 1975

J. W. Matthews (editor), EPITAXIAL GROWTH, PARTS A AND B, 1975

J. M. Blakely (editor), SURFACE PHYSICS OF MATERIALS, VOLUMES I AND II, 1975

G. A. Chadwick and D. A. Smith (editors), GRAIN BOUNDARY STRUCTURE AND PROPERTIES, 1975

John W. Hastie, HIGH TEMPERATURE VAPORS: SCIENCE AND TECHNOLOGY, 1975

John K. Tien and George S. Ansell (editors), ALLOY AND MICROSTRUCTURAL DESIGN, 1976

M. T. Sprackling, THE PLASTIC DEFORMATION OF SIMPLE IONIC CRYSTALS, 1976

James J. Burton and Robert L. Garten (editors), ADVANCED MATERIALS IN CATALYSIS, 1977

Gerald Burns, INTRODUCTION TO GROUP THEORY WITH APPLICATIONS, 1977

L. H. Schwartz and J. B. Cohen, DIFFRACTION FROM MATERIALS, 1977

Zenji Nishiyama, MARTENSITIC TRANSFORMATION, 1978

Paul Hagenmuller and W. van Gool (editors), SOLID ELECTROLYTES: GENERAL PRINCIPLES, CHARACTERIZATION, MATERIALS, APPLICATIONS, 1978

G. G. Libowitz and M. S. Whittingham, MATERIALS SCIENCE IN ENERGY TECHNOLOGY, 1978

Otto Buck, John K. Tien, and Harris L. Marcus (editors), ELECTRON AND POSITRON SPECTROSCOPIES IN MATERIALS SCIENCE AND ENGINEERING, 1979

Lawrence L. Kazmerski (editor), POLYCRYSTALLINE AND AMORPHOUS THIN FILMS AND DEVICES, 1980

Manfred von Heimendahl, ELECTRON MICROSCOPY OF MATERIALS: AN INTRODUCTION, 1980

O. Toft Sørensen (editor), NONSTOICHIOMETRIC OXIDES, 1981

M. Stanley Whittingham and Allan J. Jacobson (editors), INTERCALATION CHEMISTRY, 1982

A. Ciferri, W. R. Krigbaum, and Robert B. Meyer (editors), POLYMER LIQUID CRYSTALS, 1982

Graeme E. Murch and Arthur S. Nowick (editors), DIFFUSION IN CRYSTALLINE SOLIDS, 1984